Springer Series in Design and Innovation

Volume 17

Springer Series in Design and Innovation (SSDI) publishes books on innovation and the latest developments in the fields of Product Design, Interior Design and Communication Design, with particular emphasis on technological and formal innovation, and on the application of digital technologies and new materials. The series explores all aspects of design, e.g. Human-Centered Design/User Experience, Service Design, and Design Thinking, which provide transversal and innovative approaches oriented on the involvement of people throughout the design development process. In addition, it covers emerging areas of research that may represent essential opportunities for economic and social development.

In fields ranging from the humanities to engineering and architecture, design is increasingly being recognized as a key means of bringing ideas to the market by transforming them into user-friendly and appealing products or services. Moreover, it provides a variety of methodologies, tools and techniques that can be used at different stages of the innovation process to enhance the value of new products and services.

The series' scope includes monographs, professional books, advanced textbooks, selected contributions from specialized conferences and workshops, and outstanding Ph.D. theses.

Keywords: Product and System Innovation; Product design; Interior design; Communication Design; Human-Centered Design/User Experience; Service Design; Design Thinking; Digital Innovation; Innovation of Materials.

How to submit proposals

Proposals must include: title, keywords, presentation (max 10,000 characters), table of contents, chapter abstracts, editors'/authors' CV.

In case of proceedings, chairmen/editors are requested to submit the link to conference website (incl. relevant information such as committee members, topics, key dates, keynote speakers, information about the reviewing process, etc.), and approx. number of papers.

Proposals must be sent to: series editor Prof. Francesca Tosi (francesca.tosi@unifi.it) and/or publishing editor Mr. Pierpaolo Riva (pierpaolo.riva@springer.com).

More information about this series at https://link.springer.com/bookseries/16270

Emilia Duarte · Carlos Rosa
Editors

Developments in Design Research and Practice

Best Papers from 10th Senses and Sensibility 2019: Lost in (G)localization

Springer

Editors
Emilia Duarte
UNIDCOM/IADE - Unidade de
Investigação em Design e Comunicação
Universidade Europeia, IADE - Faculdade
de Design, Tecnologia e Comunicação
Lisboa, Portugal

Carlos Rosa
UNIDCOM/IADE - Unidade de
Investigação em Design e Comunicação
Universidade Europeia, IADE - Faculdade
de Design, Tecnologia e Comunicação
Lisboa, Portugal

ISSN 2661-8184 ISSN 2661-8192 (electronic)
Springer Series in Design and Innovation
ISBN 978-3-030-86595-5 ISBN 978-3-030-86596-2 (eBook)
https://doi.org/10.1007/978-3-030-86596-2

This Springer imprint is published by the registered company Springer Nature Switzerland AG
The registered company address is: Gewerbestrasse 11, 6330 Cham, Switzerland

Preface I

Design as a Catalyst

Design is its own domain, with its own methods and processes, as well as epistemic constraints and goals [1]. True, but with the proliferation of signs that form different interactions across the possible ecosystems of our daily lives, it is necessary to equate and study our internal and external relationships with them [2–7].

It is about this abounding growth of languages that we must reflect upon and not only about their interpretation, but also about their production and, above all, about the symbiotic attraction that these languages have within the environments and ecosystems in which we live. Design is thus a catalyst not only for production, but mainly for interpreting different languages. Consequently, I strongly believe that design is itself a language and, as any language, it needs a medium and a process to manifest itself. The creation of these languages, which may be universal, global, or local, can be forms of criticism themselves. This proves that design is above all a way of thinking.

The way we think and generate ideas could be thought of in terms of a teleportation machine, which travels at a dizzying speed through parallel worlds that are being built, and at the same time, slowly we are appreciating and looking critically at these worlds. This speed is actually the sum of several speeds, that of the brain, the hand, and the imagination of those who gaze at the drawing, those who read the processes, and those who read the solutions. We have to be patient. We have to be responsible. But above all, we need to have a critical attitude, to allow ourselves to be carried away by the "teleportation machine." The designer must regard himself as a critic.

I believe that the genesis of creativity is precisely this ability to travel, to "teleport" oneself and foresee something that appears as an alternative which, through conscious optimization, might end up being the solution. This critical attitude must be the main ingredient of design schools. Designing starts with criticism. Designing starts with teaching design.

It starts with the way we move and (g)locate in everyday life, in society. This society we inhabit is guided by self-fulfilling prophecies, where the starting point is a set of false definitions. In turn, that false definitions give rise to new behaviours making the initial conception true. When was the last time we used our memory? When was the last time we relied on a digital assistant? And what about thinking? Thinking without help?

We think in two dimensions, without depth, without volume, and without critical dimension. We do not think enough; we criticize even less. This is the genuine struggle of design: changing the world. Changing the mind-sets.

All of this hinders our capacity to reflect, and this limitation prevents us from making projections in time. And it is these temporal features that transport us to the future, for new scenarios and for new worlds in art, in culture, and in economy.

How do we manage to absorb timelessness if languages cannot project what could be, what could have been, what is and what was, what will be or what actually happened?

Our thinking archetypes have failed, and our ways to act tend to fail too. In the field of thought, the models, concepts, and arguments that we traced as a society served for a certain time, but we were unable to deal with our actions as human beings. We created scenarios and strategies, but we were incapable of abandoning them when we needed it the most.

We should and have to rethink the relationship between justice and doing, between the material goods, the resources and the ways in which we produce and enjoy artifacts. It is necessary to rethink the natural as the opposite to the artificial. To solve all this, we must strive for a daily life stimulated by critical thinking, by activist creators who transform our normality and our daily life because this is ruined.

In the 1980s and 1990s, sociologists such as Ulrich Beck, Anthony Giddens, and Scott Lash introduced the concept of reflexive modernity and pointed to the gradual fading of traditions and, consequently, the disappearance of the conventions that governed common daily life. The absence or continued camouflage between good and evil, between right and wrong, and between natural and artificial settles in and persists. We are facing a crisis of conventional ways because we decided, governed, and implemented everything wrong.

The industrial world has largely deviated from aesthetic creation, content to imitate, using substitute materials that allow mass production at low cost. We are already beyond a turning point. I would even say that we have already crossed the Rubicon and there is no going back. In fact, we need to go back, but looking forward.

Risking to contradict myself, if designers were responsible for creating the world we live in, maybe it is up to us to save it, now that we have entered the urgency of the possible.

Design "was" fundamentally concerned with making, but it is time to think about how design education should focus on an appropriate knowledge about business, scientific methods, and technologies, and how people interact with the latter, beyond developing advanced skills in drawing and prototyping [8].

Norman and Mayer [9] also note that design education should emphasize this flexibility not only to act within other contexts of action, but also to develop the discipline itself.

If I were allowed to be intrinsically lyrical, these thoughts about design are what allows it to partake in every area of human activity and what allows it to become really interdisciplinary and this is the most "designerly" [10] idea we should have concerning everyday life.

Lisbon, Portugal Carlos Rosa

References

1. Buchanan R (2009) Thinking about design: an historical perspective. In: Meijers A (ed) Philosophy of technology and engineering sciences. pp 409–453. Elsevier North Holland, Amsterdam; London; Boston
2. Santaella L, Nöth W (1998) Imagem, cognição, semiótica, mídia. Iluminuras, São Paulo
3. Santaella L (2012) Semiótica aplicada. Cengage Learning, São Paulo
4. Nöth W (1997) Representation in semiotics and in computer science. Semiótica, Berlin, vol 115, pp 203–215
5. Nöth W (2013) Semiótica visual. Tradução de Rodrigo Morais. Tríade Comunicação, Cultura e Mídia—Revista do Programa de Pós-Graduação da Universidade de Sorocaba ISSN 2318-5694. Sorocaba: Universidade de Sorocaba, vol 1, no 1, pp 13–40
6. Colapietro V (2014) Peirce e a abordagem do self: uma perspectiva semiótica sobre a subjetividade humana. Intermeios, São Paulo
7. DeTienne A (2012) Time and the flow of signs: semiosis and chronogony. VI Advanced seminar on Peirce's philosophy and semiotics, São Paulo, no 15, ano XV, Livro 15, pp 39–57. Co-edição: Centro Internacional de Estudos Peirceanos, Programa de Estudos Pós-Graduados em Comunicação e Semiótica, Programa de Pós- Graduação em Tecnologias da Inteligência e Design Digital and Pontifícia Universidade Católica de São Paulo
8. Hernández-Ramírez R, Morais R, Rosa C (2021) Foundations, research, and transdisciplinarity: re-shaping education for the 21st century at a school of design and technology in Portugal. In: Martins N, Brandão D, Moreira da Silva F (eds) Perspectives on design and digital communication II. Springer Series in Design and Innovation, vol 14. Springer, Cham. https://doi.org/10.1007/978-3-030-75867-7_17
9. Meyer MW, Norman D (2020) Changing design education for the 21st century. She Ji J Des Econ Innov 6:13–49. https://doi.org/10.1016/j.sheji.2019.12.002
10. Cross N (2006) Designerly ways of knowing. Springer-Verlag, London

Preface II

Design is everywhere, and its importance is undeniable. However, the social dynamics, the diversity of practices, the multiplicity of fields of action and research, as well as some controversies require a continuous exercise of debate and critical reflection from the community, focused not only on the past and present, but, above all, looking to the future.

Taking on the role of an active promoter of this discussion at an international level, UNIDCOM/IADE—Research Unit in Design and Communication has, since 2003, organized an international conference called, since 2011, *Senses & Sensibility*. And it is, precisely, the best articles from the 10th edition of this conference, *Senses & Sensibility '19: Lost in (G)localization* that make up this work. The intention behind the creation of this book is, therefore, to offer material for reflection on design, its role, and contribution to society, according to various perspectives and frameworks.

Although the conference took place at the end of 2019, when little was yet known about the new virus discovered in Wuhan, China, and that would become the cause of the COVID-19 pandemic, the theme of this edition was strangely prophetic. It challenged the community to discuss what role design was playing in the (un)balance between global and local approaches that were beginning to clash in an increasingly fluid world. What direction should be taken next to address the complex societal problems threatening humanity?

To organize the conference, academics from all over the world were invited to promote debates on major contemporary structuring design and design research agenda themes, crossing borders between knowledge and areas, roughly following Buchanan's [1] four orders of design. This collaborative exercise resulted in twelve tracks, leading to the seven parts making up this book: design for communication and branding; design for new materials and new manufacturing technologies; design for interaction; design for health and well-being; design for education; design for culture; and design for society.

It is our ambition that this book may be useful not only in the academic sphere, in particular within the scope of design teaching and research, but that it may also be useful to design practitioners and all stakeholders, from different sectors of society, active in this vast disciplinary area. Finally, we also hope that it will be the catalyst for a timely and necessary reflection on design's role and impact on society.

We are grateful to the many people not only who helped us to make the conference happen, but also who contributed to the organization of this book.

Thank you all so much for your involvement, commitment, and dedication. And, above all, thank you for believing in the power of design.

Lisbon, Portugal Emilia Duarte

Reference

1. Buchanan R (2001) Design research and the new learning. Design Issues 17(4):3–23

Contents

Design for Interaction

Design for Health and Wellbeing

Design for Education

Design for Culture

Design for Society

Design for Communication and Branding

The Relevance of Communication Design and the Undeniable Power of Brands

Daniel Raposo, Fernando Oliveira, and Catarina Lélis

Abstract Communication design can well be framed as the heartland of effective communication. Brands are the result of a communication design exercise and their impact on lifestyles, either in terms of behaviour change or adaptation, ways of communicating, or one's own perception of the human condition and the world, is well established. On the other hand, research fulfils the purpose of questioning, understanding, and transforming the world. This text presents a set of articles that open different perspectives on how brands can engage with ground-breaking communication design and how they transform the world we live in.

Keywords Branding · Communication design strategy · Brand · Brand language

Communication has been, since always, the greatest means for humans to define social connections, create emotional bonds, understand problems and identify solutions, establish principles and limitations in life, which is tailored to take place, for this species, in society. On the other hand, communication has also been the major responsible for misunderstandings, disagreements, inconsistencies and tensions between humans. For design, this is a central, instrumental and twofold quest: to augment

---

D. Raposo (✉)
CIAUD, Lisbon School of Architecture, Universidade de Lisboa, Rua Sá Nogueira, 1349-063 Lisboa, Portugal
e-mail: draposo@ipcb.pt

RETHINK, Research Group On Design for the Territory/Polytechnic Institute of Castelo Branco, Av. Pedro Álvares Cabral, nº 12, 6000-084 Castelo Branco, Portugal

F. Oliveira
UNIDCOM/IADE - Unidade de Investigação em Design e Comunicação, IADE – Faculdade de Design, Tecnologia e Comunicação, Universidade Europeia., Av. D. Carlos I, 4, 1200-649 Lisboa, Portugal
e-mail: fernando.oliveira@universidadeeuropeia.pt

C. Lélis
University of West London, St Mary's Road, Ealing W5 5RF, London, UK
e-mail: catarina.lelis@uwl.ac.uk

E. Duarte and C. Rosa (eds.), *Developments in Design Research and Practice*, Springer Series in Design and Innovation 17,
https://doi.org/10.1007/978-3-030-86596-2_1

communication's qualities and facilitate interaction whilst obviating its matters of contention and discrepancies. As Aakhus [1] mentions, communication design takes place when an intervention into a determined ongoing activity is made. Said intervention can be done through the creation of new techniques, artefacts, and processes or procedures, aiming at redesigning interactivity and the implicated avenues for communication. The author also states that the relationship between interaction and communication is a vital problem for design.

Interaction is one of the main concerns and strengths of design as a discipline. In fact, it is difficult to conceptualise design that is not human-centred and that does not consider the human as its main source of relevant and meaningful information. The human being is characterized by its ability to intrinsically communicate, associated with life in society and defined by a complex system of signs created together, that do not exist individually, depending on the interaction of people whilst members of a community. The value of things and the judgment humans make of people, objects and entities is based on a repertoire of cultural symbols that lives in us, in our minds, and that is constantly growing [11].

Created and tailored either by experience or artificially, signs fit into models or networks that provide them with a certain context (and value) in a given culture—known as codes of use and understanding. These codes (or networks of signs) are responsible for informing and defining the creation of brands in their symbolic, visual and communicational dimensions.

Brands are the result of a communication design exercise and their impact on lifestyles, either in terms of behaviour change or adaptation, ways of communicating, or one's own perception of the human condition and the world, is well established. As a branch of communication design, brand visual identity is a conversation between a brand and its clients or users, namely when brands have such a strong presence and influence in society, in the way we, humans, learn about the world and ourselves.

In fact, brands reflect and represent our culture: they seduce and guide people to a certain lifestyle, they provide memories and stories that enliven our sense of morals and ethics and, in some cases, they actively participate in the lives of people, acting as content producers in today's society of information.

Society is defined itself by brands as these are the ones that clarify intentions and purposes… even in visionary approaches, such as in cinema masterpieces like Blade Runner, the presence of brands is a reality. Acknowledging the assumption that individuals project the future through the known patterns of the present, we live in a hypermodern reality that cannot be expressed without brands.

When one discusses the concept of Brand, one talks about a complex system that involves several sub-systems [18]. These address products or services, management, logistics, spaces, communication, among others. Within these, the visual dimension appears as the direct link between a certain brand and its external and internal audiences, materializing the brand/entity's message. And this is most due to the fact that humans are, with few exceptions, essentially, visually driven beings.

It is through our senses, especially sight, that we perceive most of the world around us, be it real or virtual, and communication design plays the role of defining these realities by embodying the languages that individuals identify with. For better or worse, designers are sculptors of culture responsible for a process that assigns meaning to the referent, paving the way for interpretation. In that sense, communication designers hold the power to influence people and change perceptions. In fact, some authors attribute to communication design the ability to reflect society's modes of thinking in a certain period of time [9]. This theory is related to the idea that design serves people and their needs in each moment, reflecting the economy, leisure trends and aspirations of individuals.

Similarly, Sudjic [20] refers to design as one of the vectors that characterize "the language of things", along with fashion, luxury and art. In this sense, design takes on a broader perspective than merely communication-related, unfolding in its other aspects, namely in reference to objects and spaces. It is through the intervention of design that the consumer society is shaped and moulded, allowing that the "things", those that we have or want, define the various frameworks/possibilities in the lives of individuals.

By saying that communication design defines reality, we certainly do not exclude other areas that are part of the construction process of messages and that efficiently and effectively contribute to visual communication. As Berstrom [3] says, visual communication is supported by some pillars that give it meaning and structure, solidifying the setup where a given message is processed and its capacity to be characterized by a mutual exchange—or, as commonly used these days: sharing. These pillars contain strategy, narrative, creativity, message and meaning, as well as the actual visual elements such as iconography, typography, colour or images.

Strategy is necessarily present because of the purpose and intention behind the thing to be communicated. Narrative too, because storytelling and its techniques create contact points between brand values and audience's values, even if this is a cultural fabrication implemented by some institutions. It is through creativity that differentiation and innovation are processed, representing the ability to step away from the hugely standardized lines of thought (heuristics) that the brain resorts to in order to simplify routine, as described by Buzan [4], Kahneman [10]. It should be noted that, within creativity, it is the strength of visual representation that facilitates decision-making processes, since it enables the visualisation of possibilities in order to find alternatives.

As for the message and the meaning, they are actually the beginning of the whole process, associating themselves to the context, purpose and strategy, allowing for analytical interpretations that go through deconstruction and understanding, generating interpretations that can be (1) denoted, through direct interpretation, (2) suggestive, through prepositions and hypotheses, or (3) connotative, which result from the combination of various cultural universes [16]. There is also a correlation between the message, the meaning, and the media, as McLuhan [14] defends, since these elements define, within themselves, a context that it is thought to be responsible for the institutionalisation of the so-called unofficial language, where certain forms of expression—such as street-art, junk culture, or even non-sense culture—would see a fit.

Regarding the visual elements themselves, iconography has an intimate relationship with the organization of classes in which signs are contained. Semiotics plays a vital role here, in the sense that it allows understanding how societies have related to signs and symbols throughout times, most of it present in facts that are essential parts in the history of writing.

In that sense, typography also represents the visual expression of the word and should be seen in both a macro and micro context, since it enhances notions of personality and functionality. Hence, it has a close relationship with visual communication as it is at its very origin, from the moment the alphabet becomes a vehicle for transposing visual messages. In brands, typography is usually the stable element [12]. Apart from cases where typography is a dynamic feature of a visual identity, most brands opt its stability and this translates into the use of few typographic elements so that the notion of visual family is more easily recognised. Even in the so-called dynamic brands [22], in which diversity characterises their visual landscape, typography emerges as a unifying element of the system, such as in the cases of EDP (designed by Sagmeister & Walsh), of the New York brand (by Wolff Olins) or the MAD museum (by Pentagram). As for colour, it speeds up the emotional element, being able to transpose the meanings to which the message is attached—although acting differently in different cultures. However, it always has a strong emotional presence in the communication process, reinforcing the message to be conveyed.

In the context of imagery, there is also much of what visual communication transports on to audiences [13]. It concerns the content and visual characteristics of photography or illustration, elements that characterize the style and tone of voice of an identity and visual communication project. This component alone certainly justifies greater attention, as it is at the origin of the discipline known as Visual Culture and which, today, encompasses the entire visual universe here described and is not just limited to the construct of image. However, image is an increasingly important trend in contemporary communication design. Crow [6] even speaks of a transition from word to image in the way of communicating, giving examples such as the case of Benetton or Forma since today we communicate so much through our mobile devices, using characters to represent concepts that, through processes such as juxtaposition or combination, even become images, as the very popular emojis. In this sense, and alluding to an aforementioned case, the visions of a world in the future projected in Blade Runner are populated with way more images than words in what seems to be constant entertainment for the public.

1 Afterword

1. Currently, society is still proliferating through a phenomenon of globalisation that tends to absorb local cultures to simplify the message of certain brands when the latter target their audiences. Such simplification removes the presence of what is genuine and unique to people, and which is still an integral part of local cultures. However, and as postmodernism rests its pillars in the past, the recovery of this heritage can generate authenticities since it maintains the diversity that seems to be an innate characteristic of the human being. The

research work that seems to have become a case study in this area is the one addressing the heritage of Sanjo [5]. In it, communication design assumes itself at several levels as there is a framing of the visual component as a kind of anthropology, recovering the heritage of the brand and its products, as well as their entire historical and social framing, in a clear relationship with Heller and Fili [9] theory already mentioned, on the relationship with societies and their thinking.

2. It is precisely in this context that two of the chosen articles fit. The "Reactivation of graphic memory and technical knowledge of the past factors of industrial competitiveness" by Nuno Coelho, since it revives the memory of letterpress and the printed object that tell a lot about the history of communication design, but which can also essentially be means of innovation, exploring the plasticity of the technique itself and even its experimental potential. In the case of "The influence of feminism on the development and branding strategies of fashion brands" by Camila Lamartine and Lililana Ribeiro, the framing of revivalism and the perception of the past also exists, focused on an area that is thematic of the visual culture itself, aimed at a phenomenon that represents not only our way of dressing but also our social function [20]. However, what seems more relevant to us is that it adds difference, insofar as feminism was marginalised and, in this sequence, its own languages were created, often with different approaches to what was institutionalised. This situation has triggered a system in itself, assuming a preponderant role in contemporary Visual Culture, which allows interpretations and simultaneously provides a contribution to new realities.
3. In our contemporary society, pictographic language is a strategy to visually represent abstract or complex concepts in an inclusive, intelligible, and integrated way.
4. In the 1920s, Otto Neurath (1882–1945), with the support of Gerd Arntz (1900–1988) and Marie Neurath (1898–1986) proposed ISOTYPE (International System of Typographic Picture Education), a pictographic language to make complex or statistical information more accessible to all. Since then, the work of Neurath, Arntz and Neurath has been a reference for guidance and information systems, namely at an infographic level. This notion of a pictogram as a graphic synthesis of an idea integrated in a visual system (with its own character and without being considered illustration) was adopted by many designers such as Otl Aicher (1922–1991) and had repercussions in projects such as ASOTYPE (Animated System of Typographic Picture Education) by Yoonsun Joo, who aimed at studying ways to animate pictograms. Pictograms are used on different occasions, changing the meaning of the referent depending on the context, strongly present in urban settings but also found in rural areas, with specific orientation and informational purposes, in products or associated with the communicational, rhetorical and symbolic intentions of brands, in physical, audio-visual and digital artefacts. The use of pictograms in the context of branded communication and information has increased in digital media, justifying the relevance of studies in understanding their effectiveness

when they are animated. It is in this context that the third article on "Moving pictograms" by Maria Diaz, Carlos Rosa and Liliana Faria is included.

5. On the audiovisual dimension, movement, rhythm, and sound are an integral part of visual communication too, as they reinforce the stream of emotion that is conveyed by the sequences of images that characterise these kinds of media.
6. This is especially true because images have the power to trigger secondary associations, opening possibilities for connotation or to the creation of scenarios that, through memory and imagination, guide the audiences' ways of thinking once these perceive them. Pragmatics alludes to this, from the perspective of semiotics, referring to the meaning and understanding of a sign based on the recipient's cultural repertoire. The same idea is reinforced by the neurosciences concept of Default Mode Network (DMN), indicating that the human brain stores structures of narrative that are activated when one hears/views oral, written and audiovisual stories [7]. In short, the syntactics of imagery, as well as the other elements of visual communication, have the potential to correlate the meaning attributed by pragmatics, giving context and meaning to the brand message, in a given culture [19].
7. Brands inhabit our collective imagination, constituting themselves as narratives that include symbolic, idealistic and emotional experiences, always determined by (and determining) a given context. It is this characteristic that allows brands to outlast the companies that create them, to transform and combine in time according to society's behaviours and symbolic repositories. In fact, there are several brand revitalisation projects from the first half of the twentieth century, in terms of code and graphic language, currently positioned as premium. Nuno Coelho's study is also a good example of this.
8. On the other hand, design and brand research fulfil the purpose of questioning, understanding and transforming the world. Research arises from the curiosity to know more, together with the need to improve and contribute to quality of life. And in this sense, we can summarise that research is a systematised way of looking for reliable and useful knowledge, in a given context.
9. Research in visual communication is a way of accelerating the sharing of knowledge within a group or community, but also a strategy to identify, prioritise and structure thoughts and data, fostering the enjoyment and understanding of complex realities. Images and diagrams, in particular, have the potential to increase the effectiveness of the investigative process, as they provide a visible form to data, concepts and methods, facilitating collaborative/participatory discussion or even learning by doing [16]. Nevertheless, it is interesting reflecting on Downtown's [8] perspective, who defines design as a way of asking, producing and knowing, the practice of design tends to be empirical, based on non-systematised research and not research-driven, as it is more an interface than a way of producing new knowledge [15].
10. In the realm of design and brands, research necessarily includes the object, its context, people, communication, and culture. Although the dimensions referred to in the previous sentence can be studied separately, they are entirely indivisible in the scope of this kind of research, since it allows knowing the state

of a problem in its dependent relation with multiple variables that, together, define the context, hence enabling the construction of solid arguments about a given situation. As Toulmin [21] points out, an argument can be used to convince and engage another person based on a certain common idea. Brands are essentially arguments that include symbols, ideologies, and feelings, which are made visible and experienceable through design.

Such transformative strength endows designers with the power to change the others and the world. In this context, the contribution of designers to improve the world we live in seems to be real. We can use our skills to seduce audiences in order to improve their consumption habits, their lifestyle and also the quality in the workplace, in travelling, in being treated, educated, and even governed. The manifesto First Things First (1964) was already an anthem to a different society where communication design should serve greater purposes and not just a consumer society [17]. Fortunately, it has been recently revisited and adapted to the current days and consumption problems we all face.

References

1. Aakhus M (2007) Communication as design. Commun Monogr 74(1):112–117
2. Walsh A (202) And Walsh creative agency. https://andwalsh.com/work/all/edp/
3. Berstrom B (2009). The essentials of visual communication. Laurence King Publishers
4. Buzan T (2001) Use your head. BBC Worldwide Limited
5. Carvalho de Almeida PAS (2012) Brand archives: the rescuing of locally specific brand imagery as a graphic design response to the globalization of visual identity. PhD thesis, University of the Arts London, United Kingdom
6. Crow D (2006) Left to right: the cultural shift from words to pictures. Ava Publishing
7. Dehghani M, Boghrati R, Man K, Hoover J, Gimbel SI, Vaswani A, Zevin JD, Immordino-Yang MH, Gordon AS, Damasio A, Kaplan JT (2017) Decoding the neural representation of story meanings across languages. Hum Brain Mapp 38(12):6096–6106. https://doi.org/10.1002/hbm.23814
8. Downtown P (2003) Design research. Royal Melbourne Institute of Technology, RMIT
9. Heller S, Fili L (1999) Typology. Type design from the victorian era to the digital age. Chronicle Books
10. Kahneman D (2013) Thinking, fast and slow. Farrar, Straus and Giroux, NY
11. Leeds-Hurwitz W (1993) Semiotics and Communication: signs, codes, cultures. Lawrence Erlbaum Associates
12. Lelis C, Leitão S, Mealha O, Dunning B (2021) Typography: the constant vector of dynamic logos. Vis Commun. https://doi.org/10.1177/1470357220966775
13. Lupton E (2011) graphic design thinking. Beyond brainstorming. Princeton Architectural Press
14. McLuhan M (2005) The medium is the message. GP
15. Muratovski G (2016) Research for designers. A guide to methods and practice. SAGE Publications
16. Noble I, Bestley R (2005) Visual research: an introduction to research methodologies in graphic design. AVA Publishing
17. O'Brien M, Dharia N, Gaydos B, Harrison S, Yezbick J, Shanley SK, Cellinese R, Binell R (2020) First things first. A manifesto—2020 edition. https://www.firstthingsfirst2020.org
18. Oliveira F (2018) Brand's identity and visual csulture. In: Raposo D (ed) Communicating visualy: the graphic design of the brand. Cambridge Scholars Publishing, pp 117–138

19. Raposo D (2018) Communicating visually: the graphic design of the brand. Cambridge Scholars Publishing
20. Sudjic D (2008) The language of things. Penguin Books
21. Toulmin SE (2003) The uses of argument. Cambridge University Press, Cambridge
22. Van Nes I (2012) Dynamic identities: how to create a living brand. BIS Publishers, Amsterdam

Reactivation of Graphic Memory and Technical Knowledge of the Past as Factors of Industrial Competitiveness—The Case Study of the Development of Cento & Vinte at Confiança's Letterpress Workshop

Nuno Coelho

Abstract In 1950, Confiança Soap and Perfume Factory created a letterpress workshop on its premises, a rare example in the Portuguese industrial panorama. However, today, the use of its letterpress workshop has fallen out of use and is in danger of being dismantled. Complementing and in coordination with a historical and theoretical research and fieldwork, a design project component materialised in the "research-production" modality (Moreira in Edifícios & vestígios: Projeto-ensaio sobre espaços pós-industriais, Imprensa Nacional Casa da Moeda, Lisbon, 2012)—a project-based research methodology—was developed, resulting in the development of a new product ("Cento & Vinte"). The intention was to recover composition and printing processes through the manual nature of letterpress, without any use of digital editing processes. Working from concrete heritage and through the study and analysis of artefacts of material culture, this project-driven research seeks to develop mechanisms for the preservation of collective identity and memory. The main objective was to be able to contribute to the preservation of graphic memory and technical knowledge of the past as factors of industrial competitiveness. The design project component was then understood as interventionist, viewing the designer as a producer of meaning and as a non-neutral subject. The intention was to assume a more exploratory and experimental character, moving away from conventional methods of contemporary product development in order to approach a more critical practice. The developed product should not be exclusively analysed on its final formal configuration but, essentially, on the formulation and development of its concept of origin. It was our intention to contribute to the preservation of the brand's memory and also to the reactivation and maintenance of its letterpress workshop—which is now in danger of disappearing.

N. Coelho (✉)
Univertity of Coimbra, Centre for 20th Century Interdisciplinary Studies (CEIS20), Coimbra, Portugal
e-mail: ncoelho@dei.uc.pt

Department of Informatics Engineering (DEI), University of Coimbra, Coimbra, Portugal

E. Duarte and C. Rosa (eds.), *Developments in Design Research and Practice*, Springer Series in Design and Innovation 17,
https://doi.org/10.1007/978-3-030-86596-2_2

Keywords Research-production · Design project-based research methodology · Letterpress · Packaging · Labelling

1 Introduction

In 1950, Confiança Soap and Perfume Factory (founded in 1894 in Braga, in Portugal) created a letterpress workshop on its premises with around 60 different typefaces and well as printing and finishing machines, ensuring the production of packaging and labelling for decades, a rare example in the Portuguese industrial panorama. Even today, Confiança continues to use on its labels the same graphic motifs which were developed mostly during the first half of the twentieth century. However, the use of its letterpress workshop has fallen out of use and is in danger of being dismantled.

Complementing and in coordination with a historical and theoretical research and fieldwork, as part of a scientific research into the history of this factory and the letterpress printing processes, a design project component materialised in the "research-production" modality [9] was developed, resulting in the development of a new product. This design project component was then understood as interventionist, viewing the designer as a producer of meaning and as a non-neutral subject. The intention was to assume a more exploratory and experimental character, moving away from conventional methods of contemporary product development in order to approach a more critical practice. We have used the term "contamination" as the result of this project-based research methodology which is intended to insert itself into the history of this producer.

With this exercise—which resulted in the creation of the product label "Cento & Vinte" ("One Hundred and Twenty")—the intention was to recover composition and printing processes through the manual nature of letterpress, nowadays widely understood as an obsolete system, without any use of digital editing processes. Working from concrete heritage, in this case the internal letterpress workshop and all the typographic elements available on the factory premises, the intention was to reinvent the memory of its brand in contemporary times. With the clear objective of preserving this graphic memory and its material heritage, the preservation of work processes stemming from the objects with a great focus on the manual aspect of human gestures was intended.

The product name was chosen to celebrate the 120th anniversary of the brand, following a similar rationale developed by the factory in previous years. In an exercise in enhancing the letterpress process and reusing existing typographic elements, the idealised composition was conditioned by the collection of typographic characters and printing machines available. The production of new engravings or any other typographic elements for the purpose of this work was not requested, reinforcing the concept of reutilisation of existing material heritage. We were interested in exploring the idea of continuity of the typographic elements already used in countless labels in past decades, but currently without any function or utility.

Through the study and analysis of artefacts of material culture, this project-driven research seeks to develop mechanisms for the preservation of collective identity and memory. With this design project component, the main objective was to be able to contribute to the preservation of graphic memory and technical knowledge of the past as factors of industrial competitiveness. The developed product, which is also now part of the factory's graphic heritage and history, should not be exclusively analysed on its final formal configuration but, essentially, on the formulation and development of its concept of origin. With this project it was our intention to contribute to the preservation of the brand's memory and also to the reactivation and maintenance of its letterpress workshop. The existence of this internal letterpress workshop makes Confiança if not a unique case, undoubtedly a rare case in the industrial panorama of Portugal in the twenty-first century—which is now in danger of disappearing.

2 Historical Context and Justification of the Practical Project

In Europe, until the mid-fifteenth century, book production was rather limited, as they were handwritten by scribes, most of whom were members of the clergy (usually then monk copyists). Books were extremely expensive objects, and only the higher classes, namely the nobility, could afford them. The information was difficult to access, not only because of the prohibitive price of books, but also because it was the clergy's responsibility to decide which books were to be reproduced.

The German Johannes Gutenberg is credited with inventing the printing press around 1450. However, it was the adaptation to the Western alphabet of a Chinese creation—movable type [2]. The technical innovation introduced by Gutenberg in Europe allowed a considerable increase not only in the number of print runs of the same work but also in the variety of published works, constituting "the first mass production system" ([8], p. 65). The press democratised access to information in such an impactful way in society then, as perhaps only paralleled in the late twentieth century with widespread access to the Internet.

For a long time, movable type was the dominant and practically hegemonic technique in the publication of books and periodicals. This mechanised form of text reproduction constituted a radical break with the typographic tradition based on (cursive) calligraphy ([8], p. 21). According to Lupton, in the letterpress printing process "the relationship between the letters of a font has become more important than the identity of each character". From the fifteenth century, "movable types of metal were the best way to transmit information" [1]. It was only in the late nineteenth century that "other methods of composing and reproducing texts gradually replaced this old form of printing" [1]. Later, with the widespread use of the offset lithographic system over the course of the twentieth century, which is "today [the] hegemonic technique in the printing industry" ([7], p. 296), the letterpress printing process is now regarded as a virtually obsolete system. This was further enhanced in the late twentieth century with the introduction of the digital printing system, which establishes a direct link between computer and printer.

Small letterpress workshops that have not modernised their printing processes either survive today with many difficulties or have even ceased their activity. Currently, the use of the letterpress printing process in printing graphic materials is rarely found and the trend indicates that it is progressively moving towards its eventual disappearance. The letterpress printing process is time-consuming and largely dependent on manual work, characteristics that lead to the preference for other more modern and automatic techniques. Other reasons for abandoning the letterpress printing process may be pointed out, such as the slight graphic differences in print quality between copies of the same print run or, mainly, due to the fact that letterpress composition does not include digital compositing processes at a time when the use of computers in the graphic industry is unavoidable with the growing demand for time optimisation of technical and human resources.

However, despite the progressive abandonment of this technique, there is also a current of reactivation and valorisation of this visual printing language with movable types, through sporadic private initiatives but also by research groups, institutions and organisations [1]. The practical work that serves as a case study of the present communication intends to fit into this current, assuming that "movable type has great importance in the configuration of history and graphic memory" [1].

Considering the case study of this communication, we understand that the letterpress printing process strongly influenced and determined the identity of the graphic language developed by Saboaria e Perfumaria Confiança (Confiança Soap and Perfume Factory) in the conception and production of the labels of its products. The technique directly influenced the aesthetic formal aspect of the graphic materials of this industrial unit founded in 1894 in Braga, Portugal. Prior to the creation and installation of the in-house letterpress workshop on its premises in 1950, its labels had also been produced using movable type by employing various letterpress workshops in Porto, a town approximately 55 km from Braga. The description of this letterpress workshop will be thoroughly analysed later in the current text.

We understand this current processes of reactivation and valorisation of the letterpress printing process in a context very similar to that described in the description and analysis of the project Serrote [13]. This designation brings together the creative talent of artist Nuno Neves and designer Susana Vilela (Nuno Neves studied painting at the Faculty of Fine Arts of the University of Lisbon and Susana Vilela studied design at the same educational institution, where they met). At the end of 2004, they began looking for letterpress workshops in Lisbon with the intention of producing "Gutenberg-like" notebooks. In this process of investigation, they found that many letterpress workshops had already closed their doors but they were able to find one who was willing to produce the notebooks designed by the couple.

"At the bottom of bug-bitten drawers we've found worn out engravings and lead types that had been used for decades to print invoices and business cards. We made a cover, and in the early days of January 2005, the first Serrote project was born: a smooth, blue-coloured notebook with the cover printed in two colours. In the following years we increased the collection of notebooks at the same time as we were receiving orders for projects of cards, notebooks and invitations" [13]. Since the beginning of the project, the Serrote duo has even been buying movable type and prints from letterpress workshops that have closed down, noting in a recent interview that their workspace is currently "stacked with lead", the metal from which movable types are made [13].

The notebooks with the signature Serrote are of limited and numbered editions, never being reprinted, characteristics that give them the status of being practically collector's items. After more than thirty different notebooks published until 2013, the date of the development of the practical project that serves as a case study here of this communication, in May 2008 they published the first Serrote book—an illustrated book about the Minho region (northern Portugal). Two years later, it was the turn of the neighbouring region of "Tra-los-Montes" (the title uses the sparsely spoken Mirandese language that can only be found in this region). In 2013 they published "Portugal a Cores" ("Portugal in Colours"), an illustrated book with artificially coloured photographs with excerpts from foreigners visiting this country in the eighteenth and nineteenth centuries, "some of them exaggerated, others prejudiced, but almost all true". Other titles published by the duo include "Nouveau Dictionnaire de Français" (2008), "Tradicional Romanceiro" (2011) and "Finlandia" (2011).

Although up to 2013 they have published two books that refer to foreign contexts (France and Finland), practically all the production of the Serrote duo (between notebooks, books, postcards, cards, invitations and various special editions) regards the Portuguese context. And because it is so local, the Serrote project created a global appetite. In addition to Portugal, in 2013 it was possible to purchase products from this brand from points of sale spread over nineteen countries. The revivalism perpetuated by this project involves the identification of the intrinsic and unique characteristics of the letterpress printing process and, essentially, its reinterpretation in contemporary times. As Silva considers the A Vida Portuguesa phenomenon (a successful shop initiated in 2004 by journalist Catarina Portas that only sells products from historic Portuguese manufacturers) to be a "meta-nostalgic project" [14], we can also apply this definition to the Serrote project. The different works produced by the Lisbon duo are short postmodern narratives of the letterpress printing process of other times.

The couple intends to counteract the use of the computer in the creative process at a time when the use of this tool by designers is unavoidable. The supposed freedom afforded by digital editing seems to constitute a condition for both creatives, and it is precisely the technical limitations of the letterpress process that stimulate them creatively. Their reinterpretation of the letterpress printing process often involves the use of ornaments and engravings in the creation of typography, as illustrated by several works done for various clients, namely the Theophile birth announcement card made in 2006 [13]. Nuno Neves and Susana Vilela identify comics, illustration and pixel art as their inspirations but also the "anonymous Portuguese graphic artists who once designed packaging for tuna cans, toothpastes, soaps, candles, antique traffic signs and store signs" [13] (Fig. 1).

Beyond Portuguese borders, another example may be quoted in the context of our practical project, like the Brazilian edition of the book "Thinking with Type" by the

Fig. 1 A Serrote notebook produced with letterpress printing processes for A Vida Portuguesa in 2008 (*image source* the author)

North American designer, curator, teacher and critic Ellen Lupton [8]. Cosac Naify published the first edition of this book in Brazil in 2006, currently considered "one of the most recommended typography works in the world, making it an excellent reference in any typography course and a basic guide for designers and students" [5]. Although the book's pages were printed using the offset lithographic system, the cover for the Brazilian edition was printed through the letterpress printing process with movable wooden type at Fidalga, a letterpress workshop in São Paulo. In this way, it was possible to produce covers with four very distinct compositions, even with some slight graphic differences between covers with the same composition in terms of print quality (slight errors), a natural characteristic of the letterpress printing process. The cover design is by Elaine Ramos with the collaboration of Flávia Castanheira and is a direct reference to the origins of the manual and mechanical process of composition and letterpress printing processes [6] (Fig. 2).

Fig. 2 Cover of the Brazilian edition of "Pensar com tipos" ("Thinking with type") by Ellen Lupton produced with letterpress printing processes at Gráfica Fidalga (*image source* https://elaineramos-estudiografico.com.br/Pensar-com-tipos)

It was our intention to carry out our practical research project, the case study of this present communication, on the same lines of reinterpretation in contemporary times of this visual printing language with movable type, understanding that the two examples explained, especially the Portuguese case, are quite representative for our attempt to contextualise and circumscribe our area of activity. Despite the importance that the Confiança factory's in-house letterpress workshop has represented in past decades, having at the same time employed three printmakers, during our fieldwork we found that it was currently underused and far from being able to contribute to the valorisation of current graphic production of labels from the Braga producer. From the various trips made to the Confiança facilities during our fieldwork, carried out in 2013, we found that the letterpress workshop was merely used for the production of labels of low relevance and of little graphic value. Similarly, it was found that this letterpress workshop was in serious danger of being shut down as soon as the then only printmaker, Álvaro da Silva Gomes, was about to retire. This reality was confirmed later at the end of 2016, and the letterpress workshop was dismantled in 2019.

The objective of this practical project was to recover the value that the letterpress workshop once had in the factory's history, giving it a central, if not fundamental, role in the process of developing new products with the Confiança brand. In this context, we recognise the value of preserving the memory of past technical knowledge as a fundamental factor of industrial competitiveness. In order to test our concept, the development of a product in which the letterpress workshop of Confiança constituted the hinge element of the entire creative process was proposed to the brand administration at the time, the company Ach. Brito. As this practical research project approached the procedural and professional practice of a design project (in which Ach. Brito took the position of client and we as a service provider), we naturally incorporated suggestions proposed by the company's management. The project process was therefore developed in close relationship with the brand owner. We incorporated methodologies of "research-production" ([9], p. 104), which implies that each research should have a practical outcome as an essay; on the other hand, the production of a project is a research in itself.

It was our concern that the practical project would have the approval of the brand administration at the time so that the resulting product would in fact be mass produced and marketed. The marketing guidelines established by the management were taken into account, even as the Confiança brand was, in 2013, in a process of repositioning itself in the market resulting from the recent acquisition by its rival and centenary competitor Ach. Brito in 2008. The product developed was scheduled to be launched in 2014, the year in which the 120th anniversary of the foundation of Confiança was going to be celebrated. The fact that the practical project, in the end, was not taken up by the factory administration at the time will be analysed in the conclusions of this text.

3 Description of the Letterpress Workshop of Saboaria e Perfumaria Confiança

The establishment of an in-house letterpress workshop at Confiança was started in 1950, the year the first printer was purchased. The purchase of the remaining two printers was phased in over a period of approximately one year. The three printers were on the current factory premises at the time of the development of the practical project, and were fully operational.

Through interviews with current workers of the factory, particularly those with more years of service, eight printmakers were identified who had worked at the Confiança letterpress workshop throughout its existence, where, at one time, three printmakers were simultaneously employed. Direct contacts were established with two of them: António Meireles, one of the main printmakers and who has been at the service of Confiança for more years; and Álvaro da Silva Gomes, the only one working at the factory in 2013 and the last person to work there, having retired in 2016.

Just before 1950, the very first equipment purchased for the letterpress workshop was a manual printer (press) of the brand Luís Maceira Builders which was not in service at the time of our research. Due to its fully manual printing system, this printer was considered obsolete. Then, the three printers mentioned above were bought: two Minerva Original Heidelberg colour printers and an Original Heidelberg Cylinder colour printer. According to reports from Confiança employees, we are led to believe that the later was the third machine of its kind in Portugal, at a time when there were only two others in the country, one in Porto serving the newspaper O Comércio do Porto and another in Lisbon serving the newspaper Diário de Notícias (Fig. 3).

In addition to the three printers, Confiança's letterpress workshop purchased a Johne Perfecta paper shearing machine; a J. Sandt AG cutting press; a second shredder; a dry relief machine; a hand press for engraving and cutting; and a gilding machine, commonly also known as a glitter machine. The letterpress workshop was complemented by a carton section consisting of six machines: an Aug. Kolbus Rahden

Fig. 3 Printers at Confiança letterpress workshop on the factory premises in Braga (*image source* the author)

Fig. 4 Letterpress drawers at Confiança letterpress workshop on the factory premises in Braga (*image source* the author)

creasing machine (acquired in 1958); a second creasing machine; two corner cutting machines; and a card guillotine. It was not possible to identify the brands of some of these machines.

In addition to the letterpress printers and cartoning machines, the Confiança letterpress workshop had a collection of movable types, blades, stripes and clichés, elements that are distributed in 61 letterpress drawers. These elements are identified in a 1958 typographic catalogue published for internal use ([4], Annex 2). In addition to these elements, during the 1990s, Confiança acquired another six new movable type sets from a Braga letterpress workshop (the name of which was not possible to confirm) which closed its doors at the time. Then, an extra sheet was added to the typographic catalogue. This catalogue does not reproduces the names (or designations) of each movable type set, instead using Confiança employee names, Confiança warehouse addresses in various cities across Portugal, product brand names and advertising slogans created, among other information (Fig. 4).

On the factory premises there is also an unidentified number of engravings purposely produced exclusively for the composition of certain labels and packaging. The exact number of these engravings is unknown, as the company has never, throughout its history, carried out a systematic process of inventorying and cataloguing its graphic assets.

4 Analysis of the Typographic Catalogue and the Letterpress Drawers

In order to develop our practical research work, we decided, in the first phase, to carry out a systematic analysis of the typographic catalogue so that, in the second phase, on the Confiança premises, we would search for and collect other elements that could be included in the project. Our work would be heavily conditioned by the elements we would find. We consider these apparent constraints and limitations as a stimulus for the creative process (as mentioned in the description of the Serrote project). This approach also sought to optimise existing resources and was purposely chosen not to pursue the production of new elements. The project would thus be carried out in counter-cycle to an usual design process where it is common for the designer to determine what is produced specifically for a given project. By using typographic elements from the current Confiança estate, the same elements that gave rise to the composition of a significant part of the brand's labels and packaging, we intended to reactivate this same graphic legacy.

The existence of the internal typographic catalogue of 1958 allowed the prior identification and study of the elements available at the Confiança letterpress workshop before the development of the practical project. However, it was found that the catalogue was organised in a careless way, with the types organised by families, not in a rigorous fashion. As an example, we can indicate that sizes 8, 10, 12 and 28 of the movable type that we identify as "type B" are on one page (page 5 of the catalogue) while size 48 of the same type is on a different page (page 8 of the catalogue). In addition, no type is identified by name except a single case called "Machine Type", whose design mimics that of typewriters. In the light of these findings, we decided, in the first instance, to undertake a systematic analysis of the typographic catalogue, seeking to identify and classify the types and other typographic elements available throughout the sixteen pages of the catalogue (including the cover). This study was an essential step in the process of familiarisation with the typographic elements available at the letterpress workshop before moving in person to its facilities for the development of the practical work (Fig. 5).

The catalogue subcategorises typefaces into three distinct groups: "common", "cursive", and "fantasy". "Common" is the designation given to the roman or Didot characters, "very pronounced and bold typeface characters with liform bases" [11], and elzevires, "types used by the Dutch Elzeviers, a family of printers and booksellers" [11]. "Cursive" types are those that are reminiscent of the handwriting gesture, imitating calligraphy. Finally, "fantasy" types are all "whose traits deviate from ordinary types" [11] which, in the digital age, houses an uncountable number of type designs of the most diverse nature.

59 different combinations of type design and sizes are available in the catalogue, nine of them relating to "common" types, 13 to "cursive" types and 37 to "fantasy" types. Upon further examination, we began by identifying and grouping the types of the same family (of the same design) when only their size and/or weight varied (variation from the main family design, i.e. normal weight variation, which includes designations such as "bold", "thin", "condensed", or "italic", among others). We tried to identify the names of each type and, in the absence of correct identification, assigned them names so that they could be distinguished from each other (types "A" to "T"). In addition to types, five different blade designs are available in the catalogue ("thin wire", "half cane", "2 point black wire", "4 point black wire" and "spike"), twelve "stripe" designs and five "cliché" designs. These elements were not included in our classification as it only focuses on type design analysis.

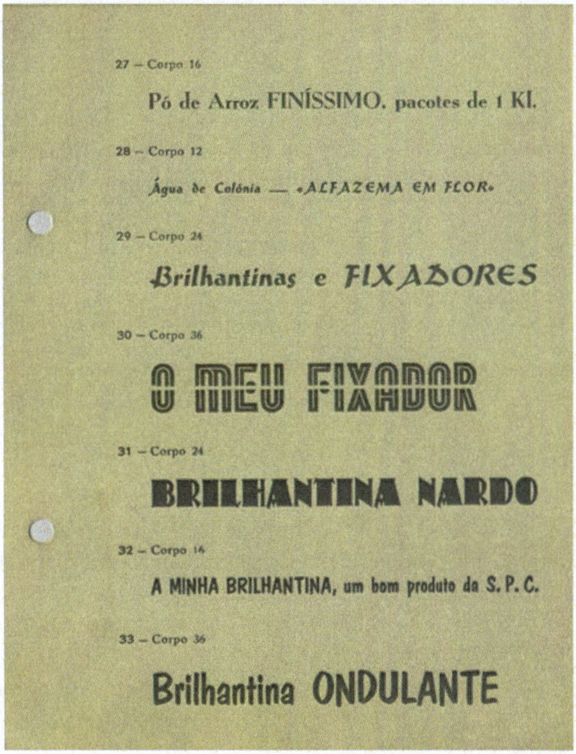

Fig. 5 One page of the typographic catalogue for internal use of the Confiança letterpress workshop (*image source* the author)

From the analysis it can be concluded that eleven different body sizes are available in the Confiança letterpress workshop, with four types of 6 points, eight of 8 points, ten of 10 points, seven of 12 points, seven of 16 points, two of 20 points, four of 24 points, five of 28 points, five of 36 points, five of 48 points and finally two of 10 quads. Point is a measurement that derives from the English metric system and corresponds to 1/72 inches. The 12-point typographic measurement is also known in Portugal as "cícero". Unlike the point that refers to the height of the character, quads is a measure that "corresponds to the normal width of the body" [12]. We need to note that this measure corresponds to 8 points in Lisbon and 12 points in Porto [11, 13].

At a later stage, carried out at the Confiança letterpress workshop, it was possible to make sure that the catalogue did in fact correspond to the reality of the letterpress drawers. On the factory premises, other available typographic elements were also identified, which were never inventoried by the company, namely engraving prints produced specifically for certain labels and packaging. In addition to typographic elements, an analysis was made of the existing printers and other machines at the Confiança letterpress workshop. This information was gathered after the many visits we made to the factory premises during our fieldwork.

Still regarding the movable types, in the absence of the correct names of the fonts included in the catalogue for internal use, we decided to proceed with their correct identification. We were interested in identifying not only the names of each of them, but also the year of their creation, their authorship and, above all, identifying the provenance of the physical elements. In the first phase of the process of identifying the fonts available at Confiança letterpress workshop, three online applications were used: "What the font", "What font is" and "Identifont". These are applications that allow one to identify fonts (typography) from scanned images submitted by users. This is a sometimes automatic process that proceeds from a mathematical algorithm and/or by answering a series of questions.

Two of these applications proved to be very limiting, even one that could not identify any of the fonts. However, when some of these applications were able to identify names, in almost all cases designations by which types are known today were suggested, not correctly identifying their original names. This is because many types created throughout the twentieth century were redesigned into digital format at the advent of the digital revolution. The switch from analogue mechanical systems to digital technology has dematerialised the fonts, transferring them from the physical format (lead or wood) to the digital format (computer software). Thus, the above-mentioned applications attributed the authorship of the fonts to the designers who carried out this digital process and not to the true authors of the original designs. The designers who converted the fonts to digital format, in almost all cases, gave them new names. Further research using search engines on the Internet and through various websites dedicated to the study of typography revealed the original names of the fonts.

Later, through a new search for the original names of each of the fonts, it was found that the authorship of two of them was attributed to Carlos Winckow, a German typographer who lived much of his life in Spain where he worked for Fundicion Tipográfica Nacional, in Madrid. In-depth research has led us to the 1955 catalogue of the Fundicion Tipográfica Nacional, digitised in full by Josep Patau (a contemporary Spanish type designer) and available on his Flickr page [10]. In this catalogue, we found a very significant number of types available at Confiança letterpress workshop, concluding that most of its movable types were acquired from this Spanish foundry (it is also curious that even in the digital age the designation "foundry" is still in use; therefore, foundry continues to designate companies that design and/or distribute fonts for computers).

Following the same research methodology, it was found that other types had been identified as being authored by Dutch type designers who in turn had developed font designs for the Lettergieterij Amsterdam foundry. The fact that we later found a typeface of precisely this origin at Confiança letterpress workshop, still packed in its original box, confirmed this information. In short, with some exceptions, the fonts at Confiança letterpress workshop have these two origins.

Thus, it can be concluded that the fonts used in the production of Confiança soap and perfume labels from 1950 (the year of the establishment of the in-house letterpress workshop) onwards, are essentially of Spanish and Dutch design and production. Besides these two findings, there was also a predominance of German designs. In addition, the fact that some of the identified fonts were created after 1950 leads us to conclude that Confiança purchased movable types after the establishment of its letterpress workshop. Obviously, all identified fonts were designed before 1958, the date of production of the typographic catalogue.

5 Briefing Concept and Activity Report

"The origin of words lies in the gestures of the body. The first fonts were modelled directly on the shapes of handwriting. However, they are not body gestures but images manufactured for infinite repetition. The history of typography reflects a continuous tension between the hand and the machine, the organic and the geometric, the human body and the abstract system. These tensions, which marked the birth of print letters over five hundred years ago, continue to energise typography today" ([8], p. 13).

The technical innovation introduced by Gutenberg introduced a mechanical process in the reproduction of communicative supports. For centuries, the first phase of the typographic process—composition—was a completely manual process, probably as lengthy as producing a copy of a handwritten copy. It was the hands of the printmakers who, letter by letter, searched for the physical movable types in the letterpress drawers to put them in the intended order. Letter by letter and, consequently, line by line, this gesture was repeated systematically until the desired composition was completed. Hands even dealt with the empty spaces of the composition, spaces that had to be filled with physical objects. Essentially, mechanisation was an attribute of the second phase of the typographic process—printing. In a time before automatic printers, it still largely depended on the body gesture needed to exert the press(ure) printing strength.

The manual nature of the letterpress printing process and also of manufacturing in general, the precise choreography of the gestures of the workers, the dexterity found over years of work, give a product unique characteristics. Even today, in some Confiança product lines, each item is individually inspected and packed by hand. This manual process is a characteristic that has been around since the beginning of Confiança, insistent in its resistance even today and constituting one of the main distinguishing factors of the brand's products.

The final packaging of certain products thus continues to follow the lengthy and delicate process of the hands of workers (essentially female workers), requiring technique, precision and dedication. "Before being involved, the soaps are subjected to a final check that ensures, among other things, the condition of the soap and the print quality of both the soap and the packaging" [3]. Manual packaging remains a trademark of Confiança, properly valued in the catalogues published by the administration. If these body gestures insist on resisting time, contrary to the tendency to mechanise the production process, we understand that the gestures of the printmakers must also be conserved, raising them to the same status of importance.

It was our intention to rescue this gesture, this body movement, as well as the value that the letterpress workshop once had in the history of Confiança. The letterpress workshop and the then only printmaker at the factory's service assumed the central, if not fundamental, role in the process of developing a new Confiança branded product within our practical design project.

Starting from a concrete heritage, in this case the internal letterpress workshop and all the movable types available on the factory's premises in Braga, our intention was to reinvent the Confiança brand memory in contemporary times. Initially observers of the brand history [4], with this project we also wanted to inscribe ourselves in its history, not only paying tribute to a graphic legacy that we understand of enormous value, but also assuming the active role that designers can play in the process of brand re-evaluation. Designers can (and should) position themselves as agents for the preservation of industrial heritage, contributing to the conservation of this memory, which can even be an important factor of industrial competitiveness.

With the clear objective of preserving the graphic memory of Confiança and its heritage, we intended that this practical project corresponded to an exercise of "contamination" of the history of Confiança. We developed this project in an immersive logic, approaching pragmatic methodologies that are close to a practice of design as a service provider activity. As mentioned earlier, in 2013, Confiança's letterpress workshop was limited to producing labels with low graphic value, resuming its production to symbolically less important labels and packaging. As stated before, we were led to believe that the in-house letterpress workshop could eventually be shut down as soon as the then only printmaker retired. This situation was due to the high costs of maintaining the letterpress workshop in operation without the expected financial return. We were, however, of the opinion that the letterpress workshop was undervalued and underused.

From dialogues with the administration and the marketing department, it was found that the letterpress workshop could not produce product labels and packaging for large retailers. The fact that each label, within the same print run through the letterpress printing process, has slight mismatches in composition and/or colour, has led to many lots being rejected by large retailers for the graphic discrepancies, which are characteristic of the letterpress printing processes. For large retailers, it is not acceptable for labels of a product lot to be slightly different in colour from the previous lot still in stock. The administration of the time thus produced almost all labels and packaging through the lithographic offset printing process to these retailers.

The product we developed was intended for a segment other than that of the mass market—the market segment that values the artisanal aspect of the products (sometimes referred as "premium" market). Any graphic inaccuracies resulting from the letterpress printing process could be seen as an added value by certain customers if this was properly announced, the same way as hand-packaged soaps are advertised. In this perspective, the label we presented as a result of our project should not be exclusively analysed by its final formal configuration but essentially as the formulation of a concept or of a system. We are convinced that the process of repositioning the Confiança brand in the market also involves the maintenance, recovery and reactivation of its letterpress workshop. With the practical project, the inclusion of a third manual/handcrafted moment in the Confiança soap production process was considered: to the hand finishing and hand packaging of soaps would now be added the letterpress composition of the label, with emphasis on the fact that the brand values the authentic quality resulting from the manual touch of its printmakers, promoted here to the quality of craftsmen (Figs. 6 and 7).

To counteract the habit of using computer-aided digital process, without taking any of the shortcuts that technology allows today, it was proposed that we would develop the practical design project on the premises of Confiança letterpress workshop, taking the time and effort necessary to achieve an unique graphic design composition, fully hand-produced. The desired gesture in our process reflects the use of an old technique that employs manual and mechanical processes. Thus, we spent long periods of time developing the project in-house with the then only Confiança printmaker, Álvaro da Silva Gomes.

In an exercise in enhancing the letterpress printing process and reusing existing movable types, the idealised composition was conditioned by the collection of types and engravings available. The production of new engravings or other typographic

Fig. 6 The hands of printmaker Álvaro da Silva Gomes composing text (using movable types) for the intended label (*images source* the author)

Fig. 7 The hands of printmaker Álvaro da Silva Gomes composing text (using movable types) for the intended label (*images source* the author)

elements purposely for this work was not requested, reinforcing the concept of reutilisation of existing heritage material. We were interested in exploring this idea of continuity of the typographic elements already used in numerous labels in past decades, but which were then without function or utility.

From the analysis of products from the same category and range in which our project was intended to be inserted, and from the indications provided by the management and marketing department, we found that the mid-range market soaps available on the market, developed by competing brands, boast graphics whose compositions are saturated and over-designed, resorting to the abusive use of "fancy" fonts and usually using glossy and shiny papers. Many of these products can be considered "pastiches", as they are not facsimiles (reliable imitations of old designs) nor are they intended to reflect a contemporary taste. The products whose labels we consider to be "pastiche" are situated in a kind of limbo, a state of uncertainty—at the same time of contemporary creation but intended to resemble those produced in earlier times, without really mimicking the aesthetics of the time. It was our aim to seriously counteract this kind of approach, in which, although it is very recurrent and common, we do not recognise any value.

The composition was therefore intended to be simple and "authentic". The number of colours was limited not only by the technique (the management asked us to use a maximum of three colours) but also to reinforce an idea of simplicity. It was decided not to use any technical embossing or special colours (such as metallised or fluorescent ones) so that the subtle feel of the ink feel on the paper—the slight low relief created by press pressure—would be most highlighted. The use of this tactile sensation was justified by the following three statements (cyclic triangular formulation):

- through the skin we feel the relief of the label;
- the label as "skin" of the soap;
- the soap, which in turn comes into direct contact with our skin.

The name chosen to designate the product was "Cento & Vinte" ("One Hundred & Twenty") in order to clearly identify its country of origin by using the Portuguese language (in the same line of thought as the Italian car brand FIAT which, since 1991, adopted the name "Cinquecento" instead of "500" for its model in production since 1957—the product is no longer a number, to directly refer today to the geographical location where it is produced). The fact that the name refers to Portugal, albeit subtly, reinforces the fact that the product as a whole, with the exception of some of the raw materials used in the production of the soap, is developed in Portugal. In addition, the inclusion of the reference "Braga, Portugal", as Confiança has always been keen to mention throughout its history, gives a precise sense of place and locality in an increasingly global marketplace. The name is also an allusion to the 120th anniversary of the foundation of Confiança, as the product would be launched during 2014. It is, therefore, a celebratory edition as Confiança had done on several previous occasions—it published a commemorative booklet for its 50th birthday and launched the "75" soap and cologne in 1969, the year in which it celebrated

Fig. 8 Various stages of the production of the "Cento & Vinte" soap label (*images source* the author)

Fig. 9 Various stages of the production of the "Cento & Vinte" soap label (*images source* the author)

75 years. The edition of the product "Cento & Vinte" would be limited and marketed exclusively during 2014 (Figs. 8, 9, 10 and 11).

Since both the Confiança brand and the two others owned by Ach. Brito (the namesake brand and Claus Porto) already produce brands whose labels are "decorative", our graphic design was based on simplicity in terms of graphic composition. We used engravings that refer to historical elements, such as the humanisation of industry, a female image created in 1904; that of the façade of Confiança's historic building at Rua Nova de Santa Cruz, built in the early 1920s; and the Confiança logo, created in the early 1950s. The different lead movable types used in the composition, in turn, date from 1950, the year of the creation of the in-house letterpress workshop (Fig. 12).

Fig. 10 Various stages of the production of the "Cento & Vinte" soap label (*images source* the author)

Fig. 11 Various stages of the production of the "Cento & Vinte" soap label (*images source* the author)

6 Conclusions

The label resulting from this practical research project, which would also form part of the Confiança valuable graphic estate, should not be exclusively analysed by its final formal configuration but essentially by the formulation and development of its concept of origin. With this project it was our intention to contribute to the preservation of the brand's memory and also to the maintenance, reactivation and reutilisation of its letterpress workshop. The existence of this in-house letterpress workshop makes Saboaria e Perfumaria Confiança if not a unique case, undoubtedly a rare case in the industrial Portuguese panorama in the twenty-first century.

However, the resulting project formulated from the concept developed did not have the expected effects on the company's management. Not only was the product (here seen as a prototype) not launched on the market, but also the reactivation of the letterpress workshop did not take place. Contrary to our beliefs, the letterpress workshop was later dismantled as a result of the sale of Confiança by the Ach. Brito

Fig. 12 The final result of the "Cento & Vinte" product (*image source* the author)

group to a company of the same sector in 2019. Regarding our case study, we came to conclude that corporate business culture is highly linked to logics of profit, not valuing the experimental and speculative characteristics in which the discipline of design often wants to be registered. We may conclude that some Portuguese historical companies, while recognising their longevity as a distinctive factor in relation to their counterparts, do not yet recognise the importance of incorporating a "design laboratory" component into their DNA that would enhance their heritage.

References

1. Aragão I et al (2012) Um estudo comparativo entre a catalogação dos tipos móveis da Editora UFPE e da oficina tipográfica da FAU USP. In: Anais do X Congresso Brasileiro de Pesquisa e Desenvolvimento em Design. Universidade Federal do Maranhão, São Luís
2. Bringhurst R (2005) Elementos do estilo tipográfico. Cosac Naify, São Paulo
3. Claus Porto (2013) https://clausporto.com/pt. Last accessed 23 June 2013
4. Coelho N (2013) O design de embalagem em Portugal no século XX–Do funcional so simbólico–O estudo de caso da Saboaria e Perfumaria Confiança (Doctotal thesis in Contemporary Art). Universidade de Coimbra, Coimbra
5. Design Colectivo: http://designcolectvo.com/tipografia/pensar-com-tipos. Last accessed 23 June 2013
6. Fidalga: https://www.flickr.com/photos/tipocracia/4464057041. Last accessed 23 June 2013
7. Fragoso M (2012) Design gráfico em Portugal: Formas e expressões da cultura visual do século XX. Livros do Horizonte, Lisbon

8. Lupton H (2006) Pensar com tipos. Cosac Naify, São Paulo
9. Moreira I (2012) Edifícios & vestígios: Projeto-ensaio sobre espaços pós-industriais. Imprensa Nacional Casa da Moeda, Lisbon
10. Patau J (2013) https://www.flickr.com/photos/pepel. Last accessed 23 June 2013
11. Portal de Artes Gráficas: https://www.portaldasartesgraficas.com. Last accessed 23 June 2013
12. Priberam: https://dicionario.priberam.org. Last accessed 23 June 2013
13. Serrote: https://www.serrote.com. Last accessed 23 June 2013
14. Silva HS (2010) Glocalness: Identidade e memória no design português contemporâneo (Master's dissertation in design). Escola Superior de Artes e Design, Matosinhos

The Influence of Feminism on the Development and Branding Strategies of Fashion Brands

Camila Lamartine and Liliana Ribeiro

Abstract Fashion and femininity have always completed each other. The fashion industry has been more connected with woman since she entered the labor market. The end of the hoop skirt, the first Chanel pants, Saint Laurent's Tuxedo, Quant's miniskirt, until more recently, a T-shirt with the slogan of activist Chimamanda Adichie, "We Should All Be Feminists," for Dior, turning to apparel as an ally of the feminine evolution. Considering the relationship between fashion and feminism, this paper aims to discuss the influence of the feminist movement in the development of branding strategies in the fashion industry, presenting feminism as a social-political movement and brand equity as a post-purchase relationship. The present study was developed through a mixed methodology, where questionnaires had open and closed questions, to obtain quantitative and qualitative data, thus analyzing the consumer and fashion brands as social agents. This study becomes relevant since it deals with feminism as a political and cultural ideological movement that has broadly emerged in the last years in the field of fashion. Besides, it helps to uncover new loyalty relationships that brands are building with their customers. Thus, from the concept of glocalization, it is possible to translate and unite in communicational tools the brand and consumer thinking, to achieve a true social transformation.

Keywords Feminism · Fashion · Branding · Glocalization · Fashion Consumption

1 Introduction

Women have never been as active in society as they are today. They set aside the domestic function—which still performs as if it were a part of their own DNA— to take the streets, the politics, the education and the labor market. According to the

C. Lamartine (✉)
FCSH—Faculdade Nova de Lisboa, Lisboa, Portugal

L. Ribeiro
IADE - Faculdade de Design, Tecnologia e Comunicação, Universidade Europeia, Av. D. Carlos I, 4, 1200-649 Lisboa, Portugal

E. Duarte and C. Rosa (eds.), *Developments in Design Research and Practice*,
Springer Series in Design and Innovation 17,
https://doi.org/10.1007/978-3-030-86596-2_3

report "Social and Employment Outlook in the World: Trends for Women 2018",[1] the overall female labor force participation rate was 48.5% in that year. They are increasingly recognized and certain in understanding of their rights. However, equality is still a utopia that seems distant in the eyes of the so-called minorities— despite the numbers expressing the contrary. In the case of Portugal according to the report of the Contemporary Portugal Database, PORDATA, 56% of the Portuguese population is female.

Due to this isonomy— which allows women an identity antonym to fragility— the social movement called feminism appears relevant. It resurfaces with new connotations beyond an organization, but as a philosophy that fits in politics and culture, seeking the equality of rights between women and men. Besides it defends the end of a society based on patriarch— a social-political system that is structured by male supremacy from "the historical seizure of power by men who appropriated the sexuality and reproduction of women and their products." (Reguant 2001 cited [8, p. 15]).

Driven by feminist movements and going beyond a mere trend, the impact of feminism in fashion has raised issues such as the objectification of the woman's body and imputed sexism in clothing, for example. This phenomenon can be explained because fashion generally turns to the female public and is characterized, as Duarte [5] explains, as a result of the manifestations of life in society, being fostered by the social context. Due to its critical and destabilizing potential, fashion was able to break patterns of representation by proposing various conceptions of subjectivity [10], including how brands dictate their ways within the fashion industry.

Branding is an essential management strategy to establish the relationship between brands and their audience since it leads them into people's daily lives giving and allowing more notoriety [3].

The present study focuses on fashion brand and consumer relationships established by branding communication and considering feminism in the contemporary Era as a political and ideological movement. Moreover, branding equity is presented as a key piece in brand identity construction. Thus, the objective is to discuss how the development of branding strategies by fashion brands is influenced by feminism, since it is through them that the consumer establishes a relationship that goes beyond the traditional sell and buy transition.

[1] https://www.ilo.org/global/research/global-reports.

2 Context

2.1 Branding

The emergence of the brand in its conceptualization dates back to Antiquity [15], where it was noticed the interest in knowing the origin, the production of a particular piece and the classification as useful or worthless, beautiful or ugly, and so on.

A brand is, therefore, something that is always cyclically under construction [15], representing what the company does and its customer's perception of it. The brand surpassed the mechanical field of production, object by object, becoming, thus, a personality immersed in people's real-life [2].

Healey [9] says that branding is a form of conflict between the producer and his consumer seeking to define what the brand is, what promise and what it wants to convey. "The brand, therefore, has to see how customers perceive the things they buy; it is not simply a feature of certain industries." [2, p. 2].

Since the beginning of nineteenth-century, consumer consumption has undergone a significant change. Possess just to possess is no longer the main reason for a purchase, especially when it comes to products that are generally not considered essential. The way of buying and form of relation with the desired object, and more than that, how to become related to the product brand, makes branding a key tool when inserted in a competitive world. "Branding is the art of creating trust; with confidence on the agenda, brands cannot be far behind." [23, p. 9].

Branding creates value-added personality when more than ever a brand must adapt to the constant changes of its consumers. Thus, knowing who your target audience actually is, knowing what this person feels and thinks, the brand can position itself by creating loyalty beyond the purchase.

Within a brand, the value of what it conveys to its consumer is one of the most important points to analyze. This is what makes brand equity one of the key concepts of branding because it allows consumers to synthesize the value experience based on the brand's history and what it targets in the future [21]. Therefore, brand equity becomes the result of the reinforcement of brand assets and the efforts involved in their construction [11], i.e., the sum of brand value plus the value added by the consumers.

In brand equity, a brand's identity is the part that aims to make it recognizable and identifiable, making it more attractive to its consumers. For Kapferer [11] there is a huge difference between identity and brand image—being the image the reception and the identity the emission. One can only happen after the other is assimilated and understood by its recipient.

Consumer interpretation is the differential and most important factor for Keller [12]. The author developed the Customer-based brand equity (CBBE) model that presents techniques and practices to influence the consumer behavior based on the understanding of customers' desires and needs in order to please them, so that they do not feel the urgency to search for another brand." The power of a brand lies in what remains in the minds of its consumers" [12, p. 49]. It is, therefore, the knowledge that the brand has about its user response to the company's marketing strategies [12].

2.2 *Fashion*

Starting from the same need for identification and communication, expressing oneself is more than intrinsic to the human being. Clothing also has expressed the connotation of segmentation between social classes, characterized periods, and continues today being a social mark of personal definition. However, according to Lipovetsky [13], fashion is not something encompassing all civilizations, it is something that emerges with the advent of conceptions of civic organization, something belonging to the modern western world.

Duarte [5] notes that fashion should be viewed as a "manifestation of life in society", based on a daily analysis that takes into account the immersion and interference of space and time, as corroborated by Palomino [19] when defines fashion as a "system that accompanies clothing and time, which integrates the simple use of clothes in everyday life with a larger, political, social, sociological context." [19, p. 3].

In the sociological context, clothing appears as something that defines us before society [5]. It is a kind of self-affirmation, thus marking the position of men and women in society, as did the bourgeois novice to stand out and establish themselves in mercantile cities.

The functionality of fashion stands for Miranda [17] as a kind of personal identification since it functions as a social tool. The growing individualistic atmosphere has placed men and women in a narcissistic process due to the search for the meaning of symbols and signs endowing them with logicality.

Human behavior has changed with modernity and postmodernity. Globalization has, on the one hand, narrowed borders and minimized differences, on the other hand, it has encouraged the individuality sought of human beings to be even deeper [17], even in the sense of a group distinction. This eventually characterizes fashion consumption as a paradigm, especially when it turns to mass consumption, and cannot infer structural separation between what it calls mass culture and elite culture [7].

The role of men and women in society has changed, and with that, their role as a consumer [22]. Society becomes to be based on symbolic consumption, on the representation of the connotative meaning of the object, and therefore the need to consume, since consumption assumes the role of identity synonym [22]. Lipovetisky [13] states that the purchase will be defined by the reason of the sign exchange value, which in turn defines the emotional character of the fashion consumer's behavior [21], linked to the feelings to which it emerges in the subconscious mind.

For Miranda [17], symbolic consumption is something characteristic of fashion brands, as they are associated with personality acquiring visibility. The differentiation sought by individuality requires variability and personalization concerning the usual product extension. A fashion brand is cheered and then consumed to the extent that it is endowed with an emotional atmosphere by subjecting the consumer to it, even before the purchase. The pleasure of owning what the piece represents as sensory, emotional, and aesthetic qualifications constitute the emotion that resides in having [21].

Fashion consumption then becomes rampant due to all of its aforementioned components that force the consumer into this endless "race". Svendsen [22] criticizes this system and says that perhaps the race will comfort us because change is always constant and satisfaction increasingly distant and unreachable—defining what is present in the social context.

2.3 Feminism

When speaking about society, we must be aware of the subjects we are looking into and the movements to which they are included. According to the Dictionary of Sociology (1990), a social movement is what comes from the collective objecting by its nature to the establishment of a new order of life. The emergence of new social movements, such as the regionalist and the feminist movement, are direct consequences of the popular rising that occurred after the Industrial Revolution.

In this sense, feminism is a social, political, and cultural movement that, by definition, seeks gender equality and rights equity based on its critical and emancipatory character [8].

Therefore, feminism can be defined as the women's awareness—as a human collective—of the oppression, domination and exploitation of which they were and are submitted, as the object of men's collective within the patriarchy, and its different historical phases. It is a movement that searches for freedom of sex, and all the society's transformations that are necessary for this purpose [8, p. 12].

Castells [4] identify feminism as a practical movement that acts from multiple discourses in defense of women's rights, framed as human beings and not being a constant objectification. "There is, therefore, a common essence underlying the diversity of feminism: the historical, individual or collective, formal or informal effort to redefine the female gender in direct opposition to patriarchalism." [4, p. 212].

According to Miguel and Biroli [16], the focus on the feminist movement on gender inequalities is a potential way to open a new (and broader) configuration of democracy that aims to a more effective way of building an egalitarian society through the structural combat of the various forms of inequalities. This is what makes feminism—beyond ideology—a commitment to society's deepest social change.

Ever since it was recognized as a movement, feminism has opposed physical and psychological dependence on men, imputed by a patriarchal society, in pursuit of legal, intellectual, economic, political, and of course social equality.

Perrot [20] states that the feminism movement behaves as a succession of various waves, spearheaded by the struggle over gender inequality during the nineteenth century with the right to vote. Already in the twentieth century it demanded female liberation through the Women's Liberation Movement (WLM), facing a confrontation for the right to know, the right to work, to have a salary and citizenship (understood as civil rights), and also to have autonomy over their own bodies.

Feminism, therefore, is immersed in diverse social organizations on several fronts, engaging with each other, agreeing or not, but always dealing with women's rights. Thus, feminist movements have gained international visibility and action—including in the United Nations (UN)—in order to discuss gender issues and assure the development of public policies for women.

According to Valente [24], feminist movements can be considered the most significant revolutionary phenomenon of the twentieth century because of how they broke certain paradigms in a deeply authoritarian and exclusionary political culture for women and other minorities. These movements question deeply about human relations and the socio-political, economic, cultural, and even sexual contexts.

The feminist struggle persists daily, as does the discrepancy between the legal equality that is obtained by the laws and supposedly guaranteed to women, and the practice that is found and faced daily in contemporary society, especially concerning socioeconomic terms. However, more plural and articulate, the feminist movement joins hands with other minorities, thus achieving an even wider range of people, countries, and supporters, dealing with issues such as harassment—moral and sexual—prejudice and wage equity, aiming that democracy stands for a society of limited liability in the credible demarcation between what is public and private.

3 Methodology

Considering the power of branding as a factor of differentiation of a fashion brand, also understanding that fashion is cyclical and momentary, this work wanted to analyze if and how feminist movements had and have the power to influence brands strategies. For this, a mixed interventionist approach was used. The mixed approach allows qualitative and quantitative research to be parallel and complementary [18]. The choice of the mixed approach is conceived by the need to understand both agents of the buying process, the consumer and the brand itself.

For this purpose, questionnaires were developed using Qualtrics platform. Before the application and dissemination of the questionnaires, a pre-test was conducted with 65 respondents over seven days at the beginning of March 2019. The goal was to assess the response time and the effectiveness of each question, aiming to verify if these questions were reliable, valid and operational—fundamental elements in the structure of this type of research [14]. After analyzing these results, it was observed that it was necessary to reformulate the survey and add two more questions in order to understand the consumer's perception of the brand.

A survey with the purpose of collecting quantitative data was directed to the consumer, and distributed through social networks—Facebook, Instagram, WhatsApp and even by email. It was held online on the Qualtrics platform from April 5, 2019, to the 20th of May of 2019. The survey was projected to select fashion consumers, looking beyond their answer about a brand of their own choice, and then presenting the feminist movement as a social movement that, in the respondents' conception, could become a social trend linked to fashion. Alternatively, feminism could be interpreted as something that fashion has "appropriated" concerning society, thus qualifying the consumer's perception of brands in general—whether they buy them or not, and how. Furthermore, was also an objective to evaluate if, in participants view, the women's movement has interfered—or not—in their purchasing decisions.

Portuguese fashion brands were the object of qualitative analysis. They were previously contacted by email, when identified the research method and considered the possibility of response. Surveys with open questions were created and shared through the Qualtrics platform and made available from April 15th until May 27th, 2019. The brands contacted had no distinguishing factor, except being Portuguese. Was analyzed how the fashion company's brand perceived the idea of women within the industry, the issue of the feminist movement as something to be worked on by the brand and what this relation would be like in a future perspective. The inquiry was addressed to the creative director of the brand or the one who considers itself responsible for the brands' communication—not being included here communication agencies or brand representation—i.e., the answers were given by people who are actually inside the company, and represent creation and communication, translating what the brand takes for branding, for DNA, and of course for itself.

4 Results and Discussion

The closed-ended questionnaires for quantitative analysis yielded a total of 122 responses. However, within the perspective of completing the survey, 110 responses were considered valid. Most respondents were female, representing a total of 82.7%, to 17.3% of the male audience, with the most relevant age group from 26 to 35 years, followed by 18 to 25 years, 49.4% and 25.9% respectively. Most respondents consider themselves fashion users, taking an interest in clothing, footwear and accessories, where 69.38% often shop at the same websites and 21.4% say they are loyal to them.

Regarding the fashion brand's DNA—i.e., what the brand shows in its construction and also in its communication activities—such as social networks, advertising and direct contact with the consumer—48% of people answered that they consider the brand's identity a determinant factor for purchasing. While 33.7% were indifferent, and 18.4% said they did not care about the instructive values of the brand. This data reaffirms the need to build good branding strategies because, as Aaker [1] says, with this strategy the brand can first genuinely identify its target audience, secondly, establish a relationship beyond purchase with the consumer, and in the end, retain it for future purchases.

To know the consumer's perception of the influence of feminist movements on fashion, six questions were made, starting with the inquiry if the feminist statement would be just a trend. Thus, through objectivity, as explained by Markoni & Lakatos (2003), measurable data captures make room for a possible generalization of results.

It was found, therefore, that 84% of respondents believe that in fashion the consolidation of women as a spokesperson in the pursuit of the imposition of their rights is not a transitory thing, but something that is here to stay, as opposed to the minority who considered the transient assertion, totaling 16%. It is clear for popular culture that the idea of a more popular feminism is an ongoing phenomenon that is not only theoretical, since it can be measured rationally [18].

Most respondents also realize that there is a certain influence of feminist movements in communicating fashion brands, either as they present themselves on their profiles, or as they communicate strategically on their media and communication channels. 32.1% states that they see a moderate amount of this influence, as opposed to 7.4% who says that they do not perceive any sensitization. This confirms what Duarte [5] presents as an expression of life in society, i.e., fashion in general. It translates the atmosphere of contemporary times.

Regarding the influence of these movements on the consumer's purchase decision, there is a clear draw on those who consider themselves heavily influenced and those who do not consider themselves influenced at all, both by 25%. Near this number, 22.5% said they perceived a moderate amount, 21.3% perceived it slightly, while only 6.3% considered themselves very influenced by the movements when making a purchase.

How the brand represents women in their communication media is marked as a decisive buying factor for 33.8% of respondents who considered it very important, followed by 23.8% who considered it slightly influential, 15% considered it largely influential, 13.8% indicated it as a moderate amount of importance and 13.8% do not consider the empowered representation of women to make a purchase decision.

Concerning fashion, 50.6% believe that there is a possibility of fashion becoming more feminist, however, there are doubts. 43.2% said yes, and only 6.2% disagree that fashion and feminism will still have a long way to go, which goes against what author Pamela Gibson (Quoted by Duarte [6]) says when she declares that feminism and fashion live a constant struggle, infrequent turns with no possibility for reconciliation.

It was still necessary to understand whether there was any relation between the answers and the fact that the respondent identifies themselves as a feminist since this was not a criterion of incidental distinction. Therefore, of all respondents, 72.84% describe themselves as feminist people, as opposed to 27.16% of respondents who do not call themselves activists in this way. This great number of self-affirmations is made possible by feminism acting in many quarters, being syncopated and resurgent [20].

The above data generally reveals that consumers perceive the influence of feminist movements on fashion brands, either through their products or mainly through their media. For them, feminism is something that will remain as a trend, increasing future buying intentions concerning the brand's interpretation of the image of women in their communication vehicles.

The qualitative survey for brands was first made available via email with a link to the Qualtrics platform. 74 Portuguese fashion brands were contacted, resulting in a total of 16 answers, with 8 valid answers from David Pereira, Katty Xiomara, João Sousa, Meam, Ariev, Josefinas, The Co.Re and Susana Bettencourt.

All brands chose to answer the questionnaire without the physical presence of the interviewer, alleging a lack of time and inadmissible appointments. However, since the questions were open-ended, they were comfortable explaining, in fact, their opinion without previously stated assertions. The questionnaire construction and the respondents answers were done in Portuguese, but translated to English to the purpose of the present paper.

Regarding the perception that brands had of women in the fashion industry, there was a conceptual difference between respondents. The view of women in fashion was seen as something objectified by some: "in *general, similar to the image that society has, i.e., as something sexist*", and still "*futile and unimportant*". However, in counterpoint, the female image associated with independence and trust is understood by some: "*Right now, the fashion world is taking women to many levels*", placing them "*always ahead of men*". In general, although fashion has been "cruel" in terms of beauty standards, between thinness and the eternal pursuit of sensuality, there is a social change fostered by the view of woman as a driving agent, as we can see: "*Before the image that was sold was the sexy, thin and tall woman and nowadays we see many campaigns in which the focus is the identity and strength of the woman. What women increasingly want is independence, they look more at what they want and need when women were once born to serve the family and to ignore their desires and dreams.*"

Regarding feminism, it is clear that most respondents do not know how to name or describe, and still discredit its strength as a social movement, arguing that the struggle should be about equality: "*I don't believe in feminist movements. I think women are very capable and it is up to each of us to ensure respect and the fight for the rights that assist us […] feminist movements tend to focus on ancillary themes.*" And also: "*Gender equality will be the most important aspect in fashion, because a woman viewed as equal as a man means the world is balanced.*" This perception contradicts what Garcia [8] establishes as feminism, since this would be a social movement that encompasses the political and cultural areas, seeking in its definition gender equality, thus not having a dissociation between such movement and what respondents perceive as equality.

In contrast, some recognize the importance of the cause: "*These days, I think the most mediatic movement was undoubtedly the MeToo that transcended the feminist frontier.*". Concerning design influence, they consider feminism a milestone: "*In my SS18 collection I was inspired by Amelia Hart, who was part of the beginning of the Suffragette movement and the first female pilot to cross the Atlantic […] Also in SS19 I spoke of Resilient individuality and acceptance of the human being as a whole.*" This corroborates with the premise that "*body liberation is fundamental because society tends to create stereotypes that we, as designers, can and should try to break.*" And also: "*Today having a feminist voice is easier, but perhaps less serious, because in many cases a woman thinks she already has everything she wants and is entitled to because she wrongly puts a negative weight on the word feminist, connoting competitively or with a need for higher quality women than men.*"

There is, however, some uncertainty regarding the ability for interference and association between the above movements and the consumer, however in the fashion industry "*the most relaxed and extravagant pieces, not in the sensual sense, have been gaining more expression. Women no longer just want to show the body but show that they are much more than that*", which can be considered a certain influence, because as Svendsen [22] explains, society now has symbolic consumption as a kind of foundation. Additionally, because of this there is a need to consume, since consuming assumes an identity character.

Brands were also asked if they, as members of the fashion industry as producers, considered that the ascendant female statement is merely a trend or something that will stand, and whether the questions raised by feminism should be discussed within fashion. In response, all brands believe that this new perception of women is something that is here to stay and that should be worked within the fashion market, as this is "*very woman-oriented [...] the discussion has to be raised here too.*" And also: "*Female affirmation is not a trend but a necessity. The standards of society are different, and they are no longer going back.*"

Regarding the influence of these movements with the branding strategies, it was questioned if the brand went through any adaptation process since most people considered feminism as something divergent from a trend. However, most respondents said they had not developed any changes in brand communication, and those who passed through changes in the past considered that this assumption is already an intrinsic part of brand DNA. A respondent said: influences existed "*through my life rather than outside tendencies. However, everything fits perfectly.*". And yet, regarding the likely impact of these movements on the brand's behavior towards its audience, most respondents may identify but are unsure, while the other vast majority described the effect these movements have had on communication with their consumer as certain.

Finally, intending to verify if fashion—as an industry—will become more feminist, since there is agreement on the permanence of female affirmation in this atmosphere, most of the interviewers have doubts about the future of fashion. 57.8% answered the statement as "*perhaps*", while 42.2% answered affirmatively the same question. One said: "*I think the world will be increasingly genderless and unprejudiced. This is the future we all need.*", as Castells [4] points out concerning the changes that must be signed in society regarding values, dogmas and ideals, as they have lost their legitimacy due to the social changes arising from feminist struggles that can increasingly embrace and aggregate people that are not represented by the current society.

5 Conclusion

In this study, feminism was presented as an ideological social movement that has links with fashion and branding. We have tried to understand to what extent this ideology can influence the branding strategies of fashion brands in Portugal, considering both the creators and the consumers.

As saw in the references reviewed, feminist movements play a key role in the struggle for the affirmation of women's rights. Thus, the research was developed based on this idea. The objective was to collect data on the perception of individuals as consumers of fashion and social agents. The data gathered expressed consumer loyalty to brands, but, above all, the attention to the role attributed to women within the fashion industry. The new consumer— more aware of their purchasing power— is also concerned with the brand identity and therefore considers the way it is built, namely in terms of the values that underlie it.

Regarding fashion brands— even though divergent positions have been identified— this study can conclude that is not yet evident a direct influence of feminist movements on the way brands relate to their audiences. There is, however, a growing concern about the subject within the industry.

The data collected also allows understanding that the way we look at brands in the Portuguese market is changing thanks to greater social openness and disintegration of prejudice and stereotypes. Brands recognize and consider positive the way women's image has changed within the fashion industry, attributing this paradigm shift to the power women have gained in society in the last years.

Despite this recognition, it appears that the awareness of feminism is still slightly present in brands. Most of them address gender equality issues, but dissociate this concept from feminism, ultimately giving the idea that feminist movements go on a distinct path. Thus, it seems that there is no understanding of feminism as a movement that seeks equality and equity, ideas that are categorically at the root of the concept.

In general, it can be concluded that the affirmation of feminism by fashion brands is not a trend, but a long-term phenomenon, which will eventually define the industry itself. In this sense, both the brand and the consumer tend to be increasingly concerned with the representation of the image of women in communicational media. As a result, we will attest a gradual interest as well in the brand identity and branding that joins the feminist and fashion elements.

The link between branding, fashion, and feminism must continue to be worked on, as there is still a long way to go before brands can establish a direct relationship with consumers concerning the values they uphold. If consumers are increasingly aware of societal problems today, fashion brands will only be able to connect with this audience if they too embody these concerns. Feminism is an example of a movement that, as part of consumers' lives, cannot be ignored by brands, but rather must be worked on by them.

Future research should focus on exploring the development of feminism within fashion brands, its conception as a transformative agent of sociability, and aspects related to purchasing relevance. From another perspective, it would be interesting to compare female and male consumers, in order to analyze the differences in customs, values, interests, thoughts, and of course, buying processes.

Given the relevance of feminism in the social and academic fields, focusing on this duality of fashion brand consumption can stimulate the development of other work in the areas of communication, branding, marketing, fashion, and gender studies.

References

1. Aaker D (2001) Criando e administrando marcas de sucesso. (3ª ed). Futura, São Paulo
2. Arnold D (1992) The handbook of brand management. The Economist Books Limited, USA
3. Cameira SR (2016) Branding + design: a estratégia na criação de identidades de marca. Editora Senac, São Paulo
4. Castells M (2003) O poder da Identidade. Fundação Calouste Gulbekian, Lisboa
5. Duarte CL (2004) Moda. Quimera, Lisboa
6. Duarte CL (2017) Moda e Feminismos em Portugal: o género como espartilho.Histórias de vida. Círculos de Leitores, Lisboa
7. Freitas RF (2019) Comunicação, consumo e moda : entre os roteiros das aparências. Comunicação, mídia e consumo 3(4):125–136. https://doi.org/10.18568/cmc.v2i4.39 [Acedido em 03.06.2019]
8. Garcia CC (2015) *Breve História do Feminismo* (3ª ed). Claridade, São Paulo
9. Healey M (2009) O que é branding? Editorial Gustavo Gil, Barcelona
10. Joaquim JT, Mesquita C (2011) Breaks in dress: articulations between fashion and feminism. DAResearch 6(8):643–659. http://doi.org/10.5965/1808312906082011643
11. Kapferer JN (1991) Marcas: Capital da Empresa. Edições CETOP, Mem Martins
12. Keller KL (2008) Strategic brand management, 3rd edn. Prentice Hall, New Jersey
13. Lipovetsky G (2009) O império do efémero. Companhia das Letras, São Paulo
14. Marconi M, Lakatos E (2003) Fundamentos da metodologia científica (5ª ed.) Atlas, São Paulo
15. Mendes A (2014) Branding: A gestão da marca (2.ª ed.) Edições IADE, Lisboa
16. Miguel LF, Biroli F (2014) Feminismo e Politica: uma introdução. Boitempo, São Paulo
17. Miranda AP (2008) Consumo de moda: a relação pessoa-objeto. Estação das Letras e Cores, São Paulo
18. Morais AM, Neves IP (2007) Fazer investigação usando uma abordagem metodológica mista. Revista Portuguesa de Educação 20(2):75–104
19. Palomino E (2003) A Moda. Publifoha, São Paulo
20. Perrot M (2017) Minha história das mulheres (2nd ed.). Contexto, São Paulo (2017)
21. Rech SR, Farias DN (2019) O branding como vantagem competitiva para marcas do setor de moda. Dapesquisa 4(6):626–630. https://doi.org/10.5965/1808312904062009626 [Acedido em 03.06.2019]
22. Svendsen L (2010) Moda: uma filosofia. Zahar, Rio de Janeiro
23. Upshaw LB (1995) Building brand identity: a strategy for sucess in a hostile marketplace. Wiley, New York
24. Valente VV (2004) Los feminismos peruanos: breve balance de três décadas. In Cevasco G (ed) 25 años de feminismo en el Peru: histórias, confluências y perspectivas (10–17). Disponível em http://www2.congreso.gob.pe/sicr/cendocbib/con4_uibd.nsf/C08CBB7DF991A3FF05257B1700675D74/$FILE/BVCI0003574.pdf [Acedido em 15.03.2019]

Moving Pictograms

Maria Diaz, Carlos Rosa, and Liliana Faria

Abstract Although pictograms and motion are part of our day-to-day life, few are the examples that tie these two topics together. The aim of our study in connection with information design was to examine the significance of adding motion to a reality that is seen as being chiefly analog. Based on theoretical concepts, case studies and the results of field research that included interviews, surveys and focus groups, we developed and tested speculative propositions for adding motion to thirteen AIGA system pictograms. The research findings have led us to conclude that the introduction of motion could be an important contribution to facilitate the interpretation of pictograms and to enable them to develop in others ways, for example, to encourage civic behaviour, enhance the content of a message and to create emotional ties. We witnessed with this research that moving pictographs are better to decode, when compared with static pictographs. We achieved this result by comparing the original AIGA pictograms with a modification that we applied to the same pictograms in order to add movement. The results are extremely relevant, because satisfaction rates are quite high, as care was taken to conduct interviews and focus groups with designers and non-designers.

Keywords Pictogram · Information design · Movement · Communication · The AIGA system

1 Introduction

Pictograms are concise, condensed and schematic forms that were first used in prehistoric times to identify, graphically, and unequivocally, different actions, processes and activities. They should not require any text to be decoded, because if they do

M. Diaz (✉) · C. Rosa · L. Faria
IADE - Faculdade de Design, Tecnologia e Comunicação, Universidade Europeia, Av. D. Carlos I, 4, 1200-649 Lisboa, Portugal

C. Rosa
UNIDCOM/IADE—Unidade de Investigação em Design e Comunicação, Av. D. Carlos I, 4, 1200-649 Lisboa, Portugal

E. Duarte and C. Rosa (eds.), *Developments in Design Research and Practice*, Springer Series in Design and Innovation 17,
https://doi.org/10.1007/978-3-030-86596-2_4

then their function is no longer relevant [9]. The word 'pictogram' derives from the joining of the words *picto* and *gram*, meaning "the painted image" and "message", respectively. This term competes with the word's ideogram and icon [2].

Information design is divided into three areas: complexity, interdisciplinary and experimentation [10]. Several authors before and after Schuller also addressed the role of information design and contributed to distinguishing this area from all others.

Considering complexity, information design can thus be seen as the transfer of complex data to a two-dimensional representation, with the aim of communicating, documenting or preserving knowledge. It deals with the relationship between facts, in order to make their interrelationships understandable, offering information in a transparent way, eliminating uncertainty. The creation of "meanings of complex information" is a task that requires information designers to take a systematic approach to the project, combining analytical, editorial and graphic elements [10]. Regarding interdisciplinarity, information design involves an interdisciplinary attitude, since the contribution of external people or marginal scientific areas to this field ended up influencing this territory, even ending up inventing new methods of visual representation, such as animation for example, which is explored in this research. This concept of interdisciplinarity is reinforced when it is observed that information design combines the benefits of graphic design, 3D, digital media, cognitive sciences, information theory, cultural sciences, and, as we can observe in this paper, animation. Interdisciplinarity is the synthesis of different partial aspects and not their juxtaposition. The information design discipline has its roots, among others, in information theory and the psychology of perception and is therefore a combination of research and design. Information design encompasses a third competence (alongside complexity and interdisciplinarity), experimentation. Therefore, technical and graphic archetypes are an important part of the design repertoire; added to the need for clear objectivity, they are often indispensable in defining information design solutions. However, it is often the case that in some cases the representation of solutions is quite subjective, and the aforementioned archetypes do not achieve their purpose. And we believe that movement can help, because it improves the relation between form and meaning.

So, with the analysis of formal elements that generate meaning in images, animation combined with graphic discourse improves decoding the messages.

While today information design cannot be defined as an independent discipline and is far from being admitted as such, in visual communication it already stands apart from other disciplines [9].

The interest in incorporating motion into pictograms is owed to the fact that twenty-first century is characterized by technologies and motion, although pictogram representation is still predominantly fixed. We know that technology plays a determining role in people's lives, increasing the presence of the digital and the disappearance of the analog.

According to Barnés [1, p. 41], a static or fixed image conveys its contents and meaning in a static way, that is, without any movement. This image can be seen for as long as one wants or needs to interpret it. A static image, however, can be dynamic, in other words, even if it has no movement and consists only of a 'frame', it can convey movement or dynamism.

The function of some pictograms, however, even if they lack movement, is precisely to give the idea of motion, that something is moving.

As regards to the motion image, Barnés [1] defines it as images that move one after the other at a specific pace.

Two key concepts should not be mixed up: "motion image" and "movement-image", both of which are mentioned by Deleuze [3]. When we refer to "motion images", they are obtained through people or things that move, whereas "movement-images" are created when the movement is obtained by moving the camera and through the movements that are created during editing, that is, whatever shows in the picture can be static.

Rancière [8] gave his summarized and objective definition of Deleuze's "movement-image" [3], arguing that the "movement-image" is an element that forms part of the natural chain created between various images.

Zunino [12] refers that Bergson explains the relation between action and movement by assuming that the origin of awareness is the movement itself, and that its function is not to represent things or objects, but rather to group all the moments while it lasts and to make important and significant progress. Each separate image, that is, static image, shows a framework that is specifically created and defined as an art, since all the parts that form a group have to be chosen in order to create the desired framework.

To create movement in pictograms, we used Deleuze's [3] editing technique, since pictograms are static fixed planes, and we edited several sequential planes to create movement.

The aim was to start from two or three complementary forms of a pictogram and create "movement-image" to effectively convey the "thing"/image/message of that pictogram.

So as to increase the knowledge of the nature of pictograms, the main aim of this empirical study was to explore the adding of motion to pictograms and to assess the possible gains resulting therefrom to real life situations. We have posed the following question: "How does the adding of motion improve the decoding of pictographic information?".

More specifically, the purpose of this empirical study is to:

Assess whether there is a need to graphically update all/some of the pictograms chosen for the project;

Analyse the need to simplify all/some of the chosen pictograms;

Check that all chosen pictograms can improve the understanding thereof with the adding of motion.

Since this is a first approach to the topic, we chose to base our work on a specific case study, namely the AIGA pictographic system, which is not only a worldwide reference that offers a broad variety of pictograms, but is also associated to transports, a relevant sector in pictographic communication.

Of the 34 pictograms that form the AIGA system we only chose those that depicted the human form, because besides being a sufficient sample size, 13 different pictograms, the sample included the most relevant pictograms for the intended objectives (Fig. 1).

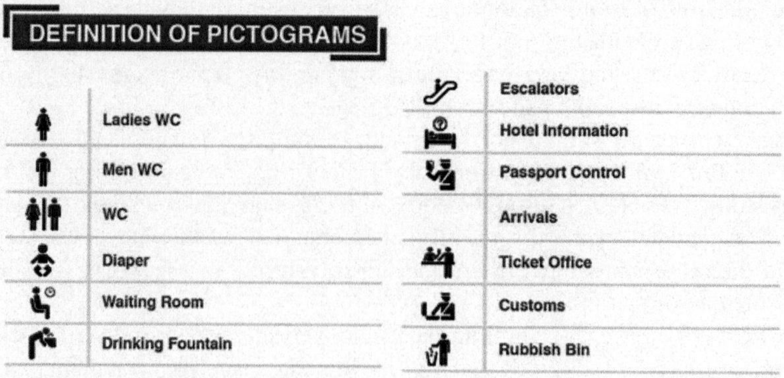

Fig. 1 Definition of the pictograms of the AIGA system

Our theoretical search allowed us to identify the best approach for producing a system of pictograms with movement.

Based on the results achieved in the three first phases of the empirical investigation—interviews, surveys and focus group, and on the study of real cases that could be a source of inspiration, we carried out thirteen movement introduction tests, one for each of the thirteen pictograms under study. We then tested the propositions developed in a fourth phase, through interviews.

Taking account of the above, the aim of this study is to contribute to a reflection and possible development of current pictographic systems, exploring new conceptual forms through the introduction of movement.

2 Methodology

To achieve the objectives outlined, the research plan developed along four main phases, following an applied research method—action, to generate knowledge that can be applied in practice. This is a quasi-experimental and exploratory study, considering that it is a first approach to the topic and our intention is to know the facts and phenomena related to the problem under study and that the research variables were not manipulated or changed [5].

In the first study, data was collected from participants not in the design area, as we hoped to find out how the chosen pictograms are interpreted by people who are not conscious of design issues, even though they use most of those pictograms on a daily basis or frequently.

The second consisted of collecting data from participants involved in design. Unlike the first phase, our aim was to examine the most sensitive side of graphic issues and image of the pictograms, and to compare their interpretation with that of the first phase group.

A focus group was organized in the third phase, in which we sought to analyse the issues identified as being the most relevant in the previous phases and to discuss possible solutions to those issues. The focus group consisted of representatives of the groups involved in the two first phases.

In the fourth phase, we again collected data from participants not in the design area, the aim being to assess the proposed pictograms with movement done in the project compared to the original pictograms.

2.1 Measurement Instruments

Different methods were chosen for data collection according to the target audience and the study phase.

During the first and fourth research phases, we interviewed people not in the design area and a second phase took place concurrently with the first phase, which consisted of a survey to people exclusively in the design area. Once the results for the first and second phase were obtained, a focus group was arranged for the third phase involving people in the design area and outside this area.

The interviews were semi-structured, taking into account that not all questions were closed-ended questions, allowing the interviewee to give his/her opinion and thoughts.

Data from the interviewees were analysed through content analysis, which, according to Bardin (2009), consists of a number of communication analysis techniques and uses systematic and objective procedures to describe the contents of messages [4].

The same script was used for all interviews, which examined the thirteen pictograms of the AIGA pictographic system chosen for the project.

The questionnaire was the measurement instrument used in the project's second phase to collect information from people in the design area, including design students, teachers or designers.

We chose to conduct a direct survey through an online questionnaire, so as to enable an efficient and effective data collection.

In the third phase, the focus group participants gathered for a round-table discussion of the topic and the most relevant results obtained in the two previous phases (phase one and two). The study was conducted in Lisbon, between Nov. 2017 and May 2018.

2.2 Samples and Data Collection Procedures

A non-probability sampling method was used in the research, typical of exploratory studies, for a cross-sectional time span.

Table 1 Characterization of participants in phase 1 of the empirical study

GENDER		AGE					PROFESSION		
MEN	WOMEN	ATÉ 10 ANOS	]10,18]	]18,25]	]25,55]	>=56	STUDENT	WORKER	OTHER
41%	59%	5%	0%	36%	41%	18%	32%	64%	4%

Table 2 Responses by participants in questionnaires

GENDER		AGE			CATEGORY			
MEN	WOMEN	]18,30]	]30,45]	>=46	LICENSEE STUDENT	MASTER STUDENT	DESIGNER	DESIGN TEACHER
31%	69%	85%	15%	0%	38%	30%	27%	5%

Table 3 Responses by participants in the focus group

GENDER		AGE		AREA		DESIGN AREA		OTHER AREA	
MEN	WOMEN	]18,30]	>=30	DESIGN	OTHER	MEN	WOMEN	MEN	WOMEN
50%	50%	87,5%	12,5%	50%	50%	50%	50%	50%	50%

Table 4 Characterization of participants in phase 4 of the empirical study

GENDER		AGE			PROFESSION		
MEN	WOMEN	]18,25]	]25,55]	>=56	STUDENT	WORKER	OTHER
57%	43%	40%	43%	17%	26%	67%	7%

Both the larger samples, used in phases one, two and four, and the sample used for the focus group were intentional samples, since the participants were chosen based on the probability and existing conditions for them to accept being part of the research (Tables 1, 2, 3 and 4).

2.3 *Suggestions for Motion Pictograms*

After analysing the results obtained in the three first research phases, we went on to apply movement to the pictograms based on Deleuze's concept of "movement-image" [3], a process that can be done through editing, that is, linking the various planes that work as static planes on their own, but when joined together create movement.

Initial sketches were required to see which solution was more appropriate and effective for each pictogram, based, as much as possible, on the current version of AIGA's pictogram. Where we found that it was more difficult to interpret the pictogram's message in the three first phases of the research, changes had to be made to the original pictogram.

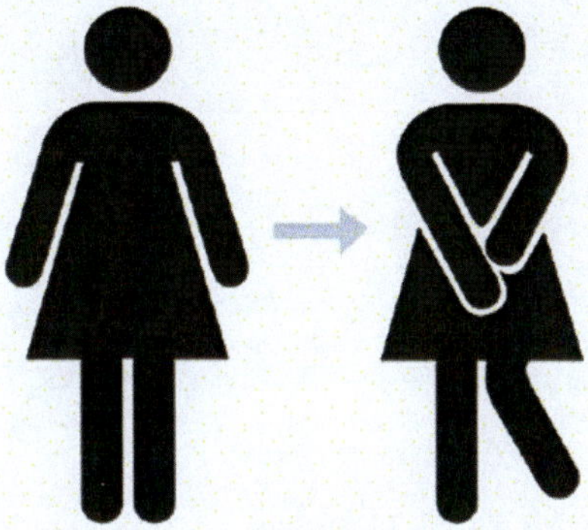

Fig. 2 Suggestion of frames for the ladies WC pictogram

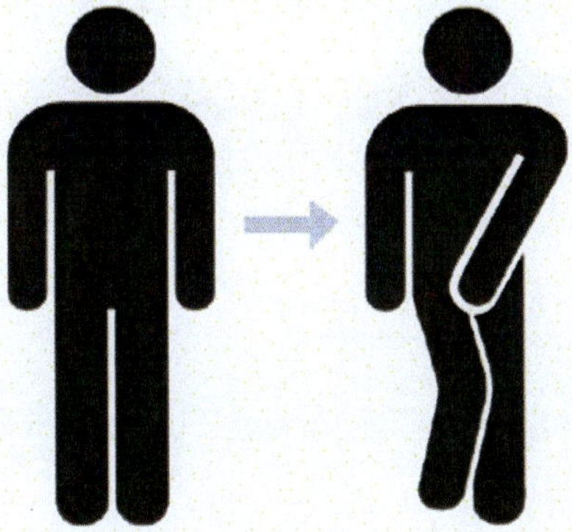

Fig. 3 Suggestion of frames for the men WC pictogram

Where the use of colour is concerned, taking into consideration the preliminary and non-exhaustive nature of the study in question, we chose to do all sketches in black and white. However, colour can be applied to any of the hypotheses.

In pictograms "WC Ladies" (Fig. 2), "WC Men" (Fig. 3) and "WC" (Fig. 4), we added only one more frame characterized by a movement that can help convey the idea of "when you have to go to the toilet/need to use a toilet".[1] In these examples, the introduction of movement was explored to enhance the message of the context to which the situation refers, humanizing and creating an emotional link.

In the pictogram for "Baby changing facility" (Fig. 5) we also added just one frame to enhance the message of the action it aims to convey.

[1] The movement introduced in the second frame of this pictogram was inspired on the pictogram representing the ladies' toilet (WC) at the Alegro Alfragide Shopping Centre, which shows that same movement.

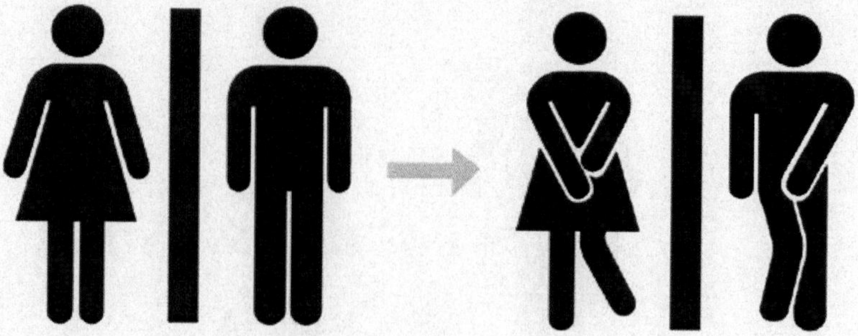

Fig. 4 Suggestion of frames for the WC pictogram

Fig. 5 Suggestion of frames for the diaper pictogram

As regards the "Waiting Room" pictogram (Fig. 6), we chose to increase the size of one of the elements (clock) and to create a sequence of five frames.

As for the "Drinking Fountain" (Fig. 7) and "Rubbish Bin" (Fig. 8) pictograms, the aim of introducing movement was to explore the association of a correspondent

Fig. 6 Suggestion of frames for the waiting room pictogram

Fig. 7 Suggestion of frames for the drinking fountain pictogram

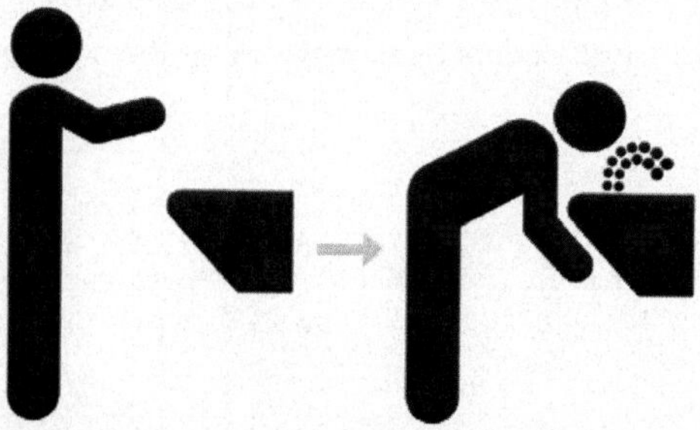

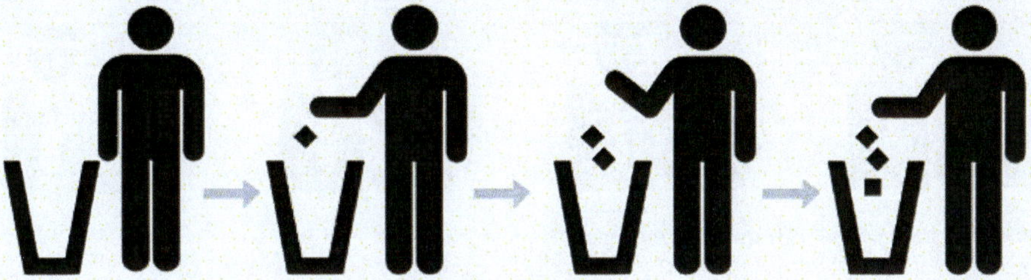

Fig. 8 Suggestion of frames for the rubbish bin pictogram

Fig. 9 Suggestion of frames for the escalators pictogram

Fig. 10 Suggestion of frames for the hotel Information pictogram

Fig. 11 Suggestion of frames for a pictogram representing passport control

civic engagement message: "use only when needed" and "throw the rubbish in the bin". The suggestions presented contained two and four frames, respectively.

For the "Escalators" pictogram" (Fig. 9), we joined three complementary frames.

The "Hotel Information" pictogram (Fig. 10), we found it appropriate to introduce movement to complement the pictogram's message with non-pictographic elements.

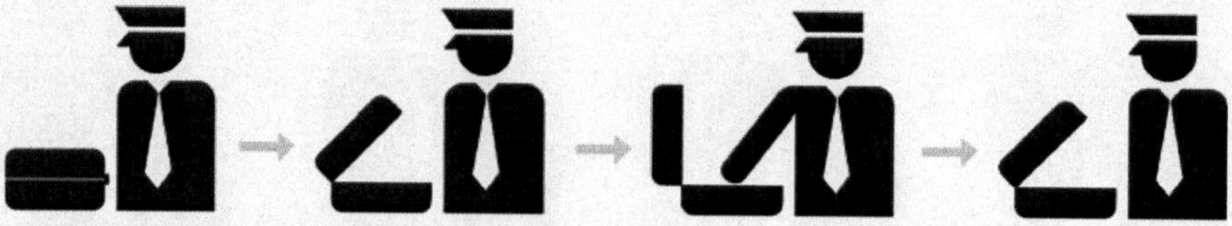

Fig. 12 Suggestion of frames for the customs pictogram

Fig. 13 Suggestion of frames for the arrivals pictogram

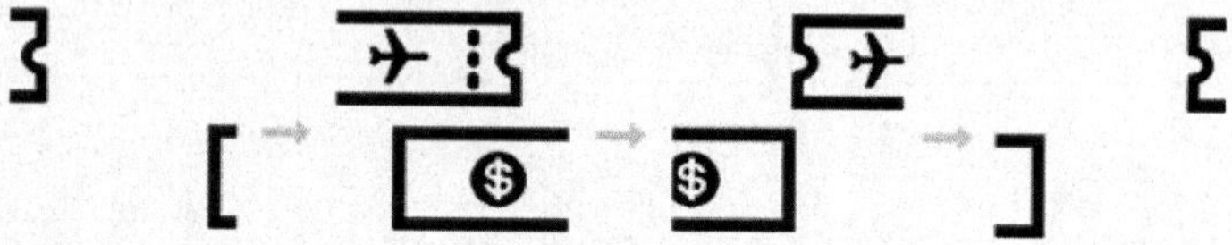

Fig. 14 Suggestion of frames for the ticket office pictogram

With regard to the "Passport Control" (Fig. 11) and "Customs" (Fig. 12) pictograms, in addition to updating the clothing of the human figure, the introduction of movement enhanced the action it intends to convey.

Our approach to the "Arrivals" pictogram (Fig. 13) was to give a perspective of space, that is, the feeling of space covered by the human figure through its different size from the first to the fourth frame.

Finally, by adding movement to the last pictogram—"Ticket Office"—(Fig. 14) we were able to replace the human figure representing the action and to update the action we intend to convey.

3 Results

In phases one and two, the percentage of answers that accurately decoded the message of the base pictograms (the interviewees understood the message) varied widely between the thirteen pictograms, with some of the situations getting all answers right and in one case no right answer.

In six pictograms (46% of all the chosen pictograms), the majority of interviewees were unable to give a right answer and, therefore, to decode the message. The results obtained were similar in phase one and two.

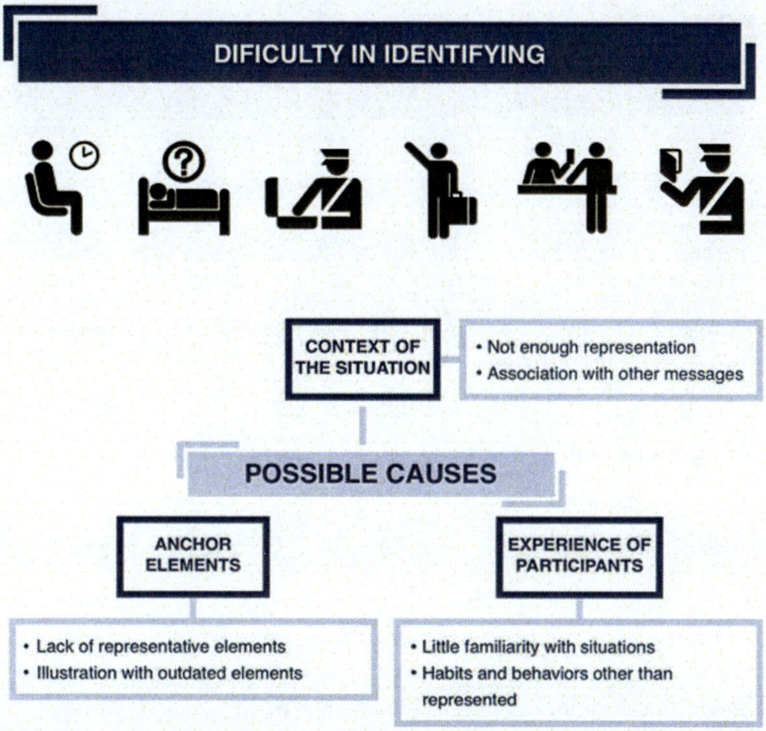

Fig. 15 Difficulty in identifying—group of reasons on three main points

The following three main problems were identified in phase three (focus group) as possible causes for not understanding the pictograms that raised more doubts, as shown in Fig. 15: (i) The context of the situation; (ii) Anchor elements, and (iii) The experience of participants.

With regard to the possible benefits in adding motion to the pictograms, all participants felt that there were significant opportunities for improvement compared to the current reality, in particular in three aspects: (i) Understanding; (ii) Visibility, and (iii) Behaviour.

Figure 16 compares the results obtained before (phase one) and after introducing motion (phase four) in the pictograms difficult to identify in phases one and two. The percentage of right answers (message decoded) increased on all situations where motion was introduced.

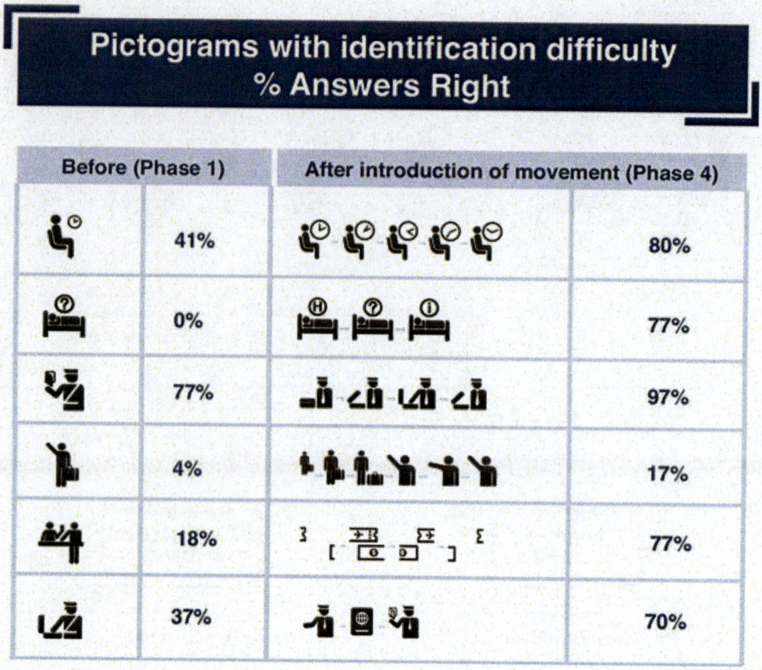

Fig. 16 Results of introducing motion and increasing the effectiveness of the message

4 Discussion

The review of literature has shown that pictography and motion are vast fields in the area of design, but studies that relate these two topics are nevertheless few.

The shortcomings found in connection with the introduction of motion in pictograms increase the interest in conducting this study.

The results obtained with regard to the objective/hypothesis—the need to redesign some of the analysed pictograms—seem to indicate that some of the AIGA system pictograms are difficult to decode and are out of date. In fact, in some cases there is a problem at syntactic level concerning its interpretation, that is., people can identify the anchor elements on their own (human figure, bag, chair, etc.), but fail to understand the message supposed to be conveyed.

As regards the second objective/hypothesis—the relevance of adding motion to pictograms—the results indicate that there is strong evidence that adding motion can be an appropriate step to increase the effectiveness of the message, although perhaps not to all pictograms.

As regard the benefits that motion can bring to the pictograms, the results point to opportunities for improvement in three aspects: (i) understanding the pictogram; (ii) visibility of the pictogram, and (iii) influence in the behaviour of the person seeing the pictogram.

Table 5 Introducing motion in pictograms—explored scenarios

SCENARIOS EXPLORED WITH THE INTRODUCTION OF MOVEMENT	
MAIN FORMS OF CONTRIBUTION	**MAIN FORMS OF APPROACH**
• Facilitate interpretation / increase effectiveness of the message	• Humanization
• Reinforce message context	• Sequential chromatic change
• Create emotional links	• Changing the size of some elements
• Encourage civic behavior	• Changing the Space Perspective
• Update processes	• Rotation with non-pragmatic elements
	• Highlight of elements

With regard to understanding, the possibility of increasing the effectiveness of the message is, without any doubt, very relevant, considering what is expected of a pictogram. As far as visibility is concerned, the benefit is also quite relevant since a pictogram will not fulfil its role if it cannot be easily seen. As for the influence on the behaviour of the person who sees the pictogram, for example, encouraging to always putting the rubbish in the designated bins, this can be a quite interesting development, through the exploration of new functional aspects in pictograms.

Table 5 presents a summary of the main contributions identified, as well as the different approaches tested. In this regard, it should be pointed out that when we ran the tests we sought to diversify as much as possible, both in terms of contributions and of forms, so that they could be an inspiring starting point for future work to optimize the message of pictograms.

In relation to the effects of "facilitating the interpretation/ increasing the effectiveness of the message", these are very positive. In fact, as shown in Fig. 16, when motion was added to the pictograms the number of right answers increased substantially in the pictograms identified in the previous phases as being difficult to interpret. With the exception of the "Arrivals" pictogram, for which results were much lower than expected (showing that the test was not enough to significantly improve the message), all pictograms tested with the addition of motion were easy to interpret for most of the interviewees.

As regards the elements added to the pictograms identified in previous phases as having no difficulty in interpretation, in particular to "strengthen the context of the message", "create emotional ties" and "encourage civic behaviour", the results obtained were also very satisfactory. As shown in Fig. 17, by exploring elements added to the base form, the effectiveness of the message not only did not decrease but in fact increased, with 100% right answers obtained in all pictograms.

We were unable to compare and complement the results obtained with other results from similar studies, as no other studies in this field have ever been carried out, according to our extensive research. We did, however, identify some aspects in the study that reflect theories and considerations of other authors that have already dealt with pictograms.

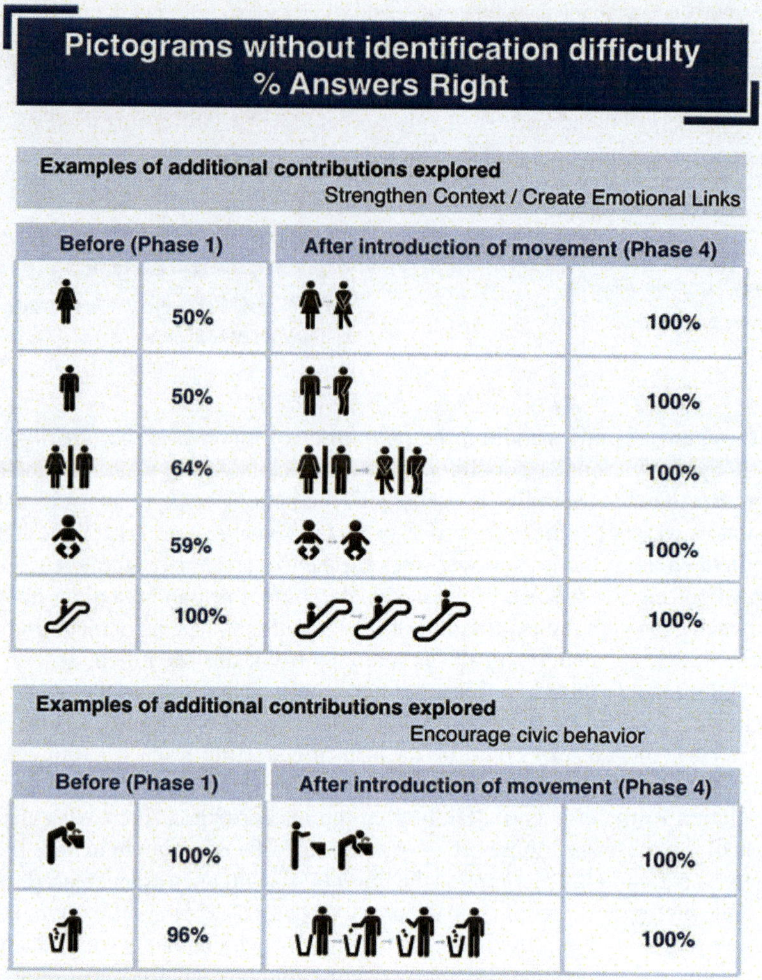

Fig. 17 Results of introducing motion in additional elements to the base form

In this context, it is considered that the three areas of improvement identified (understanding the pictogram, visibility of the pictogram, and influence in the behaviour of the person who sees the pictogram) reflect two of the three dimensions of the signage process by Morris [7].

We should also highlight the relation between the changes suggested for each pictogram and the two rules required to create this "writing style", as per Lupton [6], since throughout the project we always took the basic nature of the pictograms into consideration, giving primacy to their development to the rather than to aesthetically create new writing styles.[2]

We believe that the results obtained can strongly encourage further studies on this topic and that they reflect the need to rethink some of the AIGA system's pictograms, because although this international system is widely spread, it is quite clear that it was quite difficult to decode some of these signs.

Considering that motion is not a sufficiently explored topic in the field of pictograms, and that studies linking these two elements—motion and pictograms—are not available, we hope that this study contributes to opening up a new world of approaches and interpretations, and to the evolution of pictograms.

4.1 Limitations

In addition to the issues that seem especially relevant to design, we think it is important to mention some of the limitations we came across in our research.

First of all, we wish to comment on the method for sample selection. The fact that it is a convenience sample limits the mainstreaming of results obtained, as the factors associated with self-selection may have influenced the participants' decision to participate in the study.

Another difficulty encountered has to do with the group of instruments used, which were specifically built for the purpose (interview, questionnaire), but since they are of the self-report type, they are subject to errors of interpretation and to social desirability factors.

Limiting the study to Portugal and to the city of Lisbon is, in itself, a limitation to the mainstreaming of results to other locations, therefore its social and cultural characteristics should be taken into consideration.

5 Disclosure Statement

No potential conflict of interest was reported by the authors.

References

1. Barnés J (2006) Qué son las imágenes? Interpretaciones y aplicaciones. [What are the images? Interpretations and applications]. PhD dissertation, Universidad Pontificia de Salamanca

---

[2] According to Lupton (1989), pictograms must follow two rules: (i) reduction—determining the style of individual signs, (ii) consistency – giving a group of signs the appearance of a coherent system.

2. Bessa J (2005) Representações do Maculino e do Feminino na Sinalética [Representations of Male and Female in Signage]. PhD dissertation., Universidade de Aveiro
3. Deleuze G (2016) A Imagem-Movimento: Cinema I. [The Image-Motion: Cinema I]. Documenta, Lisbon
4. Farago C, Fofonca E (s/d) A análise de conteúdo na perspectiva de Bardin: do rigor metodológico à descoberta de um caminho de significações. Análise de Conteúd. [Content analysis from Bardin's perspective: from methodological rigor to the Discovery of a path of meanings. Conten analysis] (s.v). 1–5. Universidade Federal do Panará
5. Haro F, Serafim J, Cobra J, Faria L, Roque MI, Ramos M, Carvalho P, Costa R (2016) Investigação em Ciências Sociais: Guia prático do Estudante. [Research in Socia Sciences: Student Practical Guide]. Pactor, Lisbon
6. Lupton E (1989) Design discourse. history, theory. criticism. Reading Isotype 11(3):47–58. https://doi.org/10.2307/1511484
7. Morris C (1938) Foundations of the theory of signs. The University of Chicago Press, Chicago
8. Rancière J (2001) La fable cinematographique. [The cinematographic fable]. Paris: Le Seuil
9. Rosa C (2012) Sistemas de Informação Pictográfica-O universo dos pictogramas: métodos e procedimentos de design para obtenção de coerência formal. [Pictograpic Information Systems—The universe of pictograms: design methods and procedures for obtaining formal coherence]. PhD diss., Faculdade de Arquitetura. Lisbon
10. Schuller G (2007) Information design = complexity + interdisciplinarity + experiment. AIGA J Inf Des. New York. August 17. www.aiga.org/complexity-plus-interdisciplinarity-plus-experiment
11. Wilburg P, Burke M (1998) Information graphics:innovative solutions in contemporary design. Thames & Hudson, London
12. Zunino P (2010) Bergson: a metafísica da Ação. [Bergson: the Metaphysics of Action]. PhD diss., Universidade de São Paulo, São Paulo

Materials and Manufacturing

Design for New Materials and New Manufacturing Technologies

Pedro Oliveira, Valentina Rognoli, and Markus Holzbach

Abstract Designing new materials introduce new challenges – from the setting of materials physical characteristics to the project of their cultural identity. Such a complex challenge is being embraced by many designers in the present, however, this comes to a price. The large number of materials is already problematizing the traditional modality of direct experience upon them. For example, the traditional systems for classifying materials, commonly referred to as "material families", struggles to incorporate some new materials' unusual and contradictory properties. But also, technologies are characterized by a great dynamism and rapid evolution. This poses significant challenges but also great potential for the design of our future physical world: from the ecological to the social one's.

The human ability to shape the ecosystem is, for most of us, unquestionably evident. As has been shown, especially since the beginning of the industrial age in the nineteenth century together with the development of design as a subject, humankind is capable of actively shaping and reshaping the ecosystem and its environment. This, however, is associated with severe changes. With resource scarcity and climate change, it is time again to see the design of our environment as part of a complex overall system more clearly.

The quantification of what humanity has produced so far (anthropogenic mass) when compared to the mass of what nature currently maintains (biomass) has demonstrated that the first has already exceeded the second [4]. This is explicit proof of the overwhelming capability of humans to impose long-lasting changes on the planet.

P. Oliveira (✉)
IADE - Faculdade de Design, Tecnologia e Comunicação, Universidade Europeia. CIAUD, Lisbon, Portugal
e-mail: pedro.oliveira@universidadeeuropeia.pt

V. Rognoli
Politecnico di Milano, Milano, Italy

M. Holzbach
Offenbach University of Art and Design, Offenbach, Germany

E. Duarte and C. Rosa (eds.), *Developments in Design Research and Practice*, Springer Series in Design and Innovation 17,
https://doi.org/10.1007/978-3-030-86596-2_5

The modifying role and the scale of action of the human species gives some credit to the claim that we are already living in the Anthropocene geological era. [2].

Given dwindling resources, minimizing the use of materials and energy is becoming increasingly important.

Paradoxically, the scope of optimizing material use (the new "less is more") has induced the demand for new materials and improved traditional ones. Research into new materials and the study of their applications and impact at several levels (economic, cultural, social and environmental) has visibly increased.

New Materials and New Manufacturing Technologies are of central importance, since they create new possibilities and new solutions. Contrary to prediction, materials have not lost their significance or relevance.

The weight of what humanity has produced is overwhelming, as is the number of different materials it has developed. Attempts to quantify the number of materials currently available estimates them to be between 50,000 [3] and 100,000 ([1], p. 33), or even more. If this immense quantity of materials could be seen as pointing towards future disinvestment in research and development in this area, the evidence shows just the opposite. The pace at which new materials are being researched and developed is steadily increasing. Many areas of human development seek to optimize their actions (and solutions) in different ways. And two of them are precisely the New Production Technologies and New Materials. Both areas of development are oriented, as in the past, towards exploring new and improved performances. This is why designers now have so many materials at their disposal. And many more are being developed to meet growing manufacturing requirements and environmental necessities.

1 Opportunities and Challenges

It is now recognized that materials play a key role in the design process to satisfy technical and technological needs and the experiential dimension [7, 14] (Ashby & Johnson, 2002). The designers' ideas and purposes can only be grasped through their materialization. The playful exploration of such material approaches leads to new functional and formal impulses. We can refer here to many historical models, such as Le Corbusier, Ray and Charles Eames, Konrad Wachsmann, Jean Prouvé and Frei Otto [5]. It has also been seen that the materials themselves inspire and drive the design process [8]. The available number of materials presents both unparalleled design opportunities as well as posing new responsibilities relating, first and foremost, to environmental sustainability. In the history of design, there have been times when the available quantity of materials and related production technologies were understandable to the individual. But nowadays, to imagine such a possibility is unrealistic: there are simply too many. Furthermore, they appear in multiple performances, their identity scattered. Some important research on this has been produced in recent years (e.g. [12, 15, 18]).

Both materials and technologies are characterized by great dynamism and rapid evolution. It means that the amount of new production and the exponential speed of its generation will only widen the gap between what can be grasped and what exists—and thus, questions of form, function and meaning are also to be renegotiated. This poses significant challenges but also great potential for the design of our future environment. It is valid in terms of form, function and meaning, as well as sustainability and ecology, together with sociological and social developments.

There are expressive-sensorial qualities in materials that are difficult to describe theoretically [19, 20]. The recognition of these specific material qualities—beyond given physical properties, such as tensile strength or elasticity modulus—offers great potential for designers and is, therefore, of interest for future research. What do we learn from the texture or the smell of a material? How is it possible to exploit them to create meaningful material experiences?

Designers have started to experiment with materials. There are, in fact, many cases of them producing their own materials, as if those already available were not enough or were not sufficiently responsive to the needs of the new projects [16, 17]. Consequently, various challenges have arisen for (new?) professionals: being expected to know new materials' behaviour and properties, together with being up to date on new technologies; and also how to develop new materials by drawing on a multidisciplinary approach that involves material science, interaction and biology, to name but a few. Such a complex task increasingly requires the ability to cross the quantitative data of materials (necessary "engineering" data) with qualitative information—from sustainability to cultural associations—, dimensions in which designers demonstrate good interpretation skills.

2 Between the Taxonomies

The extremely high number of materials is problematizing the traditional modality of direct experience upon them [10]. In other words, the process of learning the physical limits and the perceptual qualities of materials is no longer compatible with direct experience. This is partially due to the speed of production. Design today is characterized by increased links with other areas of knowledge. Material quantifiable parameters are being integrated into digital models. Learning from the natural sciences and technical disciplines is being drawn on. Increasingly more work is being done on systems that are physically, chemically or biologically inspired [6]. This development can also be seen at the level of materiality. The new logic of design concepts and hybrid materials no longer correlates with materials' familiar properties.

The classic material categories seem to be increasingly dissolving. That is why Sabine Kraft wrote that the "relationship between form and material has become as diverse as it is ambiguous. A recourse to clear rules and specifications as to what can be conceived and constructed in which material and how, and what aesthetic message would be conveyed with it, is hardly possible any more—if they ever existed" ([9], p. 24; authors' translation). This leads us to ask whether the discussion about constructions suitable for the material or design ideal for the material is now an outdated approach. Today, materials can be developed for specific situations. The question of material design arises as a notable decision and development at the beginning of a design task and not necessarily a selection from a palette of pre-existing materials.

Humanity has evolved from a "Materiocentric" approach (in which materials were conceived as static—with an unchangeable set of characteristics) to an "Ideocentric" approach (in which the "purity" of ideas is no longer assumed to be compromised by material limitations). Ideas are steadily becoming less subject to adaptation in order to conform themselves to the material possibilities of production limitations. Material oriented design is increasingly turning into "design with designed materials" [5]. The traditional systems for classifying materials, commonly referred to as "material families", struggles to incorporate some new materials' unusual and contradictory properties. Many new materials are characterised by a hybridisation, which may be derived from the combination of different material "genotypes" and technologies. These material hybrids also result in hybrid properties. Formerly existing material properties are overwritten or exchanged. These informed or charged material hybrids link different contexts. Many new design solutions integrate materials and technologies whose complex structure no longer allows conclusions to be drawn about performance or function.

Metals, wood, glass, ceramics, stones, plastics are generic taxonomies whose "cultural image" is more and more outdated in the face of a much more complex material world. Therefore, in certain aspects, the inclusion of new material in any of these families involves some unavoidable stereotyping—which may very well hinder its full application potential being envisioned. Because of the current abundance of different materials and their almost inevitable complex "genotype", belonging to a traditional material family becomes more of a perceptual limitation than understanding new materials' possibilities. If for no other reason, because these nomenclatures base their members on a visual, functional, and performance identity that may be far from what they can be. In addition, the grouping of materials according to physical characteristics requires systematic corroboration in daily life that no longer occurs with the necessary consistency. That is why Manzini ([11], p.55) said that the question "What is this?" should be replaced by the question "What do I need?".

3 Research Directions

Materials emerge as cultural expressions of a social interest in overwhelming complexity and, therefore, any reflection upon materials cannot be set apart from a more comprehensive (social and cultural) framing.

A great deal of the most recent I&D involving the areas of design, materials, and production focus on the cultural dimensions simultaneously.

Many of today's approaches follow sustainability and focus on intelligence that is not always inscribed in the material itself, but how it is joined, constructively formed, and applied. New design concepts can also create entirely new relationships between form and function through the micro-scaling of new materials or new technologies. Many of today's design solutions integrate new technologies and materials in scaling and complexity that no longer allow conclusions about the performance or function of the hybrid solutions. This is also linked to the ambivalence of the new hybrid concepts. Such articulation of the new material solutions and hybridizations of design is an essential cultural mission [13].

References

1. Ashby M (1999) Materials selection in mechanical design. Butttterworth-Heinmann, Oxford
2. Bar-On YM, Phillips R, Milo R (2018) The biomass distribution on Earth. In: Proceedings of Natural Academy of Science USA 115(25):6506–6511 (2018, June 19). https://doi.org/10.1073/pnas.1711842115. Epub 2018 May 21. PMID: 29784790; PMCID: PMC6016768
3. Clauser HR (1975) Industrial and engineering materials, Wisconsin, Glencoe/McGraw-Hill
4. Elhacham E, Ben-Uri L, Grozovski J et al (2020) Global human-made mass exceeds all living biomass. Nature 588:442–444. https://doi.org/10.1038/s41586-020-3010-5
5. Holzbach M (2015) Gestalten mit gestalteten Materialien, Bauwelt. Frei nach Otto—Sieben Betrachtungen zu seinem Erbe" Issue 20, pp 26–27 (2015)
6. Holzbach M, Kracke B, Bertsch G-C (2014) Material Grove—von traditionellen Materialien zu zukunftsorientierten Materialentwicklunge (From traditional materials to future-oriented material developments), HfG Offenbach (ed.), pp 12–17, Offenbach am Main 2014
7. Karana E, Barati B, Rognoli V, Rognoli V, Zeeuw van der Laan A (2015) Material driven design (MDD): A method to design for material experiences. Int J Des 9(2):35–54
8. Karana E, Pedgley O, Rognoli V (2015) On Materials experience. Des Issues 31(3):16–27
9. Kraft S (2004) Werkstoffe. Eigenschaften als Variablen, in: Arch+ 172 pp 24–28
10. Lucibello S (2005) Materiali@design-Verso una nuova modalità di selezione su base percettiva dei materiali per il design. Editrice Librerie Dédalo, Rome
11. Manzini, E. (1989). The material of invention: Materials and design (A. Shugaar, Trans.). London, England: Design Council.
12. Oliveira, P. (2012). A Identidade Difusa dos Materiais Metálicos: Contributos para um ensino holístico sobre materiais em cursos superiores de design. (PhD Thesis), Universidade Técnica de Lisboa, Lisbon.
13. Parisi, Stefano (1), Holzbach, Markus (2), Rognoli, Valentina (1): "The hybrid dimension of Material design: two case studies of a Do-It-Yourself approach for the development of Interactive, Connected, and Smart materials", (1) Politecnico di Milano, Design Department (2) Offenbach University of Art and Design, Design Department, 2020
14. Pedgley O., Rognoli V.; Karana E. (2021) Expanding Territories of Materials and Design. In: Pedgley O., Rognoli V.; Karana E. (eds.) Materials Experience: Expanding Territories of Materials and Design. Butterworth-Heinemann: Elsevier, UK.
15. Rognoli V.; Ayala-Garcia C.; Poliini B. (2021). DIY Recipes. Ingredients, Processes and Materials Qualities. In: Laura Cleries, Valentina Rognoli, Setaal Solanki, Pere Llorach (eds.) Material Designers. Boosting talento towards circular economies. Elisava, Barcelona.
16. Rognoli V, Bianchini M, Maffei S, Karana E (2015) DIY Materials. Special Issue on Emerging Materials Experience, Materials & Design 86:692–702
17. Rognoli V., Salvia G., Levi M. (2011). The aesthetic of interaction with materials for design: the bioplastics' identity. In: Proceedings of DPPI'11, Designing Pleasurable Products and Interfaces, June 2011, Politecnico di Milano, Italy
18. Rognoli V (2010) A broad survey on expressive-sensorial characterization of materials for design education. METU J Fac Archit 27:287–300

19. Rognoli V., Levi M. (2011). Il senso dei materiali per il design, (The sense of materials for design). Milano: Franco Angeli, pp. 1–271, ISBN: 978–88–568–3230–3@@@
20. Rognoli V, Levi M (2005) Materiali per il design: espressività e sensorialità, (Materials for design: expressivity and sensoriality). Polipress, Milan, pp. 1–180. ISBN: 88-7398-012-0

Biomimetic Application Potential of *Agave sisalana* Mechanical Properties, Lightness, Resistance Strategies, and Life Cycle for Digital Fabrication

Rodrigo Araújo, Amilton Arruda, Jorge Lino Alves, Theska Soares, Tarciana Andrade, and Emília Arruda

Abstract This research seeks to understand which elements are responsible for the mechanical properties of Agave sisalana and how these properties can conform to lightweight, resilient structures and materials for bio-inspired additive manufacturing processes with the possibility of design innovation and sustainability through of multidisciplinary research involving biomimetics, biology, materials processing and 3D printing. Environmental issues, economics of matter and energy; Difficult access to biodegradable materials, lack of adaptation to the natural processes of recycling and reintegration into the natural cycle of the environment are points related to the research problem. We try to answer the problem question by aligning biomimetic processes, digital fabrication and design of bio-inspired materials. We try to answer the problem question by aligning biomimetic processes, digital fabrication and design of bio-inspired materials. It is believe that it is possible to emulate the strategies of lightness and strength of the cell wall structure of the Agave floral stem in bio-inspired digital artifacts.

Keywords Biomimicry · Agave · Digital fabrication

1 Introduction

Bio-inspired problem-solving research has focused on the development of methods and tools for the systematic use and application of natural element information. The methodological approach in biomimetics aims to study the strategies of nature, taking it as a principle and inspiration for solving design problems. Biological research

R. Araújo (✉) · A. Arruda · T. Soares · E. Arruda
PPGDesign & PPGBV, Universidade Federal de Pernambuco, Recife, Brasil

J. Lino Alves
INEGI, Faculdade de Engenharia da Universidade do Porto, Porto, Portugal

T. Andrade
CIAUD, Faculdade de Arquitetura, Universidade de Lisboa, Lisboa, Portugal

E. Duarte and C. Rosa (eds.), *Developments in Design Research and Practice*, Springer Series in Design and Innovation 17,
https://doi.org/10.1007/978-3-030-86596-2_6

favors ecological performance and allows defining metrics on the creation of bio-inspired shapes and materials [1, 2]. Possibilities given by the latest technologies, production systems and development of new structures and materials.

The field of material design as a science and technology demand for innovation with the advancement of knowledge in materials science and technology has made it possible to manipulate and create new materials with better properties for specific applications and biomimetics has strongly contributed to sustainable solutions [3]. With the advent of additive manufacturing technologies and the design of new materials for 3D printing, the possibilities for the emergence of new composite materials with different properties and possibilities for innovation in sustainability have steadily increased [4]. Biomimetics aims to bridge this gap by increasing the dimensionality of the design space by emulating strategies for bio-inspired structures, materials and bio-inspired additive manufacturing. This holistic view considers the research of biology, computing parametric and manufacturing materials as inseparable from the design dimensions, resulting in ecological artifacts from the beginning [5]. The main developments in biologically inspired structure and material solutions mainly related to the physical and mechanical properties of the material or, alternatively, to the structural characteristics of the material or structure constructed with the material. This approach is successful when relating function to the characteristics of the biological structure as well as the properties of the biological material [3].

Not only for environmental reasons, but also for technical and economic reasons, vegetable fibers have been gaining ground in the industry. Among the many lignocellulosic fibers that have been gaining ground, Agave sisalana, produced in Northeastern Brazil, has technical, economic and environmental advantages. It has extreme value, mainly due to its excellent mechanical properties, presenting mechanical behavior similar to synthetic fibers in relation to tensile strength [6]. In addition, the plant stem has strategies of lightness and resistance, these functions provided by the combination of structural characteristics and material properties. The study and emulation of the structure and function of cell walls combined with the possibilities of development of a bio-inspired material is the focus of this article. The study of cell structure and the materials that make up the walls of Agave fibers presents strategies of lightness, resistance and mechanical properties with potential for biomimetic application to structures and bio-inspired material.

2 Methodological Approach

A first time related to the research literature on the themes; parallel to the microscopic investigation and abstraction of the agave cell wall lightness and resistance strategies through the analysis and production of scanning and transmission electron micrographs. Another moment related to the emulation of agave lightness and resistance strategies through modeling processes of parametric structures with application in artifacts. Another step is to study the properties of the natural model for the development of a biodegradable composite material, produced with the elements

Fig. 1 Biology to design (Biomimicry Thinking—DesignLens) [1]

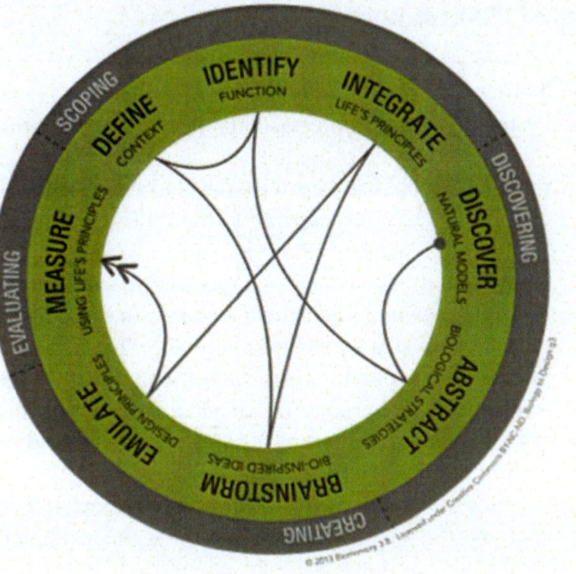

that make up the agave cell walls for the production of a filament for use in additive manufacturing 3D printers. It is a multidisciplinary research, permeating the field of biomimetics, biology, design, digital manufacturing processes and material handling. We used an approach called Biology to Design (Biomimicry Thinking—Biomimicry DesignLens) following the phases according to the need of the project when the process is configured from a biological inspiration and seeks to give direction to the development of some project or artifact and/or bioinspired materials [1] (Fig. 1).

The process begins in the phase of discovery of the natural model where is had prior knowledge of the strategies of lightness and resistance. The second moment allocated in the abstraction phase of biological strategies, seeking an application area for Agave strategies. Following is the identification of the necessary functions, later the definition of the scope of the research context (definition of the problem). The next step is to create bio-inspired ideas for emulation in lightweight, resilient, energy-optimized structures, including the study of a bio-inspired 3D printing material. Then comes the incorporation of the principles of life, which are principles of sustainability. The emulation phase represents the principles, patterns, strategies and functions found in nature that can inspire design. The last step is related to the materialization and prototyping of ideas. The process evaluation should verify compliance with the life principles listed in the process at the beginning of the project/research. Iterative replication cycles imply better results.

3 Biomimicry

Through biomimetics, we are learning to emulate natural forms, processes and ecosystems to create more sustainable designs. Imitating these more refined designs for the planet can help humans move toward technologies that consume less energy, reduce material use, reject toxins, and function as a system for creating life-friendly conditions [1, 2]. According to literature [7], bio-inspiration using insights into the role of biological systems in the development of new engineering concepts has already established itself as a successful and rapidly growing field of science. Possibilities are given by the latest technologies, that is, production systems and development of new structures and materials. The area of material design as a science and technology demand for innovation with the advancement of knowledge in materials science and technology has made it possible to manipulate and create new materials with better properties for specific applications and biomimetics has contributed greatly in this area for sustainable solutions [7].

Bio utilization (use of biological raw material) is feasible as long as it favors the construction of an objective aiming at the sustainability of the project, adapting to local realities and presenting benefits (or reduction of damages) to the ecosystem in its use, in effort to meet certain biologically inspired principles. This approach is successful when relating function to the characteristics of the biological structure as well as the properties of the biological material [3]. Sustainability is a strong point in the development of innovations by making direct use of biological materials solutions as they are sustainable in themselves. In fact, biological systems made from biological materials do not create waste or irreversible damage to the ecosystem. On the contrary, they are of great relevance as they enrich and sustain the ecosystem in which they operate. In addition, biological structures provide a wide range of properties with minimal use and flow of materials and energy, and generate fully recyclable products.

Biomimetic artifacts inserted in digital fabrication using bio-inspired material provide excellent biological strategies and models, allowing to elaborate new questions and answer questions about the relationship of cellular structure and its materials. It is precisely the transfer of function from biology to artifacts that allows biomimetics to emulate and test hypotheses of the biological sciences; otherwise there is a danger of blindly copying or imitating design principles without further knowledge of the actual functions of the forms and composition of the natural model.

3.1 Bio-inspired Structure–Function-Material Relationships and Digital Fabrication

The design of many successful products can be considered as the result of a successful relationship with the natural forms and the phenomena-functions contemplated in them [8]. Structure–function relationships can also provide a fertile platform for cooperation between engineers and biologists. This relationship is inherent in both biological and engineering thinking and may be a common denominator between the two disciplines [9].

In 1978, Bonsiepe defines the morphological analogy as the experimental search for elaborate models of the translation of structural and formal characteristics to be transposed into projects [10]. Thus, authors [11] state that this kind of analogy seeks to study and analyze why the natural form, the interrelationships of its geometry, observing and understanding its textures, paying attention to the characteristics of the form, parts and components, details of some part at macro or microscopic level, as well as for their structural forms. The idea that function fits form or structure is one of the basic design principles in nature and well accepted in both biology and design literature.

Innovative ideas are emerging from research on systems and natural properties that do not always translate only in appearance and aesthetics, but that the natural form also favors the gain in efficiency [11]. Regarding functional morphology, the principle of form and function is the first and oldest of the strands of the development of artifacts based on natural organisms and focuses on the relationship between biological forms or structures and their functions [12]. Figure 2 (left) illustrates how highly efficient structures in nature can provide structural principles that can be applied to reduce weight by up to 70% in structures such as boat rails, automotive pillars and bicycle frames. This research presents concepts such as asymmetrically radiated joints and automatic shell and volume reinforcement, where the size and geometry of each cell is adapted to the load. Smaller, more closed cells in which the load is greater becomes larger and more open cells, where smaller loads are applied [13].

According to authors [12], some scientific observations of nature have acted in both the macroscopic and microscopic fields. Technical implementations within the macroscopic dimension have been successful in observing and using avail-able techniques, this works especially well when the desired function is more closely related to its shape or structure and less to the forming material. Its technical realization in a non-biological material does not change that, the same goes for structures of various formal configurations.

The ELiSE Company has developed a bio-inspired parametric structure where automatic hardening of casings and volumes is directly related to the size and geometry of each cell and is tailored to the load—for example, smaller, more closed cells

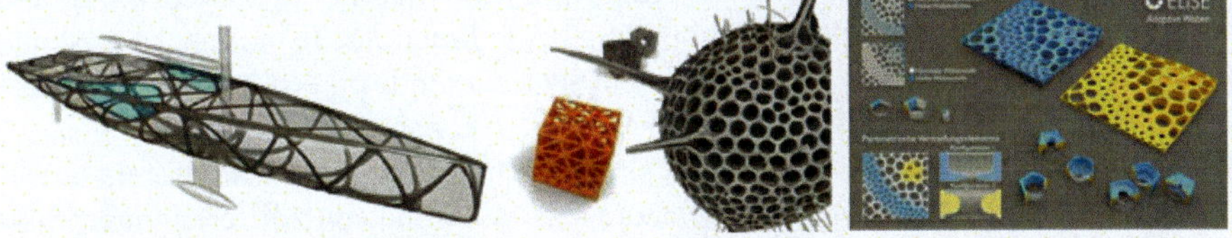

Fig. 2 ELiSE projects at (https://www.compositesworld.com/articles/ibex-2017-show-report) - Leidenfrost [13]

Fig. 3 Above: https://natitinov.files.wordpress.com/2016/01/corkscrew-willow-xylem-sem-80201781-l.jpg?w=656 (2012); http://sciencewise.anu.edu.au/articles/timbers (2007); below: https://www.engineering.com/3DPrinting/3DPrintingArticles/ArticleID/7905/New-Composite-Resins-Could-Lead-to-Larger-Stronger-3D-Printed-Structures.aspx (2014)

where the load is larger into larger and more open cells where loading is minimal [13].

The (Fig. 3-above) presents scanning electron micrographs of the light and resistant cellular structure of Balsa wood. The (Fig. 3-below) presents a bioinspired approach in the mechanical properties of balsa wood, the emulation of the structure and materials represent the transfer of function of the lightness and resistance strategies, together with the mechanical properties of the balsa materials to digital fabrication. The 3D printing material is a composite composed of resin and carbon fibers to maintain structural strength without excessive weight [14].

Biological inspiration should aim to extract nature's good designs and implement them in a way that adds value and functionality to our mechanical designs. Authors points to a basic principle of patterns of tubular structures present in nature. Such structures are characterized by a hollow cylinder, rod or tube. The hollow cylinder appears in bird feathers, flower stalks, bamboos and reeds, grain stalks, insect limbs, and long human bones, such as the femur, most of the houses of the earth's dwellers have a tubular design. The hollow cylinder provides stability against bending and deformation and is adjusted to resist bending in all directions. In the case of repeated pipe structures, each pipe in the arrangement acts as a single pipe, distributing stresses throughout the structure optimizing resistance to mechanical stress [9].

Although biological and technological functions are derived from different terminologies, structures are visual and therefore less subordinated to different interpretations. Structure–function patterns, in particular, can abstract nature's design solutions to various problems. These frameworks have inspired a new generation of innovative technologies in the science, engineering and design community. As well as the microscopic structure of cells/fibers present in plant tissue of Agave sisalana [15].

3.2 *Bio-inspired Materials and 3D Printing*

In the product development process, with 3D printing, high part complexity can be achieved at no additional cost while still making use of raw material efficiently and more sustainable material choices. Recent developments in 3D printing with regard to the use of natural materials have been analyzed [16]. According to the authors, biological printing materials, such as wood filament, require careful use of printer and media usage specifications for the most efficient and economical printing result. Wood printing filaments require very fine fiber particles to ensure a smooth printing process without nozzle blockage.

Different ways of incorporating the characteristics and properties of wood in materials for bio-inspired additive manufacturing are described [16], such as the development of bio-based filaments (Fig. 4). 3D related to biological materials will occupy the market niche, for example for more sustainable and complex format products. Agave sisalana fibers, as well as wood fibers, have application potential because they have properties and benefits for the market and the environment.

4 Agave sisalana

Brazil is the world's largest producer of Agave (or Sisal), accounting for about 70% of the world hard fiber market. Agave sisalana is the most commercialized species, common in the northeastern region of Brazil. It is an exotic and invasive plant of dunes and sandbank of the Brazilian coast, an introduced species, commonly found in several states. Harms the establishment and development of native flora species and does not provide food for local fauna [17] (Fig. 5).

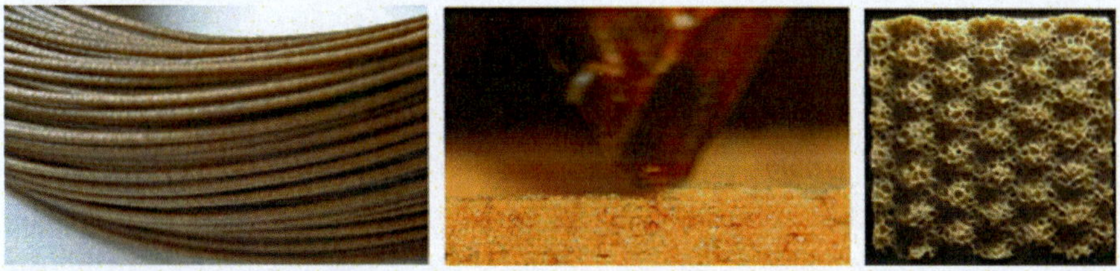

Fig. 4 Left and center images: https://tobuya3dprinter.com/expect-3d-printing-wood/#prettyPhoto (2015); Image on the right: @pa_Hugron (2018)

Fig. 5 Picture of Agave sisalana and her floral stem. Image on the right: Agave fibers grinded from the floral stem that can use as reinforcement for 3D printing materials, Authors (2019)

The plant has a life cycle that can range from 7 to 10 years, according to the Invasive Species Specialist Group.—ISSG (Global Invasive Species Database) (2019) [18]. According to literature, the floral scape can reach from six to eight meters in height. Because it is a monocarpic plant, it blooms only once during the growing season, dying later. At the end of their useful life, these materials are collected and reconfigured by other organisms, repeatedly recycled with the energy of the sun [19].

The plant stem has no commercial value comparable to leaf fiber. The ideal state for use as a raw material is when the plant dries and dies naturally, ending the life cycle of seven to twelve years. Authors explains that this way there is no deforestation, on the contrary, the removal of the environment in this region becomes a beneficial practice for the local biome, because it is an invasive species that does not serve as food in this ecosystem [17].

Sisal fibers stand out for their widespread domestic, industrial and, more recently, reinforcement of polymer composites [20].

From an anatomical point of view, sisal fibers are structural cells whose function is to support and stiffen the leaves and stem. When compared to other natural fibers, Sisal fibers have superior strength and good durability [21]. As the fibers come from the leaves and are also present in the pseudostem of the plant, their chemical constitution is basically formed by the same compounds present in the leaves, having in its chemical composition cellulose, hemicellulose, lignin, pectin and waxes [15, 22, 23] (Fig. 6).

Studies by several authors who estimated the percentage of the elements that make up the sisal fiber. It pointed out that the fibers can contain 65.8–73% cellulose, 12–13% hemicellulose, 9.9–11% lignin and 0.8–2% pectin [6].

Sisal fibers have a lignocellulosic chemical composition, they influence fiber resistance [25]. Lignin influences the structure, properties, morphology and flexibility of lignocellulosic fibers. Cellulose, on the other hand, is the polymer that gives plant fibers excellent breakage and elongation properties. In this way, the lignocellulosic chemical composition directly interferes with the strength of the fiber [6].

Among the most relevant elements for fiber, calcium stands out for being a structural component of the cell wall, considering that the wall resistance established by its composition based on cellulose, hemicellulose and lignin contents, present in the

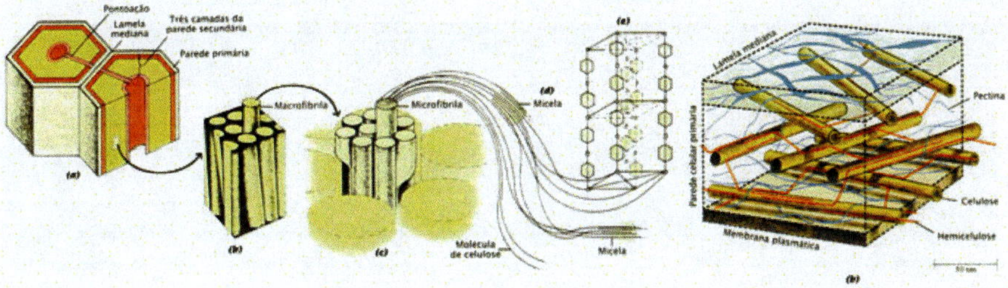

Fig. 6 Cell wall composition. Adapted from Taiz and Zeiger [24]

composition of the plant cell walls. Some authors report that P and K should considered nutritional components of great relevance when considering the fiber strength since phosphorus tends to increase the fiber length and potassium the amount of cellulose. Increments to these elements are important for greater fiber resistance, reflecting improvements in fiber length, length uniformity, and thickness [6].

Sisal fiber reinforced composites stand out for their high impact strength and good tensile and flexural strength properties. This attributed to the fact that sisal fiber has one of the highest values of modulus of elasticity and mechanical strength among natural fibers [23]. Comparative studies were made with vegetable fibers and polymeric fibers, including sisal and other natural fibers and polypropylene (PP) fibers. Studies have shown that sisal fibers have a higher modulus of elasticity, consequently greater rigidity than vegetable fibers, such as coconut bagasse and sugar cane, as well as polypropylene fibers. The breaking and elongation resistance also related due to the intermolecular forces between the cellulose chains [6].

In this sense, the authors state that sisal fibers can replace the fiberglass used to reinforce polymer composites in the manufacture of parts produced by various manufacturing processes, such as injection molding, lamination, resin transfer molding, among other utilities such as application in digital fabrication and 3D printing due to high mechanical strength and lightness [6]. The investigation of the chemical and mechanical properties of agave fibers has potential for emulation of a bioinspired material in the strategies of lightness and resistance, to be used as input for 3D printers having biodegradation conditions.

4.1 Plant Anatomy

The microscopic dimension allows the observation of nature's structures at advanced levels of detail. A recurring example represented by the plant cellular anatomy of plants in general, which reveals compacted bundles of differentiated vessels and cells. The geometric arrangement and compacted integration produce a complex, strong and flexible structure [12]. All cells have a structural role in addition to other

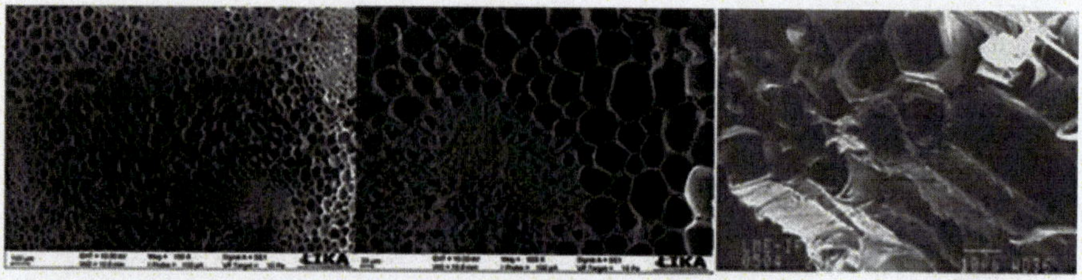

Fig. 7 SEM of Agave sisalana fibers showing tubular structures and geometric patterns. *Sources* Left and Center Images: Authors (2019); image on the right: cross-sectional view of sisal fiber Martins et al. [26]

functions. Figure 7 shows SEM images of Agave sisalana showing such a geometric arrangement.

The functions of plant systems depend on the structures, the structural form of the various tissues, such as mechanical properties, etc. May have different formats: polyhedral; cylindrical or spherical, but in general, are multifaceted isodiametric cells. It has multiple faces, that is, many sides having approximately the same dimensions. The cell wall that delimits a cell may also vary in thickness, ornamentation and frequency of holes, etc. Despite this morphological diversity, cell walls commonly classified into two main types, primary and secondary. Primary cell walls are typically thin. Secondary cell walls are thicker and more resistant than primary cells, xylem cells, such as those found in wood, are notable for having highly thickened lignin-reinforced secondary walls [24].

4.2 *Potential for Applying Agave Properties*

A study of the cell wall structure of the agave plant stem tissue was carried out to understand how their strategies of lightness and resistance are presented. It was found that the agave plant has such biological strategies. Using microscopy techniques, one can see the structural organization of the arrangement of lignocellulosic cells present in agave plant tissue (Fig. 8).

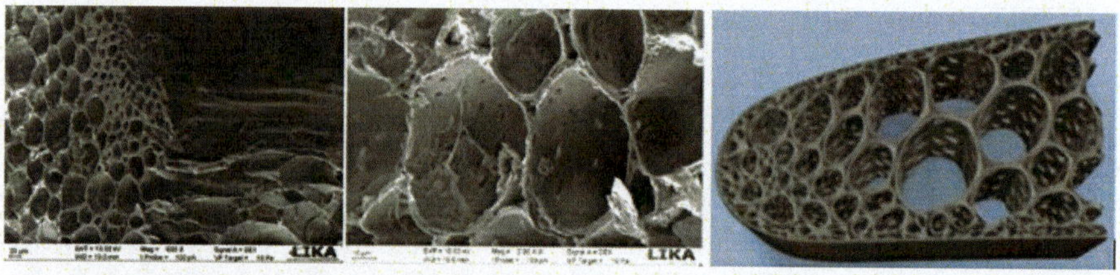

Fig. 8 Left and center, SEM images from agave. Image at right: digital parametric modeling of agave's lightness and resistance strategy. Authors (2019)

SEM images of Agave's cellular structure illustrate the variation of different cell organizations in successive hierarchies. Cells have a structural role in addition to other functions. The fibers have as one of the main functions, to support the vegetable, arranged in the form of bundles or strands, scattered throughout the primary body of the plant. Can come in many forms. The fibers that arranged in overlapping rows are elongated spindle cells that have thicker walls. These fibers are supporting cells responsible for stiffness and flexibility properties. The orientation of the fibers deposited in parallel is one of the significant factors for the mechanical properties of the plant. This structural organization results in greater tensile strength [15].

The cell wall structure of Agave plant tissue presents the strategies needed for lightness and resistance functions with optimization of matter and energy. Among other factors, the deposition of lignin (structural natural polymer responsible for the stiffness of plant cells) occurs in minimal amounts to provide the plant with mechanical support and strength. These properties applied in the development of a bio-inspired generic structure can directed to a range of artifacts that require light and resistant structures.

A study [15] was carried out in 2015 with the objective of developing a biomimetic structure inspired by the configuration of lignocellulosic cell walls of agave, for the emulation of biological strategies in materialized artifacts. Digital modeling, parametric design, and 3D printing processes (Fig. 8) enabled alignment with Agave's growth and development principles, which deposit raw material and energy by utilizing the principle of resource maximization while maintaining its mechanical properties efficient for its functions. Figure 8 shows the potential emulation of the formal/functional structure of Agave cell walls. Further investigation of the materials that make up the cell walls of agave/sisal fibers helps to understand which elements are responsible for the mechanical properties and how these properties they can be formed into a lightweight and biodegradable lightweight material for additive manufacturing processes and innovation in sustainability.

5 Conclusions

Biomimetic research and application is an effective means for bio-inspired innovations following nature's model. Understanding why the agave floral scape features strategies of lightness and endurance was the starting point. Therefore, it was necessary to investigate at a micro scale the cellular structure that make up the floral tassel of the plant through the study of biology in plant anatomy. Information about Agave's cellular anatomy can greatly contribute to achieving optimum design of lightweight, resilient structures with low energy and matter consumption. Based on these data, the choice of agave as a natural element and source of inspiration is justified in line with the principles of biomimetics and sustainability.

References

1. Baumeister D (2014) Biomimicry resource handbook. A seed of best practices. Ed. Biomimicry 3.8—Missoula, Montana
2. Benyus J (2003) Biomimicry: innovation inspired by nature. Ed. HarperCollins, New York
3. Byrne G, Dimitrov D, Monostori L, Teti R, Houten F, Wertheim R (2018) Biologicalisation: biological transformation in manufacturing. CIRP J Manuf Sci Technol 21:1–32. https://doi.org/10.1016/j.cirpj.2018.03.003
4. Daya T (2017) Facilitating sustainable material decisions: a case study of 3D printing materials. Ph.D. thesis. UC Berkeley Electronic Thesis and Dissertations
5. Oxman N, Ortiz C, Gramazio F, Kohler M (2015) Computer-aided design. Elsevier, vol 60
6. Carneiro J (2017) Caracterização de acessos de sisal usando descritores da planta e da fibra. Tese doutorado. Programa de pós-graduação em recursos genéticos vegetais - Universidade Estadual de Feira de Santana - Feira de Santana
7. Müller R, Abaid N, Boreyko J, Fowlkes C, Goel A, Grimm C, Jung S, Kennedy B, Murphy C, Cushing, N, Han J (2018) Biodiversifying Bioinspiration. Bioinspiration & Biomimetics, vol 13(5), IOP Science
8. Hsuan-na T (2017) Design: Conceitos e Métodos. Ed. Blucher, São Paulo
9. Cohen Y, Reich Y (2016) Biomimetic design method for innovation and sustainability. Springer International Publishing. https://doi.org/10.1007/978-3-319-33997-9-Switzerland
10. Bonsiepe G (1978) Teoria y Práctica del Diseño Industrial. Editorial Gustavo Gilli, Barcelona
11. Soares T, Arruda A (2018) Novas estratégias da biomimética: as analogias no biodesign e na bioarquitetura. Mix Sustentável, vol 4(1), Florianópolis pp 73–82
12. Gleich A, Pade C, Petschow U, Pissarskoi E (2002) Potentials and trends in biomimetics. Springer, New York
13. Leidenfrost D (2015) Development of a nature inspired hull structure for a 46m sailing yacht. Master thesis. Department: Naval Architecture and Ocean Engineering. Bremen
14. Compton B, Lewis J (2014) 3D-printing of lightweight cellular composites. Harv Univ Camb, Adv Mater 26:5930–5935. https://doi.org/10.1002/adma.201401804Cambridge
15. Araújo R (2015) Biomimética e artefatos para ambientes aquáticos – estratégias de leveza e resistência inspiradas na estrutura celular do agave. Ed. Novas Edições Acadêmicas, Omni Scriptum Publishing Group - Beau Bassin
16. Wimmer R, Koddenberg T, Steyrer B (2015) 3D Printing and wood. In: International conference wood science and engineering in the third millennium—ICWS, https://doi.org/10.13140/RG.2.1.2483.6563 Brasov
17. Ulysséa M, Silvestro M (2010) Pranchas de Surf Ambientalmente Corretas, Utilizando Madeira da Agave, Espécie Exótica e Invasora de Vegetação Nativa Brasileira. 1º Congresso de Inovação, Tecnologia e Sustentabilidade - Santa Catarina
18. ISSG—Global Invasive Species Database. Species profile: Agave sisalana. http://www.iucngisd.org/gisd/speciesname/Agave+sisalana last accessed 2020/04/20
19. Silva O et al (2008) Cultivo do Sisal no Nordeste Brasileiro, Circular Técnica n. 123, Embrapa, Campina Grande
20. Gondim T, Souza L (2009) Caracterização de Frutos e Sementes de Sisal - Circular técnica 127 - Embrapa, Campina Grande
21. Sapuan S, Leeniea A, Harimib M, Bengb Y (2006) Mechanical properties of woven banana fiber reinforced epoxy composites. Mater Des 27:689–693
22. Mwaikambo L, Ansell M (2002) Chemical modification of hemp, sisal, jute, and kapok fibers by alkalization. J Appl Polym Sci, 84(12):2222–2234. https://doi.org/10.1002/app.10460
23. Martin A, Martins M, Mattoso L, Silva O (2009) Caracterização química e estrutural de fibra de sisal da variedade Agave sisalana. Polímeros: Ciência e Tecnologia 19(1):40–46
24. Taiz L, Zeiger E (2004) Fisiologia vegetal. Ed. Artmed, Porto Alegre

25. Holanda E (2013) Morfologia e propriedades mecânicas da fibra de sisal unidirecional e em sobreposição de compósito com resina epóxi. Universidade Federal do Rio Grande do Norte, Natal
26. Martins M, Kiyohara P, Joekes I (2004) Scanning electron microscopy study of raw and chemically modified sisal fibers. J Appl Polym Sci 94:2333–2340

From Agricultural Waste to Microbial Growth and (G)Local Resilience

Nitzan Cohen, Emma Sicher, and Seçil Uğur Yavuz

Abstract Agriculture and food production play essential roles in defining the identity of a territory. While agrarian products are value generators for a region, the waste produced in the process is a mere byproduct with no or low value. However, in nature, the concept of waste does not exist since the residue of an organism is the nourishment for another. Being inspired by nature, the concept of 'small is beautiful' (Schumacher in Small is beautiful. Economics as if peoplemattered. Blond and 7 Briggs, London, UK [4]) can be reached not only through focusing on local production but also radically changing the scale, considering microorganisms as alternative production systems. As such, biotechnological developments can bring us to a new concept—'micro is beautiful'—which aims at converting outputs of food-related production into inputs for new production possibilities through the aid of microbial activity. Today, we witness emerging design initiatives working on living matter, such as fungi, bacteria or algae, aiming to create new sustainable materials and production models. Working on this new field of design, we present the research project InnoCell, as an example of a possible micro/macro production system based in South Tyrol in Northern Italy, which valorises a local resource (agricultural food waste) through microbial intervention and aims at creating (g)local resilient production processes. It is an interdisciplinary project focusing on the potential of locally grown microbial cellulose and its diverse 'co-products'. This microbial upcycling of local resources would provide 0 km production alternatives for various materials produced far away, enhancing territorial values, and spreading the knowledge to other localities through an open-source production system.

Keywords Growing design · Microbial cellulose · (g)local

N. Cohen (✉) · E. Sicher · S. Uğur Yavuz
Faculty of Design and Art, Free University of Bozen-Bolzano, Piazza Universitá 1, 39100 Bolzano, Italy
e-mail: nitzan.cohen@unibz.it

E. Duarte and C. Rosa (eds.), *Developments in Design Research and Practice*, Springer Series in Design and Innovation 17,
https://doi.org/10.1007/978-3-030-86596-2_7

1 Introduction

By living as nomads, once we were carrying a few belongings that could safely return to nature. After becoming sedentary, we have started cultivating and producing goods from local or even from far-away resources, and therefore we had to figure out how to manage waste that is the result of our consumption habits. Over time, these dynamics resulted in the imbalanced relationship of the current throw-away society with nature. Goods are continuously manufactured through processes of over-production and consumption. This causes incredible amounts of waste and resources are inevitably exhausting. Therefore, there should be an urgent change in how waste is perceived and how this new awareness can give rise to new ways of dealing with it. This could be possible by thinking in a micro-scale to recognize the value of local resources and dynamics, but at the same time reflecting on economic, social, and environmental developments and crisis happening on a global scale to advocate for a systematic change [1].

Agriculture and food production play essential roles in defining the identity of a territory: our perception of 'local' is generally related to the resources that a geographical area offers. However, waste generated from agricultural practices is often seen as by-product having a lower value than the main product. Focusing on a micro-scale by involving microorganisms in the production process could help us to turn this down-cycling process into a closer-loop where each product of the production has equal value and continuously transforms into new matter as it happens in nature. In natural systems, there is no concept of waste, as a residue of an organism becomes the nourishment of another [2].

Can micro-organisms help us to design new ways of creating closed loops in our production processes? Can food scraps turn into 'co-products' through fermentation?

This paper addresses these questions proposing a (g)local model aiming at a 'co-production' of substances based on local food-system byproducts. We introduce the ongoing research project 'InnoCell' as a case study that takes advantage of a fermentation process through the aid of a hybrid colony of microorganisms (SCOBY) to enhance the current value of South Tyrolean food-related 'waste'. Moreover, it addresses the 'growth unit' concept, which will be an open-source microbial production system to spread the knowledge about this innovative practice globally.

2 Nature as Source for Resilient Insights

2.1 *Micro is Beautiful*

The current status of the environment is the result of 3.8 billion years of biological adaptation to the changing planet conditions [3] therefore, the secret for survival in terms of forms, processes and ecosystems lies around us. The way organisms and natural elements adapt to changing conditions suggests nature as a perfect ally to provide insights and principles to be mimicked and adjusted for human needs. Similar to nature, where the concept of waste does not exist, our current production models should follow the same path, considering all substances generated in a process as equally valuable.

In current food production systems, while a specific product is the desired outcome, the other resulting substances are often called by-products or waste with a lower value. However, the raw resources can be considered as 'meta-products', capable of being transformed into a variety of possible 'co-products' prolonging their life cycle. Indeed, through the paradigm of 'meta-product to co-products', anything produced within a process will have equal value and could safely return to nature. This model can work if we change how we transform matter, shifting from mechanical or chemical transformation to biological by following the rules of life where macro-organisms mutually depend on other micro-organisms. The concept of 'small is beautiful' [4] can be reached not only through focusing on local production but also radically changing the scale, even considering microorganisms as production systems. Can biotechnological development bring us to a new concept—'micro is beautiful'?

Today, we have many bio-design initiatives working with living matter. Specifically, in the growing-design field, designers cooperate with existing fungi, bacteria, algae and hybrid colonies to create novel materials for sustainable alternatives to existing matter and production processes [5]. When materials are constituted or produced by living organisms, design transcends its traditional borders and teams with experts of other scientific fields [6]. The outcomes often result in being speculative and highly holistic seeing designers playing active roles in understanding growth processes and finding effective solutions to be integrated into existing or future hypothetical systems.

2.2 Biodesign for Local Resilience Case Studies

Growing-design [7] is a branch of bio-design in which diverse organisms like bacteria, fungi or algae are used and embedded into solutions for new products, processes and systems [8] aiming at efficient ecological integration. Designers take part in interdisciplinary teams offering their creative skills in envisioning scenarios and developing proofs of concept for emerging biotechnologies and bio-inspired approaches. Moreover, they have often a hands-on approach in which they actively become 'cultivators' [9] by creating the optimal conditions for the microorganisms to grow and to be later implemented into diverse applications and realities.

In past decades, artists and designers have been crossing borders to other scientific disciplines resulting in more symbiotic approaches and integration of natural processes in their practices. One of the first examples of the intersection between art and biology is 'Victimless Leather' [10] in 2004 presenting a provocative discussion piece, a 'semi-living' jacket developed by an interdisciplinary team of artists and scientists. The jacket raises questions about the moral implications of animal-based leather. Few years later 'Synthetic Aesthetics' [11] brought together artists, designers and synthetic biologists to foster discussion and generate innovative and speculative proposals concerning the potential of synthetic biology. The outcomes were equally surprising and promising, envisioning products and scenarios for more sustainable and utopic futures. One of the proposals is 'Packaging that creates its content' [12] using bacteria as manufacturing systems forming a probiotic cup releasing its nutrients into the liquid when hydrated.

Currently, a broad scope of design initiatives and activities promote open-source knowledge transfer about this emerging practice. Materiom [13] is an open-source material recipe archive utilising the advantage of algae, mycelium, and other compostable sources to accelerate the use of resources and biomaterials for circular economy. Besides, microbial processes show the potential not only to substitute other environmentally inefficient production dynamics but also to produce substances as geographically close as possible to their final utilisation. For instance, Malai [14] is an activity aimed at social empowerment through bio-design, by working with local farmers and suppliers in Southern India. It makes sustainable fashion accessories with bacterial cellulose which is grown using waste coconut water collected from the local production. Similarly, Evoware [15] is an Indonesian start-up which produces edible seaweed-based packaging and food wrappings. It provides an alternative to plastic packaging proposing water-soluble items which would not only positively impact the economic situation of local seaweed farmers but also provide additional environmental value to the urban society in Indonesia.

2.3 South Tyrol and Sustainable Relevance

Focusing on the local setting of the InnoCell project, South Tyrol[1] shows a general sensibility towards sustainable innovation displaying engagement on different levels to preserve its ecosystem. Proinsect [16], an innovative running project, is using the black soldier fly maggots in the process of converting urban organic waste into humus and practically eliminate almost all the residues of the process. Moreover, the Faculty of Design and Art of the Free University of Bozen-Bolzano is highly engaged in (g)local, social, and ecological issues with the master's degree in Eco-Social Design and design projects developed in the bachelor program. One example among these projects is 'New Menu' [17] which focused on food waste, unpacking the potential of discarded or under-used local resources. The students opened valuable discourses between tradition and potential innovation through partnership with local farmers and food-production companies.

These examples provide some insights about how to possibly speculate and create more eco-compatible solutions acting on a local scale, starting from the resources present in the territory. The next section depicts project InnoCell with its novel

[1] South Tyrol region is in the North of Italy, a part of the Tyrol area.

perspective to enhance the value of food waste in South Tyrol by creating potential co-products obtained through microbial intervention.

3 Starting Point: From Peel to Peel

Although food consumption is a short-term activity, the related packaging is often made of long/er-lasting materials, such as plastic, paper, and diverse composites. The experimental design project 'From Peel To Peel' [18] was born by questioning this paradoxical situation. The project aimed to find a locally crafted alternative to disposable food packaging and disposable tableware by employing a compostable material (microbial cellulose) obtained from the fermentation of local food 'by-products' through a symbiotic culture of bacteria and yeasts (SCOBY). Microbial cellulose (MC) provided an alternative to enhance the relationship between content and container. Another critical point was to envision the integration of possible production systems into the current regional setting.

3.1 SCOBY: A Hybrid Character

In-depth research led to identifying the possible ways of converting by-products into useful substances. Algae, fungi, and SCOBY were compared—also to paper and plastic—in terms of characteristics, production methods and manufacturing journey. SCOBY (also called Kombucha mother) was chosen for its capability to transform sugar-rich food byproducts into two valuable co-products both compatible with edible and non-edible purposes: namely a fermented liquid and a fiber-rich substance: microbial cellulose. The SCOBY is mostly known for edible-related purposes. Namely, the Filipino dessert Nata de Coco is constituted by chunks of SCOBY cooked into coconut syrup. However, it is commonly used to ferment green and black tea mixed with sugar, generating a probiotic beverage: Kombucha tea. This naturally enriched drink dating back to 220 B.C. has recently gained a high (g)local potential given by its capability to transform by-products which we consider as waste into valuable substances with a strong regional focus. Indeed, renewable resources 'naturally' call for local uses [1]. The conversion process happens through fermentation in which the SCOBY is immersed into a container filled with a nourishing liquid for 12–21 days. During this time—as it happens in any other fermentation like beer brewing—the 'mother' consumes the sugar and reproduces itself through symbiotic processes eventually generating layers of microbial cellulose floating on the surface of the liquid, as seen in Fig. 1.

This type of cellulose is purer than the one extracted from trees and appears to have a smaller environmental impact since it is synthesized directly from food-related by-products. For these reasons, it could provide a regenerative low-km solution to substitute imported materials and ingredients used on a local level. Indeed, this novel

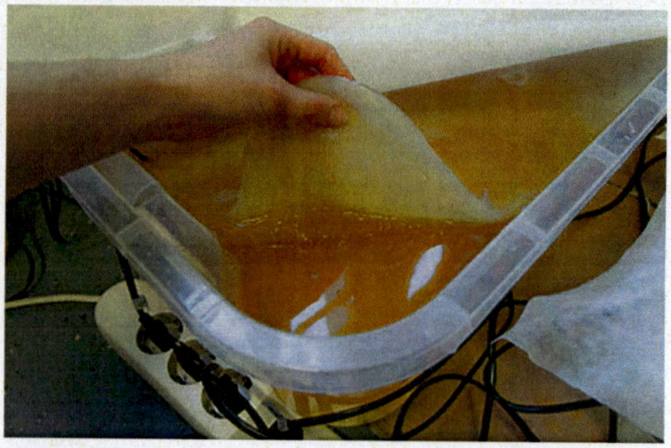

Fig. 1 Sicher. E. Microbial cellulose layer is grown from apple byproducts (2017)

process unpacks a variety of novel substances (co-products) to be used in different fields. In particular, it appears to be suitable for medical, cosmetic, biotechnological, food (source of indigestible fibers), novel material and compound, and even for extraction industry applications [19–22].

SCOBY was introduced into the design world by fashion designer Suzanne Lee [23] who wanted to provide an alternative scenario by raising the discourse about the polluting leather industry. She developed a collection of items of clothing in which she used microbial cellulose generated from tea fermentation to substitute more common textiles. However, microbial cellulose is highly hydrophilic, which means that it absorbs water becoming jelly-like and not suitable for clothing. The 'Biocouture' project was, in fact, a conversation piece. From that point in 2011 on, several design students [24–26] and professionals started tinkering, experimenting, and speculating [27] with the material to discover its potential. From Peel To Peel is one of these projects which brought the discourse to another field by focusing on food packaging.

3.2 Microbially-Produced Packaging

The project developed on two parallel levels: the material underwent an MDD (Material Driven Design) approach [28] to practically understand its capabilities and limits for food packaging and disposable tableware applications while the second level extended the practical findings towards a systematic introduction of the material into the local system.

MC was grown from local by-products such as scraps of apple, beetroot, potato, grapes, blueberries, hops, and mixed vegetables immersed into containers filled with a water-based solution and with SCOBY mothers. The fermentation happened in static conditions. The grown wet layers of MC were dried and used for further experimentation. They were treated in several manners acquiring different shapes

Fig. 2 On left: Sicher. E. Samples of microbial cellulose dyed with fruits and vegetable juices (2017), on right: Sicher. E. Sample of shiny dyed Microbial cellulose, (2017)

Fig. 3 Sicher. E. Dried apple packaging made of apple peels cellulose (2017)

and shades enabling various material applications as sheets or as substances (see Fig. 2).

The final models were representations of food packaging (Fig. 3) and disposable tableware (Fig. 4) with a familiar semantics, clear cuts, and bright shades. A particular focus was given to local food-producers proposing packaging for their specific products. Similarly, disposable trays and cups were developed for street food vendors and for hypothetical events in order to enhance the territorial identity through direct and local uses of the material.

A systemic approach was used to investigate the dynamics of the current system to speculate hypothetical ways of integration of MC production into the region of South Tyrol. In particular, the journey of food by-products was analysed from their collection from producers to their disposal into specialised centres. Visits were made to staple producers, political representatives, waste-transportation services and waste-management facilities to outline regional patterns around agri-food production systems. After framing current dynamics, two scenarios were developed proposing; first, an MC production plant integrated to the fermentation centre in charge of disposing of the urban organic waste of the city of Bolzano/Bozen (see Fig. 5). As

Fig. 4 Sicher, E. Street food plate made by beetroot cellulose (2017)

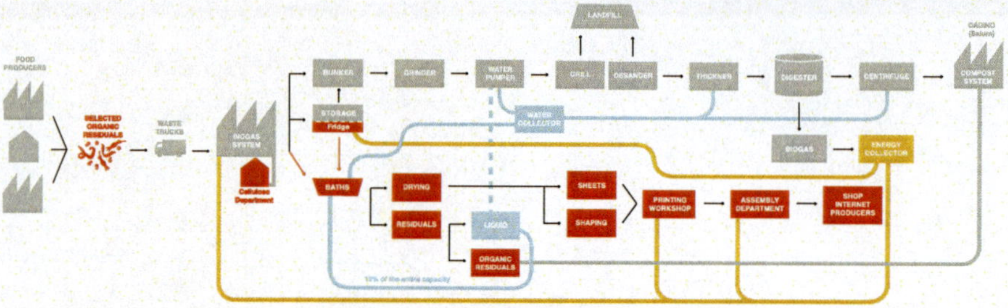

Fig. 5 Sicher, E. MC production plant integrated into organic waste fermentation plant (2017)

discussed with the plant's responsible, the current facility and the hypothetical MC production appear to be mutually nurturing.

On the other hand, the second proposal displays a scheme for an independent MC production plant (see Fig. 6).

These systemic investigations unpacked the potential of MC as a catalyser for virtuous biological cycles. The fermentation process would take advantage of both the fermented liquid and the generated microbial cellulose without leaving possibly

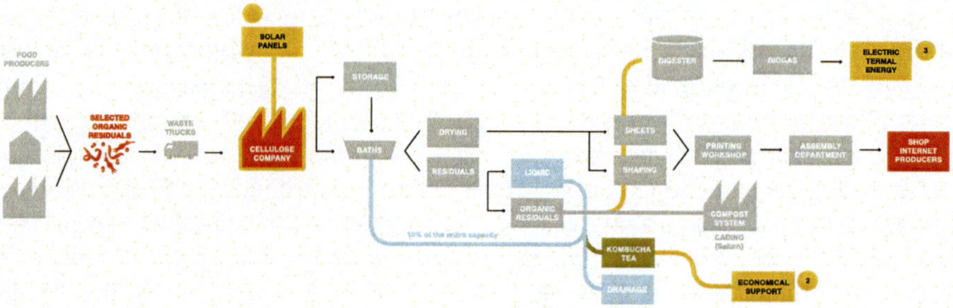

Fig. 6 Sicher, E. Independent MC production plant (2017)

any waste. In this way, local assets would create an industrial ecology [29] and be actively engaged in territorial value creation. These assumptions provided a ground for the following research project addressed in the next section.

4 InnoCell

In this paper, we present the project InnoCell, as an example of a new possible microbial production system based in South Tyrol. It would, on the one hand, valorise local agricultural food 'waste', on the other hand, it aims to propose a (g)local model which can be influential and diffused onto other local realities.

South Tyrol, the setting of the project, is the biggest apple growing area of Europe. Solely in 2018, 910.767 tons of apples were produced regionally [30] and sold as fruits or transformed into other edible products generating tons of by-products. This amount is generally down-cycled into fertiliser and animal feed.

The founding principle of InnoCell is the potential of transforming the scraps of the production of apple-related products into valuable substances. The apple is considered as 'meta-product' generating a variety of 'co-products' through transformation processes—like juice, dry apples—and through microbial intervention—such as kombucha apple juice and compostable microbial substances—for edible and non-edible purposes.

4.1 Methodology

The ongoing interdisciplinary project is led by the design team, together with food technology scientists. The partners from the food technology field are in charge of analysing the fermentation process with a growth-efficiency goal. Therefore, they selected the most efficient strains of microorganisms and defined the ideal recipe for the nourishing liquid based on apple scraps. Moreover, they will determine the most efficient parameters for the process of fermentation and for the production of the SCOBY co-products.

The design team explores the capability and potential applications of the microbial substances through practice-based design research. A DIY (do-it-yourself) growth unit is being developed in order to produce SCOBY co-products, enabling further tinkering and experimentation for possible applications. Moreover, future scenarios with a strong (g)local focus will be built through co-design workshops with partners.

4.2 *DIY (Do-It-Yourself) Growth-Unit Concept*

As MC introduces a relevant discourse towards the low-km production of substances, the consequent step would be to diffuse the knowledge of SCOBY fermentation to other localities to create novel product solutions. A growth-unit concept is being developed with the aim of making an easy-to-replicate system in order to enable anyone, anywhere to independently perform the process. The most commonly used growing technique is to let the SCOBY grow on the surface of the liquid in a static way, differently in our DIY growth-unit a dynamic system called rotating disk bioreactor was built based on the previous comparative studies done by other scholars and biotechnologists [19, 31–33] to produce an optimal amount of yield. The prototype is currently being tested with the aid of food technology partners. The goal is to create a unit to be simply replicated elsewhere with standard components and minimal assembly interventions. The unit would generate microbial cellulose substances and a fermented liquid simultaneously by using diverse food by-products coming from the local agricultural and/or food production.

5 Project Pattern: Global and Local Relevance

A relevant finding given by the ongoing research is that a systematic approach is essential when proposing novel product and production solutions. In fact, as industrial products cannot be designed as single entities, also when developing novel substances and matter, it is fundamental to consider the entire ecosystem around its life-cycle. Dynamics concerning raw materials, transformation processes, use, disposal, links, and touchpoints among stakeholders need to be investigated into current systems in order to envision meaningful scenarios of integration and/or coexistence. Indeed, desk research and design-research tools like stakeholder maps, service maps, system maps would help to visualise and to deal more efficiently with complex structures. Moreover, field research would further ground and enhance the designer's assumptions through visits to stakeholders' facilities and advice from experts. In this way, the systematic approach would support the research hypothesis to be verified based on the existing realities, aiming to a real impact.

In analysing the project's path, a relevant pattern was framed (see Fig. 7): while project 'Biocouture' by Suzanne Lee raised the discourse of an alternative production in a global dimension, project 'From Peel To Peel' had a strong locally-oriented focus. Instead, project 'InnoCell' merges the two perspectives (local and global) and provides novel production solutions based on local waste, on the other hand, it aims at carrying this know-how to other localities through the open-source DIY growth-unit.

This can be identified as a generic pattern: a global-focused principle can be taken to a local level and be enhanced with a (g)local perspective. Other projects could follow a similar path unpacking and highlight promising (g)local potentials.

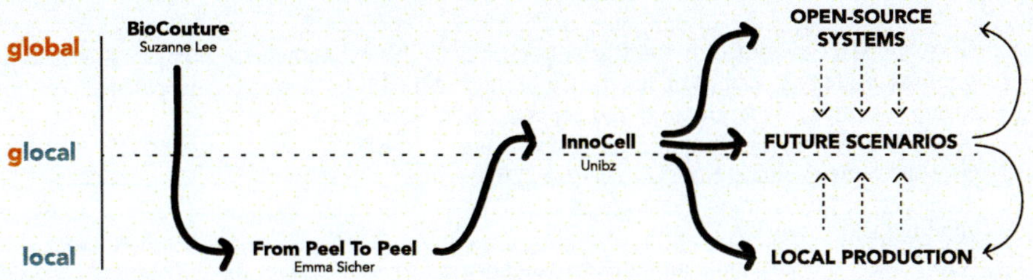

Fig. 7 Project pattern

6 Final Assumptions

This paper reflects on the potential of SCOBY as a catalyser for resilient production from a (g)local perspective. The symbiotic colony of microorganisms is capable of transforming discarded food by-products into highly valuable substances for edible and non-edible purposes and applications. This can be possible if we adopt the formula of 'meta-product to co-products' which we above mentioned. The fruitful husbandry approach between microorganisms and designers as 'cultivators' is deepened by our project InnoCell enhancing the discourse around the (g)local potential of SCOBY co-products as drivers for enhanced territorial value creation.

References

1. Manzini E (2012) Resilient systems and cosmopolitan localism—the emerging scenarios of the small, local, open and connected space. CNS Ecologia Politica
2. McDonough W, Braungart M (2002) Cradle to cradle: remaking the way we make things, 1st edn. North Point Press, New York, United States
3. Baumeister D, Tocke R, Dwyer J, Ritter S, Benyus JM (2014) Biomimicry resource handbook: a seed bank of knowledge and best practices. Montana: Missoula, United States
4. Schumacher EF (1973) Small is beautiful. Economics as if people mattered. Blond and Briggs, London, UK
5. Camere S, Karana E (2017) Growing materials for product design. In: Alive. active. Adaptive proceedings. Delft University of Technology international conference on experiential knowledge and emerging materials, Delft: TU Delft Open, June 19–20, pp 24–37
6. Antonelli P (2012) Vital design. In: Myers W. BioDesign. Nature, science, creativity, 1st edn. Thames & Hudson, High Holborn, UK, pp 6–7
7. Ciuffi V (2013) Growing design. Abitare 531:110–111
8. Myers W (2012) BioDesign. Nature, science, creativity, 1st edn. Thames & Hudson, High Holborn, UK, pp 8–9
9. Collet C (2017) Grow made textiles. In: Alive. Active. Adaptive. Proceedings. Delft University of Technology international conference on experiential knowledge and emerging materials, Delft: TU Delft Open, June 19–20, p 27
10. Victimless Leather, https://tcaproject.net/portfolio/victimless-leather/, last accessed 2020/05/06
11. Ginsberg AD, Calvert J, Schyfter P, Elfick A, Endy D (2014) Synthetic aesthetics: investigating synthetic biology's designs on nature. MIT Press, United States Cambridge

12. Ginsberg AD, Calvert J, Schyfter P, Elfick A, Endy D (2014) Synthetic aesthetics: investigating synthetic biology's designs on nature. Cambridge: MIT Press. United States, p 174
13. Materiom, https://materiom.org/, last accessed 2020/05/06
14. Malai, http://made-from-malai.com/, last accessed 2020/05/06
15. Evoware, https://evoware.id/, last accessed 2020/05/06
16. Proinsect, https://www.eco-center.it/it/attivita-servizi/ricerca-993.html, last accessed 2020/05/06
17. New Menu, https://issuu.com/idmcreativeindustries/docs/catalogue_cover_compressed__1_/70?fbclid=IwAR19JajNWptFgI2oQx0g2KA6DhIPJod-gN9FwT4pTk7JluiD49h1e69TbR0, last accessed 2020/05/06
18. Sicher E (2017) BA thesis: from Peel To Peel—an experimental design approach for packaging. UNIBZ, Bozen-Bolzano. Available online: https://ubz-primo.hosted.exlibrisgroup.com/primo-explore/fulldisplay?docid=39UBZ_ALMA_DS21151949370001241&context=L&vid=UNIBZ&lang=it_IT&search_scope=All&adaptor=Local%20Search%20Engine&tab=default_tab&query=any,contains,From%20peel%20to%20peel&offset=0, last accessed 2020/05/06
19. Chawla PR, Ishwar BB, Shrikant AS, Rekha SS (2009) Fermentative production and applications of microbial cellulose. Food Technol Biotechnol 47(2):107–124
20. Revin V, Liyaskina E, Nazarkina M, Bogatyreva A, Shchankin M (2018) Microbiology cost-effective production of bacterial cellulose using acidic food industry by-products. Braz J Microbiol 49(1):151–159
21. Dourado F, Gama M, Rodrigues AC (2017) A review on the toxicology and dietetic role of bacterial cellulose. Toxicol Rep 4:543–553
22. Cacicedo ML, Castro MC, Servetas I, Bosnea L, Boura K, Tsafrakidou P, Dima A, Terpou A, Koutinas A, Castro GR (2016) Progress in bacterial cellulose matrices for biotechnological applications. Biores Technol 213:172–180
23. Lee S (2020) Grow your own clothes. https://www.ted.com/talks/suzanne_lee_grow_your_own_clothes, last accessed 2020/05/06
24. Rykkelid E (2020) Growing products. https://growingproducts.tumblr.com/, last accessed 2020/05/06.
25. Janusz R (2020) MakeGrow Lab. https://www.makegrowlab.com/, last accessed 2020/05/06
26. Amato E (2020) Ponto biodesign. https://www.facebook.com/events/908029682970073/?notif_t=plan_reminder&notif_id=1588777207130982/, last accessed 2020/05/06
27. Dunne A, Raby F (2020) UMK, Biocars. http://unitedmicrokingdoms.org/bioliberals/, last accessed 2020/05/06
28. Karana E, Barati B, Rognoli V, Zeeuw van der Laan A (2015) Material driven design (MDD): a method to design for material experiences. Int J Des 9(2):35–54
29. Fiksel J (2003) Designing resilient, sustainable systems. Environmental science & technology. Environ Sci Technol 37(23):5330–5339
30. ASTAT, https://astat.provincia.bz.it/downloads/Siz_2018(2).pdf, p 37, last accessed 2020/05/06
31. Bungay RH, Serafica G, Mormino R (1997) Environmental implications of microbial cellulose. Stud Environ Sci 66:691–700
32. Pa'e N, Zahan KA, Muhamad II (2011) Production of biopolymer from acetobacter xylinum using different fermentation methods. Int J Eng Technol IJET-IJENS 11(5):90–98 (2011)
33. Bungay H, Pajuelo M, Hogg T, Vasconcelos I (2015) Lowering costs for microbial cellulose, pp 1–8

Design for Interaction

Should Technology Be [just] Delightful?

Rodrigo Hernández-Ramírez, Liene Jakobsone, Thomas Behrens, and Teresa Chambel

Abstract Our world has been steadily populated by a growing number of artificially "a-live" objects and smart devices, over the last decades, with varying degrees of autonomy, and seeming to become better at guessing and responding to our needs, offering pleasurable or delightful experiences, and making us more willing to trust them. Nevertheless, such development and understanding of ubiquitous smart technologies are not without criticism and challenges. The more users regard technology in magical terms, the more they ignore the complexities of the underlying infrastructure enabling their devices. However, these come with vulnerabilities that expose users to various undesirable situations, threatening their privacy and forcing them to deal with complications, with potential unforeseen social and economic consequences. This is an important reason why Human-Centered Design needs more than being empathic towards the user; it needs to assume a clear ethical stance. Designers should learn to say "no" and ask "why" while examining the role technology plays in our lives and asking what types of new relations we want to establish with it, and ultimately be equipped to ask about our human project for the future and the role a humanistic design approach should play in it. So, Should technology be [just] Delightful? This introductory chapter addresses this fundamental question while commenting on the contributions of the other chapters selected for this Design for Interaction section of the book.

R. Hernández-Ramírez (✉) · T. Behrens
UNIDCOM/IADE - Unidade de Investigação em Design e Comunicação, IADE - Faculdade de Design, Tecnologia e Comunicação, Universidade Europeia, Avenida D. Carlos I, 4, 1200-649 Lisbon, Portugal
e-mail: rodrigo.ramirez@universidadeeuropeia.pt

T. Behrens
e-mail: thomas.behrens@universidadeeuropeia.pt

L. Jakobsone
Art Academy of Latvia, Kalpaka blvd. 13, Riga 1050, Latvia
e-mail: liene.jakobsone@lma.lv

T. Chambel
LASIGE, Faculty of Sciences, University of Lisbon, 1749–016 Lisbon, Portugal
e-mail: mtchambel@ciencias.ulisboa.pt

E. Duarte and C. Rosa (eds.), *Developments in Design Research and Practice*, Springer Series in Design and Innovation 17,
https://doi.org/10.1007/978-3-030-86596-2_8

Keywords Ubiquitous smart technology · Digital interactions · Complexity · Privacy · Human-centered design · Self-reflection · Disruption · Usefulness · Empathy · Ethics and trust

1 Introduction

Over the last decades, our world has been steadily populated by a growing number of artificially alive ("a-live") objects with varying degrees of autonomy, ranging from programmable vacuums and smart thermostats to virtual assistants—e.g., "Alexa" or "Siri". Each year, these smart devices seem to become better at guessing and responding to our needs, offering pleasurable experiences, and making us more willing to trust them. These smart objects (including mobile phones) perform as well as they do, not because they are "intelligent", but due to the expansion of mobile communication systems based on satellites, cell sites, and WiFi hotspots, coupled with ever-increasing computing power has made our physical environment more friendly towards them. More humans online, faster and more efficient computing, more sensors communicating with each other, more RFID tags, and more actuators mean more opportunities for collecting data about every aspect of human behaviour [4]. This accumulation of data has enabled otherwise purely syntactical—and therefore semantically incompetent—systems to be considered "smart". Data processing is the backbone of this ubiquitous computing.

The main idea driving this wave of automation is the belief that technologies work best when they recede to the background of our experience [18] and that we should take advantage of current developments in computation to enliven everyday objects. Seen in this way, ubiquitous computing is portrayed as a form of "enchantment". As such, technology could offer information, entertainment or help people accomplish specific tasks, but all while getting out of people's way, behaving unobtrusively, without forcing them to think about what they are doing. Enchantment relies on aesthetics, playfulness and people's affective responses to turn smart devices into "partners in ubiquitous meaning-making" ([10], p. 738) for people.

Such an understanding of ubiquitous smart technology is not without criticism. The notion of enchantment is arguably underpinned by a "magical" portrayal of technological design that prevents users from confronting potentially unsettling questions about their devices [17]. Issues such as the type of labour involved (ethical or exploitative?); the provenance of the materials (do they come from conflict zones? Are they ethically produced?); fabrication and disposal (are they environmentally friendly?); and privacy (user's data is sold to third parties?) are eschewed.

"Magical thinking" combined with technological efficiency prevents users from pondering the negative consequences that technologies might bring. The more users regard technology in magical terms, the more they ignore the complexities of the underlying infrastructure enabling their devices. Magical thinking also eschews many of the constraints and frictions imposed by real life, ignoring that they exist for some reason and may have a purpose.

A common trend in ubiquitous computing seeks to make everyday human movement more fluid by automating physical barriers. For example, smart door locks at homes or offices that automatically open as users approach thanks to either biometric information (e.g., digital print) or an RFID tag. Some solutions even have suggested enhancing these door-locks with long-distance geo-location and predictive algorithms that can anticipate the time a given person might be likely to approach the door. The problem, however, is that all of these functionalities would require vast amounts of data about users, information that would otherwise remain hidden or would only be accessible through conversation or keen observation. Thus, by removing the friction of opening a door, ubiquitous computing would also have to remove the frictions that enable critical human dynamics.

This frictionless, utilitarian understanding of technologies not only brings an arguably false sense of augmented agency but also eschews three crucial problems:

1. that technologies are not simply extensions of human capacities, but are also phenomenological mediators which influence human experience;
2. that in order to predict users' needs, technologies, need to spy on users;
3. that what is often perceived as prediction is, in fact, induction of needs or actions.

The latter case is when an application encourages the user to do a certain thing, she had not thought about doing, such as buying a coffee cup as suggested by the iPhone's "time and location" function.

Moreover, consumer-level ubiquitous computing does not merely work out of the box. For example, most services require a smartphone, to accept the terms of service, manage preferences and updates, and program and monitor the devices' behaviour. To properly function, these devices presuppose that all the systems in the network in which they are embedded can exchange data. In reality, the implied seamless interoperability is difficult to achieve for both commercial and technical reasons (patents or petty cooperative rivalry). Consequently, there is no universal standard governing how different devices should communicate over a network. Thus, even the simplest device will enter an ecosystem of "balkanised" operating systems, software protocols, versions, frameworks, packages, and dependencies [7], all of which contribute to making the system more vulnerable to errors and third-party attacks. These vulnerabilities expose users to various undesirable situations, threatening their privacy and forcing them to deal with complications, such as component obsolescence, bugs, and unsafe data handling that cannot be avoided as ubiquitous computing systems continue to grow in complexity.

Despite these shortcomings, technocentric biases continue to thrive, mainly because "perfection, prediction and seamlessness" as the central (and sometimes only) goals behind technological design [2]. The problem, however, is that the promises of seamless, flawless technological advances never come, and, instead, end-users are left to deal with the unforeseen social and economic consequences of these technologies. Moreover, most of these consumer devices are now developed following some version of Human-centered Design approach (HCD) to User Experience (or at least developers claim to do so). While these methodologies are right to put users' needs at the centre of their concerns, HCD still tacitly argues that technologies should be "ready-to-hand" (to use a Heideggerean term) and disappear from users' consciousness to better perform their role. Hence, devices should blend into the background, appearing almost unnoticed as quiet servants whenever needed. This view is problematic because it perpetuates a utilitarian understanding of our relationship with human and non-human agents and because it lacks an explicit ethical stance.

From a methodological standpoint, HCD should be one of the best means to keep the consequences of poorly implemented automation at bay. However, to do so, HCD cannot continue to be a mere epistemic facilitator—i.e., a means to make the complexities of technological systems understandable. HCD needs more than being empathic towards the user; it needs to assume a clear ethical stance towards technological development and, arguably, the courage to enforce that stance. Designers should learn to say "no" and ask "why" when dealing with ubiquitous computing; they should question whether specific processes require to be automated and whether interfaces should be hidden for the sake of making technologies delightful. Designers should critically examine the role technologies play in our lives and ask what types of new relations we want to establish with our devices. Ultimately, designers should ask what our human project for the future is and what role a humanistic design approach should play in it.

2 Design for Technology and Digital Interactions

HCD advocates take human–technology relations for granted, that is, that interfaces should be "ready-to-hand", as well as epistemic facilitators, but, as we saw earlier, this often requires invasive and privacy-threatening compromises. Consequently, as our dependency on smart devices increases, so do the chances that they end up calling the shots, distorting and constraining our behaviour and our physical and conceptual environments to further accommodate us to them instead of the other way around [4]. The danger is that instead of establishing healthy dialectic relationships, we end up adapting to their "needs" only "because that is the best, or sometimes the only, way to make things work" ([4], pp. 252–253). Designers should, therefore, always consider what would be the actual role a given technology is going to play in peoples' lives.

Regardless of these concerns, however, from the point of view of the user, there is undoubtedly an experiential benefit from disappearing interfaces, as we can see in the chapter, "A Semiotic and Usability Analysis of Diegetic UI: Metro—Last Light", where Doval, Almeida, and Nesi discuss how integrating the user interface into game's art & narrative (a diegetic interface) can increase the feeling of immersion and thus provide a more pleasurable game-play experience. Furthermore, immersion can also serve a methodological purpose, as Fransreb et al. show, while exploring the potential of VR for carrying out UX research, though there is open debate concerning the scalability of the results. Thus, it is necessary to carry out more research and perhaps bring a deeper and broader understanding of users' interaction with technology.

Furthermore, when the mediation capacities of technologies are not only accounted for but also positively used, as Seçil Uğur Yavuz shows in "MASAL: Bridging between two cultures through storytelling with an interactive e-textile toy", designers make the enhancement of the relationship between humans the explicit end-goal of technological systems. Thus, here is a case in which the interface is not something that disappears but visibly mediates human relations. These chapters show the emergence of new design frameworks which, through creative speculation, challenge the dominance of HCD and show the need to reevaluate the tenets of this methodology.

3 Design for Self-reflection, Imagination and Disruption

The rapid technological development that has taken place during the previous century has enabled significant progress and improved the living conditions in certain parts of the world considerably. However, due to corporate interests, poor management, or even lack of awareness, this progress has also resulted in severe damage to the environment and deterioration of the lives of those less privileged. Furthermore, design has been too much focused on plain problem-solving innovation as its primary goal, forgetting that "innovation, as a concept, does not embed any social and ecological value sets, nor does the 'design thinking' approach to problem solving" ([1], p. 6). Even the 'human-centered' aspect of design often remains on a level of physical and emotional satisfaction of the user, without acknowledging any ethical considerations the product might require, not to mention the disregard of indirectly affecting also the non-humans.

In his chapter, "Design Delight: An Experiential Quality Framework", Omar Sosa-Tzec decodes and analyses design products' experiential qualities, which cause a sensation of delight. As he notes, until recently, this has primarily been a matter of concern for marketing researchers and practitioners, as they have realised that satisfying customers' needs were insufficient to persuade them to buy something. At the same time, it is widely acknowledged that design has historically been, and still is, employed as a powerful tool for marketing. The effect is not achieved by improving products' functionality, nor by making them more ecologically and socially responsible. If previously it was about the looks and trends, then now it is all about the experiences, and delight is among the most pleasurable ones. Sosa-Tzec describes how delight can be created through design, or in other words, what emotions a design product needs to provoke in its user in order to be perceived as delightful. However, the most crucial issue considering the current state of affairs is the purpose of causing delight. Apart from the most evident and historically accustomed reason—marketing—it can also become part of a larger-scale solution to consumerism. Sosa-Tzec remarks that delight, as a positive emotion, might have the power of altering our thinking and actions. Nevertheless, he also urges the designers to adopt a critical perspective and become ethically aware when pursuing "design delight", as this concept rejects thoughtless deployment of products and consumerism.

Luckily, he is not the only one calling for an ethical stance. In the past decades, we have witnessed an emergence of various attempts to question the very basis of design and its tradition, which has for so long contributed to the global environmental and social issues we are currently facing.

Tackling not only technologies but all sorts of design products, one of the most established results of such endeavours is the Critical and Speculative design approach, which has been gaining design community's attention since the 1990s. It is intended to address both designers and society and aims to challenge narrow assumptions and preconceptions and disrupt the status quo. By use of fictional and provocative design proposals in the form of objects and their use scenarios (often depicted in photographs and videos), Critical designers attempt to raise awareness, expose bias and provoke discussions on the role that products play in everyday life ([16], p. 94).

Another approach conceived with a similar intention is to seek ways in which design can contribute to a systemic disruption of the status quo and aim at a different "design culture" ([12], p. 58), originated in the 2010s and is known as the Transition design. It acknowledges that we live in transitional times and is based on the premise that there is a "need for societal transitions to more sustainable futures and the belief that design has a key role to play in these transitions" [9]. Similarly, advocates of this design approach are convinced that the whole society, including designers, needs to reconsider its expectations and assumptions and set new goals accordingly to achieve the necessary change. They urge designers to revise their mindsets and adopt a new paradigm in design to discontinue the ongoing "defuturing" ([5], p. IX).

The two approaches described above have things in common. They pose a strong emphasis on the future, along with an invitation to deliberately shape it according to our needs, rejecting the conformist attitude, re-learning to dream again, and setting goals that serve society instead of the industry. Both approaches also have an awareness-raising ambition and emancipatory goals, as they offer space for disciplinary and individual self-reflection ([13], pp. 389–395), which should result in a humbler and more inclusive mindset, that accepts all kinds of human 'strangeness' as well.

Also, the Disruptive Design Method [1], central to the discourse of Cristina Carmelo Gomes's chapter, "Smart Urban environments," is based on similar premises and aims at similar goals: challenging the status-quo, emphasising social and ecological values, and overcoming one's own bias. Gomes discusses the arguable success of transforming Lisbon into a smart city—a process in which, according to her, the city's peripheries have not undergone sufficient improvement of their public urban environments.

This is partly because the traditional city planning processes are complex and slow; hence they often deliver obsolete results already at the moment of implementation. Nevertheless, their methods are outdated—they replicate the existing organisational models, which are inherently unsustainable. They aim at technological innovation, which solves practical problems, but unfortunately does not tackle the social and ethical issues: neither the historically rooted nor the ones caused by this innovation.

Active citizen involvement in the form of co-design and participative processes is needed to gain insights and confront the users with the multitude of options and activate their imagination. New design methods, such as the above described, are very welcome as well, as change towards the sustainable, human and livable future "will not occur of itself; it can only occur by design" ([6], p. 45).

4 Ethical Technology and the Question of Trust

The chapter "Age Ratings for Tabletop Games' Usage in Brazil—analysis and suggestion of new criteria" by João Léste and Claudia Mont'Alvão, formulated the research question:

> Are age indications in tabletop games adequate to children's abilities and limitations? Moreover, how will children deal with the frustration of not accomplishing a game's objective?

The authors hypothesise that current tabletop games publishers do not follow adequate cognitive-oriented criteria and, therefore, the game's age rating is not adequate for children's learning and development skills.

When we purchase a tabletop game for children, we usually trust the recommended age restrictions, but what exactly is that trust based on? Chemistry sets and science kits have a long history of causing accidents, so what parameters will consumers look for before trusting a specific product? The question of trust can be extended to all interactions between humans and artefacts, but from a certain age onwards, learning and development skills are no longer tied to a specific age.

In a growing trend, companies realise that every aspect of their organisation disrupted by technology represents an opportunity to gain or lose trust. They approach trust *not as a compliance or public relations issue* but as a business-critical goal to be pursued.

Historically, scholars have viewed distrust and trust as opposite ends of the same theoretical construct or as functional equivalents. More recently, scholars have begun to identify them as theoretically different constructs [8, 14], defining distrust as "confident negative expectations regarding another's conduct" ([11]: 439). Researchers note that human perceptions are multi-dimensional; trust and distrust can, and do, coexist simultaneously [11]. "Trust and distrust are contraries, not contradictories. To say that it is not the case that one trusts another person is not to say that one distrusts him" [8]. Thus, distrust is not simply an absence of trust [11] but a proactive expectation of harm [3].

Seen from this perspective, trust becomes a full-scale undertaking to ensure that the many dimensions across an organisation's technology, processes, and people are working in sync to earn and maintain the high levels of trust expected by their many stakeholders. Consumer distrust can be particularly destructive [15], and the prospect of self-driving cars or health related services, products or policies (so relevant and debated in our current pandemic situation) are but a few examples that trust is a multi-stakeholder design challenge and represents a critical value that sets apart the responsible from the irresponsible and the corrupt, in a world that faces multiple crises. Many questions like the following are circling "ethical technology" and ethical dilemmas when making decisions on how to use disruptive technologies. "In a burning building, how would a robot that has been designed to rescue human beings from the fire have to look and act like, if it was to earn their trust?" "In case there are people of varying age groups, race and gender, whom should the robot rescue first?" The technological focus might be too narrow in this context, and it might become critical for designers to address the question of trust as a critical factor for success.

We need to evaluate what "user needs" mean nowadays, perhaps think from an ecological standpoint and this implies re-thinking (even re-designing) what humanism means in the 21st and coming centuries. Perhaps the shifting nature of the relationship between humans and artefacts via interfaces requires an accompanying adjustment in design practice, where the focus shifts from the output to the outcome. This question, in turn, raises the question of what are the preferred outcomes and who should define them? Are designers professionally equipped to engage the new ethical dilemmas that inevitably accompany the present and future technologies?

References

1. Acaroglu L (2017) Disruptive Design Method Handbook (E-Book). Disrupt Design LLC. https://www.creativityandchange.ie/wp-content/uploads/2017/06/Disruptive-Design-Method-Handbook-by-Leyla-Acaroglu-Jan-17.pdf
2. Bleecker J, Foster N, Girardin F, Nova N (2017) #TUXSAX: user experience will be as Shitty as expected. Retrieved April 26, 2021, from Near Future Laboratory website: http://tuxsax.nearfuturelaboratory.com/
3. Cho J (2006) The mechanism of trust and distrust formation and their relational outcomes. J Retail 82(1):25–35
4. Floridi L (2012) The road to the philosophy of information. In: Demir H (ed) Luciano Floridi's philosophy of technology: critical reflections, pp 245–271. https://doi.org/10.1007/978-94-007-4292-5_13
5. Fry T (1999) A new design philosophy: an introduction to defuturing. University of New South Wales Press, Sidney
6. Fry T (2009) Design futuring: sustainability, ethics, and new practice. Berg, Oxford, New York
7. Girardin F (2019) When automation bites back. Retrieved April 26, 2021, from Near Future Laboratory website: http://blog.nearfuturelaboratory.com/2019/01/16/when-automation-bites-back/
8. Govier T (1993) An epistemology of trust. Int J Moral Soc Stud 8(2):155–174
9. Irwin T, Kossoff G, Tonkinwise C (2015) Transition design provocation. Des Philos Pap 13(1):3–11
10. Jenkins T, Bogost I (2014) Designing for the Internet of Things. In: Proceedings of the extended abstracts of the 32nd annual ACM conference on human factors in computing systems—CHI EA'14. https://doi.org/10.1145/2559206.2578879

11. Lewicki RJ, McAllister DJ, Bies RJ (1998) Trust and distrust: new relationships and realities. Acad Manag Rev 23(3):438–458
12. Manzini E (2015) Design in the transition phase: a new design culture for the emerging design. Des Philos Pap 13(1):57–62
13. Mazé R (2009) Critical of What? In: Ericson M, Frostner M, Kyes Z, Teleman S, & Williamsson J (eds) The Reader. Berlin: Sternberg Press.
14. McKnight D, Choudhury V (2006) Distrust and trust in B2C e-commerce: do they differ? In: Proceedings of the international conference on electronic commerce (ICEC'06), Fredericton, Canada, 2006
15. O'Malley L, Prothero A (2004) Beyond the frills of relationship marketing. J Bus Res 57(11):1286–1294
16. Raby F (2008) Critical design. In: Erlhoff M, Marshall T, Erlhoff M, Marshall T (eds) Design dictionary: perspectives on design terminology. Basel: Birkhäuser, pp 94–96
17. Selinger E (2015) Too much magic, too little social friction: Why objects shouldn't be enchanted. Retrieved April 27, 2021, from https://lareviewofbooks.org/article/much-magic-little-social-friction-objects-shouldnt-enchanted
18. Weiser M (1999) The computer for the 21st century [Reprinted from Scientific American, 265(3):94–104]. ACM SIGMOBILE Mob Comput Commun Rev 3(3):3–11. https://doi.org/10.1145/329124.329126 (Original work published 1991)

User Experience of Real and Virtual Products: a Comparison of Perceived Product Qualities

Danny Franzreb, Alexander Warth, and Kai Futternecht

Abstract Virtual reality (VR) enables immersive experiences where users can explore products in the virtual world. In the last few years VR technologies have developed so much that the idea of substituting experiences with real products through those with virtual representation is growing, even with all inaccuracies of the computer-generated experience. If these virtual experiences are adapted as real, how huge will be the opportunities to increase the effectiveness of product development processes? For example, what if we can virtualize main parts of prototyping because we can get valuable feedback on a virtual representative? Thus, understanding the possibilities of VR to create products and evaluate their user experience (UX) is gaining in importance. While previous research analyzed the UX of interactive products, only few studies have investigated the UX of VR-product representations compared to the real world. In order to get a first step towards understanding the UX of virtual products, we conducted a product test that compares the UX of real products with their representations in virtual reality. In an empirical study, our subjects judged three different products with respect to their hedonic and pragmatic qualities by evaluating them through the AttrakDiff 2 questionnaire. The results from our user study revealed that there are in fact no significant differences in the user experience for two out of the three products. However, one object was rated significantly better in virtual reality compared to the real counterpart. Furthermore, qualitative user statements have been analyzed, showing that the users either valued or disliked different aspects of the products depending on whether these products were perceived in real life or VR: In summary, our study shows that there is already a high level of comparability between real and virtual product experience. At the same time, it reveals some question approaches that still have to be answered when experiencing virtual product representatives as if they were real.

Keywords User experience · Usability · Virtual reality · Prototyping · Testing

D. Franzreb (✉) · A. Warth · K. Futternecht
University of Applied Sciences Neu-Ulm, Neu-Ulm, Germany
e-mail: danny.franzreb@hnu.de

A. Warth
e-mail: alexander.warth@hnu.de

E. Duarte and C. Rosa (eds.), *Developments in Design Research and Practice*,
Springer Series in Design and Innovation 17,
https://doi.org/10.1007/978-3-030-86596-2_9

1 Introduction

For years, Virtual Reality (VR) has been experiencing an enormous boost in development, thanks to an increasing amount of technical innovation and it is predicted to continue to grow [28]. Large companies, such as Facebook with the acquisition of Oculus VR, are investing billions in the further development of VR hardware as well as software [3]. This trend is strengthened by the development through further complementary innovations, such as 360-degree video cameras, VR compatible development environments such as Unity [34] or Unreal Engine [35], as well as an unchanged social and entrepreneurial movement to make the knowledge of the world available to all people.

Advancement of VR innovation has impelled a worldwide race to recognize and exploit the next significant market opportunities. VR hardware manufacturers and software engineers are collaborating with specialists of several different areas such as product designers, architects or healthcare practitioners to create the next generation of interactive products and services [40].

Innovation is crucial for businesses nowadays to be successful in the global market. Many companies are trying to gain a competitive advantage towards their competitors by applying new technologies and processes to their engineering design practices. Opportunities in industrial design processes arose because the maturity of VR element technologies has increased greatly [11]. VR opens new possibilities for a faster, cheaper and more effective process and they appear in all stages of product design, such as ergonomics, product life cycle, manufacturing, etc., whereby the biggest potential seem to occur during the early stages of product development. Although computer modelling processes are currently used at different stages of the product life cycle, building a physical, fully functional prototype makes the product development process very expensive and time consuming. New technologies, such as virtual reality applications, are needed to empower the industry with a faster and more efficient way of creating products as they are an ideal tool for simulating prototypes with authentic interactions [33].

In this study, we investigate whether there are differences in the user experience of three different products when users perceive and interact with them. In order to achieve this goal, we conducted a product test that compares the user experience of real products with their representations in virtual reality. The fidelity of the computer-generated experience was still far from that of the real-life ones, thus one of the hidden interests behind the study was whether the level of fidelity that current VR experiences offer would already generate a comparable experience to real-life products in terms of the measurable UX-factors.

2 Background

This paragraph relines the topics User Experience, Virtual Reality and Prototyping and connects them to the background of our study.

2.1 User Experience, Hedonic and Pragmatic Quality

Over the last decades, design professions have become increasingly engaged with UX. Various different attempts at providing theories and UX frameworks have been made [7]. In contrast to early UX research, which was more focused on task efficiency and other pragmatic qualities, user experience nowadays also considers quality attributes like self-actualization [19] and the fact that the consumer experiences hedonic qualities while interacting with a product. In Bruno and Muzzupappa [9], the authors compared the usability of a microwave in virtual reality with the real-life version of the product. The results of this test showed that there was no significant difference between the usability of the same product in VR compared to reality. However, none of the previous research conducted has tested if the user experience of a product is different if it is seen and interacted with in virtual reality compared to reality.

2.2 Virtual Reality

Virtual Reality is the experience to perceive a computer-generated environment as if it is real. It is a powerful tool that provides an advanced immersive medium which can deliver high-impact messages and engage potential consumers [5]. According to Villani et al. [38], the experienced presence of some VR applications might even surpass reality in some occasions, depending on the quality of the application as well as the virtual environment.

Currently, there are examples of VR use which indicate that VR could be a next step in evolving product development processes or that it is already in use for it.

The whole process of assembly planning in product development is a crucial step for many companies. The process that describes how different parts will be put together, function and also how they appear has proven to be a major cost factor for the development of a product [8]. VR technology has proven to be an efficient tool to simulate such advanced interactive product models by also providing different kinds of sensations, such as haptic, auditory and visual [33]. Virtual CAD models allow product designers to import prototypes into virtual environments during early as well as later design stages. As a result, companies are able to perform evaluations based on the virtual simulations instead of building a real product prototype.

Especially formal design changes can be incorporated easily using virtual applications. Haptics, collision detection and interactions between objects can be simulated by using natural human motions. Furthermore, the concept to visualize realistic behavior and examine human interactions makes for example virtual assembly prototypes ideal for analyzing assembly related problems [33]. In order to create a reliable product model in VR, the evaluation environment must be able to simulate real world scenarios regarding physics, interactions and their behavior and properties [12]. Seth et al. therefore expanded existing definitions of VR in regards to product design and described it as "the capability to assemble virtual representations of physical models through simulating realistic environment behavior and part interaction to reduce the need for physical assembly prototyping resulting in the ability to make more encompassing design/assembly decisions in an immersive computer-generated environment."

Virtual reality applications are often used in the automotive sector to experience [17] or even drive the desired car in the virtual environment [21]. Nissan also enables consumers to even design their own Nissan car in VR [29]. Companies such as L'Oreal and Unilever [16] have created virtual reality store simulations to enable better shelf layouts and therefore increase sales [31]. Companies like IKEA [22] or Lowe's [26] use VR applications to design and explore living rooms and the furniture within and even share it on YouTube to increase the social experience of the customer. In the tourism sector, travel agencies and hotels use virtual reality to market and manage destinations [30]. These so called "virtual tours" are often panoramic pictures or videos that do not allow free navigation. Theme parks are also an increasing market for VR entertainment. Various theme parks around the globe already offer attractions that are available via VR and AR technology [18].

These various examples show that VR has actually got very far in increasing the visual quality of its generated content. In contrast, they also show that the main focus in VR has been the visual stimuli and that the integration of other sensorial channels such as haptic feedback or smell is still in its infancy. Nevertheless, VR already gives many people an immersive experience with a measurable effect on their UX.

2.3 Prototyping

Prototypes of products are simplified representations of design ideas before the final artefacts are created. The most common usage of a prototype is as a "learning tool" and can be used at any stage of the design process to find issues and explore new ideas [13]. Especially for innovative, novel concepts, the creation of prototypes, i.e. prototyping, is an important prerequisite for the evaluation of user experiences. Already in the early phases of product development, assessments of the success potential of a product idea can be obtained. Insights into promising and less promising ideas as well as possible technical realizations help to reduce development risks and thus costs. In addition to their use in evaluation, prototypes also serve to demonstrate design decisions and possible design alternatives. Thus, prototyping offers great potential from a company's point of view and is a major activity in product development [15]. Prototypes do not need to be perfect. They range from low-fidelity drawings to high-fidelity models at several levels, all designed to communicate and explore the quality and the design of the product [13]. Prototypes modelled with virtual reality technology can be seen as experience prototyping [10]. It describes the simulation of a product experience through involvement of the consumer. As in other prototyping approaches, the focus is also on the question of how a product idea plays out on the experience level, whereby usability aspects and questions of interaction design can also be examined [15].

3 Considerations for the Study Design

In order to compare the perceived user experience of real products and virtual representations, a comparable test set-up was created in reality and in VR.

3.1 Virtual Environment

It was important for this test to achieve a natural and comfortable interaction for the user, especially if the participant has limited or no previous experience with VR technology. The virtual environment was designed to create a high level of immersion for the user. Therefore, the virtual room was built similar to the real office in which the participants were located while using the VR devices. The VR setup used in this experiment was the HTC VIVE Headset [39] and the associated controllers. The virtual environment was created with the Unity Game Engine [34]. After all objects and products had been modeled with Blender [6] and Fusion 360 [4], they were imported into Unity to create the virtual set-up.

3.2 Products

As already mentioned, the product test contained three different products that are independent from each other. The choice for the three products was based on the following parameters: The objects should be part of the same environment, the interactions should be just physical, the level of interaction should be different from low to higher complexity and the objects should not be already known by the participants.

Fig. 1 Screenshot of the lamp in VR

The selected objects where part of a new lab-inventory which was introduced months after this study:

Lamp
The first object is an interactive lamp created and designed at a university. It can be moved upwards and downwards by grabbing and dragging the lamp in the desired direction. The light can be activated as well as deactivated via the black motion sensor on one of the sides. In addition, the LED light bars can be rotated in 360°. It is the object with the highest grade of interactive complexity (Figs. 1 and 2).

Convertible Table
The second product is a convertible table that can be used either as a standing table or as a normal desk. To change its purpose, the table just has to be turned over its wheel axis (Figs. 3 and 4).

Miura Barstool
The third product is a barstool designed by Konstantin Grcic. It is the object with the lowest functionality and intractability (Figs. 5 and 6).

Testing Room
For the testing we chose an office room inside the university. The choice was made because the three objects are used for agile work in office context. Additionally, the whole setup enabled a rational and undisturbed testing process with the experience in one room and the questionnaire taking place in the next one.

The main reason that the VR-twin of the room was designed as similar as possible to the real one was to have as few side effects on the VR-participants' experience as possible when they used the VR-Headset; it should feel as if there was no change in their situation (Figs. 7 and 8).

Fig. 2 Photo of the lamp

Fig. 3 Screenshot of the table in VR

Fig. 4 Photo of the table

Fig. 5 Screenshot of the barstool in VR

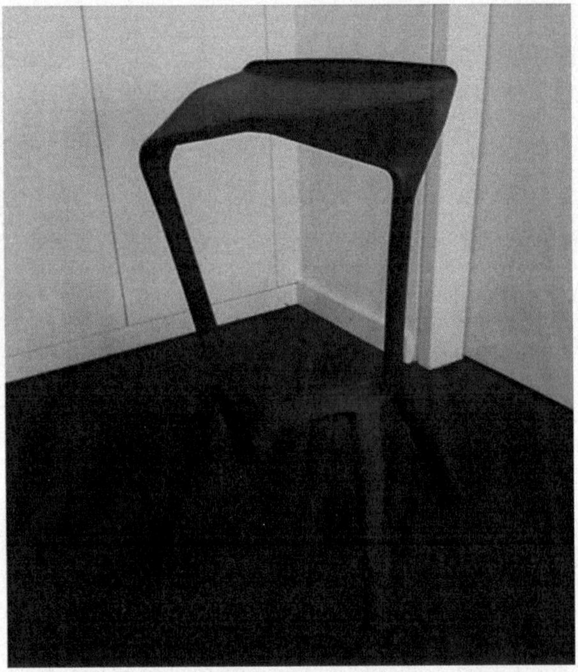

Fig. 6 Photo of the barstool

Fig. 7 Screenshot of the testing room in VR

3.3 Questionnaire

The research methodology of this survey questionnaire has a mainly quantitative focus, but also contains some qualitative questions [1, 24]. The first page of the survey has been created to collect the necessary demographic data of the participants. The

Fig. 8 Photo of the testing room

participants were asked to provide information about their gender, age, profession and annual income.

It is common practice to evaluate the user experience of interactive products via questionnaires. The User Experience Questionnaire (UEQ) [27] and the AttrakDiff 2 [20] are two of the most used tools to evaluate interactive products by questionnaires. Both constructions rely on the theoretical framework of user experience that distinguishes between pragmatic quality, hedonic quality and perceived attractiveness of a product. Furthermore, both questionnaires use a format of a seven-stage scale with two opposite adjectives on each end. A remarkable difference between both questionnaires is the selection of word pairs. The AttrakDiff 2 questionnaire has proven to be a strong and reliable source of measurement of perceived awareness of products as well as for user experience measurement with a strong background in research [2, 23, 32], and for that reason we selected it to evaluate pragmatic quality, hedonic quality and perceived attractiveness.

For each product, the AttrakDiff 2 questionnaire [20] was used to measure the perceived awareness of hedonic quality—identification, hedonic quality—stimulation, attractiveness as well as the perceived pragmatic quality. The questionnaire has proven to be a strong and reliable source of measurement of perceived awareness of products as well as for user experience measurement [2, 23, 32].

The questionnaire consists of twenty-eight 7-point adjective pairs. The word pairs represent strong contrasts and can be subdivided into further rating levels. In addition to the AttrakDiff 2, the participants were asked to rate on a 5-point scale how likely it was that they would buy the product for themselves, or recommend it to friends or colleagues. Furthermore, two qualitative questions were asked at the end of each product evaluation. The participants were tasked to indicate a fair price for each product and state what they liked and disliked about the objects.

For the virtual reality evaluation, two additional pages were added to the questionnaire to rate the VR-application, the devices and the general attitude of the participant towards VR and new technologies. Thus, the first page contained questions about the user acceptance of the virtual reality technology used in the product test. The Technology Acceptance Model [14] was used for this task and included questions about perceived ease of use [14, 36], perceived usefulness [25], attitude towards using the technology [37] and the behavioral intention to use it [37].

The second page therefore contained questions about the attitude towards new technologies, the intention to use and buy VR technology and the general usage of existing VR technology.

4 Method

The test was conducted as a participatory and summative evaluation in the rooms of the University of Applied Sciences Neu-Ulm. In VR we tested 33 (average age 23; 15 male, 18 female) and in the real environment we tested 31 (average age 24; 15 male, 16 female) students of the University of Applied Sciences Neu-Ulm. The target group was deliberately chosen homogeneously in order to exclude as many target group-specific disruptive factors as possible. All participants tested out the same three products and evaluated them afterwards. Pre-test instructions were similar for every participant in the VR test as well as the real product experience. Therefore, a guideline was created that contained all the important indications for the students. At the start of the experiment, everyone was told that they were participating in a product test and would be asked to evaluate three products afterwards via a questionnaire.

In order to learn about all the functions of each product, the participants were introduced to a scenario and had to carry out specific tasks for each object (four tasks for the lamp, four tasks for the table and two tasks for the barstool) that were described in the scenario.

In addition, the VR students were taught how to grab and move objects within the virtual environment (Fig. 9). In order to create the same prerequisite for every

Fig. 9 Screenshot of the teaching exercise

participant, an additional exercise was created to teach the user how to grab and control objects inside the virtual world. The users were tasked to sort colored balls into the according bowls. After achieving that, the instruction test was disabled and access to the three products was activated to the user.

The users were instructed to take as much time as necessary for each product as they had to evaluate the usability as well as their impression of each product afterwards. It was also mentioned that they were completely safe within the experience and could not do anything wrong inside the room. After the users experienced all the products, they were asked if they thought that they have seen enough of every product to evaluate it, otherwise they could look at a specific product again. After every participant had mentioned that they had seen enough, the product test was finished and they were asked to evaluate their product experience. The participants were taken to another room to rate the three products via the associated questionnaire. They were tasked to answer the questions spontaneously without thinking too much about it.

5 Results

To verify the reliability of the AttrakDiff2 subscales, Cronbach's Alpha was calculated for the four subscales for every product. Usually, an Alpha-Coefficient of >0.7 is considered a good indicator for the reliability of a scale. In our case, every resulting value of each product is above this threshold and therefore we can derive that the four subscales of the Attrakdiff2 questionnaire show a high reliability for our data.

5.1 Lamp

Figure 10 highlights the mean scores of the lamp for the four subscales pragmatic quality (PQ), hedonic quality—identification (HQI), hedonic quality—stimulation (HQS) and attractiveness (ATT). Analyzing the results by means of t-tests shows only slight and insignificant differences between the real and the VR outcomes. In the pragmatic quality, the real product is ranked slightly better than the VR counterpart ($M_{Real} = 4.6$; $SD_{Real} = 0.85$; $M_{VR} = 4.5$, $SD_{VR} = 0.80$; $t = -0.259$; $p = 0.800$). In the hedonic quality identification (HQI), the virtual reality object was perceived slightly better than the real product ($M_{Real} = 5.1$; $SD_{Real} = 0.4$; $M_{VR} = 5.3$, $SD_{VR} = 0.7$) $t = 0.7$, $p = 0.5$. This applies also to the hedonic quality stimulation ($M_{Real} = 5.5$; $SD_{Real} = 0.5$; $M_{VR} = 5.7$, $SD_{VR} = 0.6$) $t = 0.4$, $p = 0.7$, and also to the overall attractiveness of the products ($M_{Real} = 5.3$; $SD_{Real} = 0.3$; $M_{VR} = 5.4$, $SD_{VR} = 0.3$) $t = 1.1$, $p = 0.3$. Figure 9 shows the extended profile of word pairs with the mean values for both scenarios.

The balanced user experience of the lamp in both scenarios is also reflected in the intention to buy or recommend the product. 58% of the VR participants were

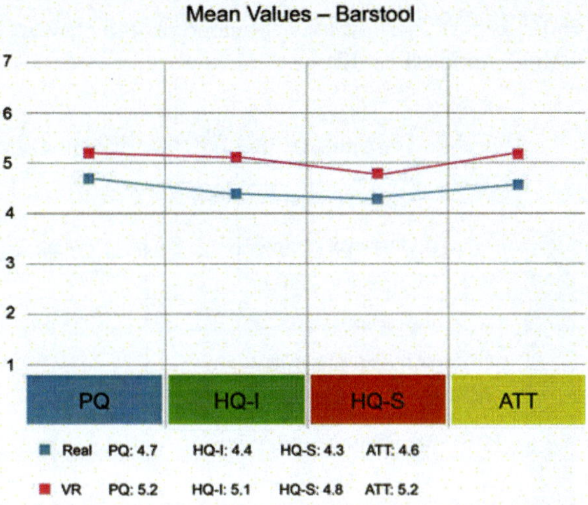

Fig. 10 Mean values for the lamp (PQ—pragmatic quality; HQ-I and HQ-S—hedonic quality individual and stimulation; ATT—attractiveness)

likely to buy the product to a fair price, whereas 53% that participated in the test of the real product answered with the same outcome. In addition, 58% (VR) and 63% (real) were likely or extremely likely to recommend the product.

However, the individual feedback of the lamp pointed out differences in the awareness of the product. The positive aspects of the object were similar in both cases. The participants rated the lamp as modern, futuristic and innovative. They enjoyed the form and style of the object and liked the rotatable LED lights. The negative aspects varied in both scenarios though. Regarding the real-life object, attendees complained about the sound that appears when the object was adjusted in height. Furthermore, several students thought it was cumbersome that the sensor of the light was difficult to find when the lamp was in a higher position. On the other hand, participants of the VR object mostly pointed out that it could be sometimes difficult to grab the LED lights with the controller, especially when the object was moving/turning.

5.2 *Convertible Table*

Figure 12 highlights the mean scores of the convertible table for the four subscales. Similar to the lamp, an analysis of both outcomes shows no significant difference between both versions of the table. In the pragmatic quality, the real product is ranked slightly worse than the VR counterpart ($M_{Real} = 5.3$; $SD_{Real} = 0.9$; $M_{VR} = 5.4$, $SD_{VR} = 0.9$) $t = 0.09$, $p = 0.9$. In the hedonic quality identification, both objects are ranked equally ($M_{Real} = 4.4$; $SD_{Real} = 0.4$; $M_{VR} = 4.4$, $SD_{VR} = 0.5$) $t = 0.05$, $p = 0.9$. This applies also to the attractiveness ($M_{Real} = 4.4$; $SD_{Real} = 0.5$; $M_{VR} = 4.4$, $SD_{VR} = 0.4$) $t = 0.0$, $p = 1$. In the hedonic quality stimulation, the real object is ranked slightly better ($M_{Real} = 4.0$; $SD_{Real} = 0.4$; $M_{VR} = 3.8$, $SD_{VR} = 0.3$) $t = -1.2$, p

= 0.2. Figure 11 shows the extended profile of word pairs with the mean values for both scenarios (Fig. 12).

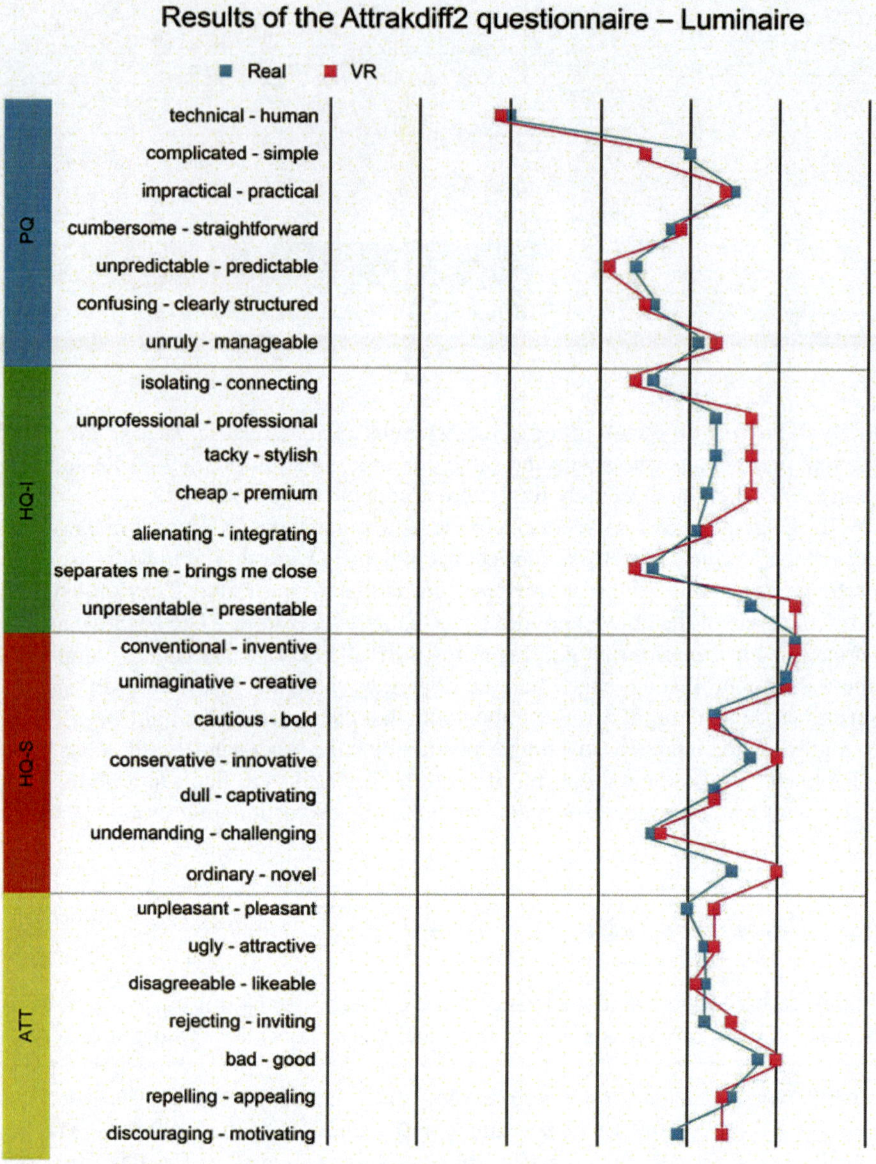

Fig. 11 Results of the Attrakdiff2 for the lamp (PQ—pragmatic quality; HQ-I and HQ-S—hedonic quality individual and stimulation; ATT—attractiveness)

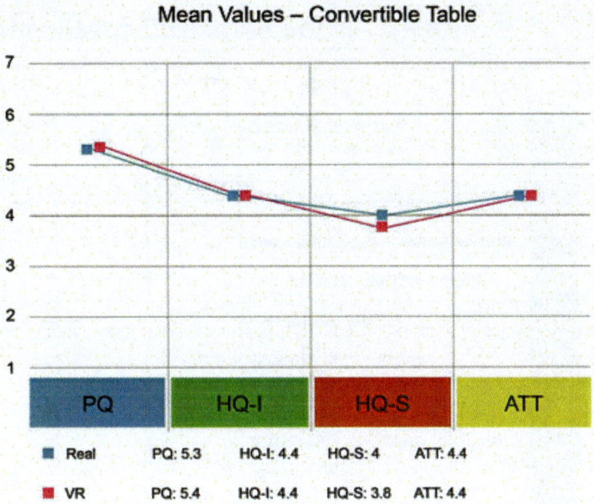

Fig. 12 Mean values for the table (PQ—pragmatic quality; HQ-I and HQ-S—hedonic quality individual and stimulation; ATT—attractiveness)

Similar to the lamp, the balanced awareness of the table is also reflected in the intention to buy or recommend the product in both scenarios. 15% of the VR participants are likely to buy the product for a fair price, whereas 23% that participated in the real-life product test answered with the same outcome. Furthermore, 30% (VR) and 33% (real) are likely or extremely likely to recommend the product.

The individual feedback this time, in contrast to the feedback about the lamp, points out that the participants perceived the product similarly in VR as well as in real life. Positive aspects about the product were that it is practical, simple and extremely space saving. On the other hand, the attendees mostly criticized the cold design of the object and mentioned that it was not aesthetic enough.

5.3 *Miura Barstool*

Figure 14 highlights the mean scores of the barstool for the four subscales. In contrast to the other two objects, the results of the barstool show significant differences between the two scenarios in the scales HQI and ATT. In the pragmatic quality, the VR product achieved a higher score than the real object ($M_{Real} = 4.7$; $SD_{Real} = 0.5$; $M_{VR} = 5.2$, $SD_{VR} = 0.6$; $t = 1.6$; $p = 0.1$). This applies also to the hedonic quality identification ($M_{Real} = 4.4$; $SD_{Real} = 0.3$; $M_{VR} = 5.1$, $SD_{VR} = 0.4$; $t = 3.4$; $p = 0.005$), and also to the hedonic quality stimulation ($M_{Real} = 4.3$; $SD_{Real} = 0.4$; $M_{VR} = 4.8$, $SD_{VR} = 0.6$; $t = 1.8$; $p = 0.09$), Finally, the VR product also achieved a much higher score in the attractiveness ($M_{Real} = 4.6$; $SD_{Real} = 0.3$; $M_{VR} = 5.2$, $SD_{VR} = 0.2$; $t = 3.7$; $p = 0.003$). Figure 13 shows the extended profile of word pairs with the mean values for the barstool (Figs. 14 and 15).

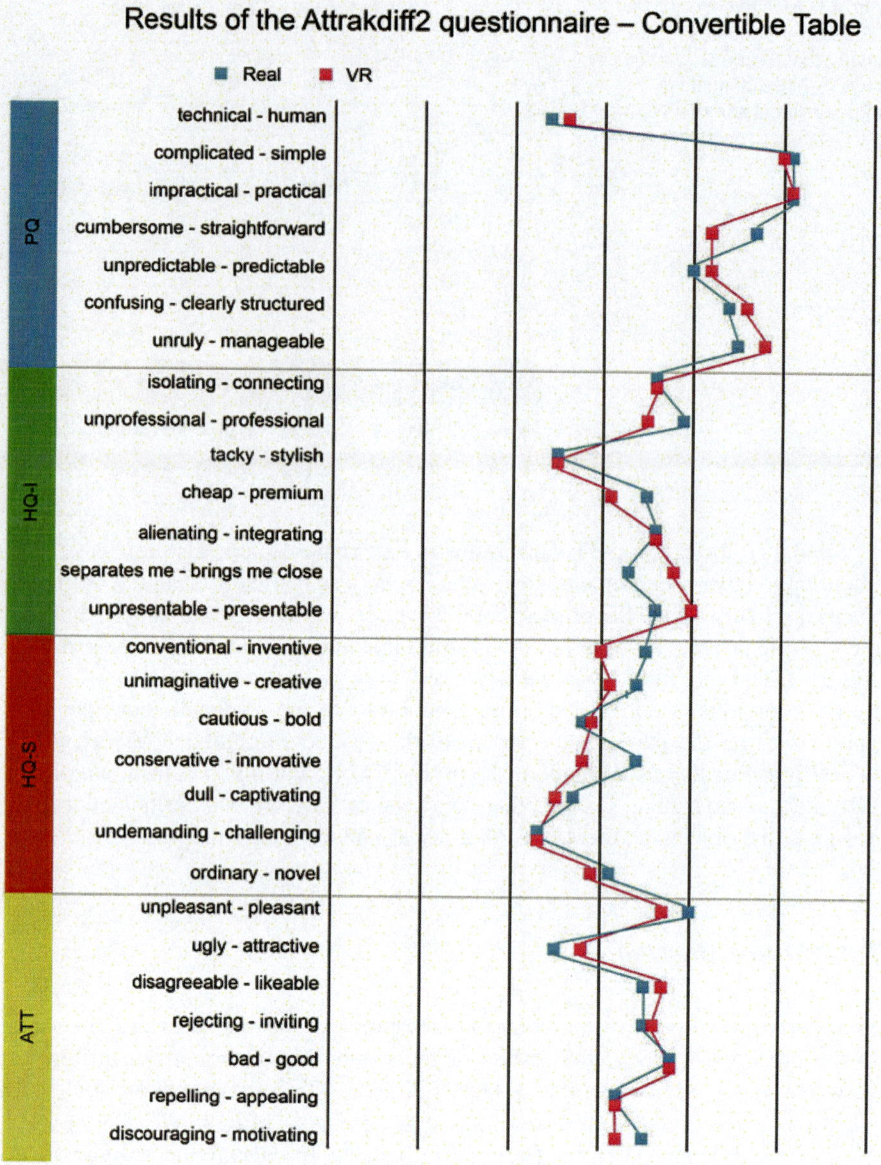

Fig. 13 Results of the Attrakdiff2 for the table (PQ—pragmatic quality; HQ-I and HQ-S—hedonic quality individual and stimulation; ATT—attractiveness)

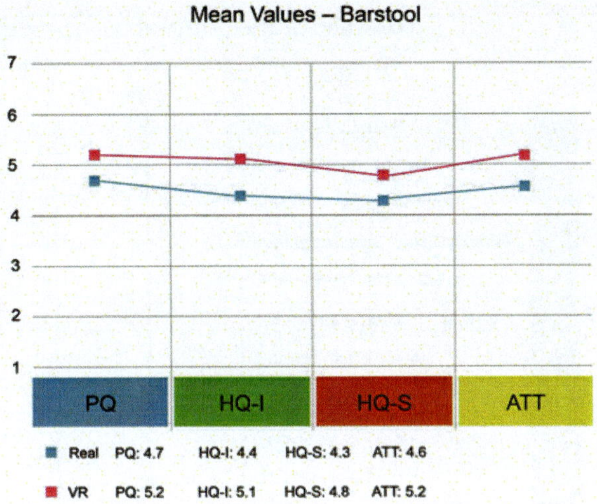

Fig. 14 Mean values for the barstool (PQ—pragmatic quality; HQ-I and HQ-S—hedonic quality individual and stimulation; ATT—attractiveness)

The VR favored awareness of the barstool is also reflected in the intention to recommend and buy the product. 51% of the VR participants were likely to buy the product at a fair price, whereas only 20% that participated in the real product test answered with the same outcome. In addition, 42% (VR) and 23% (real) were likely or extremely likely to recommend the product.

The individual feedback showed several differences between the two tests. The virtual object was described as an uncomplicated object with an attractive design. Almost no negative aspects were mentioned for the product in VR, only some students mentioned that it might look a bit uncomfortable but it was hard to assess. The perception of the real product was more focused on the comfort of sitting on the barstool instead of the design. While a few participants still mentioned that it looks quite aesthetic, many criticized the material and judged the shaky way the barstool behaved.

6 Conclusion

A first result of the study is that there is no significant difference in the evaluated experience of the lamp and the convertible table in virtual reality compared to the real-life objects. Both products were rated nearly the same (maximal difference 0.2) in the scales of pragmatic quality, hedonic quality identification, hedonic quality stimulation and attractiveness. Qualitative feedback of the table has, in fact, shown that the participants liked and criticized the same aspects of the product in both scenarios.

On the other hand, the evaluation of the lamp, which is a more complex and interactive product, demonstrated that the users valued the object in both representations

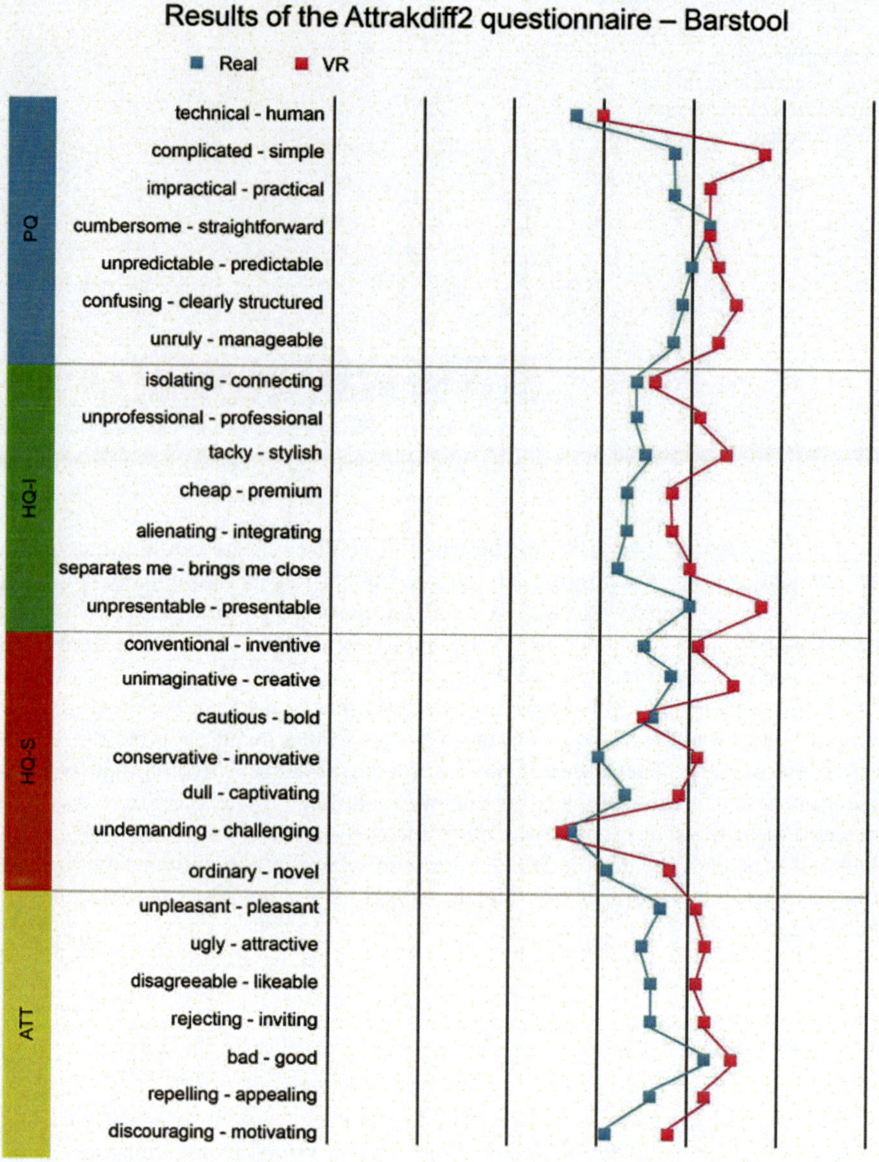

Fig. 15 Results of the Attrakdiff2 for the barstool (PQ—pragmatic quality; HQ-I and HQ-S—hedonic quality individual and stimulation; ATT—attractiveness)

nearly the same, but criticized different aspects in both cases. While the hover-function to enable the light was the only possibility in VR to activate it, it was not seen as a problem. However, participants of the real-world test found it cumbersome to activate it when the sensor was outside the range of vision.

A second finding is that the barstool was evaluated better in VR compared to the real counterpart in all four subscales (minimum difference 0.4; maximum difference 0.7). Furthermore, the results showed significant differences in the scales of attractiveness and hedonic quality identification. The qualitative individual feedback reflects the stated outcome. While most of the VR participants only pointed out the positive aspects of the barstool, students who tested out the real object criticized the material and described it as uncomfortable and shaky.

The positive results of the technology acceptance model can be an indicator that the VR technology and the associated product test in VR did not influence the perception of the products in a negative way. On the other hand, the virtual reality application did not seem to cause unusual side effects that would influence the students to rate products better in VR as there are no significant differences in two of the three products.

These results give a first glimpse of the possibilities VR can offer designers for virtualization of certain steps in a development process. It can be stated that the evaluation of these product representations in VR led to a comparable UX with the real products when they had a comparable set of features and interactions in both spaces, even though the fidelity of the virtual space differed.

7 Limitations and Further Research

In order to provide an immersive experience, one of the chosen limitations of the study was that participants experienced all three products at once and evaluated them afterwards. In this set-up participants have to remember which object felt which way. Even though the testing time for all products was short and the order was randomized, through the post hoc evaluation participants might have rated all products more similarly because the experience that they remembered was mixed.

It might also be argued whether the obtained results would have been different if the VR device were more immersive than the current setup. The current grabbing mechanics are via the VIVE controller. Newer and more immersive technology like VR haptic gloves could achieve a more natural and therefore more immersive way of interacting with the objects in virtual reality. Thus, it would be interesting to increase the fidelity of the VR application. The current set-up was built by one person with a limited amount of resources in VR. The whole experience could be even more immersive with a team of VR experts and the necessary budget.

Another factor that should be investigated in future studies is varying target groups. In both cases, only students participated in the product tests. The average age was 23 (VR) and 24 (real) which does not represent the majority of the general population. It might be argued that the results could have been completely different with an older average age. As a result of testing only students, the average annual income was in the bracket between 0€ and 5000€. This could affect the answer whether the participant would be likely to buy the product.

Even though the study could show certain design aspects that could be simulated with digital VR prototyping tools, an additional question would be what design aspects still need a real prototype to evaluate the UX, especially since the results of the barstool evaluation pointed towards a difference in perceived functional qualities for objects where some of their interactions cannot be simulated in VR.

The results of this study show the potential for product prototyping and evaluation in VR, but there is a need for further research to exactly identify the product parameters which lead to comparable UX-evaluations in VR.

References

1. Agee J (2009) Developing qualitative research questions: a reflective process. Int J Qual Stud Educ 22(4):431–447
2. Albrecht U-V, Folta-Schoofs K, Behrends M, Von Jan U (2013) Effects of mobile augmented reality learning compared to textbook learning on medical students: randomized controlled pilot study. J Med Internet Res 15(8)
3. Alex H (2017) Mark Zuckerberg reveals that Facebook paid more than we thought for Oculus VR. https://www.businessinsider.de/facebook-actually-paid-3-billion-for-oculus-vr-2017-1?r=US&IR=T. Accessed 27.03.2019
4. Autodesk (2019) Fusion 360 free 3D CAD/CAM design software for students, educators, and academic institutions. https://www.autodesk.com/products/fusion-360/students-teachers-educators. Accessed 14.03.2019
5. Barnes S (2016) Understanding virtual reality in marketing: nature, implications and potential
6. Blender (2019) Open Source 3D creation. Free to use for any purpose, forever. https://www.blender.org/. Accessed 14.03.2019
7. Blythe M, Hassenzahl M, Law E, Vermeeren A (2007) An analysis framework for user experience (UX) studies: a green paper. Towards a UX Manifesto 1 (140):6
8. Boothroyd G, Dewhurst P (1987) Product design for assembly. Boothroyd Dewhurst Incorporated
9. Bruno F, Muzzupappa M (2010) Product interface design: a participatory approach based on virtual reality. Int J Hum Comput Stud 68(5):254–269
10. Buchenau M, Suri JF Experience prototyping. In: Proceedings of the 3rd conference on designing interactive systems: processes, practices, methods, and techniques, 2000. ACM, pp 424–433
11. Choi S, Jung K, Noh SD (2015) Virtual reality applications in manufacturing industries: past research, present findings, and future directions. Concurr Eng 23(1):40–63. https://doi.org/10.1177/1063293x14568814
12. Chryssolouris G, Mavrikios D, Fragos D, Karabatsou V (2000) A virtual reality-based experimentation environment for the verification of human-related factors in assembly processes. Robot Comput-Integr Manuf 16(4):267–276
13. Coughlan P, Suri JF, Canales K (2007) Prototypes as (design) tools for behavioral and organizational change: a design-based approach to help organizations change work behaviors. J Appl Behav Sci 43(1):122–134
14. Davis FD, Bagozzi RP, Warshaw PR (1989) User acceptance of computer technology: a comparison of two theoretical models. Manage Sci 35(8):982–1003
15. Diefenbach S (2017) Psychologie in der nutzerzentrierten Produktgestaltung. Springer, Berlin

16. Fleming M (2018) How virtual reality is helping brands overhaul the shopping experience. Marketing Week. https://www.marketingweek.com/2018/06/27/virtual-reality-enhance-shopper-experience/. Accessed 14.03.2019
17. Framestore (2019) Volvo reality. https://www.framestore.com/work/volvo-reality. Accessed 14.03.2019
18. Guttentag DA (2010) Virtual reality: applications and implications for tourism. Tour Manage 31(5):637–651
19. Hassenzahl M (2018) The thing and I: understanding the relationship between user and product. In: Funology 2. Springer, pp 301–313
20. Hassenzahl M, Burmester M, Koller F (2003) AttrakDiff: Ein Fragebogen zur Messung wahrgenommener hedonischer und pragmatischer Qualität. In: Mensch & Computer 2003. Springer, pp 187–196
21. Honda P (2019) Proctor Honda virtual test drive. https://www.proctorhonda.com/ff-honda-virtual-reality-test-drive-tallahassee-fl. Accessed 14.03.2019
22. IKEA (2019) Virtual reality—into the magic. https://www.ikea.com/ms/en_US/this-is-ikea/ikea-highlights/Virtual-reality/index.html. Accessed 14.03.2019
23. Israel JH, Wiese E, Mateescu M, Zöllner C, Stark R (2009) Investigating three-dimensional sketching for early conceptual design—results from expert discussions and user studies. Comput Graph 33(4):462–473
24. Jick TD (1979) Mixing qualitative and quantitative methods: triangulation in action. Adm Sci Q 24(4):602–611
25. Karahanna E, Agarwal R, Angst CM (2006) Reconceptualizing compatibility beliefs in technology acceptance research. MIS Q 781–804
26. Labs LsI (2019) New ways to start home improvement projects. http://www.lowesinnovationlabs.com/visualization. Accessed 14.03 2019
27. Laugwitz B, Held T, Schrepp M Construction and evaluation of a user experience questionnaire. In: Symposium of the Austrian HCI and Usability Engineering Group, 2008. Springer, pp 63–76
28. Merel T (2016) Augmented And Virtual Reality To Hit $150 Billion, Disrupting Mobile By 2020. https://techcrunch.com/2015/04/06/augmented-and-virtual-reality-to-hit-150-billion-by-2020/?guccounter=1. Accessed 01/02/2019
29. Nissan (2019) Tech Drive 360 VR. https://www.nissanusa.com/experience-nissan/news-and-events/nissan-intelligent-mobility-360-vr-experience/. Accessed 14.03.2019
30. Prideaux B (2002) The tourist as a metaphor of the social world. The Cybertourist 1(6):317
31. Rutgers-Online (2014) Virtual reality: changing the way marketers are conducting. http://online.rutgers.edu/resources/infographics/virtualreality/. Accessed 10.12.2018
32. Schrepp M, Held T, Laugwitz B (2006) The influence of hedonic quality on the attractiveness of user interfaces of business management software. Interact Comput 18(5):1055–1069
33. Seth A, Vance JM, Oliver JH (2011) Virtual reality for assembly methods prototyping: a review. Virtual Reality 15(1):5–20
34. Unity (2018) Real-time solutions. Endless opportunities. https://unity.com. Accessed 14.03.2019
35. Unreal (2018) Make something unreal. https://www.unrealengine.com/en-US/what-is-unreal-engine-4. Accessed 14.03.2018
36. Venkatesh V (2000) Determinants of perceived ease of use: integrating control, intrinsic motivation, and emotion into the technology acceptance model. Inf Syst Res 11(4):342–365
37. Venkatesh V, Morris MG, Davis GB, Davis FD (2003) User acceptance of information technology: toward a unified view. MIS Q 425–478
38. Villani D, Repetto C, Cipresso P, Riva G (2012) May I experience more presence in doing the same thing in virtual reality than in reality? An answer from a simulated job interview. Interact Comput 24(4):265–272
39. Vive (2019) A VIVEPORT subscription is the best way to experience over 330 VR APPS and games. https://www.vive.com/eu/product/. Accessed 14.03.2019
40. Zaid S (2016) The very real growth of virtual reality. https://www2.deloitte.com/bh/en/pages/about-deloitte/articles/treading-water/the-very-real-growth-of-virtual-reality.html. Accessed 01/02/2019

MASAL: Bridging Between Two Cultures Through Storytelling with an Interactive E-textile Toy

Seçil Uğur Yavuz

Abstract Stories have an important role in children's cognitive and emotional development, enhancing their imagination and helping them to make sense of real-life situations. Today, by embedding digital technologies into physical spaces and objects, the storytelling experience can be enhanced by embracing interactive characteristics in order to support children's imagination and give rise to open-ended play experiences. Masal is an interactive storytelling stuffed toy constructed with e-textiles and an open-source technology for children to co-create stories. It aims at fostering creative thinking and empathy towards a different culture through a collaborative construction of a story. This paper addresses the design and prototyping process of Masal and the storytelling sessions done with three groups of participants, varying from children (3–8 years) to adults. It represents the qualitative results obtained from these sessions reflecting on how an interactive soft toy can enhance children's experience and imagination in storytelling, and above all can create a bridge between two cultures through co-creation of a story.

Keywords e-textiles · Interaction design · Storytelling · Design for children

1 Introduction

Telling a story is an important act of children's daily life since they explore and make sense of the world through storytelling. They experience real-life situations through role-playing and game setting, where they are the builders of countless scripts based on facts and fiction. Today, digital technologies have brought storytelling into a new level in which children are overwhelmed with limitless options flowing from their screens. However, screen-based devices often make them passively consuming stories, rather than actively engaging with the construction of the story which can broaden their imagination. Thanks to physical computing, today it is possible to bridge between the digital and physical world in order to support children's creativity

S. Uğur Yavuz (✉)
Faculty of Design and Art, Free University of Bozen-Bolzano, Bolzano, Italy
e-mail: secil.uguryavuz@unibz.it

E. Duarte and C. Rosa (eds.), *Developments in Design Research and Practice*,
Springer Series in Design and Innovation 17,
https://doi.org/10.1007/978-3-030-86596-2_10

and expressive abilities. According to Lampe and Hinske [1] multimedia supported children toys can increase the creativity of fantasy play. There are projects using interactive technologies embedded into space to make children tell their own stories more engagingly [2, 3]. Besides using spatial dimension for interactive storytelling, hand-held objects integrated with interactive properties can also stimulate children to create new play experiences [4, 5]. Stuffed toys that have been used in storytelling as playing companions by children for centuries, today can be embedded with sensors and actuators thanks to the use of e-textiles. E-textile field has introduced new possibilities to combine electronics with fabrics through creating soft circuits with conductive fibres and yarns [6]. The use of e-textiles presents new construction modalities of interactive toys not only for new play experiences, but also for learning electronics [7]. As e-textiles are mostly connected with craft practices, DIY culture and open-source technologies, it extends the possibilities to enable people to construct their own toys based on their own needs.

Masal is an interactive soft toy constructed with e-textiles in order to answer a specific need: introducing a foreign culture to children through storytelling. The toy allows children to create a story together with other children while learning about a new culture through this constructed narrative. Masal means "fairytale" in Turkish, therefore its first usage was for an occasion to introduce the Turkish culture to the Italian children. This paper addresses the design and prototyping process of Masal and the test sessions done with children and adults. It represents the qualitative results obtained from these sessions reflecting on how an interactive soft toy can enhance children's experience in collaborative storytelling, and above all can create a bridge between two cultures through co-creation of a story.

2 Designing Interactive Storytelling Tools for Children

2.1 Gains and Pains of Digital Natives

Digital natives, as a term, was first coined by Prensky [8], referring to the children who were born in the digital age, having a close interaction with computers and the Internet from the very first moments of their life. Being born and grown up in the digital era has many advantages, but on the other side brings also its consequences. One of the most used devices by children are smartphones and tablets that provide limitless possibilities of playing games, watching videos or staying in contact with others through a screen. Due to the fact that screen-based digital devices are meant to be used personally, children tend to use digital technology individually rather than in collaboration with others [9]. It is often seen that children sitting next to each other with their smartphones, but not even talking or having a physical interaction between them. Turkle [10] defines this situation as "flight from conversation" which can bring consequences in social interaction modalities, such as lack of empathy or lack of sensibility towards another person. Moreover, although some digital application or games offer possibilities to be creative and active, in most of the cases children passively consume the digital content overflowing from their screens.

If technology is used more properly, it can turn those pains to gains for Digital Natives. Screen-based digital content can be extended towards the physical world through tangible user interfaces (TUIs) [11]. Hence, this type of new interfaces can give rise to different scales and modalities of usage going beyond the limitation of the screen dimension in order to allow multiple users to collaborate and consequently communicate. Besides, the passive action of receiving information can turn into creating content and integrating one's imagination into it. This gives a pro-active role to the user that is enabled by the technology in order to create open-ended possibilities distributed in the space and the objects within the space. The child tangible interaction (CTI) framework presented by Antle [12] gathers five themes related to different aspects of tangible systems for children: ranging from spatial, perceptual, behavioral, semantic and collaborative dimensions. These dimensions show how digital technologies could be embedded in space and tangible objects for creating new interaction modalities for children and guide designers to create new tangible systems.

2.2 *Merging Physical and Digital Storytelling*

Children merge facts and fiction in a play setting writing a script based on their "needs and desires" [13]. They continuously create stories day and night to make sense of real-life situations. Activities related to stories are one of the most important elements of childhood, ranging between listening, creating and role-playing. Fantasy play which often embraces role-playing, storytelling and animating toys and objects helps children to enhance their imagination while improving their cognitive and language skills [2]. On the other hand, in the digital realm, stories are generally transmitted through videos that often do not give children an active role in imagining, rather imposing a ready-made fantasy. Moreover, today as a result of the Internet of Things, the screens have been extended to virtual assistants which could assist us through vocal responses. There are also applications for children to listen to stories told by the virtual assistants activated just by saying: "tell me a story" [14, 15].

Besides these applications, we witness a considerable integration of digital technology into children toys, namely "interactive toys". Interactive toys differ from traditional toys by the integration of sensors and actuators continuously transforming inputs into a variety of outputs (auditory, tactile, visual, etc.) to stimulate child play [16]. Today, there are many interactive toys already in the market offering children the experience of an interaction extending the fantasy play to another level in which the toy as well participates actively in the play. It is not only animated by the child's own will, but also stimulates the game with new inputs and provides an open-ended structure that continuously providing with impulses.

There are projects aiming at creating interactive storytelling experiences for children by the integration of digital technologies into objects and spaces. While Cassell and Ryokai [2] developed a space-based storytelling which provides the possibility to record and replay children's stories on an interactive mat called StoryMat, KidsRoom turns the child's bedroom into an interactive narrative environment in which multiple kids can play [3]. Besides the space-based interactive storytelling systems, there are also physical objects—interactive story tools augmented with digital technologies. For instance, StoryTech uses stuffed toys integrated with RFID tags generating computer-based graphics and characters to encourage children to tell stories [17]. Moreover, Arnall [18] introduces the prototypes of toy-like objects integrated with RFID tags triggering media playback on a TV. In these examples, the toys are functioning in the physical realm and working as keys to activate the virtual content shown on the screen in order to create an interactive storytelling experience. Besides, there are also examples in which the interactive toy can record voices and replay them for creating a story or animating the toy within the story [19, 20]. Going beyond the whole integration of digital technologies with physical objects, Cuddly represents an interesting example of how a mobile phone can be integrated into a stuffed toy in order to stimulate interactive play using the sensors and actuators already existing in the phone [21].

Moreover, Augmented Reality (AR) is also another method applied in children products in order to stimulate interactive storytelling. For instance, the Textales project uses AR technology to turn the patterns on the blanket into virtual images which help children and their parents to tell bedtime stories [22].

Starting from the concept of the Internet of Things (IoT), several examples to make connected toys, so-called The Internet of Toys (IoToys) have also emerged by introducing new play practices merging online and offline features [23]. Being connected to the Internet makes the toys linked to the other toys or to the applications—therefore to other people—in order to create new interaction modalities for multiple users in long or close distance. For instance, Vai Kai is a smart wooden toy that allows children to create an open-ended game or story in which the toy can connect with other toys via the Internet by the help of connected sensors and actuators [5].

These examples show us that there is a tendency for almost two decades to merge digital and physical play by using various technologies in order to stimulate children's imagination, to encourage them to collaborate and to create new storytelling experiences based on interactive spaces and objects.

3 E-textiles for Stuffed Toys

3.1 DIY E-textiles—Opening and Softening the Black Box

Terms like DIY, hacking or crafting have been entering to electronics with the maker movement unboxing the complexities of the knowledge and making it more democratized by enabling people to develop their own technologies [24]. One of the fields that supports this development of making technologies open and easy to tinker with is the e-textile craft field. By integrating conductive yarns or fibres, textiles can sense, react and connect, through becoming a sensor, an actuator or even an antenna. Although e-textile firstly appeared in high technological applications in various sectors such as healthcare, protection or communication, later it has become strongly linked to DIY culture, embracing the 'crafting' aspects of textiles. This possibility of 'making textiles behave' encouraged many practitioners doing 'contemporary craft' to merge traditional textile making techniques with electronics to create interactive soft goods [25]. There are many communities, makerspaces, online platforms dealing with e-textile craft, making and sharing this tacit and explicit knowledge in order to grow new hybrid craft techniques merging digital and physical aspects [26–28]. This type of collective making activities needs a place to bring people together physically and create a social gathering while crafting new e-textile ideas. There are also Fablabs which are specialized for this kind of activities and becoming incubators for e-textile craft practices [29–31].

The field of e-textile craft gives rise to the projects that can help people to create their own interactive soft goods to answer their specific needs. Moreover, using open-source technologies (hardware and software) provides possibilities to personalize the interaction based on new configurations of input–output and share the instructions or codes with others to be replicated anywhere by anyone.

3.2 E-textile Toys for Children

Since textile materials are often used for children products due to their comfort and tactile properties, e-textiles provide diverse possibilities to create new interactive children products that can still have the qualities of the textile material, but also can embrace interactive behaviors, such as sensing, responding and connecting. One of the first applications of e-textiles for children toys was the interactive soft toy, called Spookies providing various input and output possibilities in order to create an open-ended play experience [32]. Embedding electronics into stuffed toys has also opened a new field to create toys for children with special needs in order to create sensorial feedback for calming down or to trigger spontaneous and collaborative play [33, 34].

There are also examples of e-textile toys that embrace advanced integration of electronics on textiles such as through embroidering conductive yarns for building touch sensors [35] or even directly weaving the textile with a pattern corresponding to different sensors on an interactive puppet [36]. Moreover, there are also several examples, like DIY instructions for children to build their own interactive stuffed toys [7] or even hack their own soft toys to make them interactive [37].

4 MASAL—An Interactive Soft Toy for Collaborative Storytelling

4.1 Previous Work

This paper represents a smart storytelling soft toy as a second result of a practice-based design research—CO-STORY—funded by Free University of Bozen-Bolzano. The project aims at exploring new storytelling smart tools that can trigger children's imagination to build and experience stories through tangible objects integrated with digital technologies. While these artifacts provide tangible interaction with digital data, they introduce a new concept of storytelling through making children creators of the story.

As a first stage of the project, a co-design methodology was applied in order to create a smart object together with 8–15 years old children in various classroom settings. The result of the co-design process was an interactive capsule-like object that was used by the same children who participated in the workshops as a tool to contain and reveal their stories embracing specific emotions [38].

In the second stage of the project, the focus was given to younger children (3–8 years old) with a different method in which the children were not "design-partners" [39] in the design process, but they were involved later in the testing phase.

4.2 Inter-Cultural Storytelling

The project was based on an existing storytelling activity that is organized by Studio-Comune periodically in Bolzano, Italy. In these storytelling sessions, families and children (aged 3–8) gather to tell and listen to stories in a different language. The aim is to get to know a different culture through the language in a narrative form. The storytellers are invited to tell a story in their own language by using instruments like storybooks, illustrations or even body language.

Masal—an e-textile stuffed toy—was designed to be used in this existing inter-cultural storytelling event as an instrument to tell a story in a different language. As a difference from other instruments, it allows the participants to be active creators of a story that narrates sequence of events happening to a stranger coming from a different country.

4.3 Who Is Masal?

Masal was born due to the fact that the author of this paper was asked to tell a story in Turkish in one of the inter-cultural storytelling sessions organized by Studio-Comune. Therefore, its name—Masal comes from the authors country of origin,

Turkey, meaning "fairytale" in Turkish. It allows children to create a story together with others while learning about a new culture through this constructed narrative. Masal, the main protagonist of the story cannot talk in Italian but can express visual signals in order to help the participants to construct its story. With a LED display embedded in the stuffed toy, it shows random letters and facial expressions to give hints on what it wants to tell. The story is recorded on a paper-board that allows children to write down new foreign words learned within the story.

All participants sit in a circular setting around the paper board which is divided into eight sections for eight participants plus the initiator. The storytelling session starts with the initiator introducing Masal and describing the context of the story. Then, Masal is given to the first participant who interacts with it by closing its both hands to get a letter and by putting its hand on its heart to see an emotional expression (surprise, sadness, happiness) on its face. Based on these two inputs, the first participant creates a keyword and the first piece of the story. The initiator writes down the keyword in both languages (Italian and Turkish). Then, Masal passes from one participant to another while each participant creates a part of the story. At the end of the session, the initiator tells the whole story in Turkish while emphasizing the words written on the paper board. When the participants recognize the Turkish words written on the board, they raise their hands.

4.4 Construction of Masal

Masal is embedded with an open-source technology—Micro:bit [40] which is used as both sensor and actuator connected with touch sensors created with conductive yarns and conductive wool. The main material used in the construction of Masal was wool felt fabric embroidered with conductive yarns and felted with conductive wool.

The conductive parts embedded on the main body functions as touch sensors that are programmed to activate various visual outputs that are shown on the LED display of the microprocessor (see Fig. 1).

The microprocessor is attached on the head area of Masal, in order to display the emotional signals exactly on its mouth. The programming is done in Micro:bit's own online coding platform which works with visual blocks of codes enabling users to program the microprocessor in an easy way. The platform provides also the digital means to share the codes created by the user and make them open to be used by other users. Therefore, the codes could be shared with others to be adapted and reused for the storytelling sessions with other languages, for instance by using a different alphabet, or adding other types of emotional expressions besides the three expressions: surprised, happy and sad.

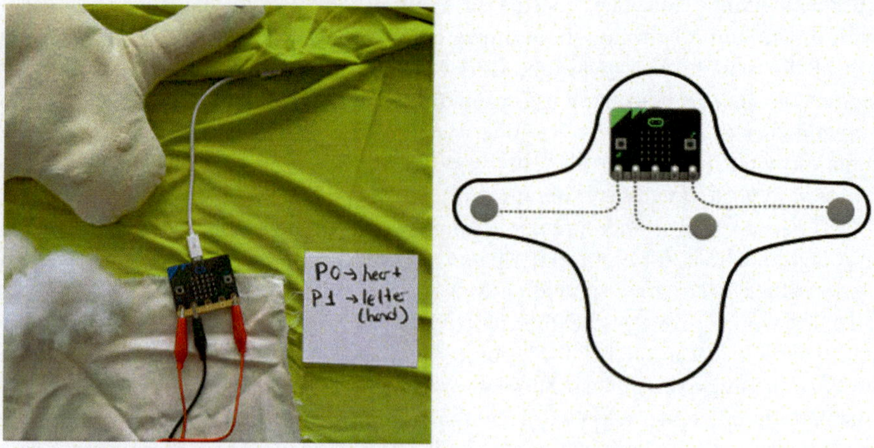

Fig. 1 Construction process of Masal with Micro:bit

5 Storytelling Sessions with Children and Adults

The first test was done with the first version of Masal using a different interaction modality (shaking: showing the letter, closing hands: showing the emotions) (see Fig. 2). However, this interaction modality was not found pleasurable, since the shaking movement of the toy was not so gentle. In this prototype, the inbuilt accelerometer of Micro:bit was used as a sensor to detect the shaking movement. Besides Masal, there was a paper board with a circular shape divided into eight

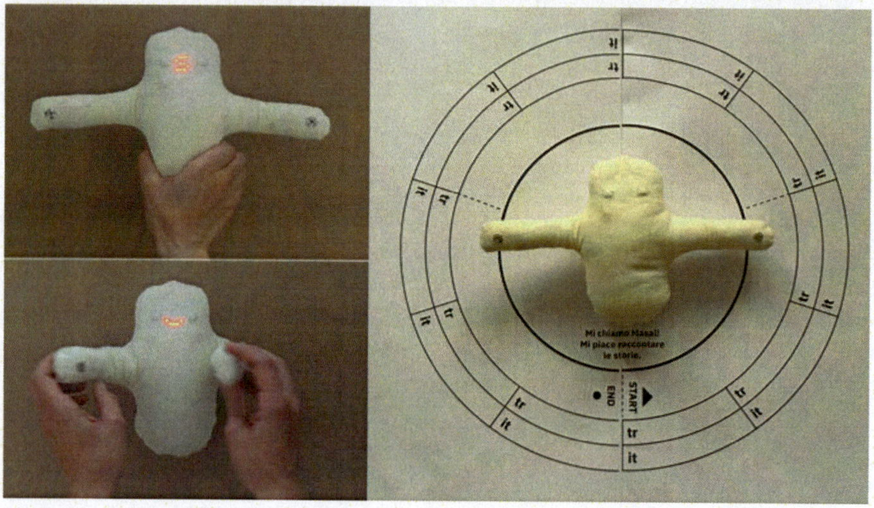

Fig. 2 On left: first version of Masal, on right: Masal on the first paper board

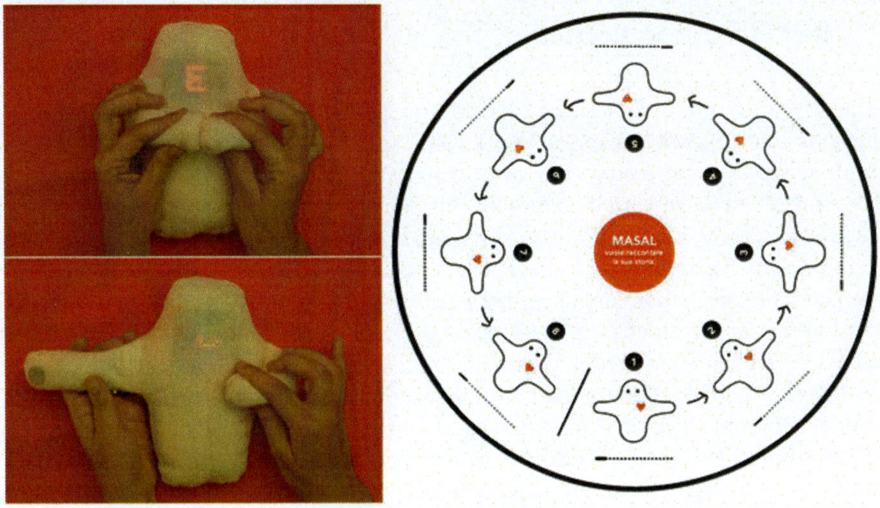

Fig. 3 On left: second version of Masal, on right: redesigned paper board

sections in which participants wrote the Italian and Turkish words with the facial expressions indicated in the toy (see Fig. 2).

In the first test session, parents accompanied their children (3–8 years old) and they were helping the children to come up with the words since some of them did not know how to read and write. The involvement of parents brought a different dynamic into the storytelling session enhancing the collaborative aspect of the activity.

In the second session, Masal was updated with its new interaction modality (closing hands: showing the letter, touching its heart: showing the emotions) (see Fig. 3). The participants were eight children (3–8 years old) accompanied by their parents. In this session, the paper board was redesigned to make the recording action easier with a more visual representation (see Fig. 3). After the storytelling session, for each participant a stuffed toy without electronics was given to create a personalized Masal. The participants later brought these toys home to continue the storytelling in an analog way.

In the third session, Masal was used to create two stories in Turkish and French by eight adults who had different cultural backgrounds. In this session, the initiator was not any more the author of the paper, but two participants from France and Turkey. While participants were creating the stories by using Masal, the author was a passive observer.

6 Results and Reflections

The three storytelling sessions conducted with children and adults gave rise to various insights about dynamics happening in a collaborative creation process. Masal, as seen from its shape, is an abstract character which does not have any specific features or form related to its gender or cultural background. Therefore, it can embody different characters easily based on the story context. This openness also comes from the digital part which allows it to have a variety of letters fitting to any language. For instance, for the Italian language there are certain letters which are not often been used as a primary letter of a word, such as H. Therefore, the coding of the digital content can be personalized based on the language spoken in the storytelling session. Although the interactive toy had an initial name Masal, it has changed during the sessions, became "Masala" as an Italian character and "Jean" as a French character. It embodies a new name and new character based on the story initiator. This openness which is both physical and digital enables the users to have a wider imagination on story creation based on their needs and desires.

During the storytelling sessions, it was observed that beyond the storytelling action superior discussions about similarities and differences of the two languages were spontaneously taking place based on the words chosen for the story. These discussions were mostly done by the parents or the adult group, but reflected with children to understand intersections between cultures.

Another observation was that children were developing emotional attachment towards Masal, such as kissing, hugging, caressing it during the sessions. The participants were getting excited to know how Masal was feeling through revealing its facial expressions. Ten days after the storytelling session, the children who participated in the first session were asked to draw Masal, and send the photo of the drawing to the author/conductor of the workshop. The drawings were very similar to Masal with all the details. This shows that it was a memorial moment as they remembered it after a certain period. Therefore, this observation gave rise to the idea to create smaller and non-digital versions of Masal to be given to the participants of the second session after the storytelling activity. In this second session, after the interactive storytelling with digital Masal, they could personalize their own mini Masals by coloring it with textile paint and take them home. One of the parents reported that her child was playing with Masal at home and creating stories with the same technique of letters and emotions. This shows us how technology can act as an enabler to give the first curiosity and experience, which then later can be applied in an analog setting to foster imagination and collaboration. In the storytelling session, the participants interacted with the digital Masal, but in the end, at home they had similar interactions with their personalized non-digital Masals.

After the second session, some parents asked for a workshop to build their own Masal to learn how to craft their own interactive toy. Due to the fact that the technology used in building the toy was open-source, the parents were interested in being involved in the making process. They commented that Masal could help their children to develop empathic skills and it could be played at home between parents and their children. The open-source technology and craft techniques used in the creation of Masal can make it replicable and allow others to personalize the usage by simply changing the code and the variables. Garzotto et al. [41] poses the question of how to make interactive storytelling tools and methods long-lasting and reaching to larger user groups, going beyond the experimentation. Today, open-source DIY technologies and platforms can allow us to overcome this problem and make small scale projects reach larger user groups in the world. By uploading the blueprints and the instructions of how to build it, the toy can be replicated anywhere in the world.

Although each participant decided what happens to Masal during the storytelling session, the group dynamics were very impactful on the story creation. The stories were constructed with the consent and collaboration of each participant. Children became co-authors of the stories, and therefore experienced a different way of storytelling, shifting from mere consumers of stories or single authors of them to co-creators.

7 Conclusions

This paper addresses a practice-based design research and an iterative design process of building and testing an interactive storytelling toy. Integration of e-textiles in children toys can introduce new play modalities, enhance the sensorial experience and facilitate the construction of a DIY game that can answer the specific needs of a user group. Open-source technologies and maker communities facilitate this type of practices enabling various users to invent new tools that are accessible for everyone. Masal is an interactive soft toy constructed with e-textiles in order to answer a specific need: introducing a foreign culture to children through collaborative storytelling. The toy allows children to create a story together with others while learning about a new culture through this constructed narrative. The results obtained from the storytelling sessions with Masal show how an interactive soft toy can enhance children's imagination and collaboration, and above all can create a bridge between two cultures aiming at having more empathy and understanding towards the other. The toy transforms into new personas linking different places, cultures and people through a collaborative storytelling session. It continuously shifts between local and global dimensions, becoming a (g)local figure. As a further application, Masal can be used in multi-cultural classroom settings or any possible environment in which there are children with different cultural backgrounds.

Acknowledgements Special thanks to Irene Visentini and Sofia Sanchez—Studio Comune (https://www.studiocomune.eu/) for the collaboration within Interracconti sessions.

References

1. Lampe M, Hinske S (2007) Integrating interactive learning experiences into augmented toy environments. In: Proceedings of the workshop on Pervasive learning 2007, Toronto, pp 1–9
2. Cassell J, Ryokai K (2001) Making space for voice: technologies to support children's fantasy and storytelling. Pers Ubiquit Comput 5(3):169–190
3. Bobick AF, Intille SS, Davis JW, Baird F, Pinhanez CS, Campbell LW, Ivanov YA, Schütte A, Wilson A (1999) The KidsRoom: a perceptually-based interactive and immersive story environment. Presence: Teleoperators Virtual Environ 8(4):369–393
4. Smart Toy, https://www.fisher-price.com/en_CA/brands/smarttoy, last accessed 2019/12/13
5. Vaikai Homepage, https://vaikai.com, last accessed 2019/12/13
6. Berzowska J (2005) Electronic textiles: wearable computers, reactive fashion, and soft computation. TEXTILE 3(1):58–75
7. Buechley L, Qiu K (2013) Sew electric: a collection of DIY projects that combine fabric, electronics, and programming. HLT Press, Cambridge, MA, USA
8. Prensky M (2001) Digital natives, digital Immigrants. On the Horizon 9(5):1–6
9. Chaudron S, Beutel ME, Donoso Navarrete Veronica, Dreier M, Fletcher-Watson B, Heikkilä AS, Kontríková V, Korkeamäki RV, Livingstone S, Marsh J, Mascheroni G, Micheli M, Milesi D, Muller KW, Niska M, Olkina O, Svenja S, Plowman L, Ribbens W, Richardson J, Schaack C, Shlyapnikov V, Šmahel DD, Soldatova G, Wölfling K (2015) Young children (0–8) and digital technology: a qualitative exploratory study across seven countries. JRC; ISPRA, Italy
10. Turkle S (2015) Reclaiming conversation, the power of talk in a digital age. Penguin Books Press, NY
11. Ishii H, Ullmer B (1997) Tangible bits: towards seamless interfaces between people, bits and atoms. In: Proceedings of the ACM SIGCHI conference on human factors in computing systems (CHI'97). ACM, New York, NY, USA, pp 234–241
12. Antle AA (2007) The CTI framework: informing the design of tangible systems for children. In: Proceedings of the 1st international conference on Tangible and embedded interaction (TEI'07). ACM, New York, NY, USA, pp 195–202
13. Vygotsky LS (2004) Imagination and creativity in childhood. J Russ East Eur Psychol 42(1):7–97
14. Amazon Echo Dot, https://www.amazon.com/Echo-Dot/dp/B07FZ8S74R?th=1, last accessed 2019/12/13
15. Google Home, https://store.google.com/product/google_home, last accessed 2019/12/13
16. Fernandes R, Yamanaka T (2019) Interactive design versus design for interaction: developing interactive play tools that promote interactions between children. In: Interactive multimedia, IntechOpen
17. Kara N, Aydin CC, Cagiltay K (2014) Design and development of a smart storytelling toy. Interact Learn Environ 22(3):288–297
18. Arnall T (2014) Exploring 'immaterials': mediating design's invisible materials. Int J Des 8(2)
19. Vaucelle C, Jehan T (2002) Dolltalk: a computational toy to enhance children's creativity. In: CHI'02 extended abstracts on human factors in computing systems (CHI EA'02). ACM, New York, NY, USA, pp 776–777
20. Sylla C, Gonçalves S, Branco P, Coutinho C (2012) T-words: playing with sounds and creating narratives. In: Nijholt A, Romão T, Reidsma D (eds) Proceedings of the 9th international conference on advances in computer entertainment (ACE'12). Springer, Berlin, Heidelberg, pp 565–568
21. Low S, Sugiura Y, Fan K, Inami M (2013) Cuddly: enchant your soft objects with a mobile phone. In: Reidsma D, Katayose H, Nijholt A. (eds) Advances in computer entertainment. ACE 2013. Lecture Notes in Compter Science 8253, Springer, Cham, pp 138–151
22. Kuusk K (2016) Crafting sustainable smart textile services. Eindhoven University of Technology, Eindhoven, The Netherlands

23. Mascheroni G, Holloway D (2017) The Internet of toys: a report on media and social discourses around young children and IoToys. Retrieved from http://digilitey.eu/wp-content/uploads/2017/01/IoToys-June-2017-reduced.pdf
24. Tanenbaum JG, Williams AM, Desjardins A, Tanenbaum K (2013) Democratizing technology: pleasure, utility and expressiveness in DIY and maker practice. In: Proceedings of the SIGCHI conference on human factors in computing systems (CHI'13). ACM, New York, NY, USA, pp 2603–2612
25. Satomi M, Perner-Wilson H (2011) Future master craftsmanship: where we want electronic textile crafts to go. In: Proceedings of the ISEA2011. The 17th international symposium on electronic art
26. High-Low Tech homepage, http://highlowtech.org, last accessed 2019/12/13
27. Kobakant website, http://www.kobakant.at/?page_id=475, last accessed 2019/12/13
28. Hertenberger A, Scholz B, Contrechoc B, Stewart B, Kurbak E, Perner-Wilson H, Posch I, Cabral I, Qi J, Childs K, Kuusk K, Calder L, Toeters M, Kisand M, Bhömer M, Donneaud M, Grant M, Coleman M, Satomi M, Tharakan M, Vierne P, Robertson S, Taylor S, Nachtigall TR (2013) E-textile swatchbook exchange: the importance of sharing physical work. In: Proceedings of the 2014 ACM international symposium on wearable computers: adjunct program (ISWC'14 Adjunct). ACM, New York, NY, USA, pp 77–81 (2014)
29. Fabtextiles homepage, https://fabtextiles.org, last accessed 2019/12/13
30. TextileLab Amsterdam. https://waag.org/en/project/textilelab-amsterdam, last accessed 2019/12/13
31. Wemake homepage, http://wemake.cc/, last accessed 2019/12/13
32. Berglin L (2005) Spookies: combining smart materials and information technology in an interactive toy. In: Proceedings of the 2005 conference on Interaction design and children (IDC'05). ACM, New York, NY, USA, pp 17–23
33. Seo JH, Aravindan P, Sungkajun A (2017) Toward creative engagement of soft haptic toys with children with autism spectrum disorder. In: Proceedings of the 2017 ACM SIGCHI conference on creativity and cognition (C&C'17). ACM, NY, USA, pp 75–79
34. Nonnis A, Bryan-Kinns N (2019) Mazi: a tangible toy for collaborative play between children with autism. In: Proceedings of the 18th ACM international conference on interaction design and children (IDC'19). ACM, New York, NY, USA, pp 672–675
35. Berzowska J, Mommersteeg A, Grueso LIR, Ducray E, Rabo MP, Moisan G (2019) Baby Tango: electronic textile toys for full-body interaction. In: Proceedings of the thirteenth international conference on tangible, embedded, and embodied interaction (TEI'19). ACM, New York, NY, USA, pp 437–442
36. Pouta E, Mikkonen J (2019) Hand puppet as means for eTextile synthesis. In: Proceedings of the thirteenth international conference on tangible, embedded, and embodied interaction (TEI'19). ACM, New York, NY, USA, pp 415–421
37. I luv my stuffie, https://makercamp.com/projects/i-luv-my-stuffie, last accessed 2019/12/13
38. Bonetti R, Cohen N, Ugur Yavuz S (2019) Enabling relationships in a co-creative process with children. In: Proceedings of RTD conference method & critique, Delft
39. Druin A (2002) The role of children in the design of new technology. Behav Inf Technol 21(1):1–25
40. Micro:bit, https://microbit.org/, last accessed 2019/12/13
41. Garzotto F, Paolini P, Sabiescu A (2010) Interactive storytelling for children. In: Proceedings of the 9th international conference on interaction design and children. ACM, pp 356–359

A Semiotic and Usability Analysis of Diegetic UI: Metro—Last Light

Guilherme Doval, Flávio Almeida, and Luan Nesi

Abstract A Game's narrative is perhaps one of the key components of creating player immersion. As technology advances, game developers increase their toolset for creating increasingly complex game worlds. The UI has a crucial role, providing the player with feedback about the various attributes and mechanics within the game. Some games sought to integrate the traditionally intrusive UI within the game's narrative and art, by the means of Diegetic UI. The goal of this research paper is to understand how integrating the User Interface into the game's art and narrative—creating what is called a Diegetic Interface—can increase the feeling of immersion for the player. To identify the processes through which meaning is observed in Diegetic UI, we've relied on the Discoursive Semiotics proposed by A. J. Greimas and to assess if these Diegetic Representations affect Usability, we employed Game Usability Heuristics proposed by Desurvire and Wiberg. The methodology proved to yield interesting results regarding the relationships between UI and Narrative as well as the Usability impact derived from such implementation in the game Metro: Last Light.

Keywords UI · Immersion · Storytelling · Game design

1 Introduction

As games become increasingly complex and realistic due to technological advancements, the subject of immersion in games becomes central to understanding how to

G. Doval
IADE – Faculdade de Design, Tecnologia e Comunicação, Universidade Europeia, Av. D. Carlos I, 4, 1200-649 Lisboa, Portugal

F. Almeida (✉)
UNIDCOM/IADE - Unidade de Investigação em Design e Comunicação, Lisboa, Portugal & LabCom, Universidade da Beira Interior, Covilhã, Portugal
e-mail: flavio.almeida@ubi.pt

L. Nesi
UNISINOS, Porto Alegre, Brasil

E. Duarte and C. Rosa (eds.), *Developments in Design Research and Practice*, Springer Series in Design and Innovation 17,
https://doi.org/10.1007/978-3-030-86596-2_11

design games for Virtual Reality. However, VR games are not the sole candidate to benefit for research into Immersion [1].

One of the main goals for story-driven games is player immersion into the game's world. Conceptually, the player stops being an outside entity and assumes the role of a fictional character, with his own set of goals, beliefs and memories [2].

The game's User Interface, as the name suggests, bridges the information gap between the player and the game, providing information regarding various status of the gameplay. In truth, some games rely on the UI as much more than just its informative function, in city manager games for example the UI is the means through which the player selects the tool he wishes to use in order to complete the set goals. The use of Diegetic UI can be seen in some form in various games such as the often-quoted example of Dead Space and discussed by various authors such as Fagerholt and Lorentzon, Azevedo et al. and Salomoni et al. [3–5].

The goal of this paper is to understand how Diegetic Interfaces can potentially increase player immersion and to assess if such implementation can prove to be detrimental to Usability.

To approach this question, we will employ the methodology first proposed by Vitorino and Serrano [6], consisting of a combination of a Semiotic Analysis, based on the work of Greimas [7] and a Heuristic Evaluation based on the work of Desurvire and Wiberg [8]. This approach is qualitative in its nature, allowing us to understand how the interface is structured in order to create meaning, by becoming part of the discourse of a game's plot or universe.

The object of the analysis is the game Metro: Last Light, published by Quicksilver and released in May 17th, 2013.

2 Theoretical Framework

2.1 Heads-Up Display (HUD)

One of the main elements of GUIs found in videogames is called a Heads-Up Display (HUD), according to Fagerholt and Lorentzon: "A head-up display, or HUD, is any transparent display that presents data without requiring the user to look away from his or her usual viewpoint. The origin of the name stems from the user being able to view information with their head "up" and looking forward, instead of angled down looking at lower instruments" [3]. In games, the term HUD refers to the method by which information is visually conveyed to the player whilst a game is in progress. The HUD is frequently used to simultaneously display several pieces of information such as the main character's health, items, and indicators of game progression and goals.

We understand HUD in videogames as any overlaid frame which conveys information regarding the player's current status, such as health, ammo, map, compass and enemy's health. The elements positioning is static, although in some cases HUD elements can be occluded when not in use and has the clear purpose of conveying information regarding the player's attributes.

The GUI is the larger system which consists of all on-screen elements, including Menus and other non-gameplay related elements that don't require the player's constant attention [3].

2.2 Diegesis and Videogames

The search for a definition of the term Diegesis apparently first arose with the postulates from Plato and Aristotle. Diegesis, from the Greek δισήγησις, means 'to narrate, set out in detail, describe' [9]. Diegesis would be aligned with poetic imitation, known as *Mimesis*, that is, they were practically the same thing. To Plato, *Mimesis* and *Diegesis* were two distinct terms: Whereas *Diegesis* stands for the narration, *Mimesis* represents the Poetic Imitation. Aristotle however saw narrative as closer to Poetic Imitation in relationship to Plato's concept [4].

Azevedo et al. further adds that through the times, these concepts were revised by many authors, such as Lodge [10], who proposes that Plato's and Aristotle's definition is exceedingly limited to classify all the variations and nuances in fictional narrative. Furthermore, it's a common occurrence where narratives exhibit a thin line between *Diegesis* and *Mimesis*. As such, it's paramount that more diverse and complex forms of classifications are proposed [4].

In games, the study of *Diegesis* in the context the Heads-Up Display in First-Person Shooters was approached by Fagerholt and Lorentzon [3], where it was proposed that Diegesis in games refers to the world in which the game's story takes place, defining whether an elements is part of the game's world and if the characters that inhabit this world can perceive it.

2.3 Discoursive Semiotics

"Semiotics seeks to determine what the text says, how and why it says, through an analysis in different forms of expressions" [6]. Greimas' semiotic theory positions itself as the theory of the processes of signification, and not the science of the study of signs.

It focuses itself on the generative process behind the creating of meaning [11].

This discipline, which was heavily influenced by the works of Ferdinand de Saussure, developed into its own form of Semiotics, aimed specifically at Discourse Analysis alongside an original method proposed by Greimas. In addition to the methodological support, an extensive terminological vocabulary is also presented, which is described in the postulate *Semiotics and Language: An Analytical Dictionary* [12]. One of such terms is 'Text', and while it might seem unusual to refer to visual representations as such, our understanding is that: "The term text is often taken as a synonym of discourse, (…) Both terms—text and discourse—can be employed interchangeably to designate the semantic axis of non-linguistic semiotics: a ritual or a ballet can be considered as either texts or discourses" [12].

The semiotic theory of Greimas is first characterized by the concept of a generative process of meaning, starting from the most simple and abstract to the most complex and concrete, and is composed of three levels, this is called the Plane of Content, where the main Discourse lies, and the Expression Plane, the externalization of the content. It is important to note that, analysing verbal and non-verbal text through Discoursive Semiotics does not necessarily imply that it is necessary to observe each text in accordance to a previously constructed structure but rather, observe the possible articulations and constructions of meaning that, in the text, result in a determined structure [13].

The Plane of Content. On the Fundamental Level, we find the basic semantics that constitute the foundation of the text's construction. It is here we find the semantic categories that order the text's content in a general and abstract fashion. This level is based on difference and opposition, in order to establish this opposition however, it is necessary the existence of common traits [14].

Lara [15] provides us with the example semantic oppositions of /humanity/ versus /divinity/, which also yields the contraries /not humanity/ and /not divinity/ (which are also contrary in comparison to each other). Besides the oppositional relationships, there also exists complementary relationships, such as the case of /humanity/ and /not divinity/, as well as /divinity/ and /not humanity/.

On the narrative level, the abstract values from the Fundamental become inscribed into objects, with which the observer can relate by Conjunction or Disjunction. The texts in this level are structured by a Canonical Narrative Schema which comprises four phases, namely: Manipulation, Competence, Performance and Sanction. [13, 15].

Lastly, on the Discursive level, the Subjects and Objects become Actors in the Discourse and the narrative begins to display the Time and Space categories, in order to transform the text into a communicative situation. The values previously embedded in objects become widespread as Themes—abstract elements which explain and instil reality—as well as Figures—or concrete elements that built the world simulacra and cover adjacent themes [13, 16].

The Plane of Expression. As we delve further into the Semiotic Theory proposed by Greimas, we find heavy influences from Saussure and his conception of the Sign as a duality between Signifier and Signified. This dyadic—i.e. two sided—model was then revisited by Louis Hjelmslev, who renamed them into the two planes discussed in this section—the Signifier and Signified becomes the Plane of Expression and the Plane of Content respectively [17].

If, in a primary phase, the Semiotic Theory focuses its efforts in analysing the content of a text, then in a second phase that focus is transferred into the Plane of Expression, which if we refer to the Sausurrian model—i.e. the Signifier—constitutes the 'sound-image' component of a Sign. Hjelmslev however, understood that the Plane of Expression represents material culture and the physical materials of a medium—images, printed words, sounds or even physical performances[18].

Furthermore, Hjelmslev adds that: "There can be no content without an expression, or expressionless content; neither can there be an expression without a content, or content-less expression" [18].

Before analysing the visual text it's important to note that, the Semiotic Analysis of strictly visual representations doesn't privilege neither the Plane of Content or the Plane of Expression, as the Dimensions—Also referred to as Plastic Formants—can also hold complementary relationships with one or more units within the Plane of Content [19].

2.4 Heuristic Evaluation

Despite this paper's perspective of games as a medium through which storytelling can be achieved and, as Hassenzahl [20] puts it, a mediator for experiences, they are still, nevertheless, software. As such, they're still bound to the principles of Usability proposed by HCI studies. One of the methods proposed to analyse software in order to isolate Usability Issues is called Heuristic Evaluation. Perhaps the most known examples of this type of Evaluation is Nielsen's [21] Usability Heuristics, broad rules of thumb—for evaluating User Interfaces in a cheap and fast manner.

For the purposes of this paper and in aiding us in answering the research question, we will rely on the work of Desurvire and Wiberg [8] where they claim that traditional heuristic evaluation fails to consider important concepts in Game Design, such as Immersion, Challenges and Entertainment. Thus, a specific set of Heuristics for games is necessary, and they introduce the PLAY Heuristics as a possible solution.

2.5 Metro: Last Light

Metro: Last Light is second instalment in a series of games developed by 4A Games and published by Deep Silver, released May 17th, 2013. The game is based around the book series Metro 2033 by the Russian author, Dmitriy Glukhovskiy. The game is set in a post-apocalyptic Moscow, following a Nuclear War that made the surface uninhabitable and forced the survivors to live in the Moscow Metro. Upon reception, the game was lauded by critics for its atmosphere, world design, story and gameplay [22, 23].

The game is classified as a singleplayer first-person shooter, and features stealth gameplay elements, enabling the player to choose between direct combat or dispatching enemies quietly.

3 Method

3.1 Object

The game "Metro: Last Light" was chosen as the object of this research paper due to its implementation of Diegetic Interfaces as well as its focus on storytelling and gameplay.

For the purposes of this paper we've opted to play the game under Ranger Hardcore, which according to the text description aims to provide the most immersive experience of the game, removing all HUD elements. This choice was made in order to assess the impact of removing on-screen elements and if this way of playing the game has direct impact on Usability and Immersion. While a comparative analysis of playing the game with the HUD versus playing the game without the HUD was considered, we believe that by starting the game without the HUD it was possible to analyse the game under the perspective of someone who has never played the game.

3.2 Materials

To conduct the analysis, the Authors resorted to utilizing a combination of Semiotic Analysis and Heuristic Evaluation as proposed by Vitorino and Serrano [6], however instead of utilizing the heuristics proposed by the authors, the PLAY Heuristics [8] was utilized due to it being an updated version of the heuristics presented in the original paper by Vitorino and Serrano.

To classify the elements according to its design space, the following categories proposed by Fagerholt and Lorentzon [3] were utilized (Fig. 1).

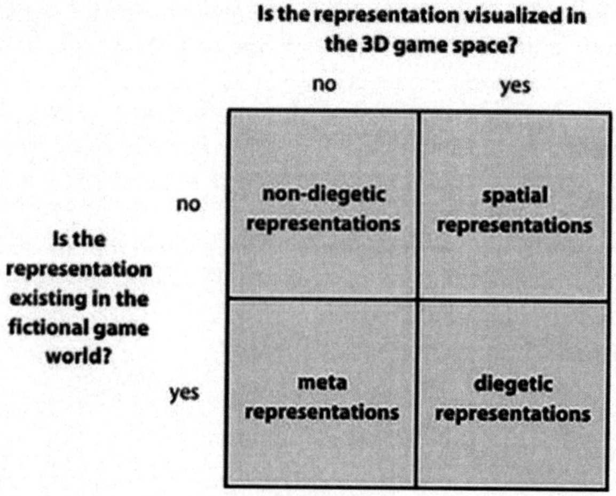

Fig. 1 Types of interfaces distinguished by their presence in the narrative and 3D space [3]

Table 1 The dimensions of the plane of expression. Adapted from [6, 15, 19, 24]

Dimensions	Terms
Topological dimension Spatial properties	High versus low Centre versus extremity
Eidetic dimension Forms and shapes	Circular versus rectilinear Uniform versus multiform Expanded versus contracted Angular versus rounded
Photochromatic dimension Light and colour	Light versus dark Monochromatic versus polychromatic Chromatic versus achromatic Opacity versus transparency Warm colours versus cool colours

For the Semiotic Analysis of the Plane of Expression, the following Dimensions were employed, with no specific hierarchy in mind (Table 1).

To conduct the heuristic analysis of the interface, the PLAY Heuristics proposed by Desurvire and Wiberg [8] were used, specifically use the Category 3, which is concerned with Usability and Game Mechanics (Table 2).

3.3 Procedure

Determining the Interface Elements. Before analysing the individual elements of the interface, the elements were outlined and separated in accordance to their function. The interface in Metro—Last Light is very straightforward, from an early gameplay segment that features a prologue to the main story, we see two elements: *A Wristwatch* on the protagonist's left hand featuring a blue LED and a digital clock, *and a Gun.* Following this opening sequence, the player assumes control of the Protagonist, which proceeds to automatically take two items from his desk: *A Journal and a Lighter*. As previously stated, the difficulty level under which the game was analysed does not feature a HUD. Upon reaching the Armory, the player is presented with the *Gas Mask*.

Four distinct items were stablished, which play important roles in the game's mechanics and gameplay, namely:

- Watch;
- Gun;
- Gas Mask;
- Journal and Lighter.

There are other interface elements which are responsible for specific game mechanics, such as a Battery system which the player must use in order to recharge its flashlight, however this system is only used occasionally and thus, we've opted

Table 2 Category 3: usability and game mechanics of the PLAY heuristics [8]

A. Heuristic: documentation/tutorial
A1. Player does not need to read the manual or documentation to play; A2. Player does not need to access the tutorial in order to play
B. Heuristic: status and score
B1. Game controls are consistent within the game and follow standard conventions B2. Status score Indicators are seamless, obvious, available and do not interfere with gameplay B3. Controls are intuitive and mapped in a natural way; they are customizable and default to industry standard settings B4. Consistency shortens the learning curve by following the trends set by the gaming industry to meet users' expectations. If no industry standard exists, perform usability/playability research to ascertain the best mapping for the majority of intended players
C. Heuristic: game provides feedback
C1. Game provides feedback and reacts in a consistent, immediate, challenging and exciting way to the players' actions C2. Provide appropriate audio/visual/visceral feedback (music, sound effects, controller vibration)
D. Heuristic: terminology
D1. The game goals are clear. The game provides clear goals, presents overriding goals early as well as short term goals throughout gameplay D2. The skills needed to attain goals are taught early enough to play or use later, or right before the new skill is needed D3. The game gives rewards that immerse the player more deeply in the game by increasing their capabilities, capacity or, for example, expanding their ability to customize
E. Heuristic: burden on player
E1. The game does not put an unnecessary burden on the player E2. Player is given controls that are basic enough to learn quickly, yet expandable for advanced options for advanced players
F. Heuristic: screen layout
F1. Screen layout is efficient, integrated, and visually pleasing; F2. The player experiences the user interface as consistent (in controller, color, typographic, dialogue and user interface design) F3. The players experience the user interface/HUD as a part of the game; F4. Art is recognizable to the player and speaks to its function
G. Heuristic: navigation
G1. Navigation is consistent, logical and minimalist
H. Heuristic: error prevention
H1. Player error is avoided H2. Player interruption is supported, so that players can easily turn the game on and off and be able to save the games in different states H3. Upon turning on the game, the player has enough information to begin play H4. Players should be given context sensitive help while playing so that they are not stuck and need to rely on a manual for help H5. All levels of players are able to play and get involved quickly and easily with
I. Heuristic: game story immersion
I.1 Game story encourages immersion (If game has story component)

not to analyse it and instead focus on the more prominent and common elements. Due to the page limit we will only showcase the analysis of the Watch element, but as the other elements were a part of the original research conducted under a master's degree thesis, we will take them into account during the conclusion.

The Watch is the one element which is consistently visible through the game with no changes in appearance, its purpose is to provide the player with two distinct status information: The current oxygen filter's duration and the protagonist's visibility to enemies, represented by a pair of the red glowing numbers on the watch and a blue led indicator, respectively.

Element Classification. The element assumes a Diegetic representation, the Watch is placed in the Protagonist's left wrist and is contextualized as part of the game's art and narrative.

Semiotic Analysis. On the Plane of Expression, the LED indicator manifests characteristics from the Photochromatic Dimension, by the oppositions /light/ versus /dark/ and /chromatic/ versus /achromatic/, they relate to the /visible/ versus /invisible/ oppositions on the Discoursive level, which lead to the /conflict/ versus /calm/ Fundamental terms, which can be found in the Plane of Content, establishing a semi-symbolic relationship. The clock's numbers also assume the /light/ versus /dark/ and /chromatic/ versus /achromatic/ in the Photochromatic Dimension of the Plane of Expression. As the numbers are displayed on what seems to be a Nixie Tube, we find the /analogue/ versus /digital/ and /decline/ versus /progress/ terms on the Discoursive level, leading us to the Fundamental opposition /culture/ versus /nature/.

The numbers on the watch continuously decline until the player replaces his oxygen filters, if he fails to do so, he will die. We find the /full/ versus /empty/ Discoursive Terms, leading us to the /life/ versus /death/ Fundamental opposition. While these oppositions are not represented on the current element's Plane of Expression, they appear on the Gas Mask element (Figs. 2, 3 and 4).

Heuristic Evaluation. The watch speaks to its function to the player, making it easy for a new player to understand the information being conveyed without having to read a manual or play a tutorial, as the game introduces the player to the watch's functionality through specific in-game sections. The watch beautifully performs as both an indicator of the player's visibility and the current air filter's duration, providing visual as well as auditory feedback—in the form of a beeping sound when the filter is running out—it does not lack any sort of feedback and thus, does not place a burden upon the player. The watch is placed on the lower centre portion of the screen, where the player can easily see looking by simply looking down during shooting portions. As a Diegetic element, it encourages the player to immerse itself into the game's narrative.

Overview. It's apparent that the watch is not hindered by the any lack of feedback from its function, its positioning in the lower-c entre of the screen attests to its importance in providing the player with crucial feedback regarding two game mechanics, Stealth and Oxygen. Not only does it provide feedback, it is also implemented into the game world in a realistic fashion, reinforcing the player's sense of immersion and conveying the Themes and Figures found in the game's discourse. In addition,

Fig. 2 While exploring the surface or in irradiated areas where the player must use a gas mask, the numbers on the watch represent the current oxygen filter's duration

Fig. 3 While close to a light source, the lit LED indicates that the player is visible

we find that while specific Diegetic elements do not express their contents visually, the content can be expressed in other elements.

4 Results and Discussion

Through the information gathered by the Analysis of Watch element, previously established, we'll now present the results in a table in order to ease the visualization of the results, it consists of the Elements Classification, the Semiotic Analysis of

Fig. 4 In a dark area, the LED become unlit, indicating that the player is not visible to enemies

the Plane of Content and Plane of Expression as well as the Heuristics in which the elements fits, followed by an overview of the Results (Table 3).

It was possible to notice during the Analysis some of the common Themes of the Game and its use of visuals to convey its Discourse, Metro: Last Light is a game about a decaying society, where Humanity's last remnants struggle to survive the desolate world left by a Nuclear War. The sense of helplessness is reinforced to the player by the game's mechanics such as the Gas Mask mechanics, where the player must scavenge the ruins of Moscow in search of filters in order to survive and complete his Objectives. Overarching Themes such as the technological stagnation, the unrelenting and uninhabitable overworld as well as the constant threat of exposure to enemies are clearly reflected in the game's Diegetic Interface and in this sense, we could say that Metro: Last Light excels at its use of Diegesis to reinforce the Player's Immersion.

However, that is not to say that this implementation is without fault. During the Analysis it was possible to understand that, while traditional First-Person Shooters mechanics did not require any type of instruction to the player, such as movements,

Table 3 Results of the semiotic and usability analysis of metro: last light

Analysis results: watch	
Element classification	Diegetic
Plane of content	Discoursive Level visible versus invisible analogue versus digital decline versus progress full versus empty
	Fundamental Level life versus death culture versus nature conflict versus calm
Plane of expression	Photochromatic dimension light versus dark chromatic versus achromatic
Usability heuristics	A1, A2, B2, C2, E1, F1, F3, F4, H5, I1

aiming and shooting, the game's specific mechanics such as the Gas Mask and Journal and Compass.

The Gas Mask is perhaps the worst offender for Usability, although the mechanic itself is very easy to understand and gives the player feedback regarding various statuses, it suffers primarily due to it having 3 keys responsible for distinct functions—Replace Filter, Clean Mask and Remove Mask—and as the mechanics are not transversal to the First-Person Shooter genre, unlike the ones found in the Gun, remembering the mapping can prove to be challenging. The lack of an indication of how many filters there are available can also prove to be a potential hurdle to be overcome by casual players, however, this can also be an incentive for exploration and resource scavenging.

One of Metro: Last Light's strongest points is its constant use of feedback to indicate the player's current status, such as the use of Meta-Perception UI in the form of condensation in the mask, providing an indication that the player's Oxygen Filter is about to expire or the LED Display on the watch which signalizes if the player is visible to enemies. It's clear that although the game might suffer significantly on Usability while being player without the HUD, if the Player manages to memorize the key mappings this can prove to be an Immersive Story-Driven experience, where the player is put in the shoes of a survivor in the post-apocalyptic underground of Moscow's Metro system, with Gameplay Mechanics, UI and Visual Storytelling that reflect the care given to it.

The Semiotic Analysis proved useful in the textualization of the UI in order to identify the elements which constitute the game's Discourse, this is useful to analyse how well the narrative's Themes and Figures are represented Visually. By combining this with the Heuristic Evaluation it was possible to see that some while some aspects of the UI suffer in Usability from its Diegetic Approach, it's a trade-off which can prove to be beneficial in storytelling and world building in games, and in this sense potentially increase the Player's immersion through storytelling rather than gameplay.

This paper seeks to further the study of Diegesis in games and the relationship between Narrative and UI in in games and suggest that more research is conducted in the use of Discoursive Semiotics in analysing UI in games.

References

1. Jennett C, Cox AL, Cairns P et al (2008) Measuring and defining the experience of immersion in games. Int J Hum Comput Stud 66:641–661. https://doi.org/10.1016/j.ijhcs.2008.04.004
2. Heussner T (2015) The game narrative toolbox, 1st edn. Routledge
3. Fagerholt E, Lorentzon M (2009) Beyond the HUD—user interfaces for increased player immersion in FPS games. Master Thesis, Chalmers University of Technology
4. de Azevedo FM, da Silva ICS, Frosi FO (2017) Game User Experience (UX): Explorando a Teoria da Diegese. In: SBC—Proceedings of SBGames 2017. Curitiba, BR, p 9
5. Salomoni P, Prandi C, Roccetti M et al (2017) Diegetic user interfaces for virtual environments with HMDs: a user experience study with oculus rift. J Multimodal User Interfaces 11:173–184. https://doi.org/10.1007/s12193-016-0236-5

6. Vitorino RF, Serrano PHSM (2017) Os Efeitos de Sentido e a Usabilidade na Interface de Call of Duty. In: SBC—Proceedings of SBGames 2017. Curitiba, BR, p 9
7. Greimas AJ (1976) Semântica Estrutural. Cultrix, São Paulo, BR
8. Desurvire H, Wiberg C (2009) Game usability heuristics (PLAY) for evaluating and designing better games: the next iteration. In: Ozok AA, Zaphiris P (eds) Online communities and social computing. Springer, Berlin Heidelberg, pp 557–566
9. Liddell HG, Scott R (1996) A Greek-English Lexicon. Oxford Univeristy Press, Oxford, UK
10. Lodge D (1984) Mimesis and diegesis in modern fiction. SPELL: Swiss papers in English language and literature 1:89–108. https://doi.org/10.5169/seals-99832
11. Dias AR (2016) A aplicação da semiótica discursiva na análise da comunicação publicitária para a divulgação da marca. / The application of discursive semiotics in the analysis of advertising communication for the dissemination of the brand. IT 394. https://doi.org/10.19132/1807-8583201637.394-412
12. Greimas AJ, Courtés J (2008) Dicionário de Semiótica. Cultrix, São Paulo, BR
13. Castro CM de, Portela JC (2018) A noção de conteúdo e de expressão no percurso gerativo do sentido. Estud semiót 14: https://doi.org/10.11606/issn.1980-4016.esse.2018.148463
14. Antonio DM (2008) O Percurso Generativo de Sentido Aplicado à Análise Documental de Textos Narrativos de Ficção: Perspectivas de utilização em bibliotecas universitárias. Master Thesis, Faculdade de Filosofia e Ciências, Universidade Estadual Paulista
15. Lara GMP (2011) A Imagem como Objeto de Ensino. CASA 9. https://doi.org/10.21709/casa.v9i1.4423
16. Fiorin JL (2010) Introdução à Linguística II: Princípios de Análise. Contexto, São Paulo, BR
17. Fontanille J (2006) The Semiotics of Discourse. Peter Lang US
18. Chandler D (2017) Semiotics: the basics, 3rd edn. Routledge, New York, US
19. de Oliveira AC (1995) As Semioses Pictóricas. Face, Revista de Semiótica e Comunicação 4:104–145
20. Hassenzahl M (2010) Experience design: technology for all the right reasons. Morgan & Claypool Publishers, San Rafael, US
21. Nielsen J (1994) 10 Heuristics for user interface design. In: Nielsen Norman Group. https://www.nngroup.com/articles/ten-usability-heuristics/. Accessed 19 Jul 2019
22. Moriarty C (2013) Metro: Last Light Review - IGN. https://www.ign.com/articles/2013/05/13/metro-last-light-review. Accessed 24 Jul 2019
23. VanOrd K (2013) Metro: Last Light Review. In: GameSpot. https://www.gamespot.com/reviews/metro-last-light-review/1900-6408240/. Accessed 24 Jul 2019
24. Greimas AJ (1984) Semiótica Figurativa e Semiótica Plástica. Significação - Revista Brasileira de Semiótica 4

Age Ratings for Tabletop Games′ Usage in Brazil Analysis and Suggestion of New Criteria

João Léste and Claudia Mont'Alvão

Abstract Currently, all the criteria regarding the age rating regulations of tabletop games, both in Brazil—by the INMETRO (Brazilian National Institute of Metrology, Standardisation, and Industrial Quality)—and in the European Union—by the European Parliament and the Council of the European Union—, are based on security standards of product usage, not taking into account the cognitive and social skills involved in the task of playing games. Therefore, this article's main objectives are: to establish a theoretical frame of analysis, based on bibliographical research, in order to define which other criteria could be taken into account when determining the age ratings; and to analyze three different games to identify if the current standards applied to them are adequate. The main document used as a reference is the BNCC (National Common Curricular Basis), which specifies what skills are expected of children from each school year—and has already been used in similar works that correlate games with cognitive skills. Considering that this is an pioneer study on the subject, only mathematical skills will be taken into account, since they are easier to assess.

Keywords Tabletop games · Age rating · Usability · Learning processes

1 Introduction

In many countries, knowing relevant information about the use of any product is considered a primary consumer's right, as clearly stated in the Directive 2009/48/CE: "Warnings which determine the decision to purchase the toy, such as those specifying the minimum and maximum ages for users [...] shall appear on the consumer packaging or be otherwise clearly visible to the consumer before the purchase" [1].

J. Léste (✉) · C. Mont'Alvão
Pontifical Catholic University of Rio de Janeiro, Rio de Janeiro, RJ 22451-900, Brazil

C. Mont'Alvão
e-mail: cmontalvao@puc-rio.br

E. Duarte and C. Rosa (eds.), *Developments in Design Research and Practice*, Springer Series in Design and Innovation 17,
https://doi.org/10.1007/978-3-030-86596-2_12

In Brazil, this rating is regulated by the INMETRO (Brazilian National Institute of Metrology, Standardization, and Industrial Quality), which takes into account most—if not only—safety risks that these kinds of products might pose to children of certain ages [2]. Likewise, in the European Union, the regulatory standards are similar, focusing mainly on safety hazards associated with physical characteristics of game parts and components, such as flammability, risk of strangulation and radiation poisoning, chemical, and electrical properties, and hygiene levels [1].

Considering consumer safety should always be a top priority [3], those criteria are undoubtedly indispensable to any product. However, it is also essential to bear in mind that different tabletop games have different levels of complexity and thus will vary in target audiences. Therefore, to thoroughly address user needs and capacity to engage with the products being purchased, it is necessary take into consideration both the children's age and stages of cognitive and social development.

Given this scenario, this paper's main objective is to analyze different bibliographical references that indicate which cognitive criteria are essential in determining the types of games and gaming mechanics that are suited for each group of children. The method is composed of two parts: first, defining a theoretical frame for analysis; then, comparing games' current age ratings to the literature review.

For the first part, the research was based on documents and literature that could relate children's competences and cognitive skills, the different stages of learning [4], and the DIKW Pyramid's hierarchy of cognitive entities [5]. By cross analyzing these references, while applying them to the process of learning any given game's rules, it was possible to identify the different topics, hierarchies, and the different levels of learning those games require their users to display. Therefore, this method offers a new parameter with which to measure any game's complexity, which can also be correlated to the school year in which children generally acquire the abilities that render those games adequately playable.

For the second part, considering the concepts and definitions mentioned above, an analysis of three games, with distinct age ratings, was carried out to investigate if they had—by chance or not—been adequately categorized. The results point out the first attempt to analyze the current standardization system and to suggest new criteria to be taken into account when deciding any game's age rating of usage, especially in Brazil.

2 The Current Age Rating for Tabletop Games

When determining the age ratings for any toys and tabletop games, there are two factors to take into account, which are almost identical for Brazil and the European Union: firstly, any product—even if it is toy-like—can only be classified as a toy, and therefore marketed to children, if its age rating is set to a maximum of 13 years old. If it is targeted to an audience of 14+ years (in this case, classified as 'adults'), it is to be considered as a collectible item. Those items then should be classified as either a model in scale; regional or decorative dolls; historical reproductions of toys; or imitations of firearms [1].

That causes, sometimes, games to be set for a target audience of 14+ years—for undisclosed reasons—, but also recommended for younger ages. As an example, it can be mentioned the reprinting of the game “No Thanks!” [6], that was initially released in 2004 by the AMIGO publisher, and the original game’s box informs that the game is adequate for children from “ages eight and up”, which was adequate for the regulations at that time. However, the Brazilian reprint, released in March 2019 [7] still maintains this recommendation, but has a second age rating in another part of the box, that targets it to audiences of 14+—therefore classifying it as a collectible item, for adults, even if it does not fit the specifications. This fact can be a contributing factor for possible buyers, especially for parents of children between the ages of 8 and 13, who are most likely not aware of this situation and, thus, have no way of determining if the game is adequate for their children or not.

Secondly, when targeted to young kids, games must take into account criteria that are related to “essential safety requirements applicable to toys, including special safety requisites in a matter of physical and mechanical properties, flammability, chemical properties, electrical properties, hygiene, and radioactivity” [1]. The main point of concern, though, is that the directive is not always specific about the age range, it only describes some points of concern, as seen on the following extract: “In addition to the instructions provided for in the first subparagraph, chemical toys shall bear the following warning on their packaging: ‘Not suitable for children under (*) years. For use under adult supervision;’.” [1].

The * symbol indicates that this type of toy’s adequate age of usage is to be specified by the producer, which is later confirmed: “It shall also be indicated that the toy must be kept out of the reach of children under a certain age, which shall be specified by the manufacturer.” [1]. The presence of this passage without further specification raises some points of concern since there is no regulation as to how each manufacturer must determine the adequate age rating.

One of the few examples of an age range being denoted is: “The ability of the users and, where appropriate, their supervisors shall be taken into account, in particular, in the case of toys which are intended for use by children under 36 months or by other specified age groups.” [1]. This extract also highlights that cognitive skills and abilities should be taken into account, even if it does not further clarify how.

Comparatively, in Brazil, the standards for the games’ age ratings are similar to those already mentioned, since the last INMETRO’s Decree [2] specifies the same criteria as the ones in the Directive 2009/48/CE [1], with the addition of “noise.” It is also a more specific document since it explicitly states that, for instance, children under the age of 5 should not be in contact with toys that contain glass parts, since at this age they are not able to properly engage with this kind of materials without running the risk of endangering themselves. On the other hand, while it states that the toys’ required abilities should be taken into consideration, it only exemplifies physical and mechanical ones, not mentioning cognitive skills whatsoever.

A research question then comes to mind: are age indications in tabletop games adequate to children's abilities and limitations? Moreover, if not, how will children deal with the frustration of not accomplishing a game's objective?

We hypothesize that current tabletop games publishers do not follow adequate cognitive-oriented criteria and, therefore, the game's age rating is not adequate for children's learning and development skills.

3 Children's Expected Skills at Different Ages

When researching games that seem to take a measurable and identifiable cognitive criteria, the only ones found were the Black Stories game series. Each version of the game contains 50 cards depicting a small parts of stories that the players must uncover by asking yes-or-no questions. On the original game [8], all stories are about murders and, therefore it is marketed for audiences of 12+, in compliance with the Brazilian National Department of Justice's Practical Guide for Content Rating [9].

Even though most versions of the game follow the same thematic trend, there are some variations—Green Stories [10], Orange Stories [11], and White Stories [12]—that depict situations that are not connected to any deaths. Those games are marketed for ages of 8+, significantly lower than the more violent version. However, the Practical Guide does not seem to be the only criteria taken into account, since the other age categories are "Audiences of 10+ years old" and "General Audiences," so there has to have been another factor to determine that the Age Rating would be 8+, but not any other age under 12. The question, then, becomes: "which other criteria were considered?" However, that information does not seem to be verifiable.

One of the few examples of cognition being explicitly considered by governmental documents when discussing age ratings for toys was the CSPC Art and Craft Safety Guide, which states: "Children's behaviors and cognitive abilities may also influence their risk. For example, children under the age of 12 are less able to remember and follow complex steps for safety procedures and are more impulsive" [13].

Other sources, such as the Kids Health Organization, also reference boardgames as being adequate for "bigger kids" as the act of playing encourages social and cognitive development: "Card games like 'war' or 'crazy eights' […] teach about strategy, turn-taking, negotiating rules, and fair play. Encourage cooperation and help the child learn to manage the emotions that come with winning as well as losing." [14].

3.1 Literature Criteria for Skills and Game Requirements

Brazilian's "National Base for Common Curriculum" [15] was used as a reference guide to evaluate which abilities children should display according to their school years (and age estimates). This document contents range from learning objectives of each school segment and child rights regarding education, thus establishing the first frame of reference to what should be expected of each of those particular user groups.

The well-known game publisher Devir recently distributed, during academic events about games and education, two documents that discuss the concept of Game-Based Learning (GBL) and how some of the publisher's games can be used as tools to help children's development—in this case, regarding the BNCC.

The first document, entitled "BNCC and Tabletop Games" [16] mainly associates cognitive and social skills with games that help children practice and develop them, focusing on Elementary and Middle School years. Each page presents a specific skill, its description according to the BNCC, and a list of relevant games for that regard.

While it offers interesting inputs for the usage of the BNCC as a reference to determine which games require and help children display different skills, it is essential to notice that is separates children in groups between 6–11 and 11–15 years. Between the starting and ending years of each group, children will undoubtedly exhibit varying skills, therefore the document can only be used to determine a rough outline, based on the starting years of each stage.

As for the second, entitled "Neuroeducation and Tabletop Games" [17], it elucidates that are more in-depth explanations of how each BNCC skill can be trained by playing specific games. The booklet also takes into account kindergarteners and high schoolers, as well as the groups mentioned in the previous document, showing a table at the end that indicates which games are more suited to each of them.

However, there are mainly soft skills being discussed, and their progression is incredibly hard to assess in a theoretical frame of analysis since they are related to psychological and social evaluations of the children's behavior. The difficulty in measuring those abilities is the reason why, in this paper, the authors decided to focus on Mathematical skills, since they are easier to evaluate.

3.2 Correlating the BNCC and Game Requirements

Once Math was specified as the skill being evaluated, it became necessary to choose which three specific skills could be compared between school years to result in the criteria able to relate to children in different age groups adequately. The skills chosen are part of Arithmetics, in which all are related to natural numbers, and present a progression of complexity, as shown below:

Afterwards, three games that are popular in Brazil were chosen, each one requiring the abilities associated with one of the three grades in Table 1; "Dr. Eureka" [18] (1st grade); "No Thanks!" [6] (4th grade); Quartz [19] (6th grade). It was also important to make sure that they require neither skills adequate to a higher grade than the one being evaluated, as it probably will not be part of the children's repertoire.

Table 1 Correlation between BNCC math skills and skills required to play a game

	1st grade	4th grade	6th grade
Learning and development objectives (Math operations)	Utilize natural numbers as quantity and order indicators in different daily situations	Elaborate and solve problems with natural numbers that involve addition and subtraction, utilizing varying strategies, while being able to estimate results	Elaborate and solve problem that involve calculations (mental or written, exact of approximate) with natural numbers, through varying strategies, with or without a calculator
	Count object quantities in groups of up to 100 units and verbally produce an answer, in situations of the individual's interest, such as games and classroom content		
Skills required to play a game	Determine the winner in a point-based game that allows up to 100 points to be made	Be able to calculate additions and subtractions between numbers, estimating the game's results	Be able to calculate multiplications and divisions, on top of additions and subtractions, to determine which alternatives are the best

4 Detailing the Games' Complexity

Even if a game only requires age-appropriate mathematical skills, it is also possible for it to be too complex for children of that age. For instance, a game that requires children to understand concepts or principles that are too complex for their age will be considered age inappropriate, even if it only demands adequate mathematical skills. For this reason, in order to accurately measure the complexity of a game, in addition to the BNCC skills, two more theoretical frames were considered, the Wisdom Hierarchy [5] and Gagné's learning levels [4], to establish criteria to analyze the rules of games. The intention was to determine what topics of understanding the children should be able to grasp in order for the game to be rendered playable.

Those criteria as less measurable in specific school years, but should offer a complementary qualitative analysis to BNCC's determination of skills for each age group, pinpointing the rating either to the higher or lower age on the spectrum.

4.1 A Subsection Sample

The Wisdom Hierarchy, also known as DIKW Hierarchy, as defined by Rowley, is composed of 4 different levels: Data, Information, Knowledge, and Wisdom. Each of them represents a different stage of mental processing, in order of complexity (Fig. 1):

Fig. 1 The DIKW pyramid represented by Rowley [5]

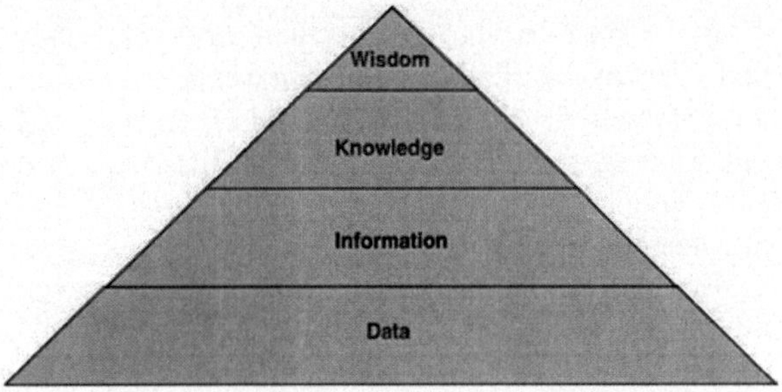

- "Data are defined as symbols that represent properties of objects, events and their environment. They are the products of observation [5]". When reading boardgame rulebooks, data would be the visual pigment of the page.
- "Information is contained in descriptions [...]. Information systems generate, store, retrieve and process data. Information is inferred from data [5]". On rulebooks, it is the meaning of the words and sentences written.
- "Knowledge is know-how, and is what makes possible the transformation of information into instructions. Knowledge can be obtained either by transmission from another who has it, by instruction, or by extracting it from experience [5]." In rulebooks, Knowledge is the answer to the question "How do you play this game?"
- "Wisdom is the ability to increase effectiveness. Wisdom adds value, which requires the mental function that we call judgement. The ethical and aesthetic values that this implies are inherent to the actor and are unique and personal [5]." In games, it is the strategy, i.e. the decision-making process of what to do and to what end.

Those four levels of mental processing can be compared to the higher levels of learning described by Gagné in his book "Conditions of Learning." The first five will not be considered in this paper since they are not consistently aligned with the DIKW Hierarchy. Therefore, only the last three will be considered since they are more indicative of complex learning processes: *Concepts*, *Principles*, and *Problem Solving*.

In this context, a *Concept* is the definition of a word, generally representing an action, object, or feeling. To grasp a *Concept*, a child would need to understand the meaning of the word that defines it, the characteristics of what is described, as well as differentiate it from similar or opposite words, e.g., for a child to understand the *Concept* of "circular" they would need to understand that it is a round shape, different from an oval. By this definition, *Concepts* are equivalent to the combination of *Data* and *Information* since they encompass the comprehension of the meaning and the characteristics of the objects they describe.

It is also important to note that some *Concepts* are composed of more than one concept, such as "risbee," that is a round object. Therefore, for a child to understand the concept of risbee, they would need to understand what is "round," what is "object," and to be able to differentiate it from a plate or shield.

As for the *Principle*, it is a situation, event or law, involving at least one *Concept*, that is always applicable to anything that it describes. To fully understand the *Principle*, the child needs to understand when it is applicable or not and to accurately predict if it is going to take place in different situations. Therefore, *Principles* are equivalent to *Knowledge*, since they represent the compilation of *Information* in order to determine general rules and how they might apply to the real world.

To grasp the *Principle* that "spherical objects will roll if you lightly kick them," the child would need to recognize that any spherical object, such as footballs, always roll in this situation, but also: that if lightly kick a box, it will not roll; if you forcefully kick an object, even if it is spherical, it will likely be thrown on the air. Lastly, when presented with the situation in practice, they should be able to predict whether the object will roll or not based on its shape. Similarly to the concepts, principles should also be divided between simple and higher ones, the latter being one that's based on two or more other Principles, such as: "even if you lightly kick a spherical object, it will not roll if it is glued to the floor."

Then, the final level of learning is the development of Problem Solving abilities, which is the application of concepts and principles to real-life situations in order to achieve a particular goal or complete a particular task. In this context, Problem Solving is similar to the definition of Wisdom and describes the process of analyzing the game's current stages, available options, and deciding which one is the adequate course of action to achieve the game's objective.

4.2 The Complexity Factors in Games

In order to contextualize the complexity factors in the games, it was necessary first to divide what the children must learn in three different categories: Game Objective (What must be achieved at the end), Tasks (How they can achieve the task); and Scoring System (How to determine if they achieved the task). Afterward, those three categories will be dissected in the concepts, principles, and problems that ought to be understood by the game players.

All those items were then applied to three different games, each of them requiring one of the skills displayed in Table 1. If the game is deemed real enough, the adequate age should be the lower one for the grade's general age range; if it is considered too complicated, for any reason, it will fall on the higher age from the age range.

4.3 Dr. Eureka

This game is about a scientist that needs help completing a series of experiments by organizing different substances in test tubes. It requires the following understandings (Fig. 2):

Fig. 2 Dr. Eureka simulated gameplay found on Blue Orange's official website, 2015 [18]

- Game Objective: Score the highest amount of points by collecting cards;
- Simple Task: move the balls between test tubes, without touching or dropping them, to match the picture on the card of the current round;
- Simple Scoring System (1st-grade skills)—each card is worth 1 point. Players need only count them and know which number is bigger to determine the winner;
- 4 Basic Concepts: Balls; Test Tubes; Card; Picture;
- 4 Advanced Concepts: Move, Drop and Touch (balls); Match (tubes and pictures)
- 2 Simple Principles: Balls have to be moved between test tubes to match the picture. Balls cannot be touched or dropped;
- 3 Higher Principles: The first one to match the picture scores the round's point. If a player touches or drop the balls, they cannot score a point during this round, even if they are the first one to match the picture. If all players touch or drop their balls, the round's point is forfeited;
- 2 Problems to be Solved: Check order, and between which tubes, the balls should be moved to match the picture?

Considering that this game's current age rating is 6+, it only requires BNCC skills from the 1st grade, elementary school level (usually 6 or 7 years old children), and since it does not require lots of previous knowledge (aside from what a scientist is) to be understood, its age rating is considered 'adequately determined.'

4.4 No Thanks!

This card game revolves around buying ad passing cards, in order to accumulate the least points, at the end of the game. It is primarily targeted to audiences of 8+ and requires the following understandings (Fig. 3):

- Game Objective: Score the lowest amount of points, by avoiding getting cards;

Fig. 3 No Thanks! Product display (retrieved from the game's Ludopedia profile), 2019 [7]

- Simple Tasks: Place a token on the open card and passing it to the next player (or) get the card with however many tokens are on it (even if 0);
- Moderately complex Scoring System (4th Grade): each card is worth the value written on it, but if any player has cards with consecutive numbers, they only score the lowest one. Add those points and then subtract the number of remaining tokens from it. The player with the lowest score wins;
- 2 Simple Concepts: Cards; Tokens;
- 4 Advanced Concepts: Get, Pass and Grouping (Cards); Natural Numbers (3–35);
- 4 Simple Principles: A player can either pass or get a card. If a player gets a card with tokens on top of it, they also get the tokens. Cards are worth the number written on them. Each token is worth −1 point, that will be subtracted off the final score;
- 4 Higher Principles: If a player gets a card with tokens on top of it, they also get the tokens. Getting cards hinders the player's chance of winning, initially. Getting cards with consecutive numbers is beneficial. Getting tokens is beneficial;
- 2 Problems to be solved: When is it worth it to get a card? Is it better to get this card to place a token on it?

This game's current age rating is 8+ but requires BNCC skills acquired during the 4th grade, elementary school (usually around 9 or 10 years old). It also demands a complex decision-making process to determine a tasks' worth, so its age rating seems to be inadequate, and should be changed to 10+, to be at a safer end.

4.5 *Quartz*

This game is about dwarfs that will mine gems for 5 days to accumulate the most wealth. It is targeted to audiences of 14+ and requires the following understandings (Fig. 4):

Fig. 4 Quartz box and components (took from author's personal copy of the game), 2020

- Game Objectives: be the richest dwarf after five workdays, by selling the most gems and getting coin;
- Tasks of moderate complexity: Mine, a gem at random; activate a card; leave the mine and collect the next reward; choose if you want to keep gems between rounds; if you choose to keep any, which gems to keep between the workdays (maximum of 2); activate which special effect when selling gems (if there's more than one available);
- Complex Scoring System (6th-grade skill): each precious (not the unstable) gem has a specific value according to its color, for which it can be sold. The player can also activate one of the following gem effects, per round: if 3 of the same gems are sold together, their value of another type of gem is doubled during the round; if 4 of the same gems are sold together, the value of 2 other types of gems are doubled for the round; if a group of 5 different gems are sold together, the player gains a bonus of 8 coins; if a group of 6 different gems are sold together, the player gains a bonus of 12 coins. Additionally, any card that remains unused is worth different values at the end (specified on each card), and any remaining "mining for dummies" book is worth 3 coins at the end;
- 5 Simple Concepts: Gems; Cards; Mine; Coins; Rewards;
- 13 Advanced Concepts: Precious, Unstable, Mine, Keep, Lose, Sell, Steam, Protect, Gain (Gems); Activate, Offensive, Defensive (Cards); Leave (Mine);
- 4 Simple Principles: During its turn, the player can choose between mining, activating a card of leaving the mine. If a player mines, they can either get a 'valuable' or an 'unstable' gem. A card can be either neutral, offensive or defensive. When all players (except one) leave the mine, the workday is over;
- 5 Higher Principles: If a player mines a second unstable gem, their cart explodes, forcing them out of the mine. If a player is forced out of the mine, the next bonus is discarded. If the player's cart explodes, they get a "mining for dummies" book, that can be used to protect against explosions, or be sold at the end of the game.

If a player activates an offensive card against another player, they can defend themselves (if they have an appropriate card). When all other players leave the mine, the last one is also forced out and does not get a bonus, but is the first to act on next workday;
- 5 Problems to be solved: Which of the actions should be chosen (each turn)? Should any gems be kept? If so, should 1 or 2 gems be kept? If so, which gems to choose? When selling gems, if it is possible to activate a gem effect, which one should be activated?

This game's current age rating is 14+ but it only requires BNCC skills acquired during the 6th grade, elementary school (usually around 11 or 12 years old), so the age rating seems to be higher than necessary.

On the other hand, the game requires many complex cognitive processes to be made than the other ones, demanding understandings of concepts and principles outside this group's daily repertoire. Therefore, the current age rating is considered to be "inadequate and should be changed to 12+" to be at the conservative end, since all safety regulations regarding the choking hazards that the gem miniatures might pose only concern children far too younger than 11 or 12 years old.

5 Conclusions, Takeaways and Next Steps

The theoretical framework proposed here set new criteria for analyzing tabletops game rules, relating them to the children's cognitive-oriented capabilities. It was the first step to analyze if the game's age rating is adequate for children's learning and development skills.

While the current age ratings on the three selected tabletop games seem to be somewhat adequate, since it takes into account important physical and mechanical properties, it lacks further clarification about some specific aspects. Only one of the games analyzed indicates an adequate age rating, suggesting that this exercise of categorization should be further applied to other games.

It may be worth noting that, differently from the safety regulations already in effect, the inadequacies of age ratings based on cognitive criteria pose considerably fewer and lighter consequences to children. Therefore, the age ratings could be given as a range of the appropriate age group, instead of a singular number. Having this range would allow parents and guardians to evaluate their children's capacities and in order to make a more accurate decision for their particular situation.

It is also essential to keep in mind that the available documentation from Devir (2016, 2018) indicates that the very act of playing the games may be a contributing factor to children's cognitive development, helping them cope with losing and not being able to perform certain skills, yet. Gagné also explains that contextualizing the learning objectives of students with a real-life objectives and interests will generally spark their interest—which means that the games may serve as an incentive for children to learn the mathematical skills required to play them.

As the next steps of this approach, it would be essential to examine if these criteria can be replicated on other countries and regions as well, using their laws and standards as guidelines. The same games should be compared to these new documents, in order to determine if the age ratings would still be the same. The results would indicate whether those age ranges vary from region to region, with two fundamental subjects being other European countries, inside and outside the European Union, and also the United States of America since they encompass the most massive tabletop games' marketplaces in the world.

Acknowledgements The authors thank the Coordination of Superior Level Staff Improvement—Brazil (CAPES)—Financing Code 001 for the support.

References

1. Council of the European Union, European Parliament (2009) Directive 2009/48/CE of the European council and the council. Official Journal of the European Union, Brussels
2. INMETRO (2016) MDIC: Portaria nº 563. Brazil
3. Senado Federal (2017) Código de Defesa do Consumidor, 2nd edn. Brasilia
4. Gagné R (1973) Como se realiza a aprendizagem. 1st edn. Ao Livro Técnico S.A., Rio de Janeiro, Brazil
5. Rowley J (2007) The Wisdom hierarchy: representations of the DIKW hierarchy. J Inf Sci 33(2):163–180
6. No Thanks! BGG profile. boardgamegeek.com/boardgame/12942. Last accessed 11 Nov 2019
7. No Thanks! Ludopedia profile, ludopedia.com.br/jogo/no-thanks. Last accessed 11 Nov 2019
8. Black Stories BGG profile. boardgamegeek.com/boardgame/18803. Last accessed 11 Nov 2019
9. National Department of Justice (2018) Content rating: practical guide. 3rd edn. Brasilia
10. Black Stories Junior Green stories BGG profile. boardgamegeek.com/boardgame/43043. Last accessed 13 May 2020
11. Black Stories Junior Orange stories BGG profile, boardgamegeek.com/boardgame/111304. Last accessed 13 May 2020
12. Black Stories Junior White stories BGG profile, boardgamegeek.com/boardgame/43171. Last accessed 13 May 2020
13. Consumer Product Safety Commission (CSPC) (2012) Art and craft safety guide. USA
14. Smart Toys. https://kidshealth.org/en/parents/smart-toys.html. Last accessed 14 May 2020
15. MEC (2017) Base Nacional Comum Curricular (BNCC). Brazil
16. Dutra JA (2018) BNCC e Jogos de Mesa. Devir Escolas, SP, Brazil
17. Sanjaume NG (2016) Neuroeducação e Jogos de Tabuleiro. Devir Escolas, SP, Brazil
18. Dr. Eureka Official Page blueorangegames.eu/pf/dr-eureka/. Last accessed 13 May 2020
19. Quartz BGG profile, boardgamegeek.com/boardgame/172881. Last accessed 13 May 2020

Design Delight: An Experiential Quality Framework

Omar Sosa-Tzec

Abstract Delight is a powerful emotion that can alter the perception of a product and its experience of use. Nevertheless, operationalizing delight for design practice and research is a complicated task. Derived from years of design analysis, *design delight* is presented as a combination of *engagement, surprise, liveliness, cuteness, serendipity,* and *reassurance.* The theoretical underpinnings, key aspects, and the six qualities of design delight are discussed. This framework posits that delight is a persuasive element of the user experience and that delightful products work as multimodal arguments that *argue by experience.* Consequently, design delight urges designers to be ethical when it comes to creating such products and to consider how these products can help users live a happy and flourishing life.

Keywords Design delight · Experiential quality · Design theory · Delightful user experience · Semiotics · Rhetoric · Aesthetics · Argument by experience

1 Introduction

Delight is broadly understood as a high degree of gratification or pleasure [1]. Many of us recognize the relevance of delight in people's lives as a result of experiencing it and how it marks events in our lives. For example, delight arises when an excited child opens a Christmas present and discovers her wish came true. As a positive emotion, delight has the power of altering our thinking and actions, and of becoming an instrument to regulate our negative emotions and pursuit of happiness [2]. Although it is possible to admit and recognize the presence of delight when it happens, the specificities of delight are difficult to describe for any person. Broadly, delight is understood as an instantaneous emotion related to joy and pleasure. However, in practice, designers need to specify what designing for delight is and how this directive differs from provoking other positive affects during the user experience.

O. Sosa-Tzec (✉)
San Francisco State University, San Francisco, CA 94132, USA
e-mail: omarsosa@sfsu.edu

E. Duarte and C. Rosa (eds.), *Developments in Design Research and Practice*,
Springer Series in Design and Innovation 17,
https://doi.org/10.1007/978-3-030-86596-2_13

This paper introduces an approach to designing for delight. Rather than scrutinizing the differences between delight and other positive emotions, this work proposes utilizing *design delight* as un umbrella term that comprises experiential qualities that are capable of creating delight—a significant moment involving high pleasure and arousal [3]. This formulation of design delight is concerned with six particular qualities, namely, *engagement, surprise, liveliness, cuteness, serendipity,* and *reassurance.* This work describes the theoretical underpinnings of this formulation, which connects with design theory, rhetoric, semiotics, and the aesthetics of interaction. Later, it presents design delight as a conceptual framework, elaborating on its key aspects and six experiential qualities. The paper closes with a brief discussion of some implications of design delight for design practice and scholarship, and an outline of next steps for its further development.

2 Delight as an Object of Study

The recognition of delight as an element of human experience and its effects have motivated academics to investigate the relationship between delight and the design of products and services. Particularly, marketing researchers have investigated the notion of delight for more than twenty years [4]. Researchers and practitioners of marketing alike realized that satisfying the customer was insufficient. Products and services should be delightful in order to become remarkable. Consequently, *customer delight* became a matter of concern for marketers. Customer delight means a higher level of satisfaction and involves a strong, positive, emotional reaction to a product or service [5, p. 314]. Marketers often associate such a reaction with a surprising or unexpected element in the experience. Pioneers of the study of delight in marketing, Oliver et al. [5] regard delight as a result of a discrepancy between what the customer expects the product to do and unexpected outcomes of the product. This formulation of delight relates to how a product and service produces a positive disconfirmation, meaning that the product or service turns out better than the customer expected. This formulation complies with the expectancy disconfirmation model [5]. Delight is a result of a positive disconfirmation of the customer's expectations in such a way that this disconfirmation activates an aroused state in the customer that is quite positive [4, p. 41]. Oliver et al. thus relate delight to encountering a pleasant surprise from the product or service [5, p. 136], providing a pleasing unexpected performance [5, p. 330]. This viewpoint, which emphasizes the relation between delight and surprise, has predominated the research on customer delight [4].

However, other marketing researchers investigate delight without the presence of surprise [6–8]. Kumar et al. [7] regard delight as a result of experiencing a higher degree of pleasure, which they refer to as joy. Kumar et al. [7] argue that joy comes in two kinds, "magic" joy and "real" joy. Magic joy is a result of a short-lived experience that involves the fulfilment of a wish or need that can change a person's situation. Real joy happens when an ongoing activity makes the person connect with some aspect of the world around her, whether this connection is mental or physical.

Dey et al. [8] argue that delight is about captivating the customer. Schneider and Bowen focus on the customer's needs, relating delight to security, justice, and self-esteem [4]. Ball and Barnes [9] broaden the notion of delight, arguing that gratitude has an impact in customer delight. Liu and Keh [10] suggest that delight without surprise is possible through novelty. Chitturri, Raghunanthan, and Mahajan regard delight as a result of meeting or exceeding the customer's desires as well as of fulfilling the customer's goals [11]. Chitturri et al. define two types of benefits that a customer gets from a product or service, namely utilitarian and hedonic. Delight occurs when the hedonic benefits promote cheerfulness and excitement [11]. Berman associates delight with events. Berman argues that delight happens as a result of fulfilling unexpected, valuable, memorable, and positive reproducible events [12, p. 132]. When surprise is disregarded, the major theme in these researchers' work is fulfillment. Delightful products or services are those not only introducing unexpected features but also capable of fulfilling the customer's needs.

Whether it involves surprise or not, delight has become a relevant matter in marketing because of its major implications in the experience of the customer. Products and services that are delightful generate a feeling of loyalty in the customer. Delight helps to create an emotional connection with products and services, and therefore, with their brands. Delight contributes to the creation of strong memories during the customer experience. Delight also works as a motivator for repurchase and promotion by word of mouth [4, 7, 11, 13–15]. Although delight plays a crucial role in customer experiences, researchers still find difficulties in its characterization [16]. This task becomes more complicated if the intention is to follow a design-oriented approach since design deals with the creation of particulars subjected not only to the circumstances of the design situation but the set of competences and expertise of the designer [17]. The idea of *designing delight* is unattainable as the design process is not only concerned with form but also with function, meaning, pertinence, and the effect of its principal outcome (hereafter referred to as the *design product*). Rather, I posit to draw on knowledge pertinent to delight that has been produced in other disciplines while considering the peculiarities of design practice to devise *design delight*: how design products engender significant moments of high pleasure and arousal during the experience of use.

3 Theoretical Underpinnings

3.1 Design as Meta-Communication

This investigation draws on Semiotic Engineering (SE), a theory of human–computer interaction (HCI) which regards the interactions between a computer system and the user as a meta-communication act in which the interface serves as the message [18]. SE defines the design of an interface as a composition of static, dynamic, and meta-linguistic signs. A successful design entails a correct interpretation of these signs.

SE is concerned with the user's semiosis, the production of meaning derived from this interpretation, which relies on abductive reasoning, the construction of plausible hypothesis from partial evidence provided by readily observable phenomena [19]. A major goal for the designer is therefore to devise the appropriate combination of signs that will lead to a consistent production of meanings for any user of a system.

The foundations of SE are crucial in this formulation of design delight. Instead of struggling with "adding more delight" to a design product, a designer can be thoughtful about the signs that not only lead to a consistent production of meaning but also have the potential for delight (as defined above), and ideally, show consistency in how they affect any user of a design product. Design delight connects meaning with high arousal and pleasure. A proper comprehension and interpretation of the static, dynamic, and meta-linguistic signs composing the design of a product are therefore necessary to experience delight. For design delight, not only does the stimuli lead to delight but also the outcome of the meaning-making process at a specific moment of the user experience.

3.2 Design as Argumentation

Design delight also draws on the notion of design as a reified argument developed by Buchanan [20, 21] and design as rhetoric by Ehses [22, 23]. These scholars connected rhetoric and design by considering that both are arts dealing with human affairs that urge taking action. The notion of *art* plays a fundamental role in design delight. From the Aristotelian perspective, an art deals with probabilities, not universal truths as science does [24]. Design delight considers designing for delight as an art. Buchanan's work accounts for how design produces lively, embodied arguments that shape people's everyday life. For Buchanan, design is a kind of rhetoric for this technological age. Such a rhetoric deals with symbols and images, physical artifacts, actions and activities, and environments and systems. Design delight applies to all of them.

Ehses situates design as a practice in service of human needs, suggesting an interesting connection between delight and fulfilling these needs. For Ehses, a visual communication design product is a composition based on signs that are intended to address a public audience being affected by a situation. Ehses builds on the intrinsic relationship between semiotics and rhetoric, regarded as the effect of these signs, to expand the notion of rhetoric and include any form of modern communication, including visual identities, navigation systems, exhibitions, websites, and fashion. This perspective underscores the connection of design delight with communication design products. Moreover, Ehses demonstrates the application of rhetorical theory for both generative and analytical purposes. Particularly, Ehses accounts for how the figures of speech, namely tropes and schemes, work efficiently for these tasks, including the creation and interpretation of *the concept* behind a design product. Design delight appears thus connected with these compositional patterns and the effect they might produce.

Buchanan and Ehses demonstrate how design is a rhetorical endeavor applicable to any media. Consequently, any design product ends up producing rhetoric (working argumentatively) regardless being visual, static, dynamic, tangible, or intangible. Design delight leverages this semiotic-rhetorical perspective of design to expand the use of the notion of sign from SE into any possible *multimodal composition,* that is, any design product. Design delight thus relates the delight that arises in an experience of use to the semantic and affective effects of the signs the designer selects to shape the multimodal patterns or features of a design product. Such a delight is a rhetorical force, an element of the user experience that can influence the behaviors, beliefs, and attitudes of the user.

3.3 Experiential Quality and Interaction Aesthetics

The concept of experiential quality gained traction in HCI as a result of a paradigmatic shift towards the notion of experience [25]. Researchers and scholars of HCI started paying attention to the experiential qualities that systems convey and how they work as elements of pleasurable experiences. An *experiential quality* is a property or trait that characterizes the experience of the user interacting with a design product. Examples of such qualities are pliability, rhythm, or fluency [26]. Formulating an experiential quality involves the understanding, critical perspective, and experiential knowledge of the one who recognizes it in a design product. Rather than offering a universal characterization of the trait, an experiential quality is formulated to serve as a conceptual tool for designers to appropriate in order to develop design judgment, and to elaborate or modify this formulation as a result of their own experience [27, p. 2].

Löwgren states that an experiential quality is neither a property of a design product itself nor a psychological or physiological property of the user. Rather, it is an effect of using the product in context [27]. A designer cannot design products with a specific experiential quality but can imagine the conditions that are conducive to the experiential quality of interest in the subsequent use of the product. Löwgren describes the goal of the designer in this regard is to increase the chances that the use of the product will be experienced in a certain way. A designer can utilize experiential qualities to formulate desirable directions for concept design in the early stages of the design process. They are also useful to identify promising candidates among a set of early design concepts [27].

Löwgren demonstrates how the formulation of experiential qualities contributes to the production of knowledge and discourse around interaction design practices, particularly, about *interaction aesthetics.* Löwgren proposes *interaction criticism* as the method for this formulation [26]. Bardzell defines interaction criticism as performing rigorous interpretive interrogations of the complex relationships between the interface and the user experience [28]. This humanistic method pays attention to the interface's material and perceptual qualities, and broader situatedness in visual languages and culture. Concerning the user experience, interaction criticism pays attention to meanings, behaviors, perceptions, affects, insights, and social sensibilities that arise in the context of interaction and its outcomes [28, p. 604]. The type of knowledge that interaction criticism produces is intermediate, meaning that the formulation of an experiential quality goes beyond the particularities of a design product but without reaching the level of a universal theory.

4 Derivation of Design Delight from Interpretive Analysis

Specifying what delight is and how it differs from other positive emotions is a daunting task. Delight manifests without notice and lasts briefly. As marketing research indicates, delight connects with surprise and unexpectedness, but not all the time. Captivation, novelty, and gratitude connect with delight as well. It is possible to note that these antecedents can be regarded as traits derived from usage besides perception, meaning that it is possible to frame antecedents of delight as experiential qualities. Instead of struggling with the impossible task of determining how to imbue delight in a product's composition, exploring qualities associated with delight is a reasonable alternative. Certainly, it is possible to frame delight as an experiential quality itself. However, this approach requires going back to the starting point in which the specificities of delight need definition.

The quality of surprise or unexpectedness represented the initial quality in this investigation as a result of a review on existing research on delight and pleasure produced in marketing, philosophy, design, and HCI. This review revealed serendipity as a trait closely connected with surprise and delight as well. After the identification of surprise and serendipity as potential experiential qualities to formulate design delight, this investigation focused on performing interaction criticism on products from different design genres and based on the theoretical underpinnings introduced above. This stage took about seven years. The objective was to gain a deeper understanding of delight based on existing products and to identify qualities in them that connect with this combined perspective involving design, rhetoric, semiotics, and pragmatic aesthetics [18, 21, 23, 26, 28]. Careful attention was paid to the appearance and behavior of the signs composing a product's design features and how these engender delight—a notion abstracted from the review—in different periods of the user experience. The abductive nature of interaction criticism led to settle a notion of delight as a remarkable moment of pleasure and arousal within an aesthetic experience. *Design delight* then emerged as a term more suitable for discussing delight from a perspective focused on human-made artifacts and their usage. Other qualities emerged. These qualities appeared significant for their frequency and the existence of research around them or related to some of their aspects. These qualities are *engagement, surprise, liveliness, cuteness, serendipity,* and *reassurance. Design delight* is the name given to the synthesis of the observations and reflections of this longitudinal analysis and that includes these six qualities.

5 Design Delight

5.1 Fundamentals

Design delight is concerned with six experiential qualities, namely *engagement, surprise, liveliness, cuteness, serendipity,* and *reassurance.* The purpose of design delight is to help a design practitioner or scholar account for how the composition of the features of a design product engender remarkable moments of pleasure and arousal within an experience of use. Design delight regards a design product as a set of features which are composed in turn by a set of functionally and semantically related signs. The features and signs shape the appearance and behavior of the product. Each sign is multimodal, meaning that it represents a conceptual composite of six elements: visual, verbal, aural, olfactory, tactile, and temporal elements. The first five elements represent the modes associated with the major human senses [29]. One or several of these elements might be sensed as more or less prominent, or even null, at all or specific times of the experience. The temporal element is the backdrop against which a sign's first five elements fluctuate at different moments of the user experience. Through this notion of a sign as a multimodal composite, design delight connects design semiotics with the field of multimodal argumentation [29], which in turn relates to rhetoric, and consequently, with the notion of design as a volitional art. Design delight therefore lies at the intersection of design, rhetoric, semiotics, multimodal argumentation, and aesthetics. A key notion in multimodal argumentation is *arguing by experience* [29], which design delight interprets as the designer's purposeful utilization of the visual, verbal, aural, oral, tactile, and temporal elements to propose and shape design features that can induce a behavioral, attitudinal, or belief change in the user through the delight provoked by these features. Particularly, design delight asks the designer to leverage engagement, liveliness, cuteness, serendipity, and reassurance to devise and construct this kind of argument.

Design delight focuses on the moments of the user experience that leave an emotional mark, acknowledging their potential for the creation of positive memories and motivation for the user to continue using the product. Design delight specifically cares about those moments that support a happy and flourishing life, referred to as a *good life.* Design delight is therefore about how design supports living a good life. Nevertheless, design delight acknowledges the gap between the designer's intent and the use of a product in context. What the designer intended to make the user feel might be different from what the user feels while using the product. There will be always a gap between the *intended design delight* and *the experienced design delight.* The ideal case is that such a gap does not exists. However, design delight embraces this gap, and acknowledges how the user's motivations and capabilities, experiential knowledge, current circumstances in the context of use, and the different ecologies of design products participating in the user's life can affect the user's semiotic and affective processes in each experience of use for a single product.

To devise the different ways a product can engender the six experiential qualities of design delight, the designer needs to understand not only the user but also how people in the context of use construe the notion of pleasure [30]. However, design delight acknowledges the value of the designer's first-hand experiences, and considers that a savvy designer knows how utilize their experiential knowledge to be empathetic with the user and have a sense of when when it would be a good time (*kairos*) for engendering the experiential qualities at different points of the user experience. As the agent in charge of constructing an effective multimodal argument, the designer's *design judgment* and *set of competences* [17] are crucial to determine why, how, and when the product's features would engender any of the six design delight qualities.

5.2 *The Six Elements of Design Delight*

Engagement takes place when a design product becomes the user's center of attention during its deliberative use. Engagement prepares the user to feel delight by creating immersion and minimizing external noise or stimuli in the context of use. Once engaged, the user can reach a state of *flow* as a result of a balanced combination between the challenges posited by the use of the product and the skills of the user [31]. Design delight considers engagement and flow closely connected. When engaged in using a product, and ideally in a state of flow, any instance of surprise, liveliness, cuteness, serendipity, and reassurance might be disruptive. Hence design delight urges the designer to the product be attentive to introduce features whose disruption is perceived as subtle or positive and that can efficiently help the user re-engage with the product or return to a state of flow. A product's engagement can be seen as its potential for captivation or attraction. However, engagement does not necessarily refer to introducing flamboyant features in order to captivate or attract the user. In design delight, the quality of engagement is mostly derived from smooth use of a product.

Surprise occurs when a feature of a product appears or behaves differently from what the user's mental model for that kind of product or similar feature dictates. The expectation inherent in this model is also a result of the formative experiences of the user with the real world, whether or not they involve using design products, which have led to the creation of *image-schematic metaphors* in the user's mind [32]. The boundary between engagement and surprise blurs when the main objective of surprise is to catch the user's attention while elevating the level of arousal in the user. Nevertheless, design delight urges the designer to utilize surprise as a differentiator that motivates the user to reflect and reconsider their current mental model and recognize how the product differs from others. The designer can consider surprise to mark starting and end points in intervals of the experience, incentivizing the user to carry on. Moreover, the designer can make use of surprise to convey some non-functional aspects that help set the rhythm of the experience [26], the product's liveliness and cuteness, or values associated with the product's brand.

Liveliness happens when a design feature conveys energy, autonomy, whim, excitement, and dynamism, either by form or behavior. Anthropomorphizing the product is one approach to liveliness. However, the designer can leverage socio-cultural conventions, including those on how pleasure operates in the context of use, to find ways to denote and connote these characteristics without the need to follow this approach. This quality complements surprise and engagement, especially when it helps the product come across alive, cognizant, intelligent, and having agency. Products involving a null or a minimal level of interactivity with the user heavily rely on features that come across swift and bold. Products involving a noticeable level of interactivity might also need to give the impression of a living creature at different points of the user experience in order to reinforce their liveliness. Either way, design delight connects a product's personality with its liveliness, and sees this connection as a factor that contributes to make the product distinct from others in the same category or that have similar features. Consequently, design delight urges the designer to leverage this connection, especially to make the user identify with the product and set the diegetic tone of the experience.

Cuteness takes place when a product's features come across not only pretty but also vulnerable, tender, innocent, harmless, or helpless. The objective of cuteness is to facilitate the user's desire to use the product and perceive it as a non-intrusive, non-judgmental, naïve companion that can help the user achieve a goal, fulfill a need, or even define their personality. A cute product leverages infantile appearance and behavior expecting to develop in the user an urge to take care and protect the product. A product's engagement and liveliness can affect or be affected by cuteness. Design delight urges the designer to be ethical concerning a product's cuteness as this quality can make the user be vulnerable and uncritical regarding the purpose and effects of the product. However, the designer can make use of this situation to induce change in the user.

Serendipity is a result of unsolicited content or functionality that produces a fortunate, grateful, valuable outcome for the user. Design delight urges the designer to leverage widely-known use patterns and image-schematic metaphors [32], as well as anecdotal evidence, to consider how the product's features can introduce such content or functionality without disturbing the user's state of flow. Moreover, design delight encourages the designer to envision serendipitous features that promote the repurposing of the product, meaning that such content or functionality enables the user to do things differently, better, or more efficiently. Design delight regards serendipity as a function of goal achievement and need fulfillment, connecting it with engagement, surprise, and reassurance. Goals can be pragmatic or hedonic. Needs can be personal or social, private or public. However, serendipity inspires the user to feel a sense of agency and that their actions are leading or can lead to a desirable future state not considered previously.

Reassurance manifests when one or several design features involved in one moment of the experience remove uncertainty, disorientation, ignorance, or anxiety in the user about events in the experience and their outcomes. Such events include the ones previous to this moment and those which could come next. The focus is on the current moment and its outcome. Reassurance is therefore concerned with how the user makes sense of their recent actions concerning the use of the product and the experience developed as a result. Design delight regards reassurance as a result of clarity, certainty, and direction, all of them directed towards provoking delight. The designer can leverage reassurance to mark milestones in the experience, expecting to re-engage the user, and hopefully, take them into a state of flow. Design delight urges the designer to utilize reassurance as a means to give closure to these milestones, especially to the end of the experience. After previous and opportune manifestations of the other qualities, engendering reassurance at the end of the experience can contribute to a positive evaluation of the product and the creation of pleasant memories around its use.

6 Discussion

Delight is a powerful emotion that can alter the perception of users about a design product and its experience of use. This papers introduces *design delight,* an approach to designing for delight, positing that this endeavor becomes more manageable when delight—a remarkable moment of pleasure and arousal within an experience of use—is regarded as a result of how the features of a design product engender one or a combination of six experiential qualities, namely *engagement, surprise, liveliness, cuteness, serendipity,* and *reassurance.* Design delight regards these features as compositions of functionally and semantically related signs that involve visual, verbal, aural, tactile, and olfactory elements over time. In this approach, this formulation is to be used as a critical vocabulary either to inspect or ideate how a product's features engender one or a combination of the six qualities. Within a moment of an experience, the boundaries among these qualities can become blurred. This approach acknowledges and embraces this fuzziness as a characteristic of the *experienced design delight.* Engagement, surprise, liveliness, cuteness, serendipity, and reassurance are not to be regarded as parameters of delight but as a conceptual framework to guide the *intended design delight* through design judgment and reflection [17], which would be informed by experiential knowledge and scholarship on the six qualities and aimed at assessing the potential for delight of a product based on its features.

Design delight seeks to bring attention to the rhetorical or argumentative character of delight and the six experiential qualities besides their aesthetic dimension. This framework urges the designer to consider the significance of this rhetorical dimension when inspecting or creating a design product. The intended design delight entails the construction of a multimodal argument [29] to affect the user's behaviors, attitudes, and beliefs. The experienced design delight carries a rhetorical dimension as well. The designer's persuasive intent can only define such a dimension partially. Unintended consequences, including some negative, might emerge from using the product. One possible unintended consequence is the appropriation of the product by the user for persuasive purposes [33]. The user might discover value in utilizing the product to induce change in themselves or others. The differences in how the designer and user might devise and utilize the product as a multimodal argument indicates a variable persuasive impact for the six qualities. The particular conditions of each experience of use affect this impact as well.

Although the complexity inherent in the intended design delight obstructs the designer's ability to foresee any unintended consequence or argumentative use of the product, they are still responsible for it. Therefore, adopting a critical perspective and becoming ethically aware are two desirable characteristics in a designer pursuing design delight. This requirement applies to design scholars investigating how existing products enable design delight as well. Both design practitioners and scholars must keep in mind that the intended design delight aims to help the user live a happy and flourishing life. Design delight rejects thoughtless deployment of products and consumerism. The critical perspective and ethical awareness included in the intended design delight is to influence design practice and scholarship and to promote outcomes demonstrating the connection between delight and living a good life. The experienced design delight attains its primary goal when it contributes to enabling a good life.

Design delight supports the notion of design as a practice focused on the creation of *ultimate particulars* [17]. Design delight also supports design criticism as a method to produce knowledge [26, 28]. Designers can utilize the six qualities described above to shape the *intended design delight* for new products. Moreover, designers can utilize them to inspect the *experienced design delight* of existing products. Both applications lead to a deeper understanding of design delight, including how these qualities help a designer *argue by experience*. Nevertheless, design delight can still benefit from empirical research to triangulate the observations that led to its formulation as well as to elicit other significant qualities from designers and users alike.

References

1. Definition of DELIGHT. https://www.merriam-webster.com/dictionary/delight. Accessed 6 Nov 2019
2. Fredrickson BL, Mancuso RA, Branigan C, Tugade MM (2000) The undoing effect of positive emotions. Motiv Emot 24(4):237–258. https://doi.org/10.1023/A:1010796329158
3. Russell JA (1980) A circumplex model of affect. J Pers Soc Psychol 39(6):1161–1178. https://doi.org/10.1037/h0077714
4. Alexander MW (2010) Customer delight: a review. Acad Mark Stud J Arden 14(1):39–53
5. Oliver RL, Rust RT, Varki S (1997) Customer delight: foundations, findings, and managerial insight. J Retail 73(3):311–336. https://doi.org/10.1016/S0022-4359(97)90021-X
6. Crotts JC, Magnini VP (2011) The customer delight construct: is surprise essential? Ann Tour Res 38(2):719–722. https://doi.org/10.1016/j.annals.2010.03.004
7. Kumar A, Olshavsky RW, King MF (2001) Exploring alternative antecedents of customer delight. J. Consum Satisfaction Dissatisfaction Complaining Behav 14:14–26
8. Dey S, Ghosh S, Datta B, Barai P (2017) A study on the antecedents and consequences of customer delight. Total Qual Manag Bus Excell 28(1–2):47–61. https://doi.org/10.1080/14783363.2015.1049146
9. Ball J, Barnes DC (2017) Delight and the grateful customer: beyond joy and surprise. J Serv Theory Pract 27(1):250–269. https://doi.org/10.1108/JSTP-01-2016-0013
10. Liu MW, Keh HT (2015) Consumer delight and outrage: scale development and validation. J Serv Theory Pract 25(6):680–699. https://doi.org/10.1108/JSTP-08-2014-0178
11. Chitturi R, Raghunathan R, Mahajan V (2008) Delight by design: the role of hedonic versus utilitarian benefits. J Mark 72(3):48–63

12. Berman B (2005) How to delight your customers. Calif Manage Rev 48(1):129–151. https://doi.org/10.2307/41166331
13. Arnold MJ, Reynolds KE, Ponder N, Lueg JE (2005) Customer delight in a retail context: investigating delightful and terrible shopping experiences. J Bus Res 58(8):1132–1145. https://doi.org/10.1016/j.jbusres.2004.01.006
14. Hsu JS-C, Lin T-C, Fu T-W, Hung Y-W (2015) The effect of unexpected features on app users' continuance intention. Electron Commer Res Appl 14(6):418–430. https://doi.org/10.1016/j.elerap.2015.06.003
15. Rust RT, Oliver RL (2000) Should we delight the customer? J Acad Mark Sci 28(1):86. https://doi.org/10.1177/0092070300281008
16. Kefalidou G, Maxwell D, Woods M, Sharples S, Makri S (2012) Is this 'Delight'? In: Proceedings of HCI, pp 1–4. Available: http://ewic.bcs.org/upload/pdf/ewic_hci12_uett_paper3.pdf. Accessed 31 Aug 2016 (Online)
17. Nelson HG, Stolterman E (2012) The design way: intentional change in an unpredictable world 2nd edn. MIT Press, Cambridge, Massachusestts; London, England
18. de Souza CS (2005) The semiotic engineering of human-computer interaction. MIT Press, Cambridge, Mass
19. de Souza CS, Leitão CF (2009) Semiotic engineering methods for scientific research in HCI. Morgan & Claypool, San Rafael, California
20. Buchanan R (1985) Declaration by design: rhetoric, argument, and demonstration in design practice. Des Issues 2(1):4–22. https://doi.org/10.2307/1511524
21. Buchanan R (2001) Design and the new rhetoric: productive arts in the philosophy of culture. Philos Rhetor 34(3):183–206
22. Ehses H (1984) Representing Macbeth: a case study in visual rhetoric. Des Issues 1(1):53–63. https://doi.org/10.2307/1511543
23. Ehses H (2008) Design on a rhetorical footing. Des Papers 6
24. Corbett EP, Roberts WR, Bywater I (1984) and others, The rhetoric and the poetics of Aristotle. The Modern Library, New York, NY
25. Harrison S, Tatar D, Sengers P (2007) The three paradigms of HCI. In: Alt. Chi. session at the SIGCHI conference on human factors in computing systems, San Jose, California, USA, pp 1–18. Available: http://people.cs.vt.edu/~srh/Downloads/HCIJournalTheThreeParadigmsofHCI.pdf. Accessed 15 Jul 2015 (Online)
26. Löwgren J (2009) Toward an articulation of interaction esthetics. New Rev Hypermedia Multimedia 15(2):129–146. https://doi.org/10.1080/13614560903117822
27. Löwgren J (2007) Fluency as an experiential quality in augmented spaces. Int J Des Taipei 1(3) (Online). Available: https://search.proquest.com/docview/921469412/abstract/CD9B7ADFBD84426EPQ/1
28. Bardzell J (2011) Interaction criticism: an introduction to the practice. Interact Comput 23(6):604–621
29. Groarke L (2015) Going multimodal: what is a mode of arguing and why does it matter? Argumentation 29(2):133–155. https://doi.org/10.1007/s10503-014-9336-0
30. Brown B, Juhlin O (2015) Enjoying machines. The MIT Press, Cambridge, Massachusetts
31. Csikszentmihalyi M (2009) Flow: the psychology of optimal experience. Harper [and] Row, Nachdr. New York
32. Hurtienne J (2017) How cognitive linguistics inspires HCI: image schemas and image-schematic metaphors. Int J Human-Comput Interact 33(1):1–20. https://doi.org/10.1080/10447318.2016.1232227
33. Sosa-Tzec O, Stolterman E, Siegel MA (2015) Gaza everywhere: exploring the applicability of a rhetorical lens in HCI. Aarhus Ser Human Centered Comput 1(1):4. https://doi.org/10.7146/aahcc.v1i1.21314

Social Smart Urban Environment: A Process that Needs to Be Disrupted to Improve the User Experience

Cristina Caramelo Gomes

Abstract This paper discusses the extent to which disruptive design can be an indisputable component of smart and inclusive solutions within the development of urban design in Lisbon peripheral dwelling areas. Driven by smart solutions to humanise the built environment, Lisbon urban centre design is changing and offers equipment and services, enhancing the people's appropriation of the urban space. Lisbon is growing as a smart place in its sustainable, economic, social and cultural dimensions whilst the dwelling areas, namely those peripheral adjoining, remain characterised by monofunctionally poor urban design solutions, where commuting possibilities (offered by the public and private transport) and proximity to employment (offered by the city centre) are the main assets. This model of planning the urban environment perpetuates itself in space and time, underestimating the development of technology, the sense of identity and belonging and, mostly significantly, overlooking a good experience for users. A new approach is urgent. It is argued that a user-oriented approach, attentive to the context, where creative solutions emerge as a contribution to a smart, sustainable and inclusive community and where the interactions between individuals and the environment, regardless of its physical or virtual features, contribute to a positive user experience.

Keywords Disruptive design · Lisbon peripheral urban areas · Smart solutions · User experience · User-centred design

1 From Design to Disruptive Design

> …design, stripped to its essence, can be defined as the human capacity to shape and make our environment in ways without precedent in nature, to serve our needs and give meaning to our lives [1, p. 5].

C. Caramelo Gomes (✉)
Lusíada University of Lisbon, R. da Junqueira 188-198, 1349-001 Lisboa, Portugal
e-mail: Cris_caramelo@netcabo.pt

E. Duarte and C. Rosa (eds.), *Developments in Design Research and Practice*, Springer Series in Design and Innovation 17,
https://doi.org/10.1007/978-3-030-86596-2_14

Academic and professional design settings establish the relationship between design and social sciences. The biunivocal relationship between design and social sciences overwhelms the functionality and/or aesthetical features of environments and products usually addressed by designers. Design has expanded its boundaries and has diverted the focus on objects, environments and communication to that onto processes, systems and services. This reality is extensively illustrated by theoretical frameworks and good practices supported and delivered by the academic research [2].

We read in the Cambridge Dictionary that design involves "make[ing] or draw[ing] plans for something… a drawing or set of drawings showing how a building or product is to be made and how it will work and look… the way in which something is planned and made". This formal definition highlights the focus on the planning, manufacture, functionality and aesthetic appearance of something regardless of the required and desirable interactions with the user.

The latest dissimilar areas of knowledge have shown a focus on the human being, developing solutions to improve human security, safety, comfort and wellbeing. Design is not excluded from this trend and the guidelines are established to encourage new fields of actuation, as well as new conceptual processes leading to concepts, such as those of user-centred design, interaction design and user experience design, among others. The choice of this trilogy has not been random; in fact, these concepts are complementary and based on similar processes. The user-centred design focuses on human heterogeneity comprehending its physical, sensory, functional and cognitive characteristics [3]. When one uses an artefact, several interactions are required. In accordance with the artefact's complexity, the required interaction can be physical and/or digital. The user's acceptance of that artefact will be closely linked to its functional performance and easiness of use. However, this rational understanding is not the key decision support. Human decisions are primarily triggered by emotional reactions. The user experience depends on time and spatial context. The way any individual perceives experience motivates emotional reactions which determine human behaviours, and, consequently, the acceptance or rejection of that artefact.

> The user experience is the totally of end-users' perceptions as they interact with a product or service. These perceptions include effectiveness (how good is the result?), efficiency (how fast or cheap is it?), emotional satisfaction (how good does it feel?), and the quality of the relationship with the entity that created the product or service (what expectations does it create for subsequent interactions?) [4, p.14].

The design of an artefact excels its physical and/or digital boundaries. It becomes the design of a context with a critical analysis of the social, cultural, economic, technical and environmental concerns; otherwise it will be the repetition of old paradigms with a new approach supported by technological development and uninformed processes.

The idea is not an update to raise the final solution for a functional performance or to guarantee an easy way of operation based on the user's technical illiteracy. The bottom line is to understand the extent to which the designed solution changes the human behaviour and contributes to a positive experience and the extent to which this experience can be conveyed into the human sense of identity, integration, comfort and wellbeing. In order to accomplish such a complex purpose a new conceptual methodology is required. From the context of its use, the function(s) to develop, the physical and (if available) digital interactions with the user, the artefact (understood as a product, communication system, built environment or services) must be conceived within a holistic approach that recurrently disrupts the ordinary status quo. A disruptive design concept contextualises, perceives and matures intricate problems to ensure sustainable solutions in their social, environmental and economic dimensions.

> Innovation, as a concept, does not embed any social and ecological value sets, nor does the "design thinking" approach to problem solving. Creating things that don't have a social, economic and environmental conscience makes no sense in the grand scheme of things [5, p.6, 7].

The disruptive design pursues a new approach to discuss the status quo in order "to make the old obsolete and the new possible, desirable and sustainable", and comprehends the following six concepts [5, p.5]:

- Everything is interconnected
- Change is constant
- The future is undefined
- All change must be sustainable
- Challenge is part of reward
- Change is interactive.

The method moves towards a holistic approach to achieve an impact balance in all the areas that the designed solution can reach. An analysis of the context in order to understand the problems that require solution, in view of the fact that the scenario is in constant economic, social and cultural change and deserve a sustainable intervention; moreover, a positive impact needs the collaboration of a multidisciplinary community which includes users as the ones that should benefit from the final result. In the end, the main goal is to offer a positive experience to all the people involved within the contextual scenario.

The disruptive design emerges as a requirement for user-centred design solutions targeting the improvement of the user experience. The aim is to conceive intended creative interferences into a pre-existed setting to boost a different outcome and create positive social change [6]. The disruptive design introduces a method developed into 3 phases: Mining—identifying and analysing the context-, exploring research approaches; Landscaping—identifying the principal elements in the context and mapping their interactions, relations, connections, impacts..; Building—exploring possible solutions raised by anterior stages, prototype, testing and repeating them [5]. As regards the peripheral urban areas, it is easy to identify a non-humanised solution. From the conceptual spatial solution to the detail of finishing materials, the urban design does not stimulate human dimensions (physical, functional, cognitive, sensorial, emotional, behavioural and social). A solution supported by these dimensions boosts a positive user experience, encouraging the sense of belonging, the will of permanence, the return and the identity of the place. A positive user experience impacts human sense of comfort and wellbeing [7].

In spite of the research in the area of knowledge of the built environment, spatial conceptual solutions, urban equipment, functionalities and finishing practices convey disrespect for the users' requirements and expectations. A new conceptual and practice design approach is urgent [8].

At the moment when the concept of smart cities invades important metropolitan areas, offering technical solutions that benefit the built environment with new equipment and services, when the exterior space is a scenario to improve human interactions, what is the contribution of the disruptive design to humanise the built environment? How can better user experiences from disruptive and creative solutions be conceived? Do the peripheral urban areas need a disruptive design? Is it possible to contribute to smart user centred environments through disruptive design?

2 Smart Understanding

The smart word emerges as a prefix to classify any artefact that shows some intelligence in its functional performance enabled by technology. Artefacts, occasionally more meaningful and often just updates of traditional ones, display a coherent and logic concept but no evidence of intelligence. The association of the smart expression to any artefact changes its symbolism and becomes a marketing parameter, where its advantages are emphasised and the contribution to a new way to develop a function or improve a user/community experience is (un)consciously forgotten to spread.

Retrieving information from the web for a consensual smart definition is an acronym which enunciates the standards to determine a set of goals, that lay significance on corporations, departments or sections; the acronyms' concept can be broadly applied and this can embrace the sustainability of cities, artefacts performance as well as the individual's professional and personal paths.

The significance of the Doran acronym is a rational way to understand the parameters required for a smart classification of any object, environment or system. As a matter of fact, despite the responsivity of an artefact to pattern uses and behaviours, it does show some improvement in its functioning, way of use or other parameters related to competition, and this improvement must be possible to measure. This is the essential parameter for smart objects and environments; otherwise they are just purposeless technology.

SMART is also an acronym defended by Doran [9] (Fig. 1).

The concept and rational use of this acronym to assess the smartness of each artefact is consistent with the process of thinking the disruptive design. It is not because an artefact has some technology or is trend in the digital market that it is considered a smart one. It must be focused on the experience of the user, responding to human requirements and desires, offering a good functional performance and a contribution for a sustainable environment.

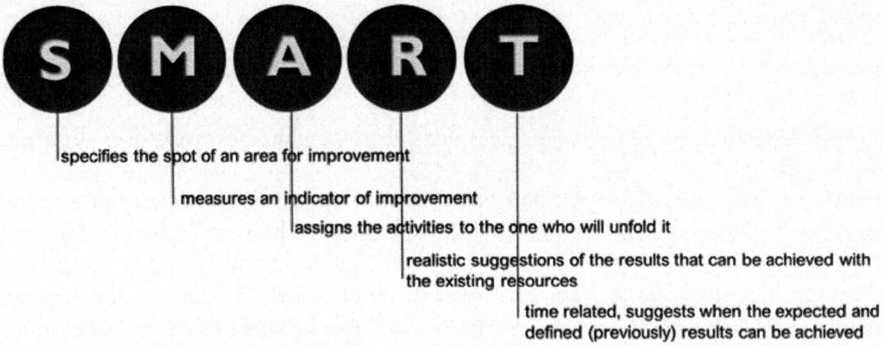

Fig. 1 SMART acronym from Doran [9]

3 Smart Solutions for Urban Environment

Since ever, and most significantly since the Industrial Revolution, technology has existed to improve the development of the built environment. The contribution of technology for the urban environment allowed an expansion of the human economic sovereignty, attracting people and, thus, depopulating the hinterland where survival resources were scarce. Technology has influenced the way people have lived and interacted, solving problems whilst new ones emerged.

The improvement of the urban environment impacted the quality of life of people, modifying the city within a continuous process of demographic gentrification, bringing people from distant geographical locations and expanding the territory to the farthest areas (social displacement), a phenomenon with an impact of the train and car (public and private) transportation with consequences. This reality has raised dissimilarities between geographical regions and has questioned the standards of human life. The expansion of cities has encouraged the construction of metropolitan areas, creating exhausting lifestyles for individuals. Services prevail in the city centres whereas the dwelling areas are mainly suburban. The distance between the city centre and suburban areas, covered by public and private transportation, entails wasting hours in commuting travel whilst it promotes unsustainable lifestyles [10].

The development of new technologies in the transition of the twentieth century to the twenty-first has promised a revolution in the way people live and interact. The emergence of the ICT (Information and communication Technologies) has offered new prospects of interacting and exchanging information, in other words, one step further to commuting people. Research (from various fields of knowledge) has pursued a desirable geographic balance, according to demographic and services distribution, reinventing the city centre and the suburban identity. Nevertheless, the gap between the academy and reality has imposed the use of technology to perpetuate existing mistakes.

The biunivocal dependence of city centres and suburban areas underpins the emergence of metropolitan centred planning on commuting, transportation and real estate speculation regardless of the user and possible interactions between the individual and the built environment, as well as the interactions between individuals within the built environment.

Literature that does not discuss its physical or virtual source, exposes different approaches and dissimilarities to the definition of the concept of smart urban environments. The first steps of smart urban environments highlighted the importance of the ICT and its impact on the daily professional and personal human routines and mostly in our way to communicate and interact with other beings, objects, and systems. The latest advances in the concept have shifted its focus toward the enhancement of the quality of life of users [11]. Thus, we privilege the definition of Thuzar (2011 apud Berardi, 2015:6) to sustain the purpose of this piece of research:

> Smart cities are cities that have a high quality of life; those that pursue sustainable economic development through investments in human and social capital, and traditional and modern communications infrastructure (transport and information communication technology) … should also be sustainable, converging economic, social, and environmental goals. (Thuzar 2011; apud Berardi 2015:6).

The urban environment needs integrated interaction, interconnection and interdependence to reward the smart label. A smart city is a place "connected, intelligent, innovative and adaptative", where smart people imagine and design smart solutions for the routines of their urban daily life.

The smart city respects the sense of place. The place's identity built up on its cultural heritage offers features that make it unique and appealing to its inhabitants and visitors (regardless of work, entertainment and tourism purposes). The initiatives in smart cities ought to be contextualized in space and time to respond to (g)local real problems and satisfy the demands of users, and not spread technology and applications based on technology development, global trends and public and/or private interests.

The smart city must be shaped by experts as well as by ordinary people: from the people who have the knowledge to develop, implement and test solutions in close collaboration with those who have to cope with their problems on a daily basis. Co-design is urgent, and the users' participation is the key. In the end, a more humanized city will emerge, which in practice means the observation of realities, identification of different uses (and inherent requirements and desires), conceptualization of solutions and implementation with close measurement results.

Beyond technology, interactions with the physical world are crucial and depend on how the built environment supports the multidisciplinary dimensions of functions and users that describe it. People, regardless of the function performed in the city, must be the most significant providers of information as well as the ones that understand the improvement in quality of life throughout the solutions implemented. For a city to be smart, it must be conceived within a human-oriented design approach.

Therefore, the aim of a smart city is to encourage and improve the quality of living of its citizens, supported by smart technology. Although the concept of the quality of life is not easy or consistent in its definition, literature review offers different definitions from the most generalist to those oriented to the problems of individuals, such as disease, while others use the term as an alternative to the concept of wellbeing. In the end, dissimilar definitions offered by several authors are not too far from Maslow's hierarchy of needs [12]; however, and because of this closeness, we choose the definition proposed by the WHOQOL Group: "…the individuals' perception of their position in life in the context of the culture and value systems in which they live, and in relation to their goals, expectations, standards and concerns" [13].

The users are the ones who benefit from a smart urban environment, shaping it all over continuous interactions. A user-centred approach adds new ways of living encouraged by the ICT and reveals the label of creativity associated with a smart urban environment. This association emerges from the education, learning and knowledge encouraged by new ways of living and interacting, thus promoting people's connections and relationships [14]. As Winters points out, smart people generate and benefit from the social capital of the city, so the smart city concept acquires a new meaning with a mix of education/training, culture/arts, and business/commerce with hybrid social, cultural, and economic enterprises [15].

4 Scenario

4.1 Lisbon

The city of Lisbon belongs to the community of European smart cities, targeting the use of the ICT as a way to improve the quality of life of its inhabitants. According to municipality documents [16], the goal of the city of Lisbon city to lure people to the city throughout energy-efficient constructions, e-mobility, smart living and smart citizens solutions, with particular attention to the elderly, a significant portion of the local demography. Social cohesion and inclusion are enhanced throughout the agendas for civic participation, rehabilitation of the urban space and the boost of efficient urban services and systems. As a result, Lisbon received The European Green Award 2020, a city with 550,000 inhabitants and nearly 40,000 daily commuters [17]. The development and use of smart and sustainable services, grounded on real-time information—gathered and shared—are smoothly accepted by users with technical familiarities.

Despite the positive impact the ICT had on users' daily routines, urban answers to ageing, accessibility and inclusion towards the pleasure of the urban spatial experience are the key input for the physical and digital interactions and, consequently, to guarantee a positive and significant user experience. The changes offered by the document on the urban environment focus on the intention of giving the urban space back to people. Lisbon smart features, such as energy efficiency, construction processes and materials, e-mobility are up to date structures that contribute to more sustainable ways of living; nevertheless, they do not disrupt the traditional procedures in a way perceived by users.

Users understand the need of more sustainable solutions although most individuals continue to perform daily routines in a traditional manner. The access to services, regardless of the offered remote solutions, continue to offer (and often require) the presential status (for workers and for customers). Nevertheless, this recognised pattern is always the way to accept and use technology that means to mimic the known reality.

Despite the presented paradox between outputs and constraints, reality shows some change in the behaviours of users. Disruption is more effective with solutions that enhance the humanisation of urban space. E-mobility demands the construction of bicycle lanes at the expense of traffic lanes and parking lots—curiously, more due to leisure than mobility requirements, people start to adhere to new ways of experiencing the city. Accessible and inclusive solutions, mostly in the busiest arteries, give more flexibility of use to everyone that presents physical, functional or sensorial limitations. There is evidence of growing living spaces added by green and/or relaxation areas, cafes and commercial spots. Some neighbourhoods economically and socially weakened, are in complex rehabilitation processes: ranging from building typologies, construction process and functions, cultural expressions, and offering new dynamics; thus, attracting more and more diversified patterns of people.

These interventions in the urban tissue are in line with the principles established by the *Carta Estratégica de Lisboa* 2010–2024 (Lisbon Strategic Document 2010–2024) [18] which is based on three main axis: agreement on the public policies about the city, centralisation of Lisbon as a world embracing capital city, and Lisbon as a cosmopolitan and neighbourhoods city.

Despite the theoretical and generalist text of the document, the time spent between policy making, planning and building, the changes shown in Lisbon's urban space convey good practices that must be continued and discussed in order to preserve a more humanized city, with respect for its identity (Fig. 2).

Lisbon city is completely different from a decade ago; yet it preserves the features that establish its own identity. Nevertheless, disruption that boosts a humanised and living city demanded also a long and intense period of gentrification.

Fig. 2 Examples of the humanisation of the urban space throughout the creation of green areas and leisure spaces. Campo das Cebolas and Largo do Intendente (Author 2020)

4.2 *Lisbon Peripheral Urban Areas*

Lisbon metropolitan area is composed of 18 municipalities, 9 on each bank of the river. Municipalities near the Tagus and the sea have the highest concentration of people, social and cultural equipment, and services. The north bank of the river is the area with the strongest economic, social and cultural context and development.

Some municipalities, due to their proximity to the city centre and/or relationship with the river, feature higher real estate values, attracting individuals from the upper classes, with higher purchasing power. These areas also display better public and private transportation and social equipment; sometimes, construction density is lower and dwelling units offer more generous areas and more qualified finishing. Living in these areas is grounded upon the identity of the place, the social and the economic status of the residents and organisations that offer services and work opportunities (Fig. 3).

The municipalities located at the eastern side of Lisbon and the ones located on the south bank of the Tagus river are the ones that, due to their characteristics, higher construction density, rational dwelling living areas, fewer services available, distance to the city centre, private and public transportation and the interior character of some of these communities, encourage lower pressure in real estate speculation, allowing and/or demanding people from lower classes.

Regardless of the possible generosity of dwelling areas, service availability and local accessibilities, there are some patterns in all the municipalities that constitute Lisbon Metropolitan Area.

Suburban areas that comprehend Lisbon Metropolitan Area are mainly the dwelling areas, known as dormitories, with a busy weekend life and quiet times during the weekdays; these areas were conceived in accordance with real estate speculation and private transportation. Green spaces are almost non-existent or an unpleasant environment, supported by some bush beds with randomly distributed benches without any shade element, regardless of their solar orientation. The benches display the same colour as the bushes, confusing most inattentive passers-by (Fig. 4).

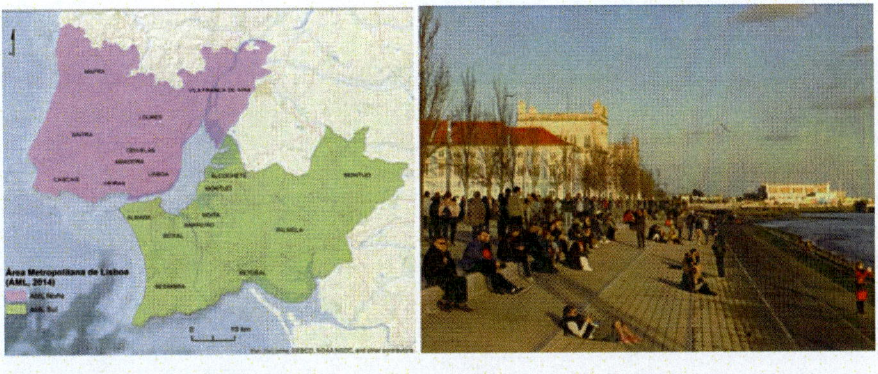

Fig. 3 Lisbon metropolitan area (DGT 2014) and Lisbon river side (Author 2020)

Fig. 4 Suburban peripheral areas, roundabouts with green area and sculpture, public space for users with small benches without shadow, green and leisure area (Author 2020)

Curiously, there are green areas, fountains and sculptures in roundabouts, some of them showing interesting art styles, but the users are unable to enjoy them; only the drivers can see them, but they must be attentive to the traffic and comply with driving rules.

Some communities, particularly those with good accesses to the highway, show considerable investment by technical and branding organisations, offering a pale reality of employment. These areas convey the investment spent to improve the image of the building, but the urban space has been neglected, as there are no green structures or entertainment areas for people, car parking is insufficient and paid, and traffic becomes particularly intense during the mornings and evenings. The traditional model has been applied without any previous strategy to overcome its weaknesses.

5 Conclusions

City planning involves a multidisciplinary cluster which addresses the urban and social tissue in accordance of their knowledge and objectives, although it converges to an optimised solution. The complexity of city planning requires (or is the traditional model that, regardless of technological advances, is still a common practice) time for observation/analysis of contextual problems, construction and implementation of solutions, which very often reveals obsolete and/or questionable results in face of contemporary dynamics. Planning decisions depend mainly on political and economic decisions, recurrently influenced by external models displaced from the spatial context and users' needs. Therefore, the implemented solutions, although well intended, expose inefficiency as they disregard conceptual considerations. Design as a discipline and the designer as a professional are crucial elements for a cluster committed to city planning. From the thinking process to a methodology towards a user-oriented solution, the design footprint seems to be required and needed more than ever. Design can go further than just providing good solutions grounded on the user and repeating worn-out models; it is time to innovate and raise solutions that respond to users as well as to planet Earth, and that is why disruptive design is needed.

The technical developments and the need of sustainable solutions promote the smartness of urban environment. Cities, more than spatial layouts, are places built up throughout social and cultural interactions, where people perform dissimilar roles searching for the improvement of quality of life in its all dimensions. A smart city is a place that is responsive to human/users' needs and wishes, independent from physical and digital responses. Technology is a facilitator to encourage collaborative actions among individuals and between individuals and organisations (regardless of their nature,) towards the improvement of the quality of life of the community. Individuals need the sense of belonging, and this is strongly connected with the place and neighbourhood relationships. The sense of belonging depends on the place's identity and the answers of physical or technical structures to real problems, within a real urban and social context. It is also enabled with the user's participation in the urban environment decisions at a community and municipality level.

Research projects promoted by the public and private organisations, as well as by the European Union, have contributed to the conceptualisation and implementation of the smart city. Service delivery, where the physical and the digital world combine to improve the efficiency of traditional activities (as well as to promote new ones), promotes new ways of human behaviour every day.

However, while the city centres of the main cities are transformed into a smart and humanised environment, peripheral areas are still developed according to traditional models where the smart character is mostly the use of technology, the humanization of the urban environment and the enhancement of a sense of belonging are forgotten. Real estate development and speculation are the major goals, commuting movements are the key and the spatial layout is designed to support traditional mobility instead of users or sustainable requirements.

The sample scenario illustrates the transformation of the big cities, and the dissimilarities between them and their suburban peripheries. A new model is required; a model aware of the dynamics stimulated by technology, with a deep insight in the context, to support broader and interrelated solutions centred on user needs and expectations, while contributing for the built environment sustainability. To conceive and implement this model, disruptive design emerges as a concept and methodology of action. City planners must be more conscious of contextual problems and base their decisions on their expertise and on the users' feedback, too. For that, public participation must be encouraged, helped and valued. Implemented solutions have a direct impact on human daily routines and individual behaviours, and users deserve a positive experience. This new model is complex, grounded in dissimilar challenges, some of them with well-known outlines. It is, however, worth the risk, if only because the traditional model is exhausted and has been proven ineffective.

To answer the questions introduced within the first section, it is possible to state that:

Disruptive design contributes to the humanization of a built environment once it has a methodology centred on the user, and the solutions conceived must impact the social, cultural, environmental and economic sustainability of the built environment. Focused on the user and aware of the context, the solutions have all the required parameters to improve the user experience. Planning cities requires a new methodological approach, particularly to pursue the improvement of the smart concept on a city's daily livings and particularly on their suburban perspectives. Knowing that a smart city is beyond the collection and the use of technological artefacts and user competencies, disruptive design is a facilitator of the conception, creation, implementation and assessment of smart solutions in (sub)urban communities.

References

1. Heskett J (2005) Design: a very short introduction. Oxford University Press
2. Frascara J (ed) (2007) Preface. In: Design and the social sciences: making connections (contemporary trends institute book 2), Kindle edn. Amazon Digital Services LLC
3. Interaction Design Foundation User Centred Design. https://www.interaction-design.org/literature/topics/user-centered-design. Last accessed 10 Nov 2019
4. Kuniavsky M (2010) Smart things: ubiquitous computing user experience design. Morgan Kaufmann
5. Acaroglu L (2017) Disruptive design method handbook (E-Book). Disrupt Design LLC. https://www.creativityandchange.ie/wp-content/uploads/2017/06/Disruptive-Design-Method-Handbook-by-Leyla-Acaroglu-Jan-17.pdf. Last accessed 15 Nov 2019
6. Kerr I, Frasca J (2019) SPURSE: innovation design workbook: disruptive innovation guide. Independently published
7. Natoli J (2015) Think first: My no-nonsense approach to creating successful products, memorable user experiences + very happy customers. BookBaby
8. Moser C (2014) Architecture 3.0: the disruptive design practice. Routledge
9. Doran GT (1981) There's a S.M.A.R.T. way to write management's goals and objectives. Manage Rev 70(11):35–36
10. Caramelo Gomes C (2004) A conceptual model to introduce telework in Lisbon. Universidade Lusíada Editora Lisbon
11. Albino V, Berardi U, Dangelico RM (2015) Smart cities: definitions, dimensions, performance, and initiatives. J Urban Technol 22(1):3–21
12. Maslow AH (1998) Toward a psychology of being, 3rd edn
13. WHOQOL Group (1998) Development of the World Health Organization WHOQOL-BREF quality of life assessment. Psychol Med 28:551–558
14. Alawadhi S et al (2012) Building understanding of smart city initiatives. Lect Notes Comput Sci 7443:40–53
15. Winters JV (2011) Why are smart cities growing? Who moves and who stays. J Reg Sci 51(2):253–270
16. Sharing Cities (2016) Building smart cities together—common solutions for shared challenges. Online. http://www.sharingcities.eu/sharingcities/smartcities
17. Smart Cities Brussels (2018) Smart City Lisbon. Online. https://smartcity.brussels/news-598-smart-city-lisbon. Last accessed 5 Nov 2019
18. CML (2009) Câmara Municipal de Lisboa: Carta Estratégica de Lisboa 2010–2024—um compromisso para o futuro da cidade. Online. https://www.lisboa.pt/fileadmin/municipio/camara/documentos/Carta_Estrategica_Lisboa_2010_2024.pdf. Accessed 10 Nov 2019

Design for Health and Wellbeing

The Future of Design for Health and Wellbeing

Louise Kiernan, Ana Correia de Barros, Teresa Cotrim, and Paul Chamberlain

Abstract Healthcare is experiencing a rapid transformation largely due to the challenges posed by rising costs, inequalities in health provision, pandemics, aging populations and the rise in non-communicable diseases. Design is now playing an important role in addressing these challenges, by applying a user centred approach to transform healthcare along with the development of new products, services systems and spaces. This special issue first describes current advances within healthcare and the role of design in the future of health care. It then introduces three papers that address design for health and wellbeing. The first paper describes design probes to configure sensory conditions to reduce pain in hospitals. The second paper, explores the use of participatory design in the co-designing of resources for knowledge-based self-reflection for people living with Parkinson's disease to enable independent living. The third paper describes a multidisciplinary research project concerned with the development and prototyping of a training tool for neurosurgery. This special issue highlights the varied benefits that design can bring to create better experiences for all stakeholder within the healthcare model. It advocates for Design to continue to be a major part of this transformation in healthcare.

Keywords Design · Design for health and wellbeing · User centred design · Participatory design · Co-design

L. Kiernan (✉)
University of Limerick, Limerick, Ireland
e-mail: lousie.kiernan@ul.ie

A. C. de Barros
Fraunhofer Portugal AICOS, Porto, Portugal

T. Cotrim
Technical University of Lisbon, Lisbon, Portugal

P. Chamberlain
SmithKline Beecham Pharmaceuticals, Harlow, UK

E. Duarte and C. Rosa (eds.), *Developments in Design Research and Practice*, Springer Series in Design and Innovation 17,
https://doi.org/10.1007/978-3-030-86596-2_15

With ageing populations, more people living with chronic conditions, Covid 19, further threats from pandemics and stretched funding, there are huge global challenges for our future healthcare. Healthcare transformation requires a shared vision between various stakeholders to create patient-centered models [6]. Design has become the focus in solving challenges in health care, and is leading to the development of new products, services systems and spaces. These can range from medical devices for surgery, wearable, smartphone and sensor based products to monitor health, digital and visual tools to support patients, systems that can be navigated smoothly, spaces that create an improved experience to virtual visits with clinicians and online pharmacies. Design is about improving people's lives. Design influences behaviour to adopt positive health care choices. It seeks to understand and incorporate all viewpoints to arrive at solutions to complex challenges.

1 Innovation in Health and Wellbeing

Healthcare has traditionally focused on diagnosis and treatment. The focus is shifting from health care to health and well-being and from treatment to prevention. There will be a greater emphasises on the promotion of healthy lifestyles, fitness, and wellbeing, prevention and early diagnosis [32]. Meskó [22] argues that the digital revolution will create an equal partnership between patients and their healthcare providers. By being able to monitor their own health at home, patients will feel empowered to take responsibility and decision making around their health. Care models will need to change the focus from acute to preventative care and well-being, empower consumers and patients to manage their own health, move towards holistic and personalized medical care and broaden the definition of health to include mental and spiritual health [12].

The severity and impact of Covid 19 has been a catalyst to prompt governments and health care providers, to respond innovatively and there is now an opportunity for design to play its part in addressing some of these challenges [12]. Future care solutions will include telehealth, remote patient monitoring, and technology enabled ways of diagnosing, monitoring, and treating patients [31]. Consumers have increasingly been placed at the centre of healthcare decision making and expect on-demand, and connected clinician-patient interactions [4]. They are more informed and showing a greater interest in managing their own health and are driving the demand for innovation in health-related products, services, and tools to track their health [12].

These innovative design solutions create an opportunity for "connected health," where individuals can collect data, remotely from a healthcare setting and then report this information to clinicians. The benefits are that the distributed gathering and monitoring of patient-generated health data (PGHD) can free up health care centres [20]. Wearable technologies such as heart rate sensors, exercise trackers and oximeters are also seeing a surge in growth. Wearable devices can provide up-to-date monitoring of high-risk patients preventing the occurrence of a major health event [29]. The wearable medical device market is expected to reach more than US$ 27 million by 2023, from just under US$ 8 million in 2017 [28]. New digital tools will continue to facilitate remote monitoring and adherence to medications allowing users to remain at home [5]. While most of these, measure vital signs, future avenues of development are in cognitive and mental health and ingestible sensors to communicate with other wearable devices to monitor the body's internals [29].

The impact of Covid 19 has also accelerated how readily consumers have adapted to digital technologies [12]. Consumers are also engaging in virtual consultations and ordering prescriptions online. Virtual visits rose from 15% in 2019 to 28% in April 2020 [13]. Video calls, phone calls, texts, and emails became an essential component in care delivery due to the pandemic to enable clinicians and patients to stay connected. Similarly, telehealth, tele-pharmacy, and virtual-hospital-at-home programs are on the rise [21]. While virtual visits are desirable for consumers, when surveyed many found it limited the quality of the interaction with the clinician pointing to the need for training tools in building virtual relationships for clinicians [12].

The need for upskilling and training of health care staff is increasing to provide clinicians with the training tools to practice at the heights of their profession [12]. This means that along with the design of medical devices, designers may also be involved in the development and advancement of simulated systems and environments for training. The overall philosophy of simulation is to develop skills for professionals and increased patient safety. The human factors aspects and training needs required for the safe and effective use of medical devices can be elucidated within the simulation environment [25]. Moreover, simulation has the potential to recreate scenarios that are rarely experienced and test professionals in challenging situations. Where there is limited access to clinical settings or limited exposure to patients who are experiencing low-frequency and high-risk situations, students might not receive the experience necessary to become confident and competent [1]. Simulation also aims to reduce both lead times and costs in the production of existing medical devices [10].

Furthermore, augmented reality (AR) and Virtual Reality (VR) are being applied to personalize medicine and create behavioural change [12]. For example, VR technology is being used to treat pain, as well as conditions ranging from anxiety and post-traumatic stress disorder and stroke to the training of surgeons and other clinicians [28]. Possible future uses of VR and AR are in training tools for medical students, calming environments for patients and the application of gamification to support patients in rehabilitation [16]. Artificial intelligence (AI) has also radically transformed health care by personalising medicine. Robots to assist nurses such as 'Moxi', Chatbots and virtual health assistants are examples of AI-based technology. AI is also being applied in precision medicine, medical imaging, drug discovery, and genomics. For example, AI can analyse multiple pathology images of different cancers to provide an accurate diagnoses and specify the best anti-cancer drug combinations to create personalised care. The AI market for health care is expected to exceed US$34 billion by 2025 (https://www.businesswire.com/news/home/20180827005149/en/) (ibid).

3D printing has revolutionised advances in health care. Current uses, include patient-specific implants, cutting guides, prosthesis and anatomical models [9]. The advantages of 3D printing is that it allows for instant bespoke rapid fix solutions to clinical problems, such as [23] where 3D printing was used to create a bespoke repair of a Percutaneous Endoscopic Gastrostomy (PEG) tube in a patient unfit for surgical replacement. Other examples are one off braces or dental implants. 3D printed medical devices can now be built to the exact specifications of a patient and patient-specific dimensions have shown greater acceptance by the body, with increased comfort and performance outcomes [11]. The capabilities of 3D printing is likely to be increased and emerging areas are in drug manufacturing and stem cell development to create organic substances such as skin and organoids [29].

Point of care testing (POCT) and diagnostics is another emerging area. Traditional models involve the transfer of a sample for testing to a lab and waiting several days for results. POCT facilitates testing the patient at multiple locations with on the spot results. The technology leverages biosensors and labs-on a-chip to integrate multiple laboratory functions into a single compact circuit that can perform diagnostics on very small samples [29].

2 The Role of Design in the Future of Health Care

Whilst we are witnessing an increasing interest in the role of design, set within the context of health there is a long history of the utilisation of design in the development of medical devices. The global market for medical devices was US$425.5 Billion in 2018 and expected to rise to US$612.7 Billion by 2025 (Insights). In building on this role, clinicians are now calling for design to help with every aspect of the healthcare system from designing medical devices to tackling pandemics, the layout of operating theatres and medical charts [14]. Designers can provide essential skill sets to healthcare. Park [24] lists five unique skills that designers can bring to healthcare; (1) problem solving and the ability to deal with ambiguity to structure problems, (2) an ability to communicate which is essential to understand the needs of others and to communicate solutions, (3) empathy and the skills to step into the shoes of those who may be anxious or suffering from chronic illness, (4) an ability to co-create with users and stakeholders and know when and where to involve them in the process and (5) creativity in challenging conventional solutions with blue sky ideas. While design alone is not the solution to all of healthcare's challenges, applying the methods of design can help make the delivery of healthcare more efficient and empathetic.

The shift in outlook towards viewing healthcare as a point of treatment for the physically unwell to the inclusion of the holistic wellbeing of the mind, body and spirit will demand a redesign of services that include the needs of consumers [12]. Fry [15] advocate that co-creation and multidisciplinary teams are essential in the redesign of healthcare services and that healthcare providers would benefit from using an iterative, user-centered and holistic approach that considers the patient experience. They state that a service design process through co-creation can challenge the hierarchy and silo-mentality that is ingrained in many healthcare organisations. This includes virtual visits, remote monitoring, prescription delivery, digital diagnostics and decision support [8].

Addressing some of the inequalities and inefficiencies of health care may also be achieved by questioning the norm of specific locations such as hospitals and GP clinics as entry points to health care. Opportunities exist for designers to consider alternative entry points. This may be achieved by targeting communities to address the drivers of health and connecting communities to a variety of health and educational services [5]. For example a more distributed healthcare service could include free-standing emergency care clinics [27].

While the future of Design within health care is promising, there are many challenges to implementing solutions. Bhavani et al. [6] state that there is a lack of evidence of whether many innovative solutions that have been implemented within healthcare actually improve outcomes and the quality of care. In response to this, the American College of Cardiology (ACC) convened a Healthcare Innovation Summit to understand the needs of various stakeholders across healthcare, including patient advocacy groups, clinicians, researchers, entrepreneurs, and industry groups. The outcome of the summit was a health policy statement that intends to guide healthcare policies, programs and innovations. The objectives are to promote patient-centric innovations that have measures and evidence of their impact on health, access, equity, costs, and outcomes (ibid). As part of a multidisciplinary team designers have a duty of care to ensure that solutions are serving the needs of the people they wish to address.

3 In This Special Issue

This special issue comprises three papers that address design for health and wellbeing. The first paper, 'Pain[off]: using Design Probes to configure sensory conditions to reduce pain in hospitals', describes how in healthcare, the contexts in which patients and caregivers find themselves in, is often complex regarding activities taking place, number and roles of people involved, number and types of objects in the environment, as well as in types of interactions. Gambera, Riccò and Duarte proposed to address a prevalent issue in these scenarios: body pain and ways to soothe it. Contrary to what is usually found in the literature on pain management, they did so through a designerly approach which embraced the challenge of considering synaesthesis rather than individual senses alone. The authors created a probe in the form of a room in which participants who were familiar with Design methods could manipulate a number of sensory stimuli until they reached the combination that felt the best to them in order to relieve pain. The authors discuss the challenges of this designerly approach, lessons learned and future work with users who are representative of a population affected by pain issues.

The second paper, 'Co-designing resources for knowledge-based self-reflection for people living with Parkinson's disease to better enable independent living' explores the use of participatory design in the Co-designing of resources for knowledge-based self-reflection for people living with Parkinson's disease to better enable independent living. Participatory designs are widely regarded as a positive way to develop and implement organizational interventions, as well as to develop products and solutions, entailing the engagement of different stakeholders and communication among them, in several activity sectors [2, 7, 26]. Healthcare is a complex and dynamic setting, where professionals and patients interact in different pathways, not always in a coordinated and complementary way. In participatory change processes, handling the interests of different stakeholders is one of the main challenges. Another one is being able to implement adequate tools regarding a holistic approach (Ibid). The paper presents an empirical study focused on Parkinson's disease, using a participatory design approach. Based on the identification of the different stakeholders needs and the problems users faced in using the services, facilitating mutual learning between participants and generating solutions. The reflexive processes lead to the integration of solutions in a final outcome: the Home Based Care Pathway. The project outcome evidences the benefits of participatory design in healthcare. It contributes to empowering patients and understanding the patients' journey to understand their needs. Finally, it shows how participatory design enhances professionals and patients to address and solve common problems.

The third paper, 'Neurosurgery Training Tool. Design as facilitator between disciplines for the improvement of medical devices' describes a multidisciplinary research project concerned with the development of a prototype training system that simulates the spinal column. The iterative development of applied design research utilises state of the art 3D printing, moulding and casting processes to simulate different material qualities of human bone and tissue. The paper highlights the increasing logistical and ethical problem of using cadavers and body parts to support learning and training of surgical techniques and the increasing need and advantages of simulated approaches. The paper argues for Design to be placed at the centre of a multi-disciplinary research environment combining both medical and technological disciplines where the adoption of User/Human-Centred Design methods can be useful in integrating information from different disciplinary perspectives. The case study also demonstrates the value of Design in creating physical tools acting as a common language to aid communication between stakeholders involved and impacted by the research. With increasing intolerance of error and the ever-increasing cost of 'medical procedures' the creation of simulated systems for testing and training are likely to be an increasing focus of research.

References

1. Aggarwal R, Mytton OT, Derbrew M, Hananel D, Heydenburg M, Issenberg B, Macaulay C, Mancini ME, Morimoto T, Soper N (2010) Training and simulation for patient safety. BMJ Qual Saf 19:i34–i43

2. Ajslev JZ, Wåhlin-Jacobsen CD, Brandt M, Møller JL, Andersen LL (2020) Losing face from engagement–an overlooked risk in the implementation of participatory organisational health and safety initiatives in the construction industry. Constr Manag Econ 38:824–839
3. Bank W (2019) Current health expenditure. Available: https://data.worldbank.org/indicator/SH.XPD.CHEX.GD.ZS
4. Barton JL, Kunneman M, Hargraves I, Leblanc A, Brito JP, Scholl I, Montori VM (2020) Envisioning shared decision making: a reflection for the next decade. SAGE Publications, Los Angeles, CA
5. Betts D, Korenda L, Giuliani S (2020) Are consumers already living the future of health? Key trends in agency, virtual health, remote monitoring, and data-sharing
6. Bhavnani SP, Parakh K, Atreja A, Druz R, Graham GN, Hayek SS, Krumholz HM, Maddox TM, Majmudar MD, Rumsfeld JS (2017) 2017 Roadmap for innovation—ACC health policy statement on healthcare transformation in the era of digital health, big data, and precision health: a report of the American College of Cardiology task force on health policy statements and systems of care. J Am Coll Cardiol 70:2696–2718
7. Burgess-Limerick R (2018) Participatory ergonomics: evidence and implementation lessons. Appl Ergon 68:289–293
8. Burill SB, Kroll D, Wheeler A, Sowar T, James (2020) Implications of the COVID-19 crisis for the health care ecosystem - gearing up for the next normal. Available: https://www2.deloitte.com/content/dam/Deloitte/us/Documents/life-sciences-health-care/us-implications-of-the-covid-19-crisis-for-the-health-care-ecosystem.pdf
9. Chae MP, Rozen WM, McMenamin PG, Findlay MW, Spychal RT, Hunter-Smith DJ (2015) Emerging applications of bedside 3D printing in plastic surgery. Front Surg 2:25
10. Ciurana J (2014) Designing, prototyping and manufacturing medical devices: an overview. Int J Comput Integr Manuf 27:901–918
11. Clinic C (2018) Cleveland clinic unveils top 10 medical innovations for 2019. Newsroom [Online]. Available: https://newsroom.clevelandclinic.org/2018/10/24/cleveland-clinic-unveils-top-10-medical-innovations-for-2019/
12. Deloitte (2021) Global health care outlook—accelerating industry change
13. Deloitte (2021) Global health care outlook laying a foundation for the future
14. Fairs M (2020). Lack of design input in healthcare is putting both patients and doctors at risk, says physician. Available: https://www.dezeen.com/2020/04/21/design-input-healthcare-risk/. Accessed 30 April 2020
15. Fry KR (2019) Why hospitals need service design. In: Service design and service thinking in healthcare and hospital management. Springer
16. Futurist TM (2020) 5 ways medical virtual reality is already changing healthcare. Available: https://medicalfuturist.com/5-ways-medical-vr-is-changing-healthcare/
17. Insights FB Medical device market size, share and industry analysis by type. Medical device/Medical device markets [Online]. Available: https://www.fortunebusinessinsights.com/industry-reports/medical-devices-market-100085. Accessed 18 April 2021
18. Jakovljevic M, Jakab M, Gerdtham U, McDaid D, Ogura S, Varavikova E, Merrick J, Adany R, Okunade A, Getzen TE (2019) Comparative financing analysis and political economy of noncommunicable diseases. J Med Econ 22:722–727
19. Jakovljevic M, Timofeyev Y, Ranabhat CL, Fernandes PO, Teixeira JP, Rancic N, Reshetnikov V (2020) Real GDP growth rates and healthcare spending–comparison between the G7 and the EM7 countries. Glob Health 16:1–13
20. Lavallee DC, Lee JR, Austin E, Bloch R, Lawrence SO, Mccall D, Munson SA, Nery-Hurwit MB Amtmann D (2020) mHealth and patient generated health data: stakeholder perspectives on opportunities and barriers for transforming healthcare. Mhealth 6
21. Mehendale RRJ (2020) Welcome to the virtual age: industrial 5.0 is changing the future of work. Deloitte Health Forward blog [Online]
22. Meskó B (2014) The guide to the future of medicine: technology and the human touch. Webicina Kft

23. O'Sullivan KJ, O'Sullivan AG, Power N, Gillick J, Dunne CP, O'Sullivan L, Linnane B (2018) Use of 3D printing to create a bespoke repair of a Percutaneous Endoscopic Gastrostomy (PEG) tube in patient unfit for surgical replacement. BMJ Innov 4:29–31
24. Park T (2020) Why healthcare needs designers. The health Care Blog [Online]. Available from: https://thehealthcareblog.com/blog/2020/01/28/why-healthcare-needs-designers/. Accessed 30 Mar 2021
25. Power D, O'Donovan K, Deasy C, Henn P (2020) Simulation test: can medical devices pass? BMJ Simul Technol Enhanced Learn 6
26. Punnett L, Nobrega S, Zhang Y, Rice S, Gore R, Kurowski A (2020) Safety and Health through Integrated, Facilitated Teams (SHIFT): stepped-wedge protocol for prospective, mixed-methods evaluation of the Healthy Workplace Participatory Program. BMC Public Health 20:1–14
27. Ratanjee V (2020) Five forces that will reshape the future of healthcare. Workplace [Online]
28. Reddy M (2021) Digital transformation in healthcare in 2021: 7 key trends. Available: https://www.digitalauthority.me/resources/state-of-digital-transformation-healthcare/
29. School of Business TGWU (2019) Which innovations are revolutionizing healthcare today? Online Healthcare MBA [Online]. Available from: https://healthcaremba.gwu.edu/blog/which-innovations-are-revolutionizing-healthcare-today/
30. Tavakoli M, Carriere J, Torabi A (2020) Robotics, smart wearable technologies, and autonomous intelligent systems for healthcare during the COVID-19 pandemic: An analysis of the state of the art and future vision. Adv Intell Syst 2:2000071
31. Taylor K, Bhatti S, Ferris K (2020) The future unmasked: Life sciences and health care predictions 2025
32. Taylor K, Properzi F, Bhatti S, Ferris K (2020) Digital transformation, shaping the future of European healthcare

Pain[off]. A Synaesthetic Design Probe is Used to Configure Sensory Conditions to Reduce Pain in Hospitals

Davide Antonio Gambera, Dina Riccò, and Emília Duarte

Abstract Hospitals are considered chaotic and stressful places. In particular, the presence of unpleasant stimuli in the environment (e.g., blinding lights, electronic noises, chemical smells) is considered one of the greatest causes of stress. All these aggressive stimuli, belonging to different sensory modalities, are processed simultaneously by the central nervous system. As a consequence, the concomitant presence of various unpleasant stimuli in healthcare environments has a negative impact on people health. This problem can be solved adopting a synaesthetic design approach, a holistic way for coordinating/manipulating all the multiple stimuli triggered by the elements present in the environment. In order to understand which types of sensory conditions could be more suitable to manipulate, we decided to run a synaesthetic design experience with design students aiming to collect data about which kinds of sensory conditions they would prefer while experiencing pain. Since pain has a more personal and intimate component, we decided to set the experimental environment as a design probe to attain the best empathic engagement. Participants could manipulate a set of stimuli (i.e., light, chromatic light, sound, vibration, texture, material density, temperature) to define sensory-based scenarios for reducing pain. The analysis of the data shows an intersubjective correspondence in associating multiple sensory stimuli in the environment to alleviate a possible condition of pain, showing a common tendency in associating the stimuli with predictable patterns. The most preferred conditions reported were: warm white light combined with a blue chromatic light, the sound of waves at a volume of 21/40%, low relief textures, high soft material density and a warm temperature of 32°. The unique exception was the not

D. A. Gambera (✉) · E. Duarte
IADE - Faculdade de Design, Tecnologia e Comunicação, Universidade Europeia, Av. D. Carlos I, 4, 1200-649 Lisboa, Portugal
e-mail: davide.gambera@universidadeeuropeia.pt

UNIDCOM/IADE - Unidade de Investigação em Design e Comunicação, Av. D. Carlos I, 4, 1200-649 Lisboa, Portugal

D. Riccò
Department of Design, School of Design, Politecnico Di Milano, via Durando 10, 20158 Milano, Italy

E. Duarte and C. Rosa (eds.), *Developments in Design Research and Practice*, Springer Series in Design and Innovation 17,
https://doi.org/10.1007/978-3-030-86596-2_16

acceptance for the TENS stimulation, case that suggests further studies with people affected by chronic pain conditions or under controlled induced pain.

Keywords Synaesthetic design · Healthcare design · Pain management

1 Introduction

1.1 Hospitals' Sensory Dimension is Still a Problem

In 1946 the World Health Organization published the WHO's Constitution.

In its first preamble health is defined as *"a state of complete physical, mental and social wellbeing and not merely the absence of disease or infirmity"* [24]. This definition should be a mantra for all those professional, including designers, that are somehow involved with healthcare systems. However, especially in healthcare design, the reality not always reflects this holistic approach. In most cases, we see that hospitals are designed around the pathology, focusing mostly on the causes that generate diseases [10]. As a consequence, the design choices are oriented to a maximization of the functionality and the minimization of the risks of infections [22]. This focus on the causes that generate disease is typical of the pathogenic approach: a successful strategy for containing the contagion during epidemics and for reducing the number of deaths for hospital-acquired infections.

Nevertheless, the pathogenic approach of healthcare design together with the emphasis on functional aspects, generated environments that are often considered stressful, uncomfortable, chaotic, painful, un-human, by a great part of the population [20]. Among the greatest causes of stress, we found that the poor sensory condition of hospitals is one of the most frequent issues. Hospitals are a source of aggressive sensory stimuli for their users: a typical hospital room is saturated by the massive presence in the environments of unpleasant stimuli such as: blinding fluorescent lights, incessant electronic noises, the smell of chemical substances, uncomfortable temperatures, etc. Moreover, all these stimuli become even more unpleasant when people are sick: in the case of illness, people sensibility to stimuli increases strongly. For example, when people have a fever, their sensibility to the temperature increases; when they have migraines their sensibility to light and sound increases so they prefer to stay in dark and silent spaces. In the same way, when people experience nausea, they are more sensitive to smells and tastes [16].

Those are some of the main important reason why the sensory dimension healthcare facilities are an urgent topic for designers that is not yet fully explored.

1.2 From a Reductionist to an Holistic Approach

Several attempts have been done aiming to solve problems regarding the sensory problems in healthcare environment, mainly for guaranteeing calm and privacy to patients and enhance their healing process. Ulrich's literature review on evidence-based healthcare design reports interesting studies regarding the improvement of patients' outcomes, such as as pain, sleep, stress, depression, length of stay, spatial orientation, privacy, communication, social support, and patients' satisfaction [22].

Great relevance has studies regarding the exposition of patients to nature, found to enhance the recovery, reduce patients' length of stay and alleviate pain [2, 21] . Other studies suggest a strong effect of lighting conditions on patients' sleep and circadian rhythms [3, 12].

In most cases the proposed solutions are mono-medial, they are focused on a single modality (sense) such as the reduction of the level of noises in a post-surgical division. To accomplish this goal, designers used to remove the uncomfortable stimuli from the environment with the use of soundproof materials for noises, dimmable lamps for lighting, etc. In this way, negative stimuli have been gradually removed from the environment, but they have not been replaced with positive ones. As a result, healthcare environments frequently are found empty, sterile, cold, boring and, in general, monotonous, because of the sensory deprivation.

Focusing on a single phenomenon is typical of reductionist approaches. In the case of sensory perceptions this approach might be insufficient because:

- There could be other concomitant stimuli that might interfere with the designed solution;
- Monomodal approaches do not consider cross-sensory interaction existing between senses of different modalities.

Subtracting stimuli from the environment could be defined as an "anaesthetic approach" (from the Greek *a*: without, *aísthēsis*: sensations). On the contrary, the Synaesthetic design approach (from the Greek *syn*: together, *aísthēsis*: sensations: literally perceiving together) proposes a model in which several positive sensations are introduced at the same time in the same space, trying to achieve the optimal configuration between the different modalities.

1.3 A Synaesthetic Design Approach for Healthcare Environments

How many times did we feel the taste of a food changing because of the presence in the environment of a terrible smell?

The importance of considering several modalities at the same time is given by the fact that human perception is never directed under a specific stimulus. As human beings we are multi-medial most of the times, only a few times we perceive with only one modality [17]. In fact, even when we deal with monomodal stimuli, what participates in the experience is not only the sensory modality directly involved (as for example taste) but also other senses (in this case smell), trying to give us more information about the surrounding environment [1].

Among the various phenomena of cross-sensory interaction we highlight here the one known as Synaesthesia, defined by Richard Cytowic as the "union of the senses" [7].

Nowadays we know that "synaesthesia" corresponds to the general name for a great set of over 80 recognized cognitive processes [6] in which the action of a sense also modifies the perception of the other sensorial systems that have been at the same time triggered by it [18]. In perception is it possible to identify common structural qualities that interest all the senses: for example, there's a vision of touch or a vision of hearing. "The hand watches, but not like eyes, and the eye touches but not like hands [...] All the senses participate in the observation" [19]. As observed by Maurice Merleau-Ponty in Phénoménologie de la perception, "Nor are these even exceptional phenomena. Synaesthetic perception is the rule" of human perception and we use synaesthetic association in our daily life" [14].

In Design these arguments have been well addressed in Synaesthetic design,—an expression defined for the first time by Anceschi et Riccò in 2000, a holistic approach and wants to focus on the whole system, rather than on the single part [1]. It has the goal of achieving the optimal configuration of objects (or environments) based upon the systematic connection between the different sensory modalities [11].

This approach is typical of a specific field of research in communication [1] and automotive design [11]. In our research, we are using a synaesthetic design approach as a strategy to improve the sensory stimuli in healthcare environments, aiming to have a positive impact on people health. In particular, the design of sensory stimuli in the environment could eventually reduce the sensation of pain of people in hospital [8]. Starting from this assumption, the hypothesis of this research is that the sensory stimuli in the environment, if adequately designed, can eventually produce analgesic effects, reducing the sensation of pain.

1.4 From a Reductionist Towards a Holistic Synaesthetic Approach

The most recent discoveries regarding processes of pain, open several possibilities for the work of designers and in particular for synaesthetic design interventions. In specific, the study of specific chronic conditions as phantom limb pain demonstrated that pain has a greater component at the level of the cortex. The analysis of particular pain chronic conditions (i.e. phantom limb pain or fibromyalgia) and the publication of the Neuromatrix Theory [13] presented a model in which pain experience could be modulated by the proceeding of sensory inputs, cognitive events, psychological stress, genetic influences) with the use of sensory counterirritation, distraction, and motivation as a strategies to alleviate pain without any pharmacological intervention [4, 5, 9, 15].

We intend to use a Synaesthetic Design approach to design concomitant stimuli in the environment able to reduce the sensation of pain. In fact, phenomena of cross-sensory interaction are found to be particularly effective with the sensation of pain.

This approach could be particularly efficient because, as designers, we know that we cannot intervene directly on the origin, but we could interfere with pain through the use of cross-sensory interactions. By assuming pain as the dependent variable of the present study, we intend to use a Synaesthetic Design approach to design concomitant stimuli in the environment able to reduce the sensation of pain in hospitals, a specific context of intervention in which the sensory stimuli are not often well designed.

In order to understand which types of sensory conditions we should manipulate in hospital, we decided to run a Synaesthtetic Design Workshop with design students in Design untitled: "Pain/off: imagining a painless hospital". The workshop aimed to collect data about which kind of sensory conditions people prefer when they experience pain.

1.5 Main Limits

This paper presents a pilot study. The main objective was to collect information about the sensory conditions people would prefer in case of pain.

The greatest limit in this sense is that subjects were not under a state of pain, they were instructed to imagine pain. Not having the possibility to deal with visceral reactions, it was not possible to obtain:

- Pain intensity measurements through visual analogue scale VAS-I
- Pain unpleasantness measurements through visual analogue scale VAS-U
- Psycho-physiological measurements (i.e., blood pressure, pulse rate, etc.)
- Other pain measurements (i.e., McGill Pain Questionnaire).

All the information gathered is based on participants previous personal and general experience with pain.

1.6 The Use of Space as a Design Probe

In our everyday life, we often use a sensory solution in case of pain. Some examples are: putting ice on a knee after a fall, turn off the lights in case of migraines, being in warm places when we have a fever, etc.

As a pilot study for our research, we wanted to understand which could be the sensory condition people would prefer in case of pain. On this purpose, we designed

Fig. 1 Participant setting chromatic lights

an environment, as a design probe, for exploring and understanding people preferences centred on empathy and personal significance [23]. In an earlier stage, we considered the use of traditional packages with design probes, to be delivered to people, but we realized that aspects as perception has to be addressed with the total involvement of participants. For this reason, we have opted for the use of a room that was configured as a design probe. The room was like a big "box" containing several stimuli to gather empathic responses from people regarding the sensory conditions they would prefer when experiencing pain. For this specific experience, we designed a specific probe for every sensory modality.

1.7 Participants

A total of 16 participants were recruited, 43.8% (7) were male and the 56.3% (9). Were female. The participants' age ranged between 20 and 56 years old. For this experiment, we were looking for people particularly comfortable with the design practice. For this reason, only design students of master (68.8%) and PhD (31.3%) degrees were recruited (Figs. 1, 2, 3, 4, 5 and 6).

1.8 Criteria of Exclusion

The participation to this experience was not allowed to people affected by:

- Epilepsy and other diseases of the Nervous Systems
- Hypertension and other cardiovascular diseases

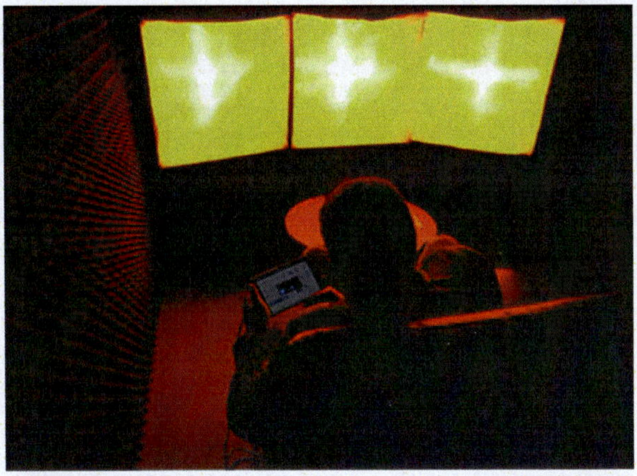

Fig. 2 Participant selecting the sound

Fig. 3 Participant applying a TENS electrode

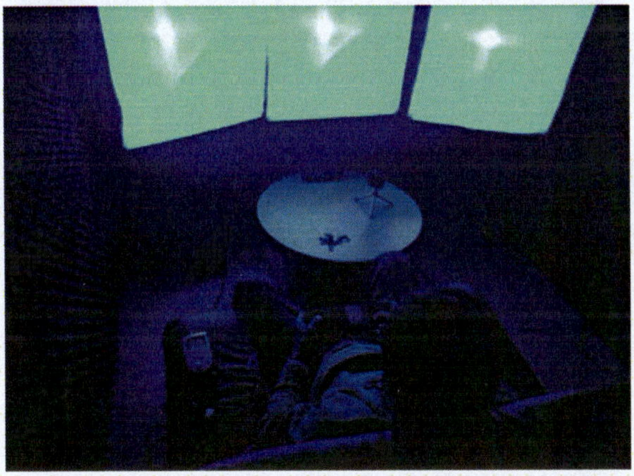

Fig. 4 Participant trying different material densities

Fig. 5 Participant trying different textures

Fig. 6 Participant trying different temperatures

- Chronic pain conditions (i.e. fibromyalgia, neuropathic pain)
- Sensory dysfunctions (i.e. colour-blindness, hearing problems)
- Anxiety.

Participants that referred to have acoustic dysfunctions (1) and anxiety (1) were excluded from the statistical analysis.

1.9 Procedure

The experience was untitled: "Pain/off: imagining a painless hospital".

A colour-blindness test was performed, and people were asked if they were feeling some kind of personal or environmental anxiety. Those that referred to such a condition were later excluded from the statistical analysis.

A set of instructions was projected, and participants were asked to configure the stimuli in the environment in the way they think would be the most appropriate for reducing pain. Participants could turn on (or keep/turn off) and set the following stimuli:

- **Light**: Intensity, Colour temperature, Chromatic Light
- **Sound**: between a set of 8 white sounds (white tv sound, a waterfall, fan, rain, storm, waves, underwater sound)
- **Vibration**: 8 level TENS intensity
- **Texture**: high relief, flat or low relief.
- **Density**: 4 materials: hard, light soft, medium soft, high soft
- **Temperature**: cold 10°, ambience 22° and warm 32°.

At the end of the setting, participants were asked if they would change something to turn the configuration more coherent with the objective of reducing pain.

1.9.1 Experimental Set-Up

The test was performed in the UX.Lab room, at IADE, Universidade Europeia, which is 12 m^2 room, without windows and acoustically isolated.

The room was equipped with:

- **3 photographic soft-boxes** with **3 Philips Hue** Color ambience RGB E27 lamps (806 lm)
- **1 Dolby Surround Audio System** 5.1 (Creative) Professional
- **TENS Electro-stimulator** (Moonssy B07X48T5L9)
- **3D printed textures** (high relief, flat and low relief)
- **4 material with different densities** (woor, Polystyrene of high, medium and low density
- **3 water baths** containing water kept constant at 10° (with a refrigeration system) 22° ambience water 32° warm water.

The place was dark, and the instructions were projected on a canvas.

2 Results

2.1 Light

The analysis of the collected data shows a greater preference for warm colour temperature (69.2%) in relation with neutral (15.4%) and cold light (15.4%). (Fig. 8). The

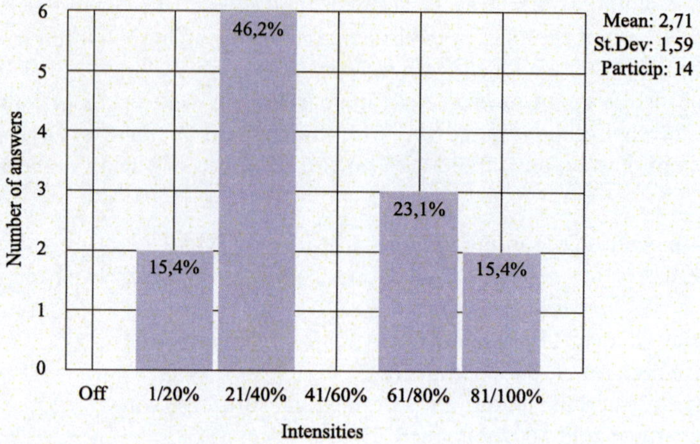

Fig. 7 Frequency of light intensities

Fig. 8 Frequency of colour temperature

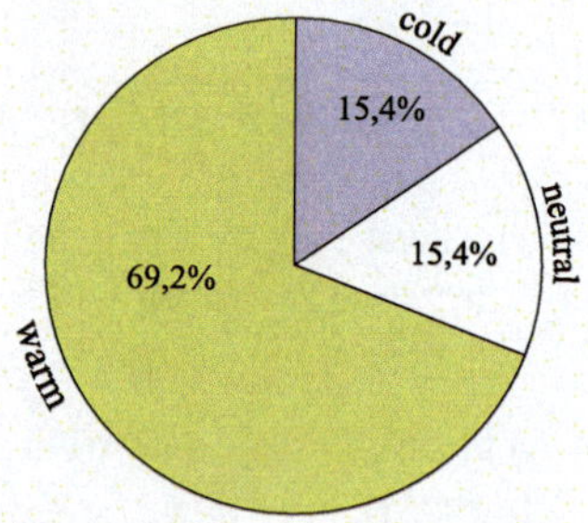

highest percentage of white light intensities is the 46.2% for the range 21–40% of intensity (Fig. 7).

The same range of intensities (21–40%) results the most preferred among the different ranges in relation with the chromatic lights (Fig. 9). Chromatic lights preferences refer a greater preference for cold colours, in particular blue (28.6%), and aquamarine (14.3%) and purple (14.3%). Yellow is the second most preferred colour with 21.4% of preferences (Fig. 10).

2.2 *Sound*

The analysis refers a preference for sound intensities included in a range between the 21/40% chosen the 50% of the cases (Fig. 11). The 35.7% of preferences is given to a range included between 41/60%. The most preferred sounds are the sound of waves (43.8%) followed by underwater bubbles (18.8%), rain (12.5%) and storm (12.5%). No other sounds have been selected (Fig. 12).

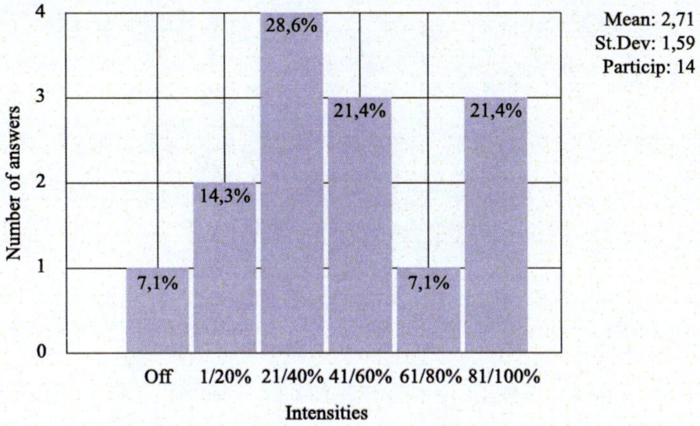

Fig. 9 Frequency of chromatic light intensities

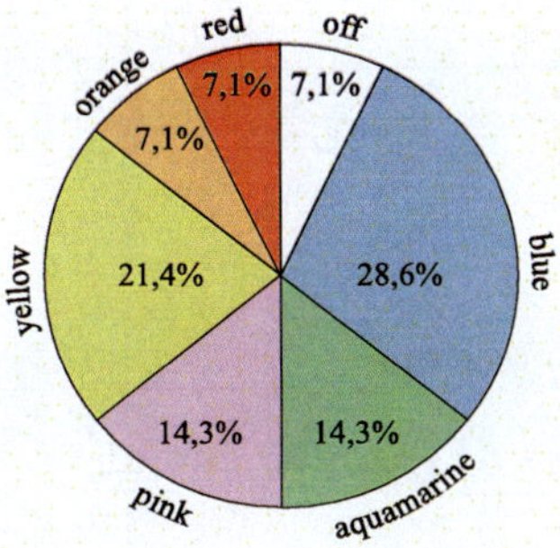

Fig. 10 Frequency of chromatic lights

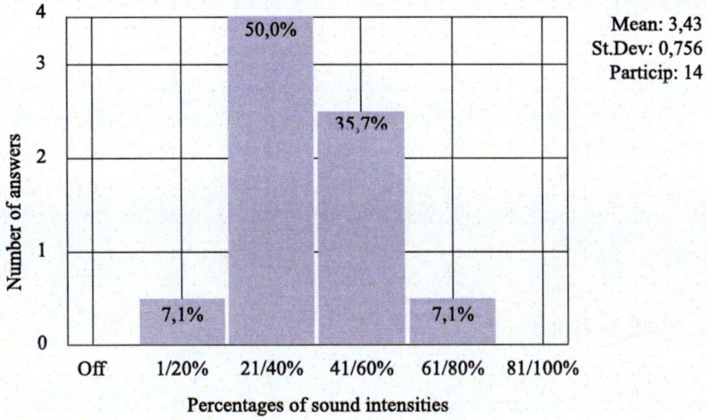

Fig. 11 Frequency of sound intensities

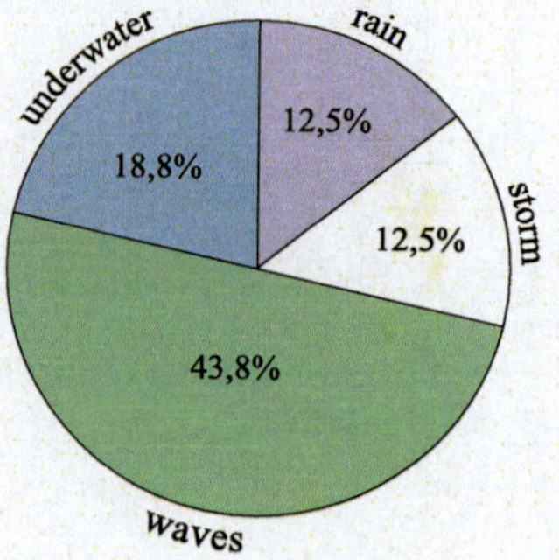

Fig. 12 Frequency of sounds

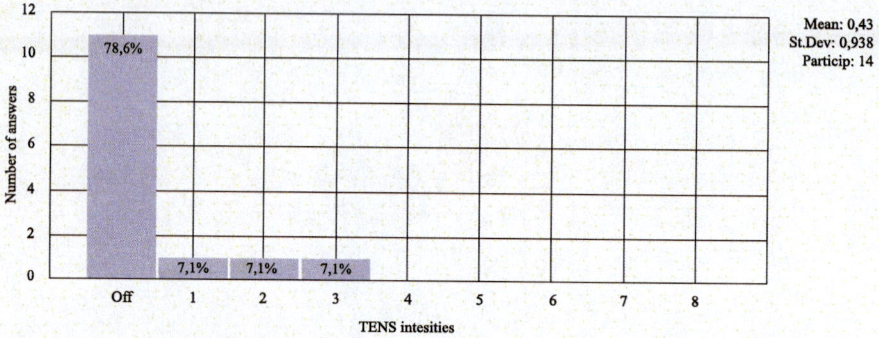

Fig. 13 Frequency of TENS intensities

2.3 *Vibration*

The data regarding vibration (Fig. 13) reported a strong preference to keep the TENS device off (78.6%) a turning off the stimulus, reported as unpleasant sensation. A premise is needed. TENS is commonly used a technique to manage pain, in particular chronic pain and its efficiency is well validated by the scientific community. Probably the test would perform differently with people affected by chronic pain conditions.

2.4 *Material Density*

Participant preference reported a frequent preference to high soft (69.2% = and medium soft materials (30.8%). No other densities have been selected (Fig. 14).

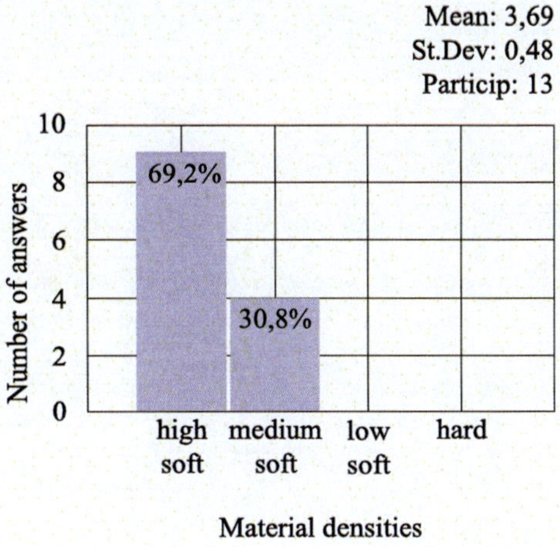

Fig. 14 Frequency of material densities

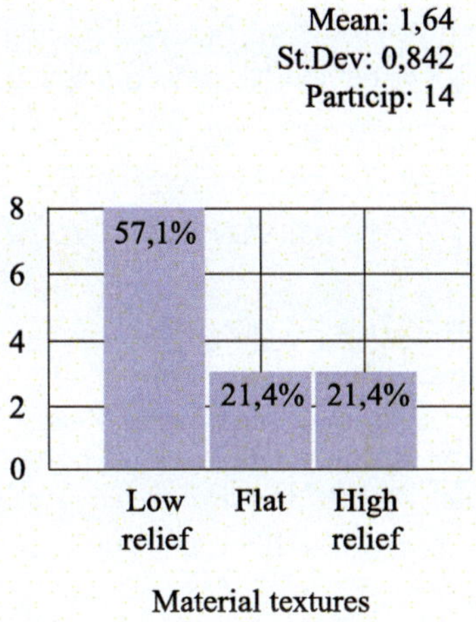

Fig. 15 Frequency of textures

2.5 *Material Textures*

The selection of the textures presents a strong preference for negative texture in low relief (57.1%) against a 21.4% of preference for flat texture and 21.4% of high relief texture (Fig. 15).

2.6 *Temperature*

The data collected show a greater preference for warm temperature at 32° (76.9%), against cold temperature at 10° (7.7%) and ambient temperature at 22° (15.4%)

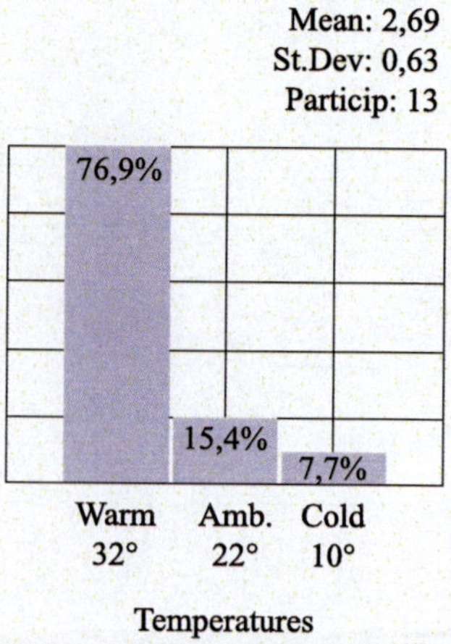

Fig. 16 Frequency of temperatures

(Fig. 16). Even thought, participants said that a cold temperature could also be a good method for reducing specific kinds of pains, (as muscular pain for example).

2.7 *Cross-Sensory Interactions*

The combination of the results collected in single modalities presented interesting associations.

Sound + chromatic lights. The association between the type of sound with the chromatic lights reported a frequent association between the blue colour and the sounds of rain, storm and waves (See Table 1). The aquamarine colour was associated to the sounds of waves and to the underwater sound. In all these cases we are in presence of associations typical of synaesthetic representations.

Sound + colour temperature. Cold colour temperatures were associated to the sounds of rain and waves; neutral colour temperatures were associated to storm

Table 1 Contingency table: associations of chromatic lights and type of sound

		Types of Sound				
		Rain	Storm	Waves	Underwater	Total
Chromatic Lights	Off	0	1	0	0	1
	Blue	1	1	2	0	4
	Bluemarine	0	0	1	1	2
	Purple	1	0	1	0	2
	Yellow	0	0	2	1	3
	Orange	0	0	0	1	1
	Red	0	0	1	0	1
Total		2	2	7	3	14

Table 2 Contingency table: associations of colour temperature and type of sound

Colour Temperature	Type of Sound: Rain	Storm	Waves	Underwater	Total
Cold	1	0	1	0	2
Neutral	0	1	1	0	2
Warm	1	1	5	2	9
Total	2	2	7	2	13

and waves (See Table 2). Warm colour temperatures were associated with rain, storm, waves and underwater sounds. The most frequent association was warm colour temperature with waves. This is also a common synaesthetic representation of sensory condition typical of a summertime experience.

Sound + temperature. The cold temperatures were associated with the sound of rain. The ambience temperatures was associated with the sounds of storm and waves, while a warm temperature was associated to rain, storm, waves and underwater (See Table 3). A frequent association as warm temperature and waves, presents a strong coherence with the association between colour temperature and sound and both are based on synaesthetic connections.

Textures + densities. Other cross-sensory interactions regarded the combination of texture and material densities (See Table 4). In this case a frequent association has been the one between low relief and high soft materials.

Temperature + colour temperature. By crossing the preferences, a significant cross-sensory association found was the combination of warm temperature with warm colour temperature (See Table 5). One of the most common synaesthetic association.

Table 3 Contingency table: associations of temperature and type of sound

Temperature	Type of Sound: Rain	Storm	Waves	Underwater	Total
Cold	1	0	0	0	1
Medium	0	1	1	0	2
Warm	1	1	6	2	10
Total	2	2	7	2	13

Table 4 Contingency table: associations of material densities and material textures

Material Densities	Texture of Material: Low relief	Flat	High relief	Total
Medium Soft	1	2	1	4
High Soft	7	1	1	9
Total	8	3	2	13

Table 5 Contingency table: associations of colour temperature and temperature

Colour Temperature		Temperature: Cold	Medium	Warm	Total
	Cold	0	0	2	2
	Neutral	0	0	2	2
	Warm	1	2	6	9
Total		1	2	10	13

Proposed Scenario.
Combining the information gathered, the most preferred conditions selected by the participant can be synthetized in the following scenario:

- **Light**: Blue chromatic light combined with warm white light
- **Sound**: waves
- **Vibration**: no vibration
- **Texture**: low relief
- **Density**: high soft
- **Temperature**: warm 32°.

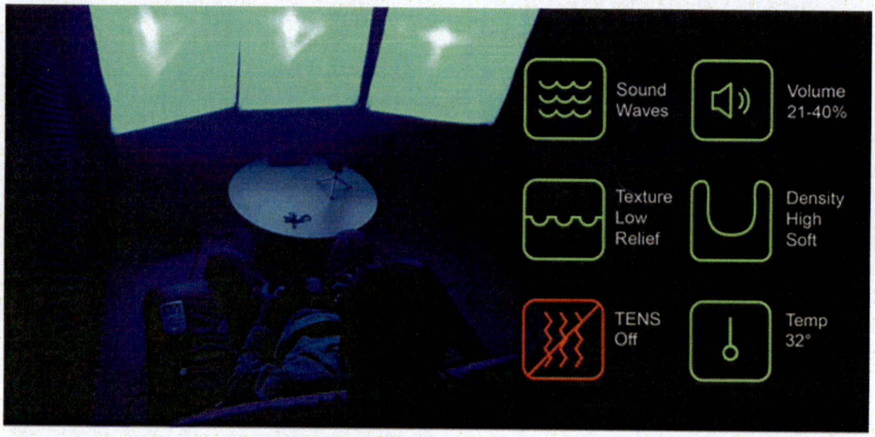

3 Discussion

The present study has been realized with healthy participants that were instructed to choose the sensory condition they would prefer in case of pain, they were not experiencing pain at the moment of the test, which is a limitation of the study. In this sense, there's the possibility they might have chosen scenarios which would render the environment more comfortable, but not necessary more analgesic and efficient for pain relief. This case was particularly evident with the TENS stimulation, a well-known counterirritation for reducing pain. The observation of the behaviour in presence of the TENS stimuli made us consider repeating the experiment with participants affected by chronic pain conditions, such as fibromyalgia or with healthy participants under controlled induced pain.

In this case it would be possible to address different levels of pain using VAS-I, VAS-U scale together with subjective measurements (McGill Pain survey) and Psycho-Physiological data (blood pressure, and hearth rate). A possible comparison with the present data and the data eventually collected with people affected by chronic pain could identify a more certain pattern that could be used in a hospital to alleviate pain.

4 Conclusions

By using the environment as a design probe, we were able to see people's preferences, regarding their own personal experiences with pain. The analysis of the data suggests a strong preference in using multiple sensory stimuli in the environment to alleviate a possible condition of pain, contrary to the anaesthetic tendency of removing stimuli in healthcare environments (Unique exception for the TENS stimulation). Even if the associations presented a predictable pattern, analysing the most prefer sensory conditions reported in the proposed scenario, it is possible to find synaesthetic relations between visual and auditory modalities in associating frequently the blue chromatic light with the sound of waves.

Acknowledgements This work was supported by the grant SFRH/BD/129136/2017 to Davide Antonio Gambera, given by the FCT, Fundação para a Ciência e a Tecnologia (Portuguese Science Foundation). This study was also conducted at the UNIDCOM, UX.Lab and supported by the Fundação para a Ciência e Tecnologia (FCT), under Grant No. UIDB/00711/20120 attributed to UNIDCOM – Unidade de Investigação em Design e Comunicação, Lisbon, Portugal (unit 711).

References

1. Anceschi G, Riccò D (2000) Research of communication design: a synesthetic approach. In: Pizzocaro S, Arruda A, de Moraes D (eds) Design plus research. Proceedings of the Politecnico di Milano conference, pp 1–7. Politecnico di Milano, May 18–20. https://doi.org/10.13140/2.1.2121.4089
2. Antonovsky A (1979) Health, stress, and coping. Jossey-bass. https://trove.nla.gov.au/work/7249690?q&versionId=10393411
3. Béphage G (2005) Promoting quality sleep in older people: the nursing care role. Br J Nurs. https://doi.org/10.12968/bjon.2005.14.4.17604
4. Boensch S (2011) Stimulation-produced analgesia: TENS, acupuncture and alternative techniques. Anaesth Intensive Care Med 12(1):28–30. https://doi.org/10.1016/J.MPAIC.2010.10.003

5. Burt RK, Korn GW (1964) Audio analgesia in obstetrics. “White Sound” analgesia during labor. Am J Obstet Gynecol 88(3):361–366. https://doi.org/10.1016/0002-9378(64)90436-3
6. Córdoba MJ, Riccò D, Day SA (2014) Synaesthesia, theorical, artistic and scientific foundations. International Foundation Artecittà. https://books.google.it/books?id=MSREBAAAQBAJ&dq=synestesia+foundations&hl=it&source=gbs_navlinks_s
7. Cytowic RE (2002) Synesthesia: a union of the senses. MIT Press. https://books.google.it/books?id=fl6wX4xzb_kC&dq=cytowic+2002&hl=it&source=gbs_navlinks_s
8. Duarte E, Gambera DA, Riccò D (2019) Beyond the five senses: a synaesthetic-design approach to humanize healthcare environments. Springer, Cham, pp 16–22. https://doi.org/10.1007/978-3-030-24067-7_2
9. Gardner WJ, Licklider JC, Weisz AZ (1960) Suppression of pain by sound. Science 132:32–33. https://doi.org/10.1126/science.132.3418.32
10. Golembiewski JA (2017) Salutogenic architecture in healthcare settings. In: The handbook of salutogenesis. Springer. https://doi.org/10.1007/978-3-319-04600-6_26
11. Haverkamp M (2013) Synesthetic design: handbook for a multisensory approach (de Gruyter W, ed.). Birkhäuser. https://books.google.pt/books?id=NfLTAAAAQBAJ&dq=michael+haverkamp+synesthetic+design&hl=it&source=gbs_navlinks_s
12. Higgins PA, Winkelman C, Lipson AR, Guo SE, Rodgers J (2007) Light measurement in the hospital: a comparison of two methods. Res Nurs Health. https://doi.org/10.1002/nur.20167
13. Melzack R (2001) Pain and the neuromatrix in the brain. J Dent Educ 65(12):1378–1382
14. Merleau-Ponty M (1945) Phenomenology of perception. Phenomenology Perception. https://doi.org/10.4324/9780203720714
15. Morosko TE, Simmons FF (1966) The effect of audio-analgesia on pain threshold and pain tolerance. J Dent Res 45(6):1608–1617. https://doi.org/10.1177/00220345660450060701
16. Nanda U (2017) A sensthetic approach to designing for health. IOP-J Interior Des 42(2):7–12. https://doi.org/10.1111/joid.12098
17. Park C, Alderman J (2018) Designing across senses. A multimodal approach to product design. O’reilly
18. Riccò D (1999) Sinestesie per il design. Le interazioni sensoriali nell’epoca dei multimedia. Etas
19. Rognoli V, Levi M (2005) Materiali per il design: espressività e sensorialità. FrancoAngeli. https://www.libreriauniversitaria.it/materiali-design-espressivita-sensorialita-rognoli/libro/9788873980124
20. Ulrich RS (n.d.) Effects of healthcare environmental design on medical outcomes. Retrieved 6 Sept 2018 from http://www.capch.org/wp-content/uploads/2012/10/Roger-Ulrich-WCDH2000.pdf
21. Ulrich RS (1991) Effects of interior design on wellness. J Healthc Inter Des
22. Ulrich RS, Zimring C, Zhu X, DuBose J, Seo H-B, Choi Y-S, Quan X, Joseph A (2008) A review of the research literature on evidence-based healthcare design. HERD 1(3):61–125
23. Wallace J, McCarthy J, Wright PC, Olivier P (2013) Making design probes work. In: Conference on human factors in computing systems—Proceedings, pp 3441–3450. https://doi.org/10.1145/2470654.2466473
24. WHO (1948) Constitution of WHO. Official records of WHO

Neurosurgery Training Tool. Design as Facilitator Between Disciplines for the Improvement of Medical Devices

Angela Giambattista

Abstract The propulsive thrust the medical sector has recently registered from the technological and methodological point of view is having substantial societal and economic impact modifying diagnostic and therapeutic practices, structures, and medical instruments. This requires that Design have to be placed at the center of a multi-disciplinary environment combining both medical and technological disciplines to develop solutions in which usability and technology coexist in a balanced manner improving treatments, diagnoses, training, and research. Design thus becomes a strategic element for stimulating incremental and radical innovation to obtain products and services that involve a multiplicity of competences and actors. Starting from these premises, the paper describes an ongoing multidisciplinary research project aiming to develop and prototype an anatomical training system of the spinal column, useful both for teaching surgical anatomy to medical students and for training experience of surgical techniques for youngsters and younger neurosurgeons.

Keywords Design as facilitator · Interdisciplinarity · Medical device development · Reality-based 3d modelling

1 Design Led Innovation in Medical Context

Design is widely recognized for years as a scientific discipline [1–3] while maintaining strong contingency relationships with other areas of knowledge, in a rhizomatic perspective that identifies it as a phenomenon total not isolated. A scientific discipline belonging to that category of phenomena that must not be examined in isolation, but always in connection to other phenomena with which they constitute a single connective tissue [4].

And it is precisely the intrinsic value of the interdependence between Design and Hard Sciences, to characterize its results, amplifying the disciplinary dimension of

A. Giambattista (✉)
Sapienza University of Rome, Rome, Italy
e-mail: angela.giambattista@uniroma1.it

E. Duarte and C. Rosa (eds.), *Developments in Design Research and Practice*, Springer Series in Design and Innovation 17,
https://doi.org/10.1007/978-3-030-86596-2_17

Design Discipline and enriching the speculative opportunities in terms of research, both qualitatively and quantitatively.

In this context straddling hybridization and scientific contamination, Design moves according to an interdisciplinary, rigorous but flexible approach [5], which presupposes coordination between knowledge and procedures, as well as the development of a common language to allow conceptual and methodological exchanges.

In this way, combined, creative and strategic modes of dialogue are triggered, aimed at the incremental and radical innovation of products, services and systems and the identification of new usage scenarios, capable of affecting a multiplicity of actors (users, businesses, organizations social, etc.) and network systems involved in decision-making, design and production processes [6].

Within this framework of scientific synergies, Medicine has also opened its arms to design-led thinking, with many using the methodology to design improved experiences and solve problems within the complex healthcare landscape [7].

In the light of continuous socio-cultural changes, such as growth and aging of the population, and considering the alternative approaches in the management of diseases as a results of result of the progress of research in the medical field, the scientific mix between Design Discipline and Medicine has evolved over time, enriching itself with new methodological models and speculative outcomes. A synchrony that has seen, over the years, different areas of application, both to improve products, technologies, and healthcare facilities, and to increase the awareness and involvement of patients within the conception, production, and consumption process.

In this scenario, Design Discipline is called to respond to new challenges and opportunities that require new design approaches for the creation of innovative products and services capable of responding to the contextual needs of all the figures involved in the healthcare process, according to a systemic approach that affects both users and operators.

However, the context of Medical Design has always been characterized by stringent constraints linked, on the one hand, to the speed and rigor of medical research, on the other, to the need to integrate with the physical, physiological and psychological factors of patients and operators [8]. The latter an aspect, not at all negligible, which aims to go beyond the traditional aspects of the mere functional requirement, in favor of a formal quality that can also incorporate the peculiarities of the user experience, the compatibility between physical-sensorial aspects and subjective components in the interaction between user and product.

The driving force that the medical sector has registered in recent years both from a technological and methodological point of view, with substantial repercussions on companies and economies, changing people's habits, structures and the way users and designers look at the artifact medical [9], guides us towards a redefinition of design priorities according to a holistic matrix that aims to consider a synchronic multiplicity of factors: functional, sensory, emotional, systemic [10].

Based on these assumptions, the discipline of Design, in addressing the design of medical devices, is therefore called to the conception and development of complex and multilevel systems, in which both the technological dimension and the experience of using products and services coexist in a balanced way with the aim of improving treatments, diagnosis, training, awareness, prevention, and research in the medical field [11].

Researches and practices led by design in the medical field thus becomes a strategic element for stimulating incremental and radical innovation [12] to obtain products and services that involve a multiplicity of competences and actors aimed at enhancing life expectancy, quality of life, diagnostics and treatment options, as well as the efficiency and cost-effectiveness of the healthcare system [13].

Starting from these premises, the paper describes an ongoing multidisciplinary research project lead by Design Discipline aiming to develop and prototype a training system that perfectly simulates the spinal column, useful both for teaching surgical anatomy to medical students and for training experience of surgical techniques for youngsters and younger neurosurgeons. In conjunction with an augmented reality software, it will allow the practitioner to experience the interaction with the physical phantom while receiving feedback and guidance in the execution of the surgical gesture. Moreover, the research project replies to the economic and organizational difficulties of the hospital structures to find useful bodies for the surgical training of neurosurgery doctors and students. Therefore, the goal is to develop an integrated system able to respond to the identified needs of operators and hospitals. On one hand, the clinical operators will have the possibility to improve their surgical skills through a dynamic tool that simulates the mechanical properties of the different human tissues and that provides a coherent feedback as for the real operations, on the other hand, the hospital structures will have the opportunity to optimize the costs of surgical training through a device, which would avoid the purchase of bodies, and that will be possible to use several times through the use of self-healing material and of easily replaceable components.

In a collaborative condition of an integrative team involving some of the most advanced competences in Sapienza University of Rome (i.e. clinical, technological, design, experimental research ones), the Design Discipline acts as facilitator between other disciplines bringing diverse experts together in a coordinated effort [14] and moving the designer role out of the traditional solo design expert role into being a design subject matter expert leading an interdisciplinary team [15].

2 The Neurosurgery Training Tool Project

2.1 Problem Identification and Opportunity Recognition

In neurosurgery, as in any procedure for which the surgical anatomy is extremely complex, and the use of microsurgical techniques is indispensable, rigorous theoretical and practical, knowledge of surgical anatomy and dissecting techniques are mandatory. For medical and surgical students, it's very important, and in some cases crucial, to practice dissection, especially as regards the study of anatomy, as well as for doctors in specialized training, mainly in surgical branches. The possibility to practice real surgery is fundamental both for the development of particularly complex interventions and in the experimentation of new approaches, techniques, and technologies.

Up today, cadavers represented a unique instrument for understanding anatomy, physiology, and pathophysiology. The lack of whole or parts of cadavers reduces opportunities to learn anatomical and surgical techniques that may be repeated in the operating theatre. The learning of surgical anatomy and technical skills during operations presents the following disadvantages: limited time at disposal, the surgical anatomical field is restricted and depends on the type of operation, the learning is influenced and limited by the pressure and psychological conditioning and by the medico-legal responsibilities. Surgical training in lab, with cadavers or models mimicking real anatomy, should be part of every residency program and a consistently available tool for senior surgeons, for the refinement of the microsurgical techniques necessary to perform the most complex [16].

However, the limited availability of cadavers for the study, scientific research, and training often force hospitals to fall back on the use of animal species or to finance training activities in foreign facilities, with an important expense of economic resources.

Particularly in Italy, hands-on courses on cadavers are limited because of bureaucratic procedures: specimens are prepared outside of Italy and provided by Departments of Anatomy of foreign universities and returned once the hands-on course is finished. Thus, many of these courses are expensive, intensive (12 h/day for 2/3 days in a row) and permit only limited working time [17].

Many alternative simulators are available for training in neurosurgery, but none of them reliably mimic the anatomy and characteristics of the spine as during live surgery. Currently, there is only one product available in the market (www.realists.de/realspine) that reproduces a model of the lumbar spine allowing realistic training for surgery, but it is limited in terms of high costs, partial immersive experience, and durability.

Recently new technologies in the field of virtual reality have tried to make up for this lack by developing a system of immersive reality with tools capable to provide active and passive resistance to the manual dexterity of the learner [18]. However, also these methodologies do not allow the training of the gesture sensitivity because they do not reproduce the physical interaction between the hand and the anatomical structures with their physical characteristics, consistency, and elasticity: a proactive feedback between hand and human tissues as essential requisite for learning the surgical gesture [19].

The Neurosurgery Training Tool (NTT) research project intends to combine the benefits of the physical feedback with the guidance of augmented reality tools moving beyond the state of the art of current simulators (Table 1) [20] and coupling physical plausibility of materials to augmented reality to provide the best of a real physical experience with the objective evaluation of an Reality-based 3d model. Besides, the three-dimensional virtual model, using surgical guides and cutting guides, allows a 3D virtual plane created on the software screen to be transferred to the operative

Table 1 Types of simulations available

Simulation	Advantages	Disadvantages	Best use
Bench models	Cheap, portable, reusable, minimal risks	Acceptance by trainees; low fidelity; basic tasks, not operations	Basic skills for novice learners, discrete skills
Live animals	High fidelity, availability, can practice hemostasis and entire operations	Cost, special facilities and per sonnel required, ethical concerns, single use, anatomical differences	Advanced procedural knowledge, procedures in which blood flow is important, dissection skills
Cadavers	High fidelity, only "true" anatomy simulator currently, can practice entire operations	Cost, availability, single use, compliance of tissue, infection risk	Advanced procedural knowledge, dissection, continuing medical education
Human performance simulation	Reusable, high fidelity, data capture, interactivity	Cost, maintenance, and down-time; limited "technical" applications	Team training, crisis management
Virtual reality surgical simulation	Reusable, data capture, mini mal setup time	Cost, maintenance, and down time; acceptance by trainees; three dimensions not well simulated	Basic laparoscopic skills, endoscopic and transcutaneous procedural skills

site and as such can be considered as an interface between the virtual plane and the physical patient.

At the same time, the NTT project explores new production opportunities by integrating three-dimensional modeling techniques in a medical context that in recent years has seen an incremental evolution thanks to the introduction of Additive Manufacturing and Virtual Representation. The world healthcare 3D printing sector is currently worth 2.3bn dollars with an annual growth rate of 26% and Medicine is set to overtake all other Industries in 3D Printing [21].

The areas of application are quite broad, extending past general medical practice and research and including surgical preparation, development of prosthetic limbs, dental application, 3D printing of tissues and organs, medication dosage and pharmacology, the manufacturing of medical tools and devices.

The field of neurosurgery has also embraced the use of these digital manufacturing technologies, because most of the surgical procedures and corresponding pathology that neurosurgeons encounter involves intricate, minute anatomical structures cannot be outwardly observed. These 3D models in neurosurgery have focused upon three main areas: the creation of patient-specific anatomical models for surgical planning, training, and education; the design of neurosurgical devices for assessment and treatment of neurosurgical diseases; and the development of biological tissue-engineered implants [22]. The translation of an image, acquired through a computerized tomography (CT), as well as a Cone Beam Computer Tomography (CBCT) or a Magnetic Resonance Imaging (MRI), in a 3d virtual model and/or in a 3d printed model, is becoming an integral component of clinical practice [23].

2.2 *Methodology*

From a methodological point of view, the research project follows the four stages of Applied Research: Definition, Design, Implementation, and Validation [24] as described next and it is currently in the second phase of development.

Stage 1—Definition Thanks to conducting information-gathering visits and observations and to the arrangement of round tables with all the expertise involved, this stage was conducted aiming to define a specific anatomical part of the vertebral column, namely the cervical section C1, C7 (subdivided into two sub-sections, vertebrae C1, C2 and C3, C7). Concerning the anatomical complexity and the number of structures that will make up the tool, an abacus of components has been defined to simplify the anatomical structure into 5 layers:

1. SKIN
   Epidermis, Dermis, Hypodermis.
2. MUSCLES
   Platysma muscle,
   Sternocleidomastoid muscle.
3. CENTRAL VISCERAL AXIS
   Trachea,
   Esophagus,
   Thyroid.
4. VASCOLOVENOUS BAND
   Carotid artery,
   Jugular,
   Nerve Cord.
5. VERTEBRAL PLAN
   Deep cervical band,
   Vertebral Column (Cervical Vertebrae, Bones, Spinal Cord).

Moreover, this step of the research was focused on understanding the State of the art of the equipment currently available through benchmarking and on identifying the medical user needs. The evaluation of possible best practices has led to the definition of the project peculiarities that were spelled out in the Designing phase.

Stage 2—Design The second stage of the NTT Project is still ongoing, and it has been focused on the development of the components according to the three specific objectives:

SO1. 3D Translation and Representation Model

Starting from the analysis of the images obtained from the Diagnostic Imaging of Computed Tomography (CT) and Magnetic Resonance (RMI), the specific aim was to determine the structural complexity of the portion to be reproduced by identifying hierarchically the technical-functional components.

Through a process of segmentation and extraction of the area of interest, these data and images, initially saved in DICOM format, were later translated into a volumetric format useful for the realization of the physical model through digital fabrication techniques. It was essential to check the quality of the mesh using customized software to avoid issues with image processing such as noises and low-contrast images.

The image segmentation process has consisted of removing all interference by highlighting the necessary parts to print, which was dividing the image into areas or categories, which corresponded to objects or parts of different objects. Each pixel of an image was assigned to one of these categories through the following selection conditions: (1) pixels in the same category with similar gray-scale values formed a connected area, (2) neighboring pixels that were in different categories had dissimilar values. Finally, the data was saved in a format recognized by a 3D printer software selecting the most used format in the standard format tessellation language (STL) (Fig. 1).

SO2. Digital Fabrication and Simulation Materials

The specific purpose in this stage was to build a table of "high performance in the use" and "high compatibility in the disposal" materials to be used through digital processes manufacturing to convincingly simulate both the physical and mechanical characteristics and the physiological sensations related to sight and touch.

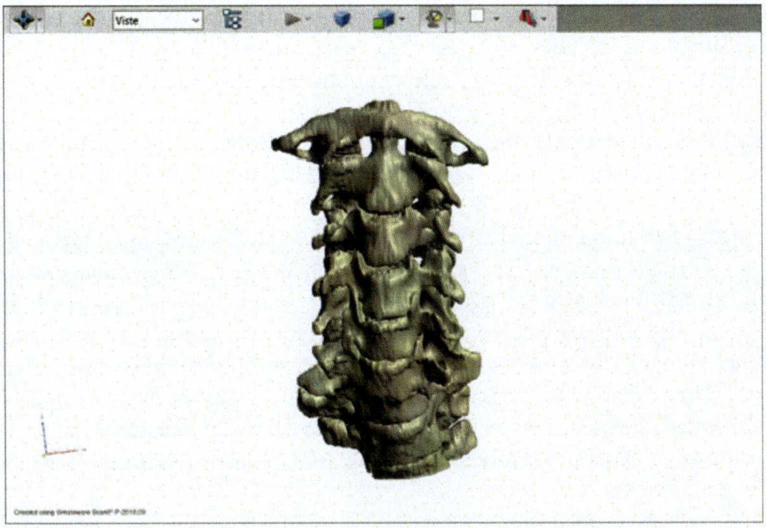

Fig. 1 Three-dimensional reconstruction of the bone part

As a result, acquired tomographic diagnostic images have been analyzed through the detection of physical properties and mechanical properties of resistance to the action of external forces or stresses.

Appropriately compared with the evidence of scientific literature about the acoustic parameters in human soft tissues [25], these data were useful for the selection of simulation materials for replication of the physical model.

The materials were selected through a literature review of studies concerning the simulation of biological tissue according to the above-mentioned abacus of components, both in the biomedical and engineering field (Table 2).

Starting from the three-dimensional model deriving from the previous phase, the materials are in a step of testing through an iterative verification process that involves the realization of the components both through direct 3D printing (as for the bone) and through the creation of molds and resin casting (as for the soft tissue).

SO3. Development of Augmented Reality Tool

The 3D virtual model featured in Objective 1, will be used to augment the current view of the operator by showing the current position of tracked surgical tools with respect to the structures (visible and hidden) included in the model. In particular, the operator will be alerted when approaching critical structures or executing non-correct surgical gestures. Navigation of the virtual model during the execution of the real surgical gesture will be obtained by a preliminary registration procedure aligning the current pose of the physical phantom to the virtual model and by real-time tracking of the surgical tools. The augmented reality coupled with the physical phantom will provide an interactive simulation system that is economically affordable, technically simple to be realized and with unlimited duration characteristics.

After the study of the components and the functionalities of the system, the second specific objective is increasing the knowledge and skills in the design field of bio-inspired, smart and collaborative artifacts. These artifacts in the medical field open to different usability and efficiency allowing operators to have immediate feedback on the vital parameters of the care procedures.

The activities in this stage aim at designing and implementing a registration procedure that, based on fiducials placed on the physical phantom in the construction phase and represented on the 3D virtual model, will align the current pose of the physical phantom with the virtual model through the use of a localization and tracking system like, e.g., the Polaris Vicra. The same system will track in real-time the pose of the surgical instruments in the hand of the surgeons. Based on their position with respect to the virtual model, the system to be developed will produce a vocal message providing feedback about the correctness of the gesture or the proximity to critical structures. The system will also include the possibility to record the data of the whole procedure with the twofold objective of (1) quantitatively evaluate the performance of practitioners; (2) setting references by using the data relative to the execution of the procedure by expert surgeons.

Table 2 Literature review of materials simulating biological tissues

Body part	Material	References
Skin	Ecoflex 00-30 (Epidermis, Dermis) Ecoflex 0010 (Hypodermis)	Sparks et al. [26]
	Standard gelatine (plus a polymer chain to avoid it dissolve with water), layer of cotton	Dąbrowska et al. [27]
	• Liquid suspensions: lipid, polymeric, and inorganic particles can be added in liquids such as water, milk or oils • Gelatinous Substances: gelatine, agar and agarose, collagens, and polyvinyl alcohol gels • Elastomers: silicones and polyurethane, polyurethanes, polyether block amides, polyisoprene, and polybutadiene • Epoxy Resin: Resin, plasticisers and diluents • Textiles: synthetic and natural leather such as Lorica® and chamois	Dąbrowska et al. [28]
	Open-celled polyurethane sponge (foam), covered by a silicone layer	Whittle et al. [29]
	Nonadhesive hydrocellular polyurethane dressing (Allevyn, #66007638, 20 × 20 cm, Smith & Nephew Medical Limited, Hull, England)	Bjellerup [30]
Muscles	Dragon skin	Sparks et al. [26]
	Elastic silicone rubber film (for ligaments) casting in SLA Viper mold	Ansari et al. [31] Li et al. [32]
	Silicone blocks SB2-50% and the SB1-60%	Hollensteiner et al. [33]
Central visceral axis	Polyurethane	Jiang et al. [34]
	Silicone rubber (curing process) (Polastosil® M-2000, Silikony Polskie Sp. zoo., Nowa Sarzyna, Poland) was hardened at room temperature using an OL-1 catalyst with the 3d Printed mold	Witowski et al. [35]
	Methacrylate or polyethylene plastic	Berman [36]
	3D printing on acrylonitrile-butadiene-styrene (mold) then coated with liquid silicone (M8012 from Asahi Kasei-Wacker Silicone, Tokyo, JP)	Mashiko et al. [37]
Vascolovenous band	Two different polylactic acid polymers: a rigid (PLA 4043D, NatureWorks LLC) and a flexible (Filaflex, Recreus, Spain) polymer	Valverde et al. [38]
	Platinum-cured silicone	Barsness et al. [39]
	Transparent resin (Connex, Stratasys, Ltd., Tokyo, Japan)	Akiba et al. [40]
	ABS + liquid silicon and hardener. Once it has hardened, the inside ABS model was melted and removed with organic solvent	Mashiko et al. [37]

(continued)

Table 2 (continued)

Body part	Material	References
Vertebral plan Vertebrae	Two-layered structure of different polyurethane (PU) foams. (Sawbones Europe AB, Malmö, Sweden) Solid rigid PU foam, density 480 mg = cm^3, compressive strength 18 MPa and cellular rigid PU foam, open cell. The block was covered with a resin to imitate the cortical layer	Fuerst et al. [41]
	Bur-sweep (Resins from Formlabs): 15% Gypsum® powder mixed with 100% Clear® Formlabs resin and 10% Castable® resin mixed with 90% Clear® resin	Hao et al. [42]
	Polymeric material including polyethylene, polystyrene and acrylic	Mire et al. [43]
Vertebral plan Bones	Epoxy resin and high-density polyethylene DuraForm PA (3D Systems Corporation, Valencia, CA), a nylon 12–based material	Adusumilli et al. [44]
	Varying ratios of multiple thermoset polymers by Objet350 Connex3™	Rose et al. [45]
	Multiple Materials of 3D printer (Objet500 Connex; Stratasys, Ltd.)	Waran et al. [46]
	Literature Review on different 3D techniques, materials, and resolution	Hoang et al. [47]
Vertebral plan Spinal cord	Elastomeric material including urethane, rubber, silicone and poly olefin	Mire et al. [43]

Stage 3 and 4—Implementation and Validation
Following the iterative nature of applied research [24], these phases, that have not yet been addressed, will concern the development of the pilot prototype and the collection and evaluation of the feedbacks on the pilot prototype.

3 Partial and Future Results

Although the NTT research project is still under development, some partial results can already be identified.

On the one hand, the ability of the Design Discipline to act as a facilitator between the various experts involved is highlighted. Through the integrated collaboration of knowledge and the sharing of a common language, it was possible to activate a synergistic creative process for the incremental innovation of medical devices in the field of surgical training according to an anthropocentric perspective that goes beyond the logic of functionality in favor of the quality of the experience as a determining factor in the project.

On the other hand, the analysis of the scientific literature review useful for the selection of materials for the simulation of human tissues for surgical training purposes can be recognized as an operational guideline for designers in the medical field.

In terms of long-term results, the research expects to have a remarkable impact first and foremost considering that this innovation will find a direct request by the national health system, starting from the two large hospitals managed by the Sapienza University of Rome: Policlinico Umberto I and Policlinico Sant'Andrea where the donation program for anatomic study and surgery training sees less than 15 donors and only up to then 5 bodies received until 2017. It is worth reminding that in Italy the practice of dissecting corpses for study purposes, scientific research and training, is quite difficult, in addition to the economic aspect of maintenance and management of bodies, due to the scarce spread of a culture of donation of bodies. This situation compels many surgeons to go abroad to learn new techniques or perfect themselves with an impact considerable in financial terms on local health resources.

In the national context, the NNT project would solve these limitations through the development of a completely synthetic system but able to perfectly simulate the physical and mechanical characteristics of portions of the human body.

Furthermore, the use of this system would have substantial economic and ethical effects also at the global level, reducing the management and maintenance costs of cadavers and the widespread use of animals in surgical practices. Besides, the interactive simulation system will be used also for developing robotically assisted procedures in the line of [48].

In terms of Future developments beyond the current state of the art, the NTT project contributes to the advancement of knowledge according to the three areas of expertise involved:

1. Neurosurgery. The anatomical/surgical training system of the spinal column developed by the project will couple, in a completely new way, the benefits of Human performance simulation and Virtual reality surgical Simulation integrating physical feedback with the guidance of augmented reality tools;
2. Bioengineering, Virtual Representation, and 3d Modelling. The process of translating diagnostic images into a 3D model, both digital and physical, will allow the expansion of this technology contributing to adapt the same process to other medical disciplines that need surgical training (i.e. orthopedics, ophthalmology, etc.).
3. Advanced and Human-Centered design for Healthcare. The phases of prototyping and usability validation and the convergence of the different disciplines involved in the project, allow moving from a Product-Centered approach to a Human-Centered approach transforming the product into a product/service with a systemic perspective.

4 Conclusion

This paper draws together findings from a very early stage of an interdisciplinary project in the sector of Medical Design.

Over the years, thanks to the growing interest of Medicine in the potential of Design approaches, from which to draw models of thought and creative practices, this field of experimentation has seen different areas of application, both to improve products, technologies, and healthcare facilities, and to increase the awareness and involvement of clinical staff and patients within the conception, production, and consumption process.

However, the context of Medical Design is still characterized by rigorous constraints linked to the speed and rigor of medical research, on one side, and to the necessity of integrating physical, physiological, and psychological factors of patients and operators, to the other. It is, therefore, essential to adopt shared tools that can contribute to the dialogue of the different expertise usually involved in the development of successful medical devices.

Based on these premises, User/Human-Centered Design (U/HCD) methods can be considered a valuable solution to integrate information from different sides and to promote cross-functional communication between the parties involved to understand a design problem/solution [49].

Following the International Standard ISO 9241-210:2010 [50], the research project here presented has adopted some of the H/UCD tools in the phases of (1) planning the process, (2) understanding and specifying the context of use, (3) specifying the user requirements, (4) producing design solution in order to prefigure shareable goals and activate common procedures.

According to Maguire [51], the H/UCD tools adopted so far by the NTT project were:

1. Usability Planning and Scoping. This strategic activity was performed by bringing together in multiple meetings all the stakeholders relevant to the device development. The aim was to create a common vision on how the Design activities could support the project objectives linking both business and usability goals.
2. Context-of-use Analysis. Through a Context Meeting, this structured method has been able to elicit detailed information about the context of use for the device as a foundation for later design activities, particularly user requirements specification and evaluation.
3. Existing system/competitor Analysis. This method was useful to identify the strengths and weaknesses of devices considered similar for some functions or completely identical to the one to be developed.
4. Brainstorming. This method was widely used in the early phases of design and it was useful for defining the three specific objectives of the project. The Brainstorming sessions were set up by bringing together design and task experts to inspire each other in the idea generation of the problem-solving process.

Considering not only the single aspect but many variables related to the different problems, all the methods used have supported the Discipline of Design in playing the decisive role of facilitator in the device development process. In such problem-solving activity, the adoption of a broader perspective and investigating the circumstances from different points of view, has allowed identifying both the direct and indirect implications, in order to develop an effective design solution and to define a wider user experience.

References

1. Simon HA (1968) The sciences of the artificial. Karl Taylor Compton Lectures
2. Archer B (1981) A view of the nature of design research. Design: Science: Method 1: 30–47
3. Cross N (1993) Science and design methodology: a review. Res Eng Design 5(2):63–69
4. Maldonado T (2003) Disegno industriale: un riesame, vol. 142. Feltrinelli Editore
5. Ito J (2016) Design and science. J Des Sci
6. Tosi F, Rinaldi A (2015) Il Design per l'Home Care. L'approccio Human-Centred Design nel progetto dei dispositivi medici. DIDA Press
7. Manley K. Healthcare design and innovation—Changing the future health experience. Retrieved 1 Dec 2019 from https://ribot.co.uk/stories/healthcare-design-and-innovation
8. Langella C (2019) Design e Scienza. Listlab, Barcelona
9. Chamberlain P, Wolstenholme D, Dexter M (2015) The State of the art of design in health: an expert-led review of the extant of the art of design theory and practice in health and social care. Sheffield Hallam University
10. Giambattista A (2017) Designing care. How design can improve medical products for a therapeutic wellbeing. Des J 20(sup1):S2158–S2167
11. Omachonu VK, Einspruch NG (2010) Innovation in healthcare delivery systems: a conceptual framework. Innov J: Publ Sector Innov J 15(1):1–20
12. Norman DA, Verganti R (2014) Incremental and radical innovation: design research vs. technology and meaning change. Design Issues 30(1):78–96
13. Varkey P, Horne A, Bennet KE (2008) Innovation in health care: a primer. Am J Med Qual 23:382–388
14. Owen C (2007) Design thinking: notes on its nature and use. Des Res Q 2(1):16–27
15. Howard Z, Melles G (20111) Beyond designing: roles of the designer in complex design projects. In: OZCHI 2011 Proceedings of the 23rd Australian computer-human interaction conference: design, culture and interaction. ACM, Canberra, Australian Capital Territory, Australia, pp. 152–155
16. Aboud E, Al-Mefty O, Yaşargil MG (2002) New laboratory model for neurosurgical training that simulates live surgery. J Neurosurgery 97:1367–1372
17. Frati P, Frati A, Salvati M, Marinozzi S, Frati R, Angeletti LR, Piccirilli M, Gaudio E, Delfini R (2006) Neuroanatomy and cadaver dissection in Italy: history, medicolegal issues, and neurosurgical perspectives: historical vignette. J Neurosurg **105**(5), 789–796
18. Kockro RA, Serra L, Tseng-Tsai Y, Chan C, Yih-Yian S, Gim-Guan C, Lee E, Hoe LY, Hern N, Nowinski WL (2000) Planning and simulation of neurosurgery in a virtual reality environment. Neurosurgery 46(1):118–137
19. Gaba DM (2004) The future vision of simulation in health care. BMJ Qual Saf 13(suppl 1):i2–i10
20. Reznick RK, MacRae H (2006) Teaching surgical skills—Changes in the wind. N Engl J Med 355(25):2664–2669
21. Capital IBIS (2016) Global HealthTech investment report. IBIS Capital, London, UK

22. Randazzo M, Pisapia JM, Singh N, Thawani JP (2016) 3D printing in neurosurgery: a systematic review. Surg Neurol Int 7(Suppl 33):S801
23. Sun L, Tian H, Wu D, Du Z (2010) Three-dimensional geometric modeling of the spine based on reverse engineering technology. In: 2010 3rd international conference on biomedical engineering and informatics, vol 3. IEEE, pp 1292–1295
24. Bickman L, Rog DJ, Hedrick TE (1998) Applied research design: a practical approach. Handbook of applied social research methods, **19** (1998)
25. Mast TD (2000) Empirical relationships between acoustic parameters in human soft tissues. Acoust Res Lett Online 1(2):37–42
26. Sparks JL, Vavalle NA, Kasting KE, Long B, Tanaka ML, Sanger PA, Conner-Kerr TA et al (2015) Use of silicone materials to simulate tissue biomechanics as related to deep tissue injury. Adv Skin Wound Care 28(2):59–68
27. Dąbrowska A, Rotaru GM, Spano F, Affolter C, Fortunato G, Lehmann S, Rossi RM et al (2017) A water-responsive, gelatine-based human skin model. Tribol Int 113:316–322
28. Dąbrowska AK, Rotaru GM, Derler S, Spano F, Camenzind M, Annaheim S, Rossi RM et al (2016) Materials used to simulate physical properties of human skin. Skin Res Technol 22(1):3–14
29. Whittle K, Kieser J, Ichim I, Swain M, Waddell N, Livingstone V, Taylor M (2008) The biomechanical modelling of non-ballistic skin wounding: blunt-force injury. Forensic Sci Med Pathol 4(1):33–39
30. Bjellerup M (2005) Novel method for training skin flap surgery: polyurethane foam dressing used as a skin equivalent. Dermatol Surg 31(9):1107–1111
31. Ansari M, Lee SK, Cho CD (2007) Hyperelastic muscle simulation. In: Key engineering materials, vol 345. Trans Tech Publications Ltd, pp 1241–1244
32. Li P, Jiang S, Yu Y, Yang J, Yang Z (2015) Biomaterial characteristics and application of silicone rubber and PVA hydrogels mimicked in organ groups for prostate brachytherapy. J Mech Behav Biomed Mater 49:220–234
33. Hollensteiner M, Fuerst D, Schrempf A (2014) Artificial muscles for a novel simulator in minimally invasive spine surgery. In: 2014 36th annual international conference of the IEEE engineering in medicine and biology society, pp 506–509. IEEE, August
34. Jiang H, Cui Y, Ma K, Gong M, Chang D, Wang T (2016) Experimental reconstruction of cervical esophageal defect with artificial esophagus made of polyurethane in a dog model. Dis Esophaus 29(1):62–69
35. Witowski JS, Jasamine C-B, Tomasz ZZ, Michal P, Jason C, Piotr M, Andrzej B (2017) 3D printing in liver surgery: a systematic review. Telemed J EHealth Official J Am Telemed Assoc 23(12): 943–947. https://doi.org/10.1089/tmj.2017.0049
36. Berman EF (1952) The experimental replacement of portions of the esophagus by a plastic tube. Ann Surg 135(3):337
37. Mashiko T, Otani K, Kawano R, Konno T, Kaneko N, Ito Y, Watanabe E (2015) Development of three-dimensional hollow elastic model for cerebral aneurysm clipping simulation enabling rapid and low cost prototyping. World Neurosurg 83(3):351–361
38. Valverde I, Gomez G, Coserria JF, Suarez-Mejias C, Uribe S, Sotelo J, Gomez-Cia T et al (2015) 3D printed models for planning endovascular stenting in transverse aortic arch hypoplasia. Cathet Cardiovasc Interv 85(6):1006–1012
39. Barsness KA, Rooney DM, Davis LM, O'Brien E (2015) Evaluation of three sources of validity evidence for a synthetic thoracoscopic esophageal atresia/tracheoesophageal fistula repair simulator. J Laparoendosc Adv Surg Tech 25(7):599–604
40. Akiba T, Nakada T, Inagaki T (2014) A three-dimensional mediastinal model created with rapid prototyping in a patient with ectopic thymoma. Ann Thorac Cardiovasc Surg, nm-13
41. Fuerst D, Stephan D, Augat P, Schrempf A (2012) Foam phantom development for artificial vertebrae used for surgical training. In: 2012 annual international conference of the IEEE engineering in medicine and biology society, pp 5773–5776. IEEE, August
42. Hao J, Nangunoori R, Wu YY, Rajaraman M, Cook D, Yu A, Shimada K et al (2018) Material characterization and selection for 3D-printed spine models. 3D Printing Med 4(1):1–17

43. Mire DA, Nycz JH, Adcox WK, Olson SW, Saladino JJ (2010) U.S. Patent Application No 12/255,014
44. Adusumilli P, McCreesh K, Evans T (2014) Development of an anthropomorphic shoulder phantom model that simulates bony anatomy for sonographic measurement of the acromio-humeral distance. J Ultrasound Med 33(11):2011–2019
45. Rose AS, Kimbell JS, Webster CE, Harrysson OL, Formeister EJ, Buchman CA (2015) Multimaterial 3D models for temporal bone surgical simulation. Ann Otol Rhinol Laryngol 124(7):528–536
46. Waran V, Narayanan V, Karuppiah R, Owen SL, Aziz T (2014) Utility of multimaterial 3D printers in creating models with pathological entities to enhance the training experience of neurosurgeons. J Neurosurg 120(2):489–492
47. Hoang D, Perrault D, Stevanovic M, Ghiassi A (2016) Surgical applications of three-dimensional printing: a review of the current literature & how to get started. Ann Transl Med 4(23)
48. Cacciotti N, Cifonelli A, Gaz C, Paduano V, Russo AV, Vendittelli M (2018) Enhancing force feedback in teleoperated needle insertion through on-line identification of the needle-tissue interaction parameters. In: 2018 7th IEEE international conference on biomedical robotics and biomechatronics, pp 79–85. IEEE
49. Veryzer RW, Borja de Mozota B (2005) The impact of user-oriented design on new product development: an examination of fundamental relationships. J Prod Innov Manag 22(2):128–143
50. ISO 9241-210:2019 Ergonomics of human-system interaction—Part 210: Human-centred design for interactive systems. https://doi.org/10.3403/30388991
51. Maguire M (2001) Methods to support human-centred design. Int J Hum Comput Stud 55(4):587–634

Co-designing Resources for Knowledge-Based Self-reflection for People Living with Parkinson's Disease to Better Enable Independent Living

Joe Langley, Rebecca Partridge, Ursula Ankeny, Gemma Wheeler, and Camille Carroll

Abstract Parkinson's disease (PD) is a complex progressive neurodegenerative disease. Individuals experience PD in a variety of ways, leading to difficulty in diagnosis, acceptance and on-going management. Service provision is complex and varied depending on provider, often with a lack of 'joined up' provision between acute hospital care and community care. This project utilised a participatory design methodology to identify patient and provider needs for PD services in the South West Peninsula (UK). A co-design approach was then used to develop tools, resources and service structures to meet these needs. The application of co-design in healthcare settings is fraught by challenges of power dynamics between healthcare professionals and 'patients'. This can create difficulties in eliciting the 'voice' of the patient, or in facilitating their genuine engagement or agency in the process. Specific features of our co-design process sought to address these issues. The tools and resources resulting from this process aimed to facilitate independent living for people with PD, yet importantly to enable them to do so from an informed position, understanding complex medical knowledge in the context of their own personal life and priorities. As such, the resources intended to facilitate both the mobilisation of complex knowledge and self-reflection. The final resource pack has recently entered a feasibility trial of 150 people living with PD.

Keywords Parkinson's disease · Home-based care · Co-design · Knowledge mobilization · Power

J. Langley (✉) · R. Partridge · U. Ankeny · G. Wheeler
Sheffield Hallam University, Sheffield S1 1WB, UK
e-mail: j.langley@shu.ac.uk

C. Carroll
Faculty of Health, University of Plymouth, Plymouth N146 8BX, UK

E. Duarte and C. Rosa (eds.), *Developments in Design Research and Practice*, Springer Series in Design and Innovation 17,
https://doi.org/10.1007/978-3-030-86596-2_18

1 Background

Parkinson's disease is the second most common neurodegenerative disorder. It is caused by death of nerve cells in the brain, primarily those producing the neurotransmitter dopamine, although other neurotransmitter systems are also affected [1]. The main symptoms of Parkinson's are related to movement (motor symptoms): slowness (bradykinesia), stiffness (rigidity) and shaking (tremor). Other aspects of movement, such as balance, speech and swallowing, are also affected. In addition to these movement related symptoms, people with Parkinson's also experience non-motor symptoms [2]. These include problems with thinking and memory, mood, bladder and bowel function, sleep and blood pressure control. The symptoms of Parkinson's vary greatly between individuals as well as with disease duration.

This individualised experience leads to difficulty in diagnosis, acceptance and on-going management. There is currently no cure for Parkinson's. Symptoms are managed through a combination of drugs and lifestyle (i.e. exercise, diet, sleep hygiene), with input from different healthcare professionals such as: Speech and Language Therapists, Dietitians, Occupational Therapists, Physiotherapists, Parkinson's Nurse Specialists and Parkinson's Specialist Consultants [3].

Drug management is complex and gets more complex with each passing year. Dosage and frequency of doses is unique to each individual and has to be empirically determined. Consultants set a 'titration' process, where dose and frequency are iteratively increased by the patient based on guidance from the consultant, until motor symptoms are optimally balanced against any side effects of the medication [3].

The provision of Parkinson's services is complex and varied. Services in the South West Peninsula, UK, cover acute and community settings, cross multiple regional and budgetary boundaries, a large geographical (urban and rural) area and have staffing pressures from unfulfilled roles, long-term sick-leave and increased patient numbers. Current provision following diagnosis is modelled on standard time-locked (usually six monthly but can be up to 2 years due to service pressures) clinic reviews that fail to meet the evolving needs of patients and their carers, and contributes to staff stress. Patients often travel long distances to attend these reviews that are often conducted by doctors who are not Parkinson's specialists, meaning the broad spectrum of disease manifestation is not explored in the consultation, resulting in problems not being identified or treated. Additionally, these reviews often do not result in any change in management.

The lead consultant for PD services in Plymouth had a vision of creating a more "joined up" service that supported people with PD when they needed support, but also enabled them to live their lives as 'normally' as possible. She reached out to the team at Lab4Living to initiate a co-design process giving them complete autonomy over planning and delivering the co-design process, securing an unrestricted educational grant from a pharmaceutical company to support the initiative.

2 Method

2.1 Participants

The participants included: People living with Parkinson's and their families and carers, Parkinson's Specialist Nurses, Community care teams, Therapy Specialists, Consultant Neurologists, Finance officer, a Primary Care physician, Parkinson's Charity representatives and representatives of local health services research groups from Exeter University and Plymouth University.

2.2 Guiding Principles of Co-Design

Within all the co-design work and collaborative work with participants, we followed a key set of principles. These were:

- Valuing different perspectives
- Adapting co-design methods and tools to the needs of the group or individuals
- Visibility and transparency
- Valuing different forms of knowledge
- Less talking, more doing.

More detailed information describing these principles and how they are practically deployed can be found in Langley et al. [4]. The way that these principles are enacted within a co-design workshop is through the use of visual- and design practice-based activities (i.e. creative activities or making mock-ups). In order to build participants' confidence with each type of activity, the skills involved are always 'practised' by setting a quick, 'throwaway' task on a trivial topic first, before asking participants to use the activity purposefully for the project. This is underpinned by the thinking articulated in Langley et al. [5], arguing that these practice tasks not only generate knowledge but additionally manage the group dynamic and hierarchies, re-distributing power and enabling equal and meaningful contribution by all participants.

2.3 Co-Design Activities

Prior to, between and after each workshop, several kinds of design studio activity were required:

1. Planning and preparation: planning the workshops and preparing activity resources and data collection tools

2. Data recording and analysis: digitally capturing all the data from the workshops after each workshop, laying out on a project wall, interrogating the data as *designers,* thinking about the problem-solution frames
3. Reflection: reflecting on the data and the workshop itself as *design researchers,* reflecting on the effectiveness of the workshop methods and tools as 'stewards' of equity and power balance between stakeholders, and as tools for knowledge creation and sharing between stakeholders
4. Making: making resources and data collection tools for the subsequent workshops, making prototypes to stimulate, provoke or build on ideas and develop them further in preparation for the next workshop
5. Communication: preparing a summary of each workshop and the studio activity between workshops to feedback to participants. Additionally, participants were asked to complete on-going data collection activities between workshops.

A second phase of four further co-design workshops was required to take four of the developed concepts and turn them into final outputs to be used in the next phase of the project—development of the Home Based Care Pathway. These are outlined below:

This second phase of co-design work for implementation was followed by a final phase of intensive design and feedback iterations as the last details of the content for piloting were developed and refined, using both digital and some physical prototypes, emailed or mailed to staff and patient representatives for them to review, try and critically respond to.

3 Results

3.1 Process

Phase 1 co-design work rapidly introduced us to the physical needs of people living with Parkinson's in the context of co-design. Our design partners gave us valuable feedback about avoiding writing activities and the need to accept big and 'wonky' writing if necessary, the use of big Lego, plenty of water and toilet breaks, accommodating frequent medication reminders and many other practical considerations to collaborating with them.

The phase 1 co-design work identified the needs of different stakeholders, specified the problems with current services (as well as areas that are working well), generated and developed ideas and prototyped them. The methods used (described in Tables 1 and 2) allowed individual reflection, the sharing of all perspectives and collaborative prioritization of key issues. Time and space was given to gradually identify root problems and shared goals of the group, and the visual nature of data collection tools represented knowledge and input of all participants equally. All of this helped to level hierarchies, engage participants meaningfully, facilitate mutual learning between participants and, in turn, generate valuable (and valued) solutions.

Table 1 First phase co-design workshops (approx. 12 months) and design tools used in each

Workshop number and attendees	Purpose of the session	Design tools used
Workshop 1 (n = 36 + 2 design researchers)	Initiating the collaboration: Setting expectations, sharing experiences and creating a set of ground rules for our co-design process	Lego serious play

(continued)

Table 1 (continued)

Workshop number and attendees	Purpose of the session	Design tools used
Workshop 2 (n = 17 + 2 design researchers)	Understanding multiple facets of the current service, including: • What are the current service experiences? • Who are the (multiple) service users, and what are their service journeys like? • How does the broad geographical 'jurisdiction' of the service affect the way it runs? • What does a Parkinson's disease trajectory look like? • What would an ideal service look like?	Service journey mapping Personas Disease trajectory models Ideal service touchpoints

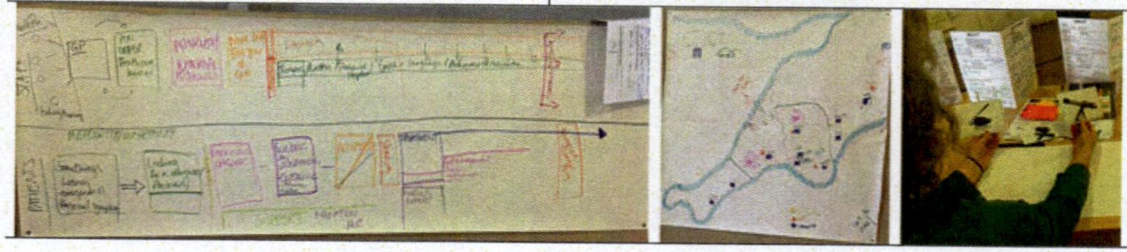

(continued)

Table 1 (continued)

Workshop number and attendees	Purpose of the session	Design tools used
Workshop 3 (n = 22 + 2 design researchers)	Ideas development, prioritisation and prototyping: Ideation, idea exploration and development, sorting and prioritisation exercises, knowledge stratification exercises (what levels of knowledge family, friends, co-workers, general public etc. might need about PD) and prototyping methods, selection of five top ideas	Ideation games Mock-ups Body storming Role plays Impact versus effort charting

(continued)

Table 1 (continued)

Workshop number and attendees	Purpose of the session	Design tools used
Workshop 4 (n = 20 + 5 design researchers)	Design 'Hack': Supported by a larger team of designers, the participants were divided into five teams and each allocated an idea and a designer. Ideas were 'fast-tracked' through prototyping, feeding back to the wider group at critical steps. Plans were drawn up for prototype refinement, production and delivery to key stakeholders for testing in context	Prototyping

(continued)

Table 1 (continued)

Workshop number and attendees	Purpose of the session	Design tools used
Workshop 5 (n = 23 + 2 design researchers)	Presentation of prototypes, testing and planning: Prior to this workshop the designers spent time developing the ideas further, creating high-fidelity prototypes to the specification set (collaboratively) in the previous workshop. These were presented to the group who generated plans for the 'what next' such as testing, implementation or further funding. This workshop was also a celebration event the project so far	Prototype presentations and role plays Party poppers

Table 2 Second phase co-design workshops (approx. 14 months) for implementation

Workshop identification	Purpose	Design tools
Workshop 1 (n = 14 + 2 design researchers)	Preparing for implementation: Combining the technology and the concepts from first phase co-design into Home Based Care Pathway. Understanding service resource requirements. Identifying the stresses of implementation, trialling and evaluation of the new pathway	Technology and concept prototypes and probes Service mapping Blue prints and visualization tools Implementation planning and stress identification
Workshop 2 (n = 18 + 2 design researchers)	Service resource mapping and stress testing: Exploring the service blueprints from previous session and adding greater detail to the resource requirements. Subsequent scenario-based (role play) stress testing of the pathway using personas	Service and resource mapping Personas Scenarios and role play
Workshop 3 (n = 30 + 1 design researchers)	Detailed co-design of patient facing data and reports: This session focused on one crucial aspect - the reports that patients will receive with data about their motor and non-motor symptoms. The group co-designed the reports to ensure they were meaningful and useful (i.e. would be able to prompt action) It was important to enable patients to make sense of this data and then to consider it in the wider context of their life, family and personal goals	Data visualisation and information presentation modelling activities Interface and report design
Workshop 4 (n = 22 + 1 design researchers)	Launch event: Presentation of the whole patient pack, 'live' user testing and feedback	Prototyping testing and user feedback

In this way, the co-design tools can be viewed as tools for *knowledge creation* and sharing, and as 'stewards' of equity.

The phase 2 co-design work was focused more on refinement rather than generation of new ideas. This involved prioritizing and consolidating the multiple concepts generated in phase 1, but also ensuring that these concepts would work in the real, complex contexts of healthcare services and people's lives. In other words, 'design for implementation.' The importance of the continued involvement of all stakeholders at this stage (unlike many other examples of co-design in healthcare) cannot be underestimated. Designerly methods (such as prototyping and role play) allow low-risk, early testing of ideas before they are too detailed, and elicit the tacit knowledge from participants' lived experiences (as service providers and users). This can 'flag' potential challenges earlier and 'fix' them prior to implementation (saving time and money, as well as reducing risk, compared to learning these lessons 'live').

Co-design processes can often be evaluated by the quality and efficacy of the outputs developed from them (as described further below). Beyond this, though, feedback from participants themselves has been positive:

> This project has had a different approach and an interactive style. It's made me work with people who I would not normally interact with. We have worked with patients and their families. They have been at the heart of everything we have discussed and addressed. It's been genuine co-design with patients.
>
> STAFF MEMBER, SW PENINSULA PARKINSON'S SERVICE.

> I've really enjoyed the exposure to different methods and ideas to support creative thinking. It's challenged me to think outside the box
>
> PERSON LIVING WITH PARKINSON'S.

The high participant retention rate (shown in Tables 1 and 2) also suggests that participants felt valued in the process by choosing to continue their involvement in the project. Travelling to, and staying for the duration of, full-day workshops can be challenging considering the symptoms of people living with PD, and demanding workloads of PD healthcare professionals. Indeed, whilst this is not often discussed in academic literature, care was taken not just in the planning of effective co-design tools, but importantly also in the creation of a comfortable, relaxed environment for participants. We have found that the design researcher often plays multiple roles in such sessions, being receptive and reactive to the mood and needs of participants as the first priority.

3.2 Outcomes

Four of the five original concepts developed in phase one[1] have been combined into one cohesive output - the Home Based Care Pathway.

The Home Based Care (HBC) Pathway includes a pack of information and tools, owned by the patient (Fig. 1), to support three key elements; self-management, triggered contacts and remote monitoring. The pack includes; a Parkinson's patient passport (Fig. 2), New service and local information (Fig. 3); a card deck to support self-reflection (Fig. 4); a self-management support and general information package (Fig. 5).

[1] The fifth concept, a media campaign which aimed to 'mythbust' common misconceptions around PD, will not be developed further in this project.

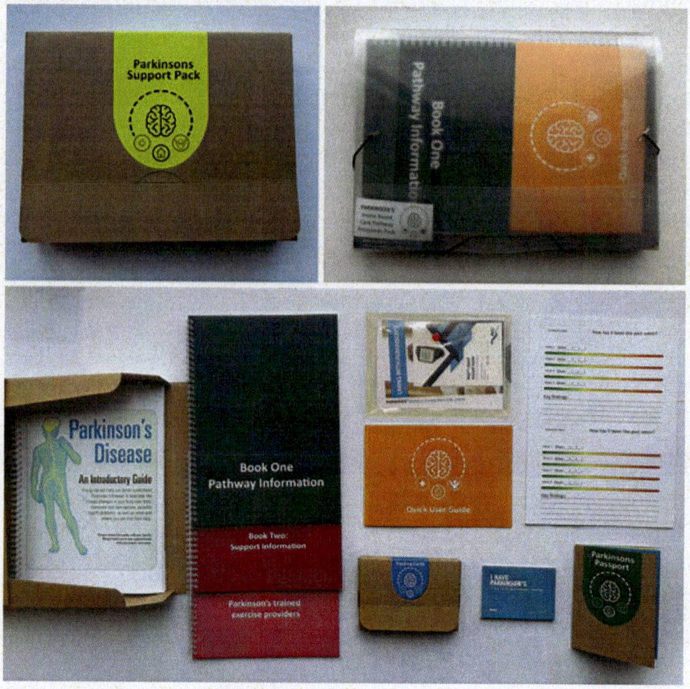

Fig. 1 The Home Based Care Pathway pack as it arrives with the patient

Fig. 2 Parkinson's patient passport

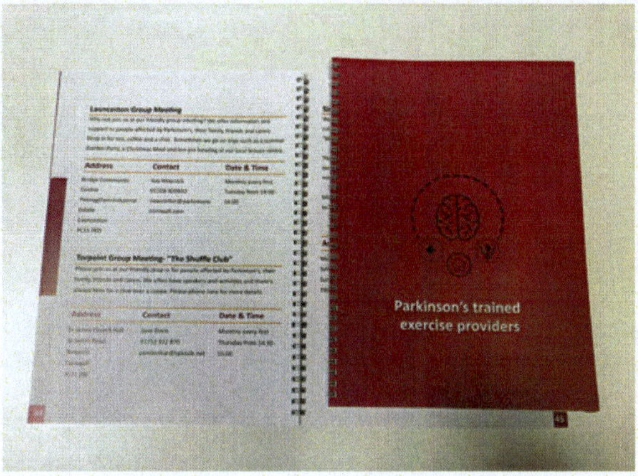

Fig. 3 New service and local information

Fig. 4 A card deck to support self-reflection

The HBC Pathway is currently being evaluated and further refined with 150 patients and three separate Parkinson's care delivery teams in Plymouth, West Devon and East Cornwall.

4 Discussion

Within healthcare, one of the biggest challenges is getting new ideas into practice [6]. Co-design or coproduction is being offered as a possible mechanism for addressing

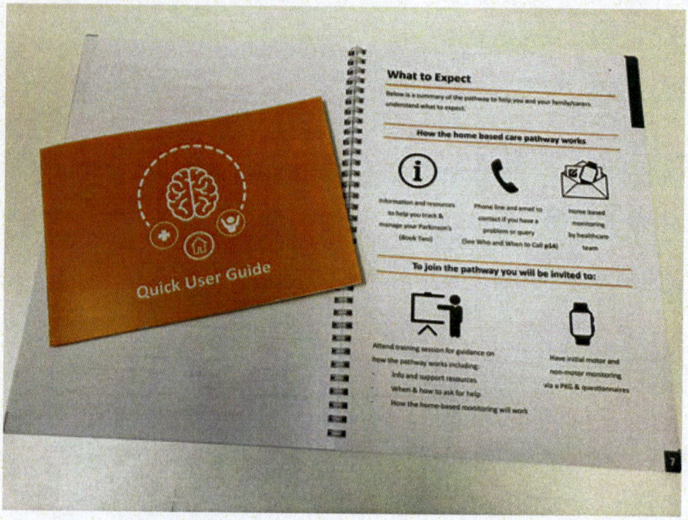

Fig. 5 A self-management support and general information package

some of the issues that cause this translational gap [7]. However, the dominant epistemology in health services research is one that privileges positivistic, quantitatively measurable data over experiential or tacit knowledge, so it is questionable whether 'real' coproduction can genuinely happen while hierarchies of evidence exist [8].

We have suggested that the creative practices of design [5], used within a co-design process, help to reduce or even remove these hierarchies, give patients agency and voice and enable the effective embodiment (within the prototypes) of the experiential knowledge of professionals and patients as well as the latest scientific knowledge of researchers and research literature. It is because the outputs of co-design processes embody these different forms of knowledge, that they become easier to implement – and reduce the translation gap [9].

Framing the design and co-design process in this way; as a knowledge creation and knowledge mobilization process, has been effective for the authors at enabling them to work in the healthcare sector, with healthcare professionals and health services researchers. It is a 'lens' that they can understand and appreciate the value of, as it relates directly to addressing a known challenge of getting research into practice.

We would encourage other designers wanting to work in this space to consider framing their work in this way, and perhaps more importantly, to all designers who use co-design, to consider their process in this way - as one of eliciting and embodying experiential knowledge of users and service providers, along with contextual knowledge and the latest research evidence from the relevant field. This can support 'evidence-informed' design outputs without stifling creativity and imagination.

5 Conclusions

The participatory design methodology described in this paper has been shown to be effective in addressing the power dynamics inherent in co-design in healthcare. By striving for equal, meaningful involvement of patients, carers and healthcare professionals, a care pathway has been designed, supported by a resource pack to empower people living with PD in the assimilation of complex medical knowledge in the contexts of their own lives. This, in turn, hopes to support their independent living from an informed position.

Acknowledgements With thanks to our design research colleagues (Chris Redford and Helen Fisher) and all our co-design partners from Devon and Cornwall. We wish to acknowledge University Hospitals Plymouth NHS Trust and University of Plymouth, who provided administrative and clinician support for the work. Finally, we also wish to acknowledge funding from Bial Pharmaceutical for the first five workshops and from The Health Foundation and Parkinson's UK for the subsequent four workshops.

References

1. Cuenca L, Gil-Martinez AL, Cano-Fernandez L, Sanchez-Rodrigo C, Estrada C, Fernandez-Villalba E, Herrero MT (2019) Parkinson's disease: a short story of 200 years. Histol Histopathol. 34(6):573–591. https://doi.org/10.14670/HH-18-073. Epub 2018 Dec 12. Review. PubMed PMID: 30540129
2. Jankovic J (2008) Parkinson's disease: clinical features and diagnosis. J Neurol Neurosurg Psychiatry 79(4):368–376
3. Armstrong MJ, Okun MS (2020) Diagnosis and treatment of Parkinson disease: a review. JAMA 323(6):548–560. https://doi.org/10.1001/jama.2019.22360.Review PubMed PMID: 32044947
4. Langley J, Wheeler G, Partridge R, Bec R, Wolstenholme D, Sproson L (2020) Designing with and for older people. In: Woodcock A, Moody L, McDonagh D, Jain A, Jain LC (eds) Design of assistive technology for ageing populations, vol 167. Springer, Intelligent Systems Reference Library, pp 3–19. https://doi.org/10.1007/978-3-030-26292-1. eBook ISBN 978-3-030-26292-1, Hardback ISBN 978-3-030-26291-4
5. Langley J, Wolstenholme D, Cooke (2018) 'Collective making' as knowledge mobilisation: the contribution of participatory design in the co-creation of knowledge in healthcare. BMC Health Serv. Res. 18:585. https://doi.org/10.1186/s12913-018-3397-y
6. Greenhalgh T, Wieringa S (2011) 'Is it time to drop the "knowledge translation" metaphor? A critical literature review.' J R Soc Med 104:501–509
7. Greenhalgh T, Jackson C, Shaw S, Janamina T (2016) Achieving research impact through co-creation in community-based health services: literature review and case study. Millbank Q 94:392–429
8. Rose D, Kalathil J (2019) Power, privilege and knowledge: the untenable promise of co-production in mental "health." Front Sociol 4:57. https://doi.org/10.3389/fsoc.2019.00057
9. Rycroft-Malone J, Seers K, Titchen A, Harvey G, Kitson A, McCormack B (2004) What counts as evidence in evidence-based practice? J Adv Nurs 47(1):81–90

Design for Education

Design Education: A Trend in the Right Direction…

Nicos Souleles, Violeta Clemente, and Naz A. G. Z. Börekçi

Abstract The chapters in this section of this publication share a common denominator. From various perspectives, they deal with design education as transcending the narrow confines of a nascent design discipline. They support the notion that designerly ways of knowing require the synthesis of cognitive skills that have relevance to a wide spectrum of contexts. Thus, they represent a trend in the right direction and confirm the shift in the epistemology of design towards a wider social role for design. The chapters deal specifically with teaching and learning issues that enhance the role of design in this new context and reveal the concerns of the authors as educators to seek instructional ways that can support this shift.

Keywords Design education · Cognitive skills · Employability · Interdisciplinarity · Life-long learning · Creativity

1 Introduction

Design education aims to provide learners with the necessary skill set as the foundation for the continuous and sustained progression of designerly practice and thinking [1]. Hands-on and project-based learning, well-grounded on theoretical and methodological frameworks that draw from practice and research and informed by field-related knowledge, are essential to the educational process. Moving on from design education to design practice, novice designers are expected to constantly develop professional knowledge through iterations of designerly practices, eventually gaining

N. Souleles (✉)
Cyprus University of Technology, Limassol, Cyprus
e-mail: nicos.souleles@cut.ac.cy

V. Clemente
University of Aveiro, Aveiro, Portugal
e-mail: catarina.clemente@ua.pt

N. A. G. Z. Börekçi
Middle East Technical University, Ankara, Turkey
e-mail: nborekci@metu.edu.tr

E. Duarte and C. Rosa (eds.), *Developments in Design Research and Practice*,
Springer Series in Design and Innovation 17,
https://doi.org/10.1007/978-3-030-86596-2_19

expertise at a level where they can in turn contribute to knowledge. Design education is how future designers gain the learning experiences that help develop the ability of agile navigation through the design process, with designerly thinking and practices at the core of the intellectual agility to be attained [17]. To keep up with the fast pace of the evolving field of design practice, learners need to be qualified with the abilities to identify opportunities in the challenges that are expected to arise during their career and respond to them through paradigm shifts in practice [8]. This requires a mindset for sustained progression as a manifestation of experience-based design expertise. It requires a permanent state of professional development.

In constantly changing social contexts where the demands and requirements of diverse consumers are moving targets, there is an increased need for design professionals with the wide perspective required to deal with multifaceted challenges. Given the increasingly complex and interdisciplinary problems humanity faces, inevitably design is either part of the problem or the solution. Due to its transformative potential, design can play a significant role in dealing with such challenges [19]. Consequently, there is a definite and growing need to invest in pedagogies that can help learners develop skills beyond learning how to think in intellectual silos [4, 13].

Based on studies carried out with experienced design practitioners, Cross [7] made three observations. Designers bring a systematic approach to exploring a problem as a whole; they frame design problems from their perspectives; and they develop creative design solutions from foundational principles, by engaging with forms and structures appropriate for the identified requirements. Design education endeavours to provide learners with the ability to be solution-oriented when dealing with ill-defined problems [9]. Also, learners need to develop the ability for synthesis based on a thorough exploration through design divergence and critical and timely decision-making through the convergence of various ideas. Creativity and evaluation are two modal shifts that parallel divergence and convergence during the design process [2], and design education strives for this primary objective. Thus, we need to determine courses of action for how learners can become aware of the complexities of a sustainable world and contribute to improvements in society.

The evolution of design from crafts, standardised methods and emphasis on symbolic and visual communications, through to engagement with organised services, stakeholder methods and complex systems and environments, came with a corresponding evolution in design education [14, 16, p. S930]. As design evolved from a practitioner's field to an academic discipline, the epistemological and methodological aspects of the discipline were acknowledged by other academic areas [5]. One can discern this characteristic among the seemingly diverse chapters in this part of the book. They represent a trend in the right direction, a problematization of how design education can foster the competencies required from design learners to succeed in an advancing, global and transdisciplinary design milieu. At the same time, these chapters fortify the academic value of design by demonstrating its uniqueness as an inherently cross-disciplinary area both in content and application.

2 The Chapters

The fundamental issue of the acquisition of the essential skills and competencies that are transferable to varied contexts is addressed in the chapter 'Teaching and Learning Soft Skills in Design Education, Opportunities and Challenges: A Literature Review'. The significance of soft skills, in general, has been articulated previously as well as the appropriate instructional strategies to foster them in Higher Education (HE) curricula, and the authors refer to a total of 65 articles on this subject. The question is what is unique to the design disciplines *vis-a-vis* soft skills, because—as the authors argue—"There are still few studies on soft skills in the context of design education". Employability emerges from this chapter as a desired graduate attribute, and to achieve it the prescribed pedagogies—not surprisingly—entail active and cognitive teaching and learning strategies. In unpredictable environments, design graduates may be confronted with diverse and multifaceted challenges that have explicit social relevance and impact. Their cognitive skills will be tested.

All this is inward-looking. One looks at curricula and attempts to prescribe the appropriate balance of teaching and learning strategies that can enhance graduate employability. In this respect, the 'Subject Benchmark Statement: Art and Design' provides a useful guide for the required skill set to be fostered in design curricula. In this pedagogical benchmark, the reference to graduates that can "apply, consolidate and extend learning in different contexts and situations, both within and beyond the field of art and design" [11 p. 9], confirms educational congruence on the objectives of design.

Another approach is to look at design curricula from an external viewpoint, from that of ex-graduates who are now design professionals. How do they sustain employability? They adopt informal life-long learning patterns, i.e. they continuously acquire knowledge outside the formal education system [15]. A sustained attitude of life-long learning can enhance employability prospects, as not all can be known about future professional challenges and a lot is learnt—informally—in situ. Estimates suggest that this form of learning comprises anywhere between seventy to eighty percent of the learning that takes place in the workplace [10, 15].

An insightful case study on how pedagogy can foster the competencies required to undertake complex design tasks is dealt with in the chapter 'Can the Pedagogy of Sheila Levrant de Bretteville be Considered a Relevant Model for Adapting Design Education to Global and Local Contexts?'. The author puts forward the proposition that Sheila Levrant de Bretteville—a multi-award-winning graphic designer with a lifelong interest in communal forms of art and feminist design—advocates through her teaching and learning the need for emancipatory education and interdisciplinary collaboration to facilitate critical thinking and social engagement.

The authors present some of the instructional tenets of Sheila Levrant de Bretteville with the argument that these promote critical thinking. The instructional value of cross-disciplinary collaborations in communities of practice is widely acknowledged and thus not a unique educational concept in her armoury of instructional tools. However, the pursuit of a subjectivist expression of design where form reveals ideology makes explicit design's political dimensions. One way to encourage politically and socially informed learners is to open a series of dialogues with them, and this Sheila Levrant de Bretteville did, often by triggering an exchange of ideas about an object that has multiple interpretations. All the different interpretations are part of the same narrative. Critical thinking, empathy, subjective and political expression are cultivated through productive tension, the acceptance of contradictory worldviews, teaching as a horizontal exchange, participative methodologies, a collaborative relation between learner and teacher and a direct connection to wider society. Her overall approach is described as the emancipative model of design education and has obvious overlaps with design activism and critical design [14].

Together with an attitude for life-long learning and the ability to apply critical thinking, creativity is another essential skill and an instructional challenge not only for design curricula. There is extensive literature on this multidimensional and seemingly abstract term and in particular how to foster it in teaching and learning [10, 15]. Triggers and factors that can promote creativity can be divided broadly into three categories: a) external causes such as the role of the environment (various external inspirations and triggers) and/or a method that can facilitate the creative process (e.g. collaboration, ideation), b) internal factors and attributes that make creativity possible (e.g. personality, inquisitiveness, attitude) and c) a combination of internal and external factors. The latter category encompasses multidimensional considerations and departs from the emphasis on individual creative self-expression; it is better placed to help prepare graduates for the professional challenges ahead.

The chapter 'Idea Generation Using the Fictionation Design Tool in an Interactive Prototyping Course for Industrial Designers' adopts an inward-looking approach, i.e. it is a case study with critical reflection from within the learning environment on how to introduce curricula learning strategies that can foster creativity. At the start of design projects, we often observe some learners who remain limited to the boundaries they embrace for themselves. They do not explore and research for further development and generation of ideas. The innovative aspect of this case study based on a product design curriculum is that it seeks to harness the collective creative ideas through a group reflection process by a community of practice that elaborates on hypothetical scenarios—*Fictionation*—that explore imaginary what-if scenarios. What if a product were designed for an extraordinary user? What would the form of the product be? How would the product's function change, and how would the user interact with the product? Beyond the potential of a collective generation of new creative ideas, one can discern that through *Fictionation*, learners need to reflect on user empathy, a skill of particular significance for participatory and co-design approaches to design.

The transference of designerly ways of knowing to other disciplines is evidence of the value of design epistemology. Cash [3, p. 97] argues that while design draws extensively from different academic areas, the reverse does not occur. One way to overcome this perceived weakness is to strengthen design education practices in ways that demonstrate value for other domains [18, p. 64]. This is the theme of the case study in the chapter 'The Value of Design Tools: Using Design Tools to Teach Psychology'. Positive psychology deals with human well-being. The author describes and analyses how a designerly approach was used to help postgraduate learners of psychology to think like designers and approach learning from a more horizontal perspective of knowledge transfer—thinking about people and their contexts.

Groups of students brainstormed, discussed, reflected and *empathised*, using a set of predetermined canvases where outcomes for each stage were recorded. They followed a series of moderated and structured stages of a cognitive scaffolding approach towards knowledge generation, and they concluded with the identification of design opportunities and a designerly-generated outcome. One can discern from this staged process the user-centred and evidence-based characteristics of design thinking, such as discovery, interpretation, ideation, experimentation and evolution [14]—the transference of designerly ways of knowing.

3 Afterword

The authors of this chapter came together not known to each other previously and not familiar with each other's work. Coming from three different institutions, speaking different languages and representing different cultures, it turned out to be rather easy to find common ground based on a shared passion for design education. This passion emerged from a common understanding of design education and practice, in which the underlying mindsets and skill sets are agreed upon and shared, and the pool of knowledge related to history, theory and methodology are available to refer to, use, improve and build upon. On the wider discussions on the current state and future of design education, we noticed it builds on a common understanding; designerly thinking and practices can be beneficial for many fields of endeavour. Academics from different locations, with different perspectives, perceptions and cultures of design education work on the terms of this common understanding, thus making collaborations across borders and disciplines possible.

We were fortunate enough to have met before the outbreak of the Covid-19 pandemic, when the writing of this chapter started. Due to the pandemic, the collaboration continued by other means to which we adapted quickly. Drawing a parallel between our collaborative effort and distant design education, we see that, in many respects, it is easy to share knowledge on practices in design education. We exchanged information, theories, methodologies, expertise and experiences. The similarities confirmed our approaches, whereas our differences set examples for alternative ways of doing things. We developed new perspectives, new tools and new measures to carry out distant design education, and this encouraged us to think about the new designerly skill sets and mindsets required for the future. Challenges in design education are opportunities for improvements in design practice. The evolution is from different paths but towards widely shared common goals. The professional patterns diverge, but the objectives converge. This was possible due to a common language that embraces the fundamentals of design education and acknowledges its trend in the right direction.

References

1. Börekçi NAGZ (2018) Design divergence using the morphological chart. Des Technol Educ: Int J 23(3):62–87
2. Börekçi NAGZ (2017) Visual thinking styles and idea generation strategies employed in visual brainstorming sessions. Des Technol Educ: Int J 22(2):1–19
3. Cash PJ (2018) Developing theory-driven design research. Des Stud 56:84–119
4. Clemente V, Tschimmel K, Vieira R (2018) A metaphoric thinking styles taxonomy to promote students' metacognitive self-reflection in design learning. In: Back to the future. The future in the past: ICDHS 10th+ 1 Barcelona 2018: conference proceedings book. Edicions de la Universitat de Barcelona, pp 679–683
5. Clemente V, Tschimmel K, Pombo F (2017) A future scenario for a methodological approach applied to Ph.D. design research. Development of an analytical canvas. Des J 20(1):S792–S802
6. Clemente V, Vieira R, Tschimmel K (2016) A learning toolkit to promote creative and critical thinking in product design and development through Design Thinking. In: 2016 2nd International conference of the Portuguese Society for Engineering Education (CISPEE), IEEE, pp 1–6
7. Cross N (2006) Designerly ways of knowing. Springer, London
8. Davis M (2017) Teaching design: a guide to curriculum and pedagogy for college design faculty and teachers who use design in their classrooms. Allworth Press, New York
9. Lawson B, Dorst K (2009) Design expertise. Architectural Press, Oxon
10. Leslie B, Aring MK, Brand B (1998) Informal learning: The new frontier of employee development and organizational development. Econ Dev Rev 15(4):12–18
11. QAA (2019) Subject benchmark statement art and design. The Quality Assurance Agency for Higher Education, London
12. Souleles N, Ferreira AM, Savva S (2019) Threshold concepts and design for social change. In: International conference on applied human factors and ergonomics. Springer, Cham, pp 80–89
13. Souleles N, Savva S, Parmaxi P (2019) Perceptions of design students on creativity and social networking sites [Paper presentation]. In: 2019 INTED conference, Valencia, Spain
14. Souleles N (2017) Design for social change and design education: social challenges versus teacher-centred pedagogies. Des J 20(1):S927–S936. https://doi.org/10.1080/14606925.2017.1353037
15. Souleles N, Savva S (2015) Sources of informal learning and undergraduate graphic design students: a grounded theory approach [Paper presentation]. In: 2015 EDULEARN15 conference, Valencia, Spain
16. Souleles N (2013) The evolution of art and design pedagogies in England: influences of the past, challenges for the future. Int J Art Des Educ 32(2):243–255
17. Tovey M (2016) Design pedagogy: developments in art and design education. Routledge, Oxon and New York
18. Voûte E, Stappers PJ, Giaccardi E, Mooij S, van Boeijen A (2020) Innovating a large design education program at a university of technology. She Ji: J Des Econ Innov 6(1):50–66
19. Wilde D (2020) Design research education and global concerns. She Ji: J Des Econ Innov 6(2):170–212

Teaching and Learning Soft Skills in Design Education, Opportunities and Challenges: A Literature Review

Ana Paula Nazaré de Freitas and Rita Assoreira Almendra

Abstract Soft skills are subjective, behavioral, and socio-emotional competences, also called "21st Century Skills", "Key Competences", "Generic Competences", and "Core Competences". They are usually related to creativity, problem solving, communication skills, empathy and team- work. They have been considered differential and necessary skills for all professions in the twenty-first century. The main objective of this study was to trace the state-of-the-art of soft skills in higher education in design. To this end, we sought to identify the importance of soft skills for design, as well as the teaching methodologies, approaches and practices adopted. A systematic literature review was conducted to synthesize the relevant academic literature about soft skills in design. A total of 65 different articles were selected, of which eleven met the predefined inclusion criteria. The results show that soft skills have been perceived as very important for employability in this area, but also as a way to improve project results, both from a functional approach as well as concerning sustainability and ethics. The most adopted teaching methods are active methods, with a constructivist approach and collaborative practices, developed in groups and in exchange milieus. In addition, 17 key competences mentioned as important in design practice were identified. However, this study has shown that there are problematic issues regarding the ambiguity of terms and the lack of precise definitions of skills, as well as the lack of tools to assess skills learning.

Keywords Soft skills · Design education · Review · Active learning

A. P. Nazaré de Freitas (✉)
CIAUD, Lisbon School of Architeture, Universidade de Lisboa, Rua Sá Nogueira, Alto da Ajuda, 1349-063 Lisbon, Portugal
e-mail: anapaulafreitas@campus.ul.pt

Universidade do Estado do Pará, Tv. Dr. Enéas Pinheiro, 2626 - Marco, Belém, PA 66095-015, Brazil

R. Assoreira Almendra
CIAUD, Lisbon School of Architeture, Universidade de Lisboa, Rua Sá Nogueira, Alto da Ajuda, 1349-063 Lisbon, Portugal

E. Duarte and C. Rosa (eds.), *Developments in Design Research and Practice*, Springer Series in Design and Innovation 17,
https://doi.org/10.1007/978-3-030-86596-2_20

1 Introduction

The configuration of the contemporary world, with the challenges of globalization, emerging technologies, and the social and economic impact of the neoliberal system, poses challenges on many fronts. In the field of design this is no different. The subject's paradigm shift, as well as the expansion, the deepening and the dissemination of its practices, mindset, solutions and culture in contemporary societies, present challenges for design education. That is because much of the teaching in the area of design is still largely related to a modern tradition that is not consistent with the changes and paradigms of contemporary times. Design is now a fundamental and inherent part of society, and the work of the designers is not limited to physical products, but has expanded to actions on organizational structures, social issues, interaction issues, services, and experiences [1].

Several authors [2–4] have identified the need to rethink and discuss design teaching models. The usual criticism is that the models are still very much tied to traditions and do not encourage more critical thinking and the awareness that designers have a great responsibility in creating a viable and sustainable future. Design education needs to be updated from a context education that must come from the economic, historical and social understanding of post-industrial society. There is a need to instigate in the students a thinking that is complex, deep and critical in order to ensure that a range of social and ethical commitments are met [4].

The changes in design teaching indicated by these authors converge to the development of some competences that arise as fundamental skills in contemporary society, both for the exercise of design and for other areas. We intend here to emphasize skills that are not quite specific to design, but which are essential to its good development. Such competences have been highlighted in several studies developed by entities thinking about education [5, 6] and that have been drawing attention to the need to teach and learn what we now call "soft skills".

Soft skills (also called "generic skills", "key skills" and "core skills") are social, subjective and emotional competences. They are transversal to various kinds of professional practice and have been increasingly pointed out as important and differentiating by education professionals, researchers and public policies agents [7]. Soft skills are different from hard skills because they are not attached to a specific profession. These competences cover a range of different and independent skills that are rarely explicitly evaluated during the educational process [8, 9]. They are related to creativity, problem solving, personal development and communication, teamwork and cooperation, emotional intelligence and leadership [10–12].

Several studies [7, 13] indicate that soft skills are positive differentials for students not only to attain their academic goals but also concerning their professional trajectory after graduation. There is a growing awareness that hard skills, even for technical positions, are not enough to succeed beyond early career [14]. There are also studies that bring evidence of the relationship between soft skills and academic performance, that is to say, they could lead to content learning [15].

Soft skills can be taught and learned in educational settings [13, 15–18]. There is research indicating that learning is improved when it happens through the study of formal subjects, namely, soft skills are best taught when they are incorporated in the learning of some other subject [13, 19]. Which means that these skills should arise and be practiced during the teaching of content related to hard skills. However, this learning happens most effectively when some active teaching approach practices are used [20].

There are still few studies on soft skills in the context of design education. Although design require skills such as communication, empathy, teamwork and multidisciplinarity, little has been explored regarding how to teach and how to learn these skills.

This systematic literature review study has four objectives. The first is to identify the importance of soft skills in design in higher education. Here we seek to verify the importance attributed by the research authors to these competences. The second objective is to describe the methodologies used in teaching soft skills and relate them to active teaching methodologies. The third objective is to identify the soft skills that are important in design education. The last objective is to trace the state-of-the-art of soft skills teaching in design education.

2 Method

2.1 Systematic Literature Review

The systematic literature review is "a review of a clearly formulated question that uses systematic and explicit methods to identify, select, and critically appraise relevant re- search, and to collect and analyze data from the studies that are included in the review." [21, p. 1]. It is a scientific method that aims to systematize studies in a concise and reliable manner.

In this research, the review focused on the literature about soft skills in design education. The study was conducted based on the Preferred Reporting Items for Systematic Reviews and Meta-Analyzes (PRISMA) approach [21].

PRISMA is a systematic review tool consisting of 27 items that aim to increase the reliability of review surveys and to reduce biases. The PRISMA flow diagram was also used to report the research developed.

2.2 Search Terms

The research was conducted using four data sources: Scopus, Web of Science, Science Direct and ERIC, which were elected due to their wide and recognized utilization, impact and approval in the field of research in education and applied social sciences.

The search action aimed for the terms "soft skills", "generic skills", "key skills", "core competences", "twenty-first century competences", "design education", and "design teaching".

The string or the Boolean search conducted was: ("soft skills" OR "generic skills" OR "key skills" OR "core competences" OR "twenty-first century competences") AND ("design education" OR "design teaching"). The objective was to find studies that had within their title, abstract or keywords one of the keywords defined in the string (since keywords are more like "key expressions", quotes were used so that the result could capture the exact expression).

The search comprised all studies published between 2004 and October 2019 and was conducted between 1 and 5 October 2019 (15 years of study).

2.3 Eligibility Criteria

The eligibility criteria adopted were:

1. Articles that address studies developed in the context of Design in Higher Education.
2. Articles written in English or Portuguese.
3. Peer-reviewed Articles or Conference Papers

2.4 Study Selection

The selection of articles followed three stages. Initially, the titles according to the eligibility criteria, then the abstract of each article selected in the first stage, and afterwards the full text of all remaining publications were analyzed. All articles considered relevant were coded as per: authors' names, date of publication, journal, type of study, term adopted to refer to the competences, the soft skills mentioned, the teaching methods identified, the importance of the soft skills listed by the authors as well as the design areas to which they refer.

The final part of the process of content analysis consisted of making connections between the information collected in the articles and the theoretical framework initially proposed. That is to say, connect the active methodologies and design education with the soft skills and, thus, trace a state-of-the-art of soft skills studies within design teaching.

We identified 72 articles in the database, seven of which were duplicated. After sorting the title and abstract, 65 articles were fully read, of which only eleven articles met the inclusion criteria. The main reason for excluding the articles in the three phases was that they belonged to another area of study, other than design. Twenty-six papers were on engineering, 21 were studies in various areas such as architecture, sustainability, business, nursing, technology, teacher training and mathematics. Another reason for exclusion was the ambiguous use of the terms. Some

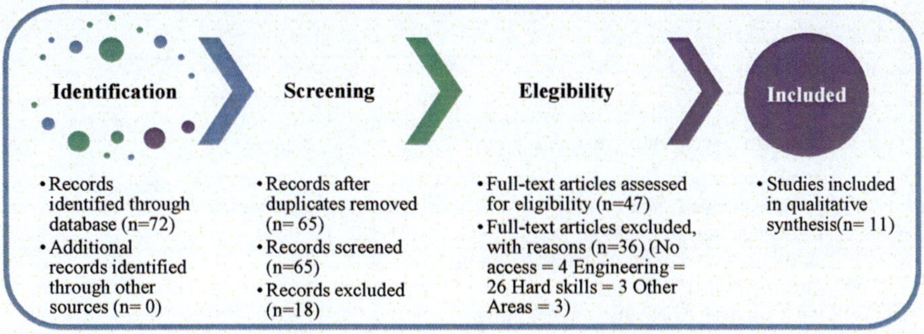

Fig. 1 Research and selection flow

works adopted the terms "core competences" and "key competences" but used these concepts to refer to hard skills. Ambiguity also influenced the research results, as many authors use the expression "design education" to refer to the action of planning the teaching, or meaning project teaching, rather than to the study about the education in the design area.

The final analysis comprised eleven articles that were synthesized according to the codes defined by the research (Fig. 1).

3 Results

Table 1 presents characteristics of the types of studies and the areas of design to which they belong and/or which are mentioned in the articles.

The most used terms in the articles were "soft skills" (n = 5); "core competences" is the second most used expression (n = 3); "twenty-first century competences" and "key skills" were used in two articles each.

Table 2 provides an overview of the competences mentioned in the included articles, and also the main concepts and themes addressed in each of the studies. The

Table 1 Characteristics of the studies included (n = 11)

Study type	n	Design field	N
Action research	2	All design areas	4
Review	1	Design ICT	1
Experimental	1	Digital media	1
Delphi-study	1	Social design	1
Mixed method	1	Fashion design	1
Experimental	1	Product design	3
Case study	2		
Survey	2		

Table 2 Soft skills mentioned and main concepts covered, by study

Studies	Soft skills mentioned	Main concepts
[22]	Originality; Creativity; Communication; Empathy; Versatility; Teamwork	Design curricula; Design education
[23]	Professionalism; Communication skills; Responsiveness; Teamwork; Multi-tasking; Innovation; Marketing skills; Business skills; Managerial skills; Commercial skills; Leadership; Participation; Curiosity; Questioning; Inquisitiveness; Persuasiveness; Presentation skills; Corporate and social responsibility; Ethics	Design education; Core competences; Transferable skills
[24]	Self-programming; Digital literacy	Self-programming; Networked society; Design for teaching and learning
[25]	Verbal communication of ideas; Teamwork; Communication skills; Willingness to learn; Flexibility; Commercial awareness; Problem solving	Digital Media; Design education; Learning and teaching model
[26]	Communication; Decision-making; Teamwork; Social interaction; Problem solving; Compromise; Negotiation; Participation; Sharing; Openness; Engagement; Investment; Reflection; Judgment; Critical question; Acceptance of diversity; Empathy; Comparison; Accountability	Design education; Social sustainability
[27]	Teamwork; Communication; Decision making; Research; Presentation skills; Collaboration; Creative problem solving; Dissemination of ideas	Learning by teaching pedagogy; Emerging technology; Design education
[28]	Ethics; Creative thinking; Self-management; Analysis applications	Core competency; Evaluation indicators
[29]	Leadership; Discrimination; Communication; Creativity; Receptivity; Adaptability; Organization; Exploration	Design competence; Design education; Design psychometrics;
[30]	Twenty-first century competencies; Creativity; Problem solving skills; Entrepreneurial skills; Collaborative skills; Creativity/innovation, Critical thinking; Information literacy; Decision making; Flexibility; Research and inquiry; Communication; Initiative; Self-direction; Productivity; Leadership; Responsibility; Collaboration; Information; Communication technology operations and concepts; Digital citizenship; Media literacy; Autonomy	Design education, twenty-first century competencies, future scenario

(continued)

Table 2 (continued)

Studies	Soft skills mentioned	Main concepts
[31]	Problem solving; Self-motivation; teamwork; Communication skills; Empathy; Team-building; Integrity; Courtesy; Responsibility	Soft skills in design education; On-line learning;
[32]	Creativity; Problem solving; Critical thinking; Collaboration; Communication	Design Education; Design Thinking Flipped Classroom

studies had a diversified scope, six of them addressed pedagogical projects carried out in learning environments that resulted in soft skills learning, one is a review article, three are about curriculum development for design courses, and one deals specifically with concepts and definitions concerning the competences and their relationship to design.

We identified 49 different competences mentioned in the articles. Some of them appear more often, namely communication skills, creativity and creative thinking, problem solving and teamwork. Table 3 presents the 17 competences that were mentioned by more than one study, in order of appearance.

Figure 2 shows the importance of teaching and learning the soft skills cited in the articles. Employability is the most mentioned, appearing in seven studies. Responsible innovation, sustainable development, as well as the impact on the quality and the results of the designer's work were also cited.

Table 4 summarizes the teaching methods identified in the studies and which, according to the authors, had good results in soft skills teaching and learning processes. The authors suggest to reproduce, whenever possible, the reality of the work environment, the use of collaborative and multidisciplinary methods, the group tasks, the experimentation, as well as encouraging the students to reflect about their ways of learning. Active teaching methodologies were also identified as "Learning by Teaching Pedagogy", "Group Teaching" and the "Flipped Classroom" methods. One design thinking method (Future Scenarios) is cited in one of the articles. There

Table 3 Overview of the competences mentioned. N = 11

Competence/Number of mentions			
Communication skills	10	Research/Exploration	3
Creativity/Creative thinking	7	Self-direction/Self-management	3
Problem solving	6	Leadership	3
Teamwork	5	Responsibility/Compromise	2
Flexibility/Adaptability/Versatility	4	Participation	2
Critical thinking/Questioning	4	Openness/Receptivity	2
Collaboration	4	Judgment/Analyses application	2
Decision making	3	Managerial skills/Business skills	2
Empathy	3		

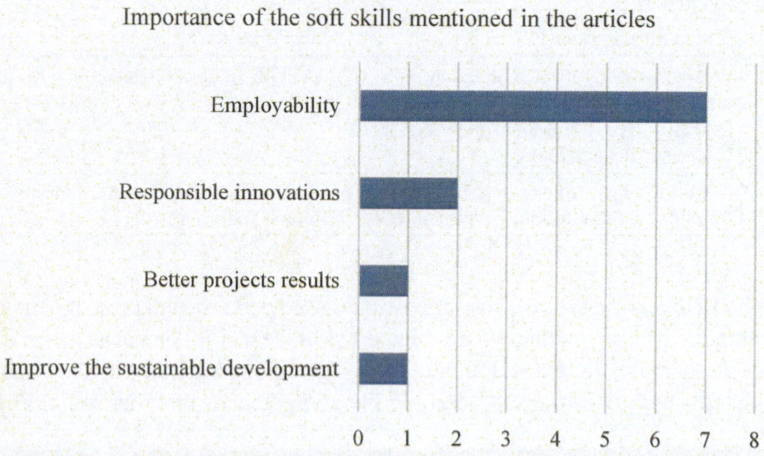

Fig. 2 Importance of the soft skills mentioned in the articles. N = 11

Table 4 Teaching methods mentioned in the articles

Active learning methods	Pedagogical practices	Design thinking methods	Preliminary studies indicated
Learning by teaching pedagogy; Group teaching; Future Scenarios; Flipped Classroom	Replicate commercial reality; Group task; Multidisciplinary and collaborative projects; Experimentation; Self-reflection	Future scenarios	Testing student's learning styles, Core competences and aptitude

is also reference to the need to identify the already existing learning styles, aptitude and skills in students, in order to develop methodologies that are more individual for teaching the competences.

4 Discussion

4.1 Main Findings

The aim of this study was to trace the state-of-the-art of soft skills teaching in design education. To this end, the importance of the soft skills mentioned in the articles and described teaching methodologies applied to the teaching of these competences were identified.

In the study on the importance of the skills, employability was the most mentioned; however, issues concerning sustainable development, project results that are more ethical, and more responsible innovations were also suggested. Although the importance of soft skills in employability in the design field is quite accepted, there is still no consensus on their importance concerning project results. There are, however, studies indicating that soft skills have a direct influence on the learning of other disciplines [20], and OECD documents highlighting the importance and necessity of developing soft skills for all professions as a way to face the many crises that are challenging the world.

Regarding the methodologies used, constructivist pedagogical approaches, more specifically those related to active teaching methods, were identified. Active teaching understands the learning processes as multifactorial, complex and unique. Active teaching can be defined as "The process of having students engage in some activity that forces them to reflect upon ideas and upon how they are using those ideas. Requiring students to regularly assess their own degree of understanding and skill in handling concepts or problems in a particular discipline. The attainment of knowledge by participating or contributing. The process of keeping students mentally, and often physically active in their learning through activities that involve them in gathering information, thinking, and problem solving [33, p. 5]". Even though many authors did not use the framework of active teaching methodologies explicitly in their work, it is quite clear by the description of the practices that they were active approaches and practices. In the literature on soft skills there is already evidence showing that the use of active teaching methods or multidisciplinary and collaborative practices conducted in groups promote the teaching of soft skills more than individual studies and traditional methods [34–37].

In this study, six articles described teaching projects set in higher education design courses that aimed to teach a subject matter content and resulted in teaching soft skills as a by-product. That is to say, these studies identified that the students can learn and develop soft skills while learning the subject matter content. This data converges with literature evidence indicating that soft skills learning takes place concurrently with the subject matter content learning [20].

The analysis of the studies also shows that soft skills are important in all areas where design can be applied, and not only those in which social skills are most required. It was also observed that some competences were more frequently mentioned than others. Skills such as communication skills, creativity, problem solving, teamwork, flexibility, collaboration, critical thinking and empathy seem to be the most important in developing design projects.

There are some problematic issues that can be pointed out: the lack of tools to evaluate the competences and of skills definitions. Some of the works indicated the students' self-assessment as a way of evaluating the learning of these competences. It is not inadequate as an evaluation tool, however, it is the only one mentioned. Sure enough, assessing the learning of these competences is a challenge, since they are subjective, linked to attitudes and behaviors and, therefore, impossible to quantify.

The other problem is the lack of systematization and clear definitions of the competences. In the study, only one article [28] presented operationalizable concepts and definitions of what they understood from each skill analyzed. This ambiguity affects the systematization of the field, generating inaccurate meanings and making analysis more difficult.

5 Future Studies

It is necessary to develop further studies in order to deepen the questions regarding the teaching of soft skills in design courses and also concerning the most important skills for design practice.

Another question that arises as necessary is whether there are competences that precede the learning of others, namely, skills precursors of others. There are studies in the literature indicating this possibility in relation to communication skills [38].

This study is part of an ongoing doctoral research that intends to deepen these questions and to develop a framework of pedagogical practices for the teaching of soft skills in higher education design. The research hypothesis is that the context of social design can bring greater opportunities for soft skills to be learned. In this review, only one of the analyzed papers was developed in the context of social design [26], so we do not discuss more broadly the possibilities that this approach may bring to the learning of these skills. However, because social design is an approach that needs critical, political and contextual training, it seems to be a productive context for soft skills teaching. Some authors have already indicated the need for new ways of teaching to meet the needs of this approach [39, 40]. Souleles [41] argues that education in design and social design will benefit from approaches that are not teacher-centered but approaches that promote the autonomy and self-construction of knowledge by students. We point out that active teaching can be one of the possibilities to meet this teaching need.

6 Final Considerations

The aim of this study was to understand what is being produced regarding the teaching of soft skills in design in higher education. The result demonstrates that it is necessary to broaden and deepen the studies, to disseminate the importance of soft skills to improve design practices and results and to update its teaching in a contextualized way with the changes of design and also of the world in the twenty-first century. The teaching of soft skills in design education seems to be one of the possible ways to face the challenges of globalization and the contemporary crises that are posed.

Acknowledgements This work was financially supported by the Centro de Investigação em Arquitetura, Urbanismo e Design - CIAUD [Center for Research in Architecture, Urbanism and Design] of the Faculdade de Arquitetura da Universidade de Lisboa [Faculty of Archi- tecture of the University of Lisbon] and also by the Universidade do Estado do Pará - UEPA [State University of Pará], Brazil.

References[1]

1. Norman D. Why design education must change. https://jnd.org/why_design_education_must_change/. Last accessed 22 Jan 2019
2. Bonsiepe G (2006) Design and democracy. Des Issues 22(2):27–34
3. Findeli A (2001) Rethinking design education for the 21st century: theoretical, methodological, and ethical discussion. Eur J Arts Educ 17(1):5–18
4. Friedman K (2012) Models of design: envisioning a future design education. Visible Lang 46:132–153
5. OECD (2009). Higher education to 2030, vol 2, globalisation, educational research and innovation. OECD Publishing, Paris
6. Delors J (1997) La educación encierra un tesoro: informe a la UNESCO de la Comisión Internacional sobre la educación para el siglo XXI, presidida por Jacques Delors. Educación y cultura para el nuevo milenio
7. Bennett N, Dunne E, Carré C (1999) Patterns of core and generic skill provision in higher education. High Educ 37:71–93
8. Fraser S (2001) Graduate attributes and generic skills at Macquarie. And Gladly Teche 1(1):1–4
9. Chamorro-Premuzic T, Arteche A, Bremner A, Greven C, Furnham A (2010) Soft skills in higher education: importance and improvement ratings as a function of individual differences and academic performance. Educ Psychol 30(2):221–241
10. Heckman JJ, Kautz T (2012) Hard evidence on soft skills. Labour Econ 19(4):451–464
11. Pereira OP, Costa CAAT (2018) The importance of soft skills in the university academic curriculum: The perceptions of the students in the new society of knowledge. Int J Bus Soc Res 7(6):01
12. Ringvold, T. A., Digranes, I. Future scenarios in general design education and 21St century competencies. In: Proceedings of the 19th International Conference on Engineering and Product Design Education: Building Community: Design Education for a Sustainable Future, 98–103, Oslo (2017).
13. Kember D, Leung D, Rosa M (2007) Characterizing learning environments of nurturing generic capabilities in higher education. Res High Educ 48:609–632
14. Laker D, Powell J (2011) The differences between hard and soft skills and their relative impact on training transfer. Hum Resour Dev Q 22:111–122
15. Barrie SC (2006) Understanding what we mean by the generic attributes of graduates. High Educ 51(2):215–241
16. Majid S, Liming Z, Tong S, Raihana S (2016) Importance of soft skills for education and career success. Int J Cross-Disciplinary Subjects Educ 2(Special 2):1036–1042
17. Rani S (2010) Need and importance of soft skills in students. J Liter Culture Media Stud 1–6
18. Santos D, Primi R (2014) Desenvolvimento socioemocional e aprendizado escolar: Uma proposta de mensuração para apoiar políticas públicas. Resultados preliminares do Projeto de medição de competências socioemocionais no Rio de Janeiro. OCDE, SEEDUC, Instituto Ayrton Senna, São Paulo
19. Hattie J, Biggs J, Purdie N (1996) Effects of learning skills interventions on student learning: a meta-analysis. Rev Educ Res 66:99–136
20. Virtanen A, Tynjälä P (2019) Factors explaining the learning of generic skills: a study of university students' experiences. Teach High Educ 24(7):880–894
21. Moher D, Liberati A, Tetzlaff J, Altman DG (2009) Preferred reporting items for systematic reviews and meta-analyses: the PRISMA statement. BMJ 338:1–8
22. *Holmlid S, Arvola M (2007) Developing a thematic design curriculum as a bologna master. In: DS 43: proceedings of E and PDE 2007, the 9th international conference on engineering and product design education, pp 63–68, United States

[1] References included in the review are marked with an asterisk.

23. *Griffiths R, Gordon B, Wilgeroth P (2007) Shaping design graduates: developing and assessing core competencies. In: DS 43: proceedings of E and PDE 2007, the 9th International conference on engineering and product design education, pp 393–398. United States
24. Levinsen KT (2011) Fluidity in the networked society—Self-initiated learning as a digital literacy competence. Electron J E-Learn 9(1):52–62
25. Fleischmann K (2011) Closing the gap: flexible and dynamic design education to better prepare digital media design students for professional practice. Des Principles Practices 5(4):213–224
26. McMahon M, Bhamra T (2012) 'Design beyond Borders': international collaborative projects as a mechanism to integrate social sustainability into student design practice. J Clean Prod 23(1):86–95
27. *Winge TM, Embry MC (2013) Fashion design podcast initiative: emerging technologies and fashion design teaching strategies. Wankel LA, Blessinger P (eds) Increasing student engagement and retention using mobile applications: smartphones, skype and texting technologies (cutting-edge technologies in higher education, vol 6 Part D). Emerald Group Publishing Limited, Bingley, pp 145–174
28. * Chiu S, Chuang L (2017) To assess the core competitiveness of taiwanese design departments students. In: Proceedings of 2017 international conference on green informatics (ICGI), Fuzhou, pp 159–162
29. Leclerc R, Horan R (2018) 'Fit' for change: measuring designer competence. Int J Des Creativity Innov 6(3–4):185–210
30. *Ringvold TA, Digranes I (2017) Future scenarios in general design education and 21st century competencies. In: Proceedings of the 19th international conference on engineering and product design education: building community: design education for a sustainable future, E and PDE 2017, pp 98–103. Oslo
31. *Katoch S, Doan V, Dadashi N (2018) User centred e-learning, opportunities and challenges: a literature review. J Interact Learn Res 29(4):481–495
32. *Canina M, Bruno C (2018) Design thinking via flipped classroom. In: Proceedings of the 20th international conference on engineering and product design education, pp 468–475. London
33. Collins JW, O'Brien NP (2011) The greenwood dictionary of education, 2nd edn. Greenwood Press, USA
34. Moy J (1999) The impact of generic competencies on workplace performance. NCVER, Australia
35. Harpe B, Radloff A, Wyber J (2000) Quality and Generic (professional) skills. Qual High Educ 6(3):231–243
36. Ballantine J, Larres P (2007) Cooperative learning: a pedagogy to improve students' genericskills? Educ Train 49(2):126–137
37. Smith C, Bath D (2006) The role of the learning community in the development of discipline knowledge and generic graduate outcomes. High Educ 51:259–286
38. van Laar E, van Deursen AJAM, van Dijk JAGM, de Haan J (2019) The sequential and conditional nature of 21st-century digital skills. Int J Commun 13:3462–3487
39. Easterday MW, Elizabeth M, Lewis DGR (2018) Design and social impact. A cross-sectoral agenda for design education, research, and practice. Des Issues 34(2):64–76
40. Manzini E (2011) Design schools as agents of (sustainable) change. In: 1st International symposium CUMULUS, pp 9–16. DRS for Design, Paris
41. Souleles N (2017) Design for social change and design education: social challenges versus teacher-centred pedagogies. Des J 20(sup1):S927–S936

Can the Pedagogy of Sheila Levrant de Bretteville be Considered a Relevant Model for Adapting Design Education to Global and Local Contexts?

Paul David Hardman and Nuno Coelho

Abstract The teaching of Sheila Levrant de Bretteville can be considered an example of an explicitly emancipative form of design education. De Bretteville's approach to design implied seeing education as a platform to engage in the prospect of reimagining and redesigning the world. Transcending the material aspects of design, de Bretteville's focus is rather on its social and political effects. Her students are encouraged to go beyond conventional notions of design studies as job training: to see their work as having social and cultural power and relevance. As such, de Bretteville's pedagogy offers clues as to how design education could adapt to the challenges of the contemporary paradigm, with the potential of resolving, or at least continuing to engage with, the contradictions implicit in design education, namely the relation of design to art and how design can be socially responsible. These issues are relevant for adapting design education to both global and local conditions.

Keywords Pedagogy · Interdisciplinarity · Collaboration · De Bretteville · Design

1 Introduction

The teaching of Sheila Levrant de Bretteville focusses on the use of participative processes and the social, cultural, political and ideological aspects of design. It is the intention of this paper to make the proposition that this approach to pedagogy is of particular relevance to the contemporary paradigm of design education, which is shifting towards emancipating forms of instruction, process focussed teaching, and interdisciplinary collaboration. We argue that design education must recognise its

P. D. Hardman (✉) · N. Coelho
University of Coimbra, Coimbra, Portugal
e-mail: phardman@dei.uc.pt

P. D. Hardman
Centre for Informatics and Systems of the University of Coimbra (CISUC), Coimbra, Portugal

N. Coelho
Centro de Estudos Interdisciplinares Do Século XX, Coimbra, Portugal

E. Duarte and C. Rosa (eds.), *Developments in Design Research and Practice*, Springer Series in Design and Innovation 17,
https://doi.org/10.1007/978-3-030-86596-2_21

intrinsically political nature and increasingly attempt to facilitate critical thinking and social engagement, concluding that the pedagogy of de Bretteville provides a relevant model for how these changes can be achieved.

It can be recognised that there are a combination of concerns manifested in recent discussions of design education that taken together constitute the contemporary paradigm, but that these are yet to be realised. These include calls for students to become fluid interdisciplinary collaborators [1, 2] who can develop awareness and responsiveness to problems for themselves [3]; to develop skills in critical thinking, which must be understood as permeating all designerly and technological behaviours and circumstances [4]; and to become informed, empathetic and culturally competent designers [5]. It can be argued that the traditional studio model is over reliant on teacher-centred pedagogy [6] and the reevaluation and adaptation of the studio model certainly seems an obvious way to make design education more suitable for this emerging paradigm, by paying attention to design's signature pedagogies such as the crit [7]. Piecemeal changes however, can only take us so far. It is also necessary to challenge the norms of design education on a more fundamental level if real change can be expected to be achieved.

It is against this background that this paper draws attention to the pedagogy of de Bretteville as an example of (what appears at least) to be a genuinely radical teaching practice. Her approach to education is based on the prospect of reimagining and redesigning the world [8] but goes beyond the physical products of design, focussing rather on its social and political effects. Students are encouraged to transcend notions of design studies as job training: to see their work as having social and cultural power and relevance, thus potentially developing their conception of 'the design entity' [9] and seeing design as something that offers a transformational experience. As such, de Bretteville's pedagogy offers clues as to how design education could develop, with the potential of resolving, or at least continuing to engage with, the contradictions that emerged at the Hochschule für Gestaltung, Ulm (HfG) [10] and that were implicit throughout the Arts and Crafts and Bauhaus eras, namely the relation of design to art and how design can be socially responsible. These issues are relevant for adapting design education to both global and local conditions.

De Bretteville studied art history at Barnard College, and graphic design at Yale University School of Art. She taught at CalArts (the California Institute of the Arts) from 1971 to 73, which she left to found an independent school for female artists (with Judy Chicago and Arlene Raven), the Feminist Studio Workshop in Los Angeles, later the Woman's Building. She then taught at the Art Institute/Parsons School of Design in 1980 (presently known as the Otis College of Art & Design). Since 1990 she has been the Director of Graduate Studies in Graphic Design at Yale School of Art. This summary of her background is pertinent to this discussion because it shows her involvement in both mainstream design education in the United States and in the Feminist movement. This paper discusses several aspects of her pedagogy, covering her overall approach to design and design education as an emancipative process; general characteristics of her teaching such as its focus on interdisciplinary and student-centred learning; specific features such as 'the object project'; her use of 'consciousness raising' in the earlier years of her teaching; and later developments in both her teaching and design practice in order to build up a fuller picture of her approach to design education throughout her career.

2 Shifting Paradigms of Design Education and Corresponding Conflicts and Contradictions

The contemporary paradigm of design education has been discussed in detail elsewhere [11], but it is relevant to draw attention to some of the main areas of contention and to note the ways in which these issues are not necessarily completely new, but to some extent are characteristic of the inherent conflicts within both design and design education. Calls for design education to encourage interdisciplinary collaboration, agency, empathy, adaptive solutions, participative methodologies, fieldwork, ethnography, peer to peer learning, and critical thinking, are familiar topics in recent design education literature and can be seen as a shift from an object-focussed paradigm to a human-centred or service-focussed paradigm. A move from fixed solutions to adaptive systems; from a concern with aesthetics and quality, to effectiveness and performance; from design as simplifying complexity, to design as managing complexity; from design as planning for making things, to design as active social/political/economic engagement; are all themes that appear to define the contemporary paradigm—and indeed the influence of human factors and design thinking methodologies seem to be capable of improving certain aspects of design practice, making it more rational, reliable, and evidence based.

Yet, in reading the passage above, it may be noted that some of these ideas do not perhaps sit so comfortably together as they may seem at first glance. For example, if the criteria for a successful design outcome is its measurable effect, what then of its ethical implications? If design processes are to become more rational, how are they to accommodate critical thinking and political engagement? It seems, after all, that a struggle between conflicting world views is maintained in discussions of the contemporary paradigm of design education, and it may be inevitable that design must continue to grapple with apparently incompatible ideas as it tries to resolve its 'wicked problems' [12].

Of all the aspirations for contemporary design education, the aspect that is possibly the most difficult to resolve and most likely to be pushed aside is critical thinking. Indeed, it has been argued that critical thinking is not possible to teach and that the wide support for the idea of critical thinking is due to its ambiguity, since theoreticians of critical thinking are divided on the fundamental questions of what critical thinking is and how it can be taught [13]. There are those who defend that critical thinking is a skill that can be applied to achieve specific results, while others would see critical thinking as an end in itself that is not so much a skill as a disposition to criticality—a willingness to question authority and challenge belief systems; an intention and capacity to be critical, rather than of having critical thinking skills. Thus, an agreement on one level about how design education should change or is changing, may mask more fundamental disagreements on a deeper level, in part due to ambiguity of terms but, more importantly, due to to differences in conflicting world views about the nature and purpose of education, and of course, the nature and purpose of design.

Design can be defined in many ways, but no definition is final. As Buchanan has noted, there are generally two kinds of definition used in the design discourse: descriptive and formal [14] and both of these serve ideological and rhetorical purposes, so we should be wary of them. Descriptive definitions tend to identify and elevate a single cause for design, such as 'design is the humanizing of technology' [15], while formal definitions leave open 'creative space' and serve a more strategic purpose in building connections in the field. Buchanan's own preferred definition that 'design is the creative human power to conceive, plan and realize products that serve human beings in the accomplishment of their individual and collective purposes', is an excellent example of an appealing and apparently universal definition of design, but that unfortunately is based on problematic assumptions. This kind of statement can easily give the impression that design is generally benevolent and that individual and collective purposes tend to be compatible. It would not be difficult to find examples to contradict either of these ideas. One has only to look at the degradation of ecosystems or the propaganda that is endemic on social networks to be reminded that design may just as easily frustrate our individual (and especially) collective purposes, and that the purposes of one group often directly damage the interests of another. We are reminded that, 'design contributes to the proliferation of large problems but offers little to their solution' [16]. A certain ambivalence towards design should therefore be encouraged—not all design is good, and even technically or aesthetically 'good' design may be ethically or politically questionable.

Even the Bauhaus, the archetypal school of design, was the site of serious conflicts about what design was and how it should be taught. The Bauhaus manifesto itself has been described as 'Janus-faced' [17] due to the way it looked forward on one side, 'let us desire, conceive, and create the new structure of the future [...] like the crystal symbol of a new faith' [18], but on the other, harked back to an idealised medieval past. More troublesome, was the struggle between the idea of free artistic expression and the requirements of mass production that pervaded the entire history of the Bauhaus [19]. Attempts, for example, to 'combine the greatest possible standardisation with the greatest possible variation of form' [20], seem intractable. A conflict between the expressionistic and the pragmatic can be perhaps be better understood as an attempt not only to reconcile art and technology, but to implement a threefold technology/art/science structure [21], reminding us that design is a necessarily an interdisciplinary field.

At the school that arguably inhereted the legacy of the Bauhaus, the HfG, an attempt was made to finally remove subjectivity from this structure, by severing design from art in favour of a rationalised scientific approach. Over the 15 years of its existence, the school moved from the Bauhaus influenced formalism of Max Bill, to the scientific operationalism aspired to by Maldonado, to a crisis of this position that ultimately led to a realisation that design must focus not on industry, but on the resolution of social and environmental issues [22] as politics returned as an indispensable element of design [23]. Maldonado had hoped that the school could produce a new kind of designer who would be characterised with the 'finesse and precision of his methods of thought and work, on the breadth of his scientific and technical knowledge, as well as on his capacity of interpreting the most secret and most subtle processes of our culture' [24]. In this development, the idea that design was applied aesthetics had been replaced by a new theoretical model, which considered design as an applied (human and social) science [21]. This idea was taken to an extreme which eventually became problematic when, 'in the name of rationality, precision and objectivity, the design process was to be purged of all nonrational framing devices, whether these were taken to be normative, ethical or political in kind' [23].

There were problems that scientific operationalism could not resolve. In 1967 Schnaidt addressed this issue in the editorial of the HfG Journal which concluded that we (as designers) must realise that we work in a field of conflicting interests and that if we want to change society, we must know it and must commit ourselves politically [25]—design could no longer be neutral. This idea is incompatible with scientific operationalism, which maintained the contradictory requirements that design be both critically exterior to, and operationally integrated within, the process of production. The ideal of the artist-designer shaping society from a position of distance had been thoroughly repudiated [23]. Design has a moral and political dimension that can be hidden or ignored, but cannot be avoided. The pedagogy of de Bretteville is relevant to this discussion, both because of the way it can be seen as addressing the issues that came to the fore at the HfG, but were not resolved—the necessity for political, social and critical design education; and because it has many characteristics that make it of interest to the concerns of the contemporary paradigm of design education.

3 Design and Design Education as Emancipation

In the 1970s de Bretteville had begun to question the hierarchical, authoritarian aspects of design and the fading modern idea that there were singular formal principles that were universally appropriate [8]. Accordingly, her pedagogical approach was influenced not only by the modernist background of her education at Yale in the sixties, but by the ideas of the feminist movement and the radical pedagogy of Paulo Freire, which had convinced her that teaching could be a horizontal exchange of information [8]. De Bretteville's teaching was part of what can be seen as a new wave of radical changes to teaching in design, which were rooted in the counterculture, including exponents such as Keith Godard, who at Yale began the first course of graphic design history; Marice R. Stein and Larry Miller, who were behind the 'anti-textbook' Blueprint for Counter Education; and Victor Papanek, author of the seminal Design for the Real World, then Dean of the School of Design at CalArts who de Bretteville had approached to start the Women's Design Program at the school [26].

The Women's Design Program gave de Bretteville the chance to begin developing her pedagogy, but it was when she broke away from CalArts with Raven and Chicago to establish the Feminist Studio Workshop that she was able to formulate it as a distinct approach [8]. An article she wrote for the journal Icographic set out her ideas clearly, including a description of her techniques, philosophy and rationale, which are explained using examples of student assignments and her own design work. What is striking in this article is how de Bretteville extends the significance of design beyond the realm of aesthetics or utility: 'The process by which forms are made and the forms themselves embody values and standards or behavior that affect large numbers of people and every aspect of our lives. For me, it is this integral relationship between individual creativity and social responsibility that draws me to the design arts' [27].

In this statement it can be seen that art is brought to the fore, not for its sublime qualities, but to highlight the emancipative potential of design, and crucially, the way in which artefacts embody values. De Bretteville highlighted several distinctions between art and design however, pointing out that design reaches a broader audience and speaks a common language [28]. Importantly, de Bretteville emphasises the way that design of the artificial world shapes consciousness and is therefore far from being a neutral activity: 'The design arts are public arts, and as such are major vehicles for forming our consciousness. Consciousness is, in turn, illuminated by communications, objects, buildings and environments. The design activity stands between us and our material existence, affecting not only our visual and physical environment but a sense of ourselves as well' [27].

Design for de Bretteville is not considered to be neutral, as in Modernism and positivistic epistemologies. Instead, design becomes explicitly political and is capable of reinforcing repressive attitudes and behaviour, a factor which led de Bretteville to increasingly question the desirability of simplicity and clarity and to associate it with repressive attitudes and behaviour. 'The thrust to control almost inevitably operates through simplification. Control is undermined by ambiguity, choice and complexity, because subjective factors in the user become more effective and the user is invited to participate. Participation undermines control' [27]. This inverts the stand point of scientific rationality, in which complexity is to be understood and controlled. For de Bretteville, complexity becomes a productive factor that can be used to challenge paternalistic hegemony. De Bretteville advocated for a form of design that would embrace ambiguity of messages and complexity of content as a way of supporting individual subjective opinions as a proactive, person-centred practice that built and strengthened individuals by encouraging the expression of new ideas and new communities [8]. In this sense, the aim of communication design is to counteract its previous success in advertising and other media. De Bretteville was aware of the contradictions of this position however, noting that design appears to be an ambiguous enterprise, since it is often difficult to know if a particular design could be used to reinforce values that the designer opposes. Her solution to this dilemma was that design should produce artefacts in which a political standpoint was explicit, and to this end even the design process had to be reimagined as a public and social process, no longer a mysterious activity only practiced by an elite.

One example of a process of creating a nonhierarchical design solution with an emphasis on its social context rather than its form, can be seen in the publication de Bretteville designed for the International Design Conference in Aspen, 1971. Participatory methods were used to produce the content, and even the reader is considered a participant in the construction of meaning, since they are required to 'create and combine these fragmented responses into their own personal picture' [27]. This reconstructs the role of the designer, not as a conduit for the clients message, but rather as one who enables the creation of meaning, and as such, de Bretteville considered her role to be a facilitator of participatory processes.

The notion that design is a proactive activity and should not be driven only by corporate service [27], demands an attitude to design education that emphasises the personal perspective of the student not as expression, but as communication. The intent is for students to produce meaningful content of their own, and this requires that they should seek out and include varied perspectives for each project. Ambiguity and complexity is encouraged and a supportive approach to education is crucial, so that students have freedom to fail, a sense of community, support, and feel that they can take chances [27]. To discuss what this could mean in practice, the following sections cover specific aspects of de Bretteville's approach to teaching.

4 Interdisciplinarity Collaboration and Subjective Perspectives

Calls for design students to become interdisciplinary collaborators are relatively commonplace, as mentioned in the introduction to this paper, but for de Bretteville interdisciplinarity is not simply about integrating technical knowledge or methodologies from different disciplines, it is seen as the ability of interpreting other perspectives:

> I believe that a productive tension comes from diverse points of view, and that students should grapple with diverse points of view for any act of design. We have given students readings from various critical perspectives, including psychoanalytic, semiotic, postmodern feminist and formalist. And we encourage them to take classes at the university, from people whose daily work is thinking from perspectives [29].

For designers to become adept at understanding different perspectives and interpreting different world-views, they become facilitators, communicators and interpreters, thus suggesting a subtle and holistic role for designers as enablers of collaboration between experts and participants in design processes.

Collaboration too is given a rich significance by de Bretteville, whose approach is comparable to human-centred fieldwork that involves engaging with communities can be related to Wenger's notion of creating communities of practice [30] as part of the learning process. Wenger has argued that Individuals learn when they engage in and contribute to practice in their community and that we should therefore engage students in actions, discussions and reflections that make a difference to the communities that they value. Thus, collaboration is important within the design community and also between designers and their audience, clients or other stakeholders.

Examples of this kinds of community engagement abound in de Bretteville's own work and in the projects she encourages her students to undertake. Her own work often involves a deep sense of place, in the early stages resembling a form of psychogeography—walking in neighbourhoods, talking to residents, immigrants, old-people, shop owners and so on, looking for connections and stories and finding ways to make these hidden narratives visible, to 'celebrate the voices and experiences of anonymous citizens, immigrants and their decendants' [31]. De Bretteville is influenced by the 'history-from-below' concept, which attempts to excavate and interpret events by focussing on the lives of ordinary people. This approach is demonstrated in her project Path of stars (1994) which developed from her involvement with the community of New Haven. She used conversations with residents to identify local 'stars', unsung heroes of the area. For example, Joseph McAlpine, a janitor who had represented the black community and had been instrumental in setting up a community relations board and a college scholarship program; and Dinah Chidsey who had lived in the area in the Eighteenth Century, and, unusually, was a female land owner who had divorced and retaken her maiden name. These lesser known but inspirational local figures were honoured by having their biographies mounted on the pavement in the style of the stars from Hollywood's Walk of Fame. Making an otherwise hidden history a permanent feature of the neighbourhood. In this example, the members of a community are collaborators with the designer in the sense that they work together to define and develop content.

5 Student-Centred Learning

De Bretteville uses the term 'person-centred' rather than the more familiar, 'human-centred' or 'student-centred', but the meaning is related. In her pedagogy, the student is given the responsibility of actively defining their own learning, emphasizing the students' desire to communicate, and focusing on what each student felt necessary to be made and said and to whom they wanted to say it [26]. De Bretteville prefers that 'the students do their own talking about their work; teachers are there to listen to the students and help them learn how best to communicate their own ideas' [31]. Students are encouraged to conduct research both directly and theoretically, but always from a personal subjective point of view, paying attention to their own thinking and to locate their emotional attraction to content. Cultural artefacts are used in class as a way of studying how the values of a society are reflected in what it produces, from its art to its tools to its detritus. Simultaneously, students are encouraged to use their interpretations of these artefacts to draw out and reflect upon their own ideas and values.

An example of this can be seen in 'the object project', which is based on the idea that forms themselves embody values. At the Women's Building, de Bretteville developed the exercise in an interdisciplinary class taught with Jivan Tabibian (a political scientist) and Ben Lifson (a photographer). It is an exercise that she has used throughout her teaching career and she continues to use the format at Yale. The exercise involves asking students to bring an object to the studio, which each individual then takes a turn to describe to the rest of the class. The discussion of these objects is a way of investigating the history, production, cultural/social impact of objects and therefore of design [32] but it is also very revealing of the students themselves. Notably, this exercise is not typical of the studio model. It is an exercise in which the discussion itself is the outcome, unlike conventional exercises that teach principles of design, or projects focussed on producing artefacts. For designers to derive content from objects upends modernist principles, as form becomes the basis for a transformative experience [26]. The concept, ‘form follows function’ is inverted, instead extracting content from form, or, to put it another way, ’form reveals ideology’.

In other instances, de Bretteville selects a text, a film, or a recording with which to confront and stimulate the students, who are challenged to identify the ideas and structures that are implicit in the given text, and to use these as the starting point for an open ended project. One year, de Bretteville chose an essay by Salman Rushdie about The Wizard of Oz, that interwove references to the original story, variations of the screenplay and Rushdie’s own experience of seeing the film for the first time. One students’ interpretation was that the film was about asking for help, and so his project revolved around this theme. De Bretteville points out that her own personal interpretation of the film centred on the wizard—she wanted the students to understand, that they should not look to the teachers as wizards; ‘that if they came to Oz, otherwise known as Yale, they will find that they already have a heart a brain and courage—that these are not what the teachers will give them; they already have them’ [31]. The story of the film was used as an analogy to refer to the role of the individual in society and to the validity of authority, in the institution of the university and beyond.

The aim of these exercises is for the students to develop an empathy based practice through interpreting the diverse perspectives of others, and crucially, by becoming more aware of their own subjective reactions: ‘It is important to me that this programme be person-centred. The students are encouraged to put and find themselves in their work; my agenda is to let the differences between my students be visible in everything they do. In most projects—not just in thesis work—it’s the students’ job to figure out what they want to say’ [33].

It can be noted that this ‘agenda’ rehabilitates the idea of subjectivity in design—not as mere self-indulgence, but as a way of accepting differences between individuals and making these differences productive. Yet, the question of subjectivity is controversial, and it can be argued that subjectivist epistemologies clash with notions of user-centred and evidence-based design [6]. Nonetheless, it is problematic to aim for design to solely follow rational methodologies if design students are to be encouraged to progressively extend the arena of possibilities within which they operate, not only to seek enduing solutions or answers, but to open up unfamiliar territory and new ideas. Thus, it can be argued that design education necessarily requires the understanding that knowledge is always partial, incomplete and contingent [34].

De Bretteville makes it clear that she does not intend to impress her own politics on students, nor her own visual style or choice of content. Instead, she claims to emphasise a pedagogy of asking, listening, reflecting, suggesting, and sustaining. Regardless of this intention, the effect of the ideas implicit in her teaching should lead her students to more critical, politically informed work.

An outcome of de Bretteville's encouragement of collaboration that reflects a powerful critical position is the work of the graphic design collective Class Action, formed by graduates of de Bretteville's course, whose self-stated goal is to influence the way public issues are understood and to motivate audiences to participate in civic dialogue [35]. An exemplary project of theirs is the Human Bodies Pamphlet, a powerful subversion of the logo of the 1994 International Design Conference in Aspen which was distributed at the conference, sparking much controversy and debate. The theme of the conference that year was 'Design and Human Bodies', which was represented in the graphic communications of the event by a representation of a nude woman, reminiscent of the collages of Matisse, designed by Ivan Chermayeff. The issue that Class Action identified was that in this image, the woman was depicted in an unnaturally prone position, in what seemed to be an image of submission or violence, and that was surely problematic as an image to represent a design conference. The collective responded by creating a pamphlet that they distributed at the event, featuring a rearranged version of the logo to change its gender and a collage that used photographic images to replaced the abstraction with a brutally figurative version [36]. The piece represented a clash not only in terms of the problems it raised about the representation of women and the myopia of designers, but also because of the way that it turned the visual communication back on itself: abstraction becomes figurative, the general becomes specific, a supposedly neutral communication is shown to be loaded with ideology.

The collective recognise that their education effected their perspective, claiming 'we have been trained to see the image making process in a certain way. There is a tradition of representing women as beautiful objects. It is part of an accepted past, certainly a part of history and therefore it does relate to how designers have been taught to look at them. We've been taught to be more critical, to question things' [36].

Critical thinking then, is strongly emphasised in de Bretteville's teaching, although perhaps there are some limitations. In the earlier phase of her career she practiced what was termed 'consciousness raising', a technique that focuses the attention of a wider group of people on some cause or condition. This activity was frequent at the Women's Building, in one schedule of classes for a fall term, consciousness-raising sessions appeared in three separate weekly time slots. In practice this meant discussions in which a group of students (and potentially teachers and other participants) would have a round table discussion on general themes such as money, power, sex, work, etc. Unlike ordinary discussions however, each participant had to speak for an equal amount of time as the others in a process that was difficult but valuable, as an alumni of the Woman's building relates: 'It was agonizing to have to talk for five minutes. We couldn't say, "Ah, skip me, I've got nothing to say." We were not allowed. Everybody had to sit there and, one by one, talk about our experiences, share things we'd never told anybody. And then listen to the other women in the group. It really was a discipline. That's where the personal became political, though, because in sharing our experiences as women, we were like, "You were raped? I was raped." We started to connect dots that we as women had never really been able to connect before' [37].

It obvious that this kind of approach would be difficult to enact in crowded contemporary classrooms and studios, and we can presume that de Bretteville did not practice such confrontational techniques in her later career. Although, it is possible to see that practicing structured group discussion, even not directly related to design, could have several benefits in terms of developing critical thinking, verbal rhetoric, confidence in presenting and discussing work, and so on. It is important to note that the role of discussion in de Bretteville's pedagogy highlights the importance of both personal experience and subjective interpretation and that these are seen as essential features of design processes.

6 Conclusion

This chapter has used the pedagogy of Sheila de Bretteville as an example of how design education could be adapted to both respond to the demands of the contemporary paradigm and to take on a more explicitly political, social and emancipative character. This form of teaching is not limited to de Bretteville of course, it has its roots in the radical movements of the 1970's counterculture, and it is important to recognise this as these ideas become more current, as Wild and Karwan argue:

> Focussing on communities instead of clients, valuing individual engagement over professional detachment, bypassing top-down hierarchy in favour of feedback and audience engagement, and expanding the visual vocabulary to reflect experience are all factors of underground design that are now an accepted part of design practice [8].

What is important is to ensure that these ideas are not stripped of their radical character as they are absorbed into general design teaching. As such, de Bretteville's pedagogy offers clues as to how design education could develop, with the potential of resolving, or at least continuing to engage with, contradictions that are implicit in design discourse. These issues are relevant for adapting design education to both global and local conditions. The pedagogy that follows this position has several remarkable features: productive tension; the acceptance of contradictory worldviews; teaching as horizontal exchange; participative methodologies; a collaborative relation between student and teacher; interdisciplinary collaboration; and direct connection to public. These characteristics make de Bretteville's pedagogy particularly relevant to address the cultural and ideological specificities required for adapting design education to simultaneously local and global perspectives.

This cursory review of de Bretteville's teaching and ideas is however, rather incomplete, based mostly on information from the 1970's and the 1990's while her career continues to the time of writing. There is a need for further investigation in order to understand the relevance of her ideas and practice in greater depth, and it would be equally relevant to know more about how her pedagogy has developed over time: whether it maintains its initial radical character and if it has taken on other aspects that may shed further light on its potential.

In conclusion, we can say that although challenging, this emancipative model of design education offers at least a partial model for how the field could change. The concerns of de Bretteville's pedagogy are relevant to many of the issues faced by contemporary design education.

References

1. Furniss L (2015) Beyond discipline: design practice and design education in the 21st century. http://files.cargocollective.com/649648/BeyondDiscipline_web.pdf. Last accessed 21 May 2020
2. Smith NK (ed) (2018) Creative collaboration: a report from creative review and dropbox. Creative Review, London. https://www.creativereview.co.uk/importance-creative-collaboration-special-report/. Last accessed 21 May 2020
3. Sadler DR (2013) Opening up feedback: teaching learners to see. In: Merry S, Price M, Carless D, Taras M (eds) Reconceptualising feedback in higher education: developing dialogue with students, Chap. 5, pp 54–63. Routledge, London. http://www.innovativeconservatoire.com/wp-content/uploads/2013/12/Sadler-%E2%80%93-Opening-up-feedback-Teaching-learners-to-see-%E2%80%93-Publication-2013.pdf. Last accessed 21 May 2020
4. Keirl S (2017) Critiquing as design and technology curriculum journey: history, theory, politics, and potential. In: Williams J, Stables K (eds) Critique in technology education. Springer, Dordrecht, pp 109–133
5. Bennett A, Vulpinari O (eds) (2011) Icograda design education Manifesto. http://www.ico-d.org/database/files/library/IcogradaEducationManifesto_2011.pdf. Last accessed 21 May 2020
6. Souleles N (2017) Design for social change and design education: social challenges versus teacher-centred pedagogies. Paper presented at the Proceedings of the 12th EAD conference. Sapienza University of Rome. https://www.researchgate.net/publication/318280459_Design_for_social_change_and_design_education_Social_challenges_versus_teacher-centred_pedagogies/citations. Last accessed 21 May 2020
7. Hardman P (2018) Navigator, tour guide, or travel companion: the role of the teacher in the territory of the crit. Paper presented at Designa 2018: territory. International Conference on Design Research. Universidade de Beira Interior, Covilhã (UBI). Instituto Politécnico de Castelo Branco. Portugal
8. Wild L, Karwan D (2016) Agency and urgency: the medium and its message. In Blaufelt A (ed) Hippie modernism. The Walker Art Centre, Minneapolis, pp 44–57
9. Davies A, Reid A (2000) Uncovering problematics in design education-learning and the design entity. In: Proceedings of international conference re-inventing design education in the university. pp 178–184. http://ualresearchonline.arts.ac.uk/620/. Last accessed 21 May 2020
10. Spitz R (2002) HfG Ulm. The view behind the foreground. The political history of the Ulm School of Design (1953–1968). Edition Axel Menges, Stuttgart, London. https://www.researchgate.net/publication/309202095_HfG_Ulm_The_view_behind_the_foreground_The_political_history_of_the_Ulm_School_of_Design_1953-1968. Last accessed 21 May 2020

11. Hardman P (2017) Framing design education within the contemporary paradigm. Paper presented at EIMAD—Meeting of research in music, arts and design. Escola Superior de Artes Aplicadas (ESART). Instituto Politécnico de Castelo Branco, Portugal
12. Rittel H, Webber M (1973) Dilemmas in a general theory of planning. Policy Sci 4(2):155–169. Springer
13. Harpaz Y (2010) Conflicting logics in education to critical thinking. Published in INQUIRY: critical thinking across the disciplines, vol 25 No 2, pp 5–17. https://yoramharpaz.com//pubs/en_thinking/conflicting-logics-in-teaching-critical-thinking.pdf. Last accessed 21 May 2020
14. Buchanan R (2001) Human dignity and human rights: towards a human-centred framework for design. Keynote Speech, Cape Town. https://www.defsa.org.za/sites/default/files/downloads/2000conference/09.%20Keynote%20Dr%20Richard%20Buchanan.pdf. Last accessed 21 May 2020
15. Morello A (2001) Cited in Buchanan R (2001) Human dignity and human rights: towards a human-centred framework for design. Keynote Speech, Cape Town. https://www.defsa.org.za/sites/default/files/downloads/2000conference/09.%20Keynote%20Dr%20Richard%20Buchanan.pdf. Last accessed 21 May 2020
16. Lange J (2011) The change we seek: an African perspective. Bennett A, Vulpinari O (eds). Icograda Design Education Manifesto, pp 90–95. http://www.ico-d.org/database/files/library/IcogradaEducationManifesto_2011.pdf. Last accessed 21 May 2020
17. Haxthausen C (2009) Lyonel Feininger/Walter Gropius: Bauhaus Manifesto, 1919. Bauhaus 1919–1933: workshops for modernity, exhibition catalogue. Barry Bergdoll B, Dickerman L (eds) MoMA, pp 64–67. https://www.academia.edu/5392236/Walter_Gropius_and_Lyonel_Feininger_Bauhaus_Manifesto._1919. Last accessed 21 May 2020
18. Gropius W (1993) Cited in Wingler, H. Bauhaus. Jabs W, Glilbert B (trans). MIT Press, Cambridge
19. Wick R (2000) Teaching at the Bauhaus. Hatje Cantz, Ostfildern-Ruit (2000)
20. Bayer H, Gropius W, Gropius I (1938) Bauhaus 1919–1928. The Museum of Modern Art, New York, p 30
21. Findeli A (2001) Rethinking design education for the 21st century: theoretical, methodological, and ethical discussion. In: Design issues, vol 17, No 1 Winter
22. Bonsiepe G (1968) Ulm 21: quarterly bulletin of the Hochschule für Gestaltung. Bonsiepe G (ed). Ulm
23. Kapos P (2018) The Ulm model. Exhibition catalogue. Raven Row, London, p 10. Retrieved 12 November. http://dasprogramm.co.uk/files/4.pdf
24. Maldonado T (1958) New developments in industry and the training of the designer. In: Quarterly bulletin of the Horschule für Gestaltung. Ulm No. 2, p 40. https://monoskop.org/images/5/50/Ulm_2.pdf. Last accessed 21 May 2020
25. Schnaidt C (1967) Architecture and political commitment. Ulm 19/20: quarterly bulletin of the Hochschule fürGestaltung, Gui Bonsiepe (ed). Ulm, pp. 26–34. https://monoskop.org/images/3/39/Ulm_19-20.pdf. Last accessed 21 May 2020
26. Berenson I, Honeth S (2016) Clearing the Haze: Prologue to postmodern graphic design education through Sheila de Bretteville. Retrieved from http://blogs.walkerart.org/design/2016/04/26/clearing-the-haze-prologue-to-postmodern-graphic-design-education-through-sheila-de-bretteville-2. Last accessed 21 May 2020
27. de Bretteville SL (1973) Some aspects of design from the perspective of a woman designer. In: Burke WB (ed) Icographic, No 6. http://dash.ico-d.org/files/icographic/ICOD_Icographic_06_full.pdf. Last accessed 21 May 2020
28. Walkup K (2011) Books in a new language (267–299). In: From site to vision: the woman's building in contemporary culture. Otis College of Art and Design, Los Angeles, p 270. http://thewomansbuilding.org/images/FSTV%20PDFs/Walkup.pdf
29. Lupton E (1993) Reputations: Sheila Levrant de Bretteville. [Interview with de Bretteville, S. L.]. para. 15. http://www.eyemagazine.com/feature/article/reputations-sheila-levrant-de-bretteville. Last accessed 21 May 2020

30. Wenger E (1998) Communities of practice: learning, meaning and identity. Cambridge University Press, Cambridge, UK
31. de Bretteville SL (1997) In: van Toorn J (ed) Design beyond design. Jan van Eyck Akadamie, Maastricht, pp 115–127
32. Winter EF (2015) Textiles teaching portfolio. https://risdcollegiateteaching.files.wordpress.com/2015/05/emily-winter-teaching-portfolio-2015.pdf. Last accessed 21 May 2020
33. de Bretteville cited inLupton E (1993) Reputations: Sheila Levrant de Bretteville. [Interview with de Bretteville, S. L.]. para. 6.
34. Danvers J (2003) Towards a radical pedagogy: notes on learning and teaching in art & design. Article Int J Art Design Educ 22(1):47–57. https://www.researchgate.net/publication/229491841_Towards_a_Radical_Pedagogy_Provisional_Notes_on_Learning_and_Teaching_in_Art_Design. Last accessed 21 May 2020
35. Class Action, Homepage (2020) https://www.classactioncollective.org/. Last accessed 21 May 2020
36. Miller JA (1995) The bodies' politics. Interview: class action. In: The American Centre for Design. Statements, Winter/Spring, pp 14–19
37. Gaulke C, Wolverton T, Maberry S, Cotton C (2018) The woman's building: animating the archives. Discussion Panel at the LA Art Book Fair, p 17. https://www.metabolicstudio.org/posts/pdf/197_the-womans-building-animating-the-archives-discussion-panel-at-the-la-art-book-fair.pdf. Last accessed 21 May 2020

Idea Generation Using the Fictionation Design Tool in an Interactive Prototyping Course for Industrial Designers

Ümit Bayırlı, İsmail Yavuz Paksoy, and Naz A. G. Z. Börekçi

Abstract *Fictionation* is a design tool developed as part of an ongoing Ph.D. research, for supporting design students in their idea generation process. This paper describes a study in which the *Fictionation* design tool has been pilot tested in an Interactive Prototyping elective undergraduate course with industrial design students. The aim was to encourage design divergence, by inviting students into a fictional world where restrictive considerations of the design process do not exist for them. The study was carried out with thirteen students using the design tool in the idea generation phase of their projects. The submissions were analyzed for their visual contents. The results show that students were able to diverge in their design exploration, offering variety and novelty in their design ideas. Divergence in the idea generation phase of the design process in design education allows the students to extend the boundaries of the design solution space, and provide variety of ideas within this space, which in return, reflect on the quality of the final design solution.

Keywords Idea generation · Design fiction · Interactive prototyping · Fictionation design tool · Design divergence

1 Introduction

The creation of a product goes through a variety of phases from an initial idea in the mind to a product at a showcase. Aspelund [1, p. 2] associates this process to a path, stating that "there is a path all designs take on their journey from the world of imagination to the world of objects." The phases that designers visit during their journey is called the product development process. Within the product development process, idea generation has a critical role for the success of the final product

Ü. Bayırlı (✉)
Pamukkale University, Denizli, Turkey
e-mail: ubayirli@pau.edu.tr

İ. Y. Paksoy · N. A. G. Z. Börekçi
Middle East Technical University, Ankara, Turkey

E. Duarte and C. Rosa (eds.), *Developments in Design Research and Practice*, Springer Series in Design and Innovation 17,
https://doi.org/10.1007/978-3-030-86596-2_22

since this is the phase in which designers generate their initial ideas that may be developed into a prevailing product of the market later on. Idea generation is the creative process of developing and communicating numerous ideas that can range from an abstract to a concrete level [2, 3], generally carried out as a visual exploration of a problem space, towards the determination of the boundaries of a solution space. For a fruitful idea generation process, visual exploration aims for divergence [4, 5], meaning to think as broadly as possible, considering many alternatives and searching for different approaches. Ideally this process ends with numerous design ideas (quantity), covering a broad span of possible means (variety), and creating new opportunities for designers (novelty) [6–8]. The whole of the generated ideas during this phase creates a theoretical space called the design solution space. Studies show that, the size of the design solution space, and variety within, affects the quality of the final design solution [6, 7, 9].

1.1 Problem Statement

Many idea generation methods and tools have been developed to support designers in idea generation and to ensure quality and quantity in the design solution space [10]. These methods differ in procedure and medium, but their common point is to provoke divergent thinking by using triggers or instructions [11]. Despite the availability of methods and tools, design students may still have problems in achieving quantity, variety and novelty in their ideas [12–14]. It is important to encourage the use of developmentally appropriate design methodologies in design education, as a strategy to both help design instructors understand the cognitive processes of students as novice designers, and also for students to gain the design expertise that sets the basis of these methods [15]. On the other hand, students need to be in a certain method mindset to be able to profit from the methods [16], meaning readiness to understand the procedure and benefits of a method, and willingness to employ it in one's process.

Designing involves complex activities, each requiring the acquisition and mastering of designerly skills [17]. Through design education students gain expertise in their field. Dreyfus [2003; cited in 18] offers a 7-level model of design expertise, these being the novice, advanced beginner, competent, proficient, expert, master, and visionary, with the ways in which a problem is perceived, interpreted, structured, and solved differing in each level. Design education helps the development of competencies from a novice learner to an expert performer [19]. In this process, design students need guidance in developing competencies in idea generation and design divergence.

1.2 Aim of the Paper

This paper presents a study conducted with advanced beginner and competent design students that tests a design tool developed with the aim of enlarging the boundaries of the solution space during idea generation. The main argument is that, due to lack of experience and expertise, design students remain fixated to the constraints they determine for themselves at the onset of a design project, limiting themselves in their design exploration. The tool presented in this paper aims for the elimination of such restrictive considerations by using the design fiction approach, and the *what-if* scenarios associated with it. This tool, named *Fictionation*, is developed as part of an ongoing Ph.D. research, and was pilot tested in this study. The paper introduces the study procedure and the design tool, and presents the outcomes based on content analysis. It finally discusses the implications of the findings on idea generation in an educational context.

2 The Study

The study has three main phases. The first phase is a needs assessment study that aims to determine the restrictive considerations of students in handling their design projects, also limiting their design exploration resulting in a lack of quantity, variety and novelty in design ideas. The second phase is the development process of the *Fictionation* design tool that aims to overcome the restrictive considerations of students and meet their needs for a free-flowing and fruitful idea generation process. The third phase consists of the testing of the *Fictionation* design tool.

2.1 Phase 1: Needs Assessment

Beginning the study, a needs assessment was conducted to determine the reasons why design students have problems in design divergence. Needs assessment is an approach for collecting knowledge and information, and learning the ability and attitude of a group, to determine deficiencies and improve the current situation [20, 21].

Observation of an Educational Design Project. A seven-week project conducted with the 3rd year undergraduate students at Middle East Technical University, Department of Industrial Design was observed. The project was on outdoor lighting unit product families, and had four stages: research, idea generation, design development, and evaluation. Within the idea generation phase, students used different idea generation methods such as the morphological chart [22], matrix [23], and switch task [24]. All the same, they were observed to constrain themselves with restrictive considerations such as cost, production criteria and material selection, limiting variety and novelty in their design ideas. These considerations should ideally be dealt in the following stages of a design process.

Adoption of the Design Fiction Approach. This observation provided the motivation to search for ways of supporting students in overcoming their restrictive considerations and encouraging a divergent idea generation process. Based on earlier literature search carried out on the topics of creativity, idea generation, and design education, design fiction was adopted as the approach in developing a tool that would provide an enriched context for this support.

Design fiction is a relatively new approach that allows for the creation of an imaginative world in which rules, limitations and constraints are defined by the designer [25]. It is similar to a laboratory where designers can analyze the features of the real world in a fictional world where there is an appropriate environment to think differently from the real world [26, 27]. To create fictional worlds, design fiction uses imaginary what-if scenarios [28]. This approach was found to have the potential of creating a fictional world where the restrictive considerations of the design process would not exist for the design students.

2.2 *Phase 2: Development of the Design Tool*

The design tool was developed based on insights gained from the observation of students during the outdoor lighting units project (students' perspective), followed by insights gained from focus group interviews carried out with research assistants involved in design studio tutoring (design educators' perspective). Focus group is an interactive method conducted with a group of stakeholders to obtain their opinions and experiences regarding a specific topic [20, 29].

Focus Group Interviews. Two sessions of focus group interviews were carried out, with a total of eight research assistants from the department. The aim was to gain an in-depth understanding of the restrictive considerations of students, and to create a fictional world with its accompanying tools in response. The research assistants had experience as tutors of undergraduate design studio courses (ranging from 2 to 6 years), as well as fresh memories from their experiences as students themselves. They were asked to make a list of issues they observed students to consider during their individual idea generation processes. Then all the lists were combined, forming a pool of 24 considerations. Participants were also asked to group the considerations under themes identified as the main issues that students took into account during idea generation. The themes obtained were form, function, and interaction. Following the sessions, participants were asked to grade the 24 considerations between one to ten according to their importance for the idea generation process, on an online form. The following ten considerations obtained the highest scores in order of importance: existing products, user, anthropometric measurements, purpose, production techniques, accustomed interactions, gravity, new technologies, cost, and sense organ.

Content and Medium of the Design Tool. A card deck was chosen as the design tool since it is easy to use and applicable without the need of a moderator. The ten considerations obtained from the focus group interviews were converted into what-if questions. The what-if questions were applied to the themes of form, function and interaction, giving a total of 30 cards. For example, for the consideration of user, the what-if question for the theme of form was "what if the product were designed for an extraordinary user, how would the form of the product be?"; for the theme of function was "what if the product were designed for an extraordinary user, how would the function of the product change?"; and finally, for the theme of interaction was "what if the product were designed for an extraordinary user, how would the user interact with the product?".

Blue was used for the form-related cards, green was used for the function-related cards, and red was used for the interaction-related cards. A logo for the tool was designed as a print at the back side of the cards (Fig. 1). The cards were printed on plastic coated paper, and the dimensions were 55 by 85 mm. A box was prepared to keep the card deck together. The design tool was named *Fictionation*, which is the combination of the words fiction and generation.

Fig. 1 Fictionation cards

2.3 Phase 3: Testing of the Fictionation Design Tool

The *Fictionation* design tool was implemented in the idea generation phase of a project carried out in the Interactive Prototyping undergraduate elective course given the department, to evaluate the effects of the tool on the variety and novelty of the generated ideas. This implementation was done as a pilot study for an ongoing Ph.D. research.

Interactive Prototyping Course. The Interactive Prototyping course offers advanced techniques of electronics and programming through building working prototypes of interactive devices and systems. By taking a hands-on approach, which means spending a lot of time on building circuits, coding, and playing with sensors and controls, the students observe how best to relate all these to the user's behaviors or environmental changes. These experimentations also allow students to test and observe the affordances of their designed systems.

The course caters design students in a variety of ways; (1) they create working prototypes for their design projects to better communicate their ideas [30], (2) they employ interactive prototyping for user testing and design thinking, (3) they are introduced to electronics: (a) understanding how an electronic product works, (b) learning which component is used for what, (c) and learning the terminology for electronic components (e.g. sensors, microprocessors, actuators, and their types), (4) they grasp the interconnectedness of products with the introduction of IoT and cloud computing, (5) they familiarize with coding and algorithm, which improves their reasoning, and helps them understand programming languages of similar kinds, and finally, (6) they develop their communication skills with people from the engineering disciplines. The course is carried out in 14 weeks. The weekly course hour is four. The phases in the course are: (1) introduction to the fundamentals of prototyping, (2) defining a viable problem, (3) idea generation, and (4) implementation and iteration.

The course requirement was the definition of a project statement on a problem area that students would identify for themselves, the development of a design project on connected devices not restricted to mainstream requirements, and the delivery of working prototypes. Students were required to employ freshly learned skills such as coding, dealing with electronics, delivering UX goals, and creating a meaningful design incorporating all these specifications, which, from their perspectives, did not seem feasible in the limited time given for an elective course. It was observed that they experienced difficulty in idea generation due to the nature of interactive prototyping, which was yet unfamiliar to them. They were reserved in coming up with original ideas that they were not sure about whether they could realize. Though, as a final output, they were able to deliver four projects, which were: (1) a toy that entertains cats, allowing owners to sleep or work, (2) an enchanted coffee machine that includes different operation modes for cultural rituals and social activities, (3) an exercise assistant that helps those who lack activity during their working/studying period, and (4) a smart water closet that encourages users to keep the toilet clean.

Implementation of the Fictionation Design Tool. There were 13 students enrolled to the course. Eleven were 3rd year undergraduate industrial design students and two were graduate architecture students. The *Fictionation* design tool was implemented on the 10th week of the course. The students were by now informed on the area and familiar with the basics of interactive prototyping, placing them at a novice to advanced beginner level of expertise. At the time of the study, the students were ready to carry out idea generation. The session was conducted in the computer lab studio, and was moderated by the first and second authors, as well as another course instructor. The study procedure consisted of four phases (Fig. 2).

In the first phase, the students were asked as a home assignment to generate ideas on their own, making sketches of 30 different ideas for their projects. The students were given two A3 size charts in portrait orientation, containing 15 slots each, on which to make their sketches (Fig. 3). They were instructed to spend an uninterrupted 90 min for this phase and bring the charts to class. The following phases were carried out in class, all together.

The class began with the second phase in which a presentation was delivered to the students on the design fiction approach and the *Fictionation* design tool, by the first

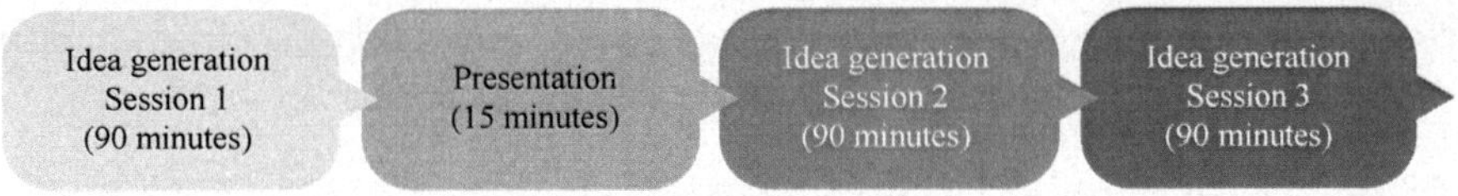

Fig. 2 Implementation process

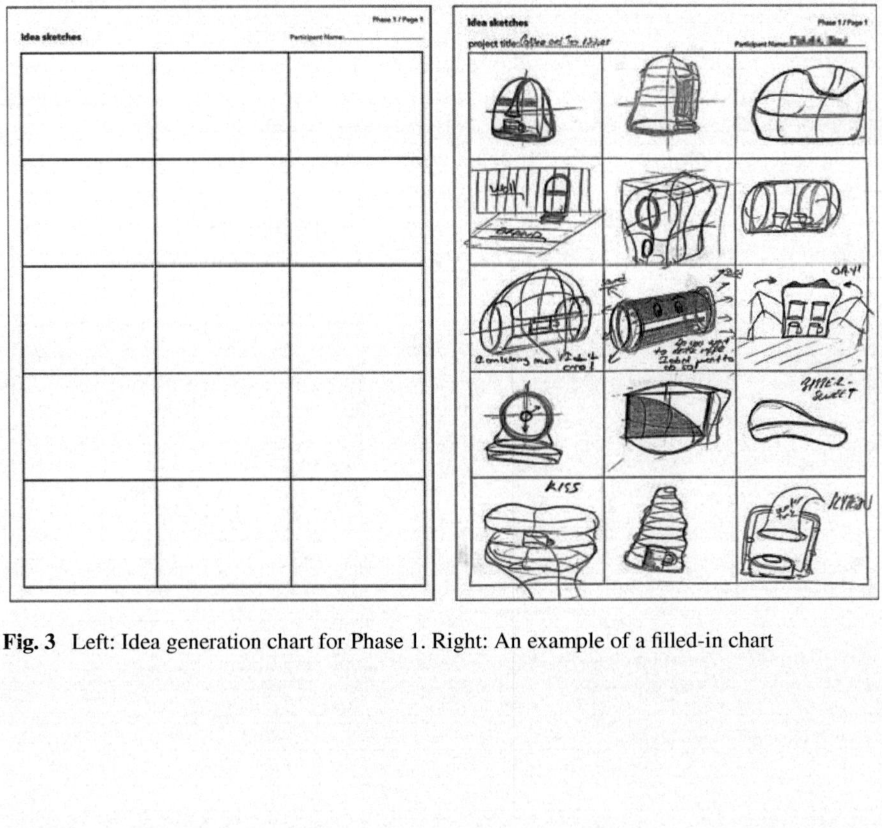

Fig. 3 Left: Idea generation chart for Phase 1. Right: An example of a filled-in chart

Fig. 4 Left: Phase 2, presentation. Right: Phase 3

author and researcher who developed the tool (Fig. 4, Left). Students were informed on the procedure of their design tasks for the day. In the third phase, students were asked to form four groups and take seats by their desks. One group of two, one group of three and two groups of four were formed (Fig. 4, Right).

Four card decks were distributed to the four groups of students. They were asked to pick ten each, out of the 30 cards. This could be a random selection, as well as a deliberate one. Students were also distributed five A3 size landscape orientation charts each, with each page containing a table of two cells, making a total of 10 cells (Fig. 5). Based on the *what-if* questions on the ten cards, students were asked to make design exploration with sketches for a fictional world. Five minutes were given for idea generation on each card and the process took 50 min in total.

In the fourth and final phase, students were asked to investigate the sketches that they made in the previous phases and process them, also generating new ones if necessary, into 30 new ideas that are applicable in our contemporary world. For this, students were again distributed two A3 size portrait orientation sheets containing 15

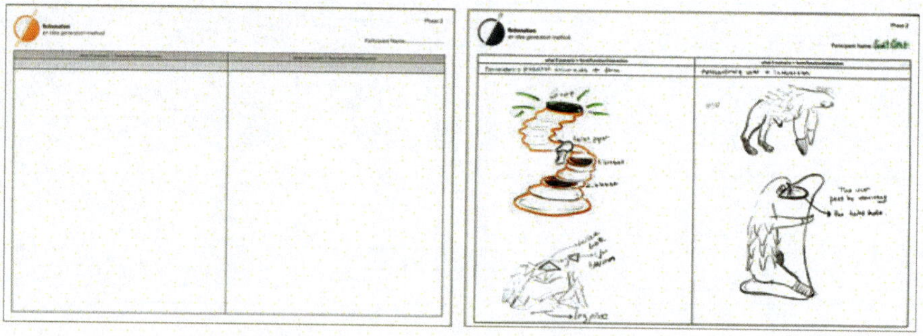

Fig. 5 Left: Idea generation chart for Phase 3. Right: Example of a filled-in chart

Fig. 6 Left: Idea generation chart for Phase 4. Right: Example of a filled-in chart

slots on each (Fig. 6). At the end of the study, a discussion session was carried out with the students to receive their feedback on the usability of the *Fictionation* design tool and its applicability for the idea generation process. Students were asked whether the tool was easy to use, easy to understand, and applicable in the idea generation process. Furthermore, students were asked to make suggestions for the development of the design tool. The researcher took written notes of the discussions.

3 Analysis and Findings

According to Corremans [31], an offered method or tool should be tested for its effectiveness, usability and applicability. The aim of the *Fictionation* design tool was to remove the cognitive barriers that students built in their minds and support them during idea generation for design divergence. Therefore, to test the effectiveness of the *Fictionation* design tool, the sketches that students made in Phase 1 without using the tool, and the sketches that they made in Phase 4 after having used the tool, were compared. The analysis included data collected from 10 of the students who participated in both Phase 1 and Phase 4. The collected data was a set of 19 charts for Phase 1 containing 256 sketches, a set of 50 charts for Phase 3 containing 127 sketches, and a set of 18 charts for Phase 4 containing 198 sketches. The sketches were analyzed using visual content analysis, which is an unbiased and empirical method for evaluating visual representations [32, 33]. The sketches were broken down into their components reflecting independent units of design ideas [34, 35] and examined separately for solutions about form, function and interaction.

3.1 Effects of the Fictionation Design Tool on Generated Ideas

Quantity. The aim of the *Fictionation* design tool was to support students in idea generation and design divergence in particular. The 10 students who submitted for Phase 1 filled in 256 slots out of 285, within 19 charts. Their number of sketches ranged between 13 and 30, with an average of 25.6. Having carried out Phase 3, in which they produced between 7 and 22 sketches for design exploration, these students submitted 18 charts for Phase 4, with 198 out of 270 slots filled with sketches. For both phases, students generated between 11 and 52 sketches, with an average of 32.1. The students were able to achieve quantity in their idea generation.

Divergence. A main observation regarding the charts produced for Phase I was that, on the second page of their sheets many students made explorations of the same ideas represented on the first page. Ideas seemed to saturate after 15 sketches, students had difficulty in diverging further, and instead they returned to previous ideas as sources for exploration. In Phase 4, the idea content increased, with students diverging in their explorations.

In terms of form, it was seen that, while many students represented only one form alternative and developed a usage scenario considering that form in Phase 1, they increased the number of form alternatives in Phase 4. For example, while in Phase 1 students generating design ideas for an interactive cat toy explored a tortoise-shell-like form, in Phase 4 they were able to offer diversified forms (Fig. 7). As another example, while in Phase 1 students generating design ideas for a coffee machine,

Fig. 7 Left: Explorations in Phase 1 concentrating on a tortoise-shell-like form. Right: Diverging form explorations in Phase 4

Fig. 8 Left: Explorations of a main function in Phase 1 concentrating on either locking the door or toilet seat. Right: Explorations of additional functions in Phase 4

explored form alternatives similar to those available in the market, in Phase 4 they were able to offer variations in forms.

In terms of function, it was observed that in Phase 1 the students focused only on the main function of the product idea. However, in Phase 4, besides the main function, students offered additional functions, also referring to novelty. For example, in Phase 1, students developing an interactive cat toy, generated product ideas that function mainly as a toy that goes around in the house for the cat to follow. In phase 4, they offered ideas in which the toy also functions as a robotic vacuum cleaner, or as a weighing scale for the cat. As another example, in Phase 1, students developing a toilet that reminds people to clean the bathroom before leaving, generated ideas that mainly focused on preventing the user from leaving if the bathroom were in a dirty state. However, in Phase 4, they generated further ideas with additional features such as personalized projections of visuals on the walls, and additional functions, such as a toilet bowl that adjusts its height according to the user, and a toilet bowl that measures the user's weight (Fig. 8).

In terms of interaction, it was observed that in Phase 4 students generated new ideas offering new interactions compared to those generated in Phase 1, again referring to novelty. For example, in Phase 1, students developing an exercise robot that detects user inactivity, generated ideas in which the user has to physically interact with the robot by touching it, and the robot moves its parts and makes noises in return. However, in Phase 4, they generated ideas in which the robot interacts with the user by activating light, producing smell or replicating the user's emotions. As another example, in Phase 1 the students developing a coffee machine generated ideas in which the coffee machine interacts with the user (e.g. considering time, counting the cups of coffee that the user makes). However, in Phase 4, the user is put in a more active position, in which he/she initiates interaction with the coffee machine (e.g. making gestures, holding) (Fig. 9).

3.2 *Usability and Applicability of the Fictionation Design Tool*

In terms of feedback, students described the tool as easy to use, and the content of the *what-if* questions as understandable. They indicated that the tool was beneficial in

Fig. 9 Left: Explorations of user-product interactions for a coffee machine in Phase 1. Right: Explorations of new types of user-product interaction in Phase 4

terms of making them think from different perspectives. They mentioned that using the tool made idea generation entertaining for them. For the purpose of developing the design tool, students offered a new consideration, *the place*, as a new *what-if* scenario in idea generation. On the other hand, the students indicated that choosing ten cards and making 30 sketches required too much effort, and suggested decreasing this number. In Phase 3, where students generated ideas for the ten cards that they chose, students were expected to make design explorations with numerous sketches for the *what-if* scenario in the designated area on the charts. However, many students made only one sketch in this area. As a final suggestion, students offered to divide this area into smaller slots.

4 Conclusion

Reflections on the Fictionation Design Tool. This paper presented the development of a design tool supporting idea generation for design divergence, and a pilot study in which the tool was implemented in an undergraduate elective course on interactive prototyping. As a result of this pilot study, it is possible to suggest improvements towards the development of the tool.

Firstly, one more consideration was added to the card deck: *the place*. With the addition of three new cards incorporating the *what-if* questions related to form, function and interaction, the number of cards in the deck reached 33. As for the design charts, the number of slots were increased from two to six, in the A4 size chart of Phase 3, in order to guide students in making numerous sketches. Furthermore, the number of slots in the A3 size charts of phases 1 and 4 were decreased from 15 to 8, in order to reduce the effort required for completing the task. Also, it was decided to ask students to choose 8 cards instead of 10.

Reflections on the Interactive Prototyping Course. It was noted through the submissions made for the course at the end of the semester that, some of the ideas generated in this session with the support of the *Fictionation* design tool were reflected on the final design solutions. One self-criticism is that, the idea generation session was executed late in the semester, and not enough time was left for students to realize their final design solutions in full. However, the implementation of the *Fictionation* design tool was the correct action to take, since the students were short of generating original ideas before, and with the implementation of the tool, they were able to generate design ideas that matched the expectations of the course to a degree. The tool is compatible with the essential objective of the course, which is to learn a non-familiar and hard-to-grasp topic, prototyping, in an entertaining way, without dealing with the actual production requirements expected in the studio courses. As a future recommendation, the *Fictionation* design tool should be implemented earlier in the semester.

Implications of the Findings on Design Education. The study carried the purpose of testing the benefits of a design tool using the design fiction approach in supporting design divergence in idea generation. It did this in the three dimensions of form, function and interaction related to product ideas. In terms of form, design divergence was observed in the explorations carried out following the implementation of the method. In terms of function, following the implementation of the method, explorations were made for the integration of additional functional features into the previously offered product ideas. In terms of interaction, user-product interactions explored in the sketches were mostly altered, suggesting changes in the types of interaction that make the product actively involved rather than having it passively wait for the physical involvement of the user. Consequently, it has been possible to suggest that using the *Fictionation* design tool allowed students to generate new ideas in general, diverging in their exploration of form, integration of additional functions, and enrichment of user experiences by reconsidering user-product interaction in mutual ways.

In the light of the student feedback and reflections of the generated ideas on the final design solutions, the *Fictionation* design tool can be evaluated as usable for the idea generation process. The developed version of the *Fictionation* design tool has been implemented in further studies carried out in the undergraduate educational setting. Currently the outcomes of these studies are being analyzed as the concluding stage of Ph.D. research.

References

1. Aspelund K (2014) The design process, 3rd edn. Fairchild Books, New York
2. Shah JJ, Kulkarni SV, Vargas Hernandez N (2000) Evaluation of idea generation methods for conceptual design: effectiveness metrics and design of experiments. J Mech Des 122:377–384
3. Innolytics homepage, What is idea generation? https://innolytics-innovation.com/idea-generation/. Last accessed 23 Nov 2019
4. Jones JC (1980) Design methods: seeds of human futures. Wiley, Chichester
5. O'Kane B (2014) Building bridges to new realms. In: Gardner S, Evans C (eds) Design and designing: a critical introduction. Berg, London
6. Christian JL, Daly SR, Yilmaz S, Seifert CM, Gonzalez R (2012) Assessing design heuristics for idea generation in an introductory engineering course. Int J Eng Educ 28(2):463–473
7. Daly SR, Seifert CM, Yilmaz S, Gonzalez R (2016) Comparing ideation techniques for beginning designers. J Mech Des 138(10):101108/1–101108/12

8. Saunders R, Pourmohamadi M (2009) Designerly ways of customizing. In: Suominen J, Piller F, Ruohonen M, Tseng M, Jacobson S (eds) Proceedings of the 5th international conference on mass customization & personalization MCPC 2009 conference. Helsinki, October 4–8
9. Römer A, Weißhahn G, Hacker W (2001) Effort saving product representations in design—results of a questionnaire survey. Des Stud 22(6):473–490
10. Shroyer K, Lovins T, Turns J, Cardella ME, Atman CJ (2018) Timescales and ideaspace: an examination of idea generation in design practice. Des Stud 57:9–36
11. Haritaipan L (2019) Towards the creation of creativity tools for real-practice: a review of 112 design tools in the market. Des J 22(4):529–539
12. Bruseberg A, McDonagh-Philp D (2002) Focus groups to support the industrial/product designer: a review based on current literature and designers' feedback. Appl Ergon 33(1):27–38
13. Sio UN, Kotovsky K, Cagan J (2015) Fixation or inspiration? a meta-analytic review of the role of examples on design processes. Des Stud 39:70–99
14. Vasconcelos LA, Crilly N (2016) Inspiration and fixation: questions, methods, findings, and challenges. Des Stud 42:1–32
15. Curry T (2014) A theoretical basis for recommending the use of design methodologies as teaching strategies in the design studio. Des Stud 35(6):632–646
16. Daalhuizen J, Person O, Gattol V (2014) A personal matter? an investigation of students' design process experiences when using a heuristic or a systematic method. Des Stud 35:133–159
17. Lawson B, Dorst K (2009) Design expertise. Architectural Press, Oxford
18. Dorst K, Reymen I (2004) Levels of expertise in design education. In: International engineering and product design education conference, 2–3 September 2004, Delft, The Netherlands
19. Cross N (2011) Design thinking. Berg, Oxford
20. McCawley PF (2009) Methods for conducting and educational needs assessment: guidelines for cooperative extension system professionals. University of Idaho Extensions, Moscow, Idaho
21. Altschuld JW, Eastmond JN (2010) Needs assessment phase 1: getting started. Sage Publications, California
22. Börekçi NAGZ (2018) Design divergence using the morphological chart. Des Technol Educ Int J 23(3):62–87
23. Korkut F, Doğan Ç (2010) My dear monster friends: matrix as an emotionally rich generative design tool. In: Proceedings of the 7th international conference on design and emotion (CD). IIT Institute of Design, Chicago
24. ID302 Industrial Design IV Studio Team (2018) 2017–2018 Academic Year Spring Semester Switch Task Class Assignment Brief. Department of Industrial Design, Middle East Technical University, Ankara, Turkey
25. Bleecker J (2010) Design fiction: from props to prototypes. In: Swiss Design Network (eds) Negotiating futures—design fiction. Paper presented at the proceedings of the 6th Swiss design network conference, 28–30 October 2010. Steudler Press, Basel, Switzerland, pp 58–67
26. Franke B (2010) Design fiction is not necessarily about the future. In: Valentine L, Borja de Mozota B, Nelson J, Merter S, Atkinson P (eds) The value of design research. Paper presented at the proceedings of the 11th international european academy of design conference, Paris Descartes University, France, 22–24 April 2015. Sheffield Hallam University, UK, Sheffield
27. Tanenbaum J (2014) Design fictional interactions: why HCI should care about stories. Interactions 21(5):22–23
28. Knutz E, Markussen T, Cristensen PR (2014) The role of fiction in experiments within design, art & architecture—towards a new typology of design fiction. Artifact 3(2):8.1–8.13
29. Kamberelis G, Dimitriadis G (2013) Focus groups: from structured interviews to collective conversations. Routledge, Oxon
30. Aprile WA, Van Der Helm A (2011) Interactive technology design at the delft university of technology—a course about how to design interactive products. In: International conference on engineering and product design education. City University, UK, 8–9 September
31. Corremans J (2011) Measuring the effectiveness of a design method to generate form alternatives: an experiment performed with freshmen student's product development. J Eng Des 22(4):259–274

32. Bell P (2001) Content analysis of visual images. In: Leeuwen TV, Jewitt C (eds) Handbook of visual analysis. Sage Publications, London, pp 10–34
33. Krippendorf K (2004) Content analysis: an introduction to its methodology, 2nd edn. Sage, Thousand Oaks, California
34. Goldschmidt G (2016) Linkographic evidence for concurrent divergent and convergent thinking in creative design. Creat Res J 28(2):115–122
35. Yilmaz S, Seifert CM (2011) Creativity through design heuristics: a case study of expert product design. Des Stud 32(4):384–415

The Value of Design Education for Other Fields: Using Design Tools to Teach Psychology

Mafalda Casais

Abstract This paper proposes the use of design tools and studio environment in psychology teaching, based on a type of outcome that is already produced in this field—interventions to support people's well-being. In a class with 24 students from a post-graduation study in positive psychology, we introduced a sequence of six canvases (persona, empathy canvas, journey mapping, design vision, well-being matrix, and a blank canvas to draw the intervention) and distributed students in multidisciplinary groups. Introducing a studio format with design tools aimed to offer a different perspective on thinking about potential patients/clients/users and contexts through an action-based, opportunity-driven setup. Results show an impactful effect, a successful production of interventions to apply in practice, and overall high levels of engagement and satisfaction. While this paper reports a single case, it proposes that this approach is worth exploring further. Its contribution is twofold: considering process and content, it introduces human-centered design thinking to an educational context that already sought it tacitly; considering format, it empowers psychology students to think like designers and approach the educational experience in a more horizontal perspective of knowledge transfer. We discuss how design tools and educational modalities might be appropriate to introduce into the education of other disciplines, still considering their specific needs and aims—like a *globalized* approach to education, which we call *Education through Design*. Also, we discuss it in the context of the future of education, from a convergency tendency perspective at a European level.

Keywords Design studio pedagogy · Globalized education · Design tools and methods · Human-centered design thinking

M. Casais (✉)
CIAUD, Lisbon School of Architecture, Universidade de Lisboa, 1349-063 Lisboa, Portugal
e-mail: mafaldacasais.ciaud@fa.ulisboa.pt

E. Duarte and C. Rosa (eds.), *Developments in Design Research and Practice*,
Springer Series in Design and Innovation 17,
https://doi.org/10.1007/978-3-030-86596-2_23

1 Introduction

Designing for the future of education requires design education. [1]

Pedagogical instruments originating from design—i.e., design methodologies, the studio classroom format, design tools—are of great value for the education of design disciplines (communication design, product design, engineering, architecture, etc.). Among the reasons for this are the fact that designers think visually [2–4], and that design is learned in the process of designing [5].

Introducing design tools (instruments that help in *doing* something), together with design methods (recipes for knowing *how* to do something), is potentially interesting in the educational context of fields outside of design. The reason for this is that often the goal of other domains of knowledge and practice is also to create interventions for a specific target audience. These *designerly ways* of teaching that originate from the design field can bring in depth understanding of people and contexts, and help in the development of situated interventions for specific aims, which is a goal of fields like psychology.

In this paper, we report a case in which design tools conveying design methods were used in a psychology class. Specifically, we introduced a set of canvases with different methods to understand people and their contexts, and to help create something for their well-being. In addition, and inherent to the use of these tools in groups, a studio environment was set up in the classroom. This allowed diverse teams to form and to explore the material progressively, in close connection with the teacher. The response to the introduction of this didactic material was evaluated from the point of view of the students, through a questionnaire.

Specifically, we present a study with 24 students from a post-graduate course in applied positive psychology, describe the *designerly* materials that were introduced, and discuss the questionnaire results. Also, we discuss the resulting insights in the context of the future of education, from a convergency tendency perspective, as observed by the streamlining of higher education at a European level—with formats such as ECTS credits or the Bologna cycles, proposing an *Education through Design* approach.

1.1 Teaching Positive Psychology Using Positive Design Thinking

Positive psychology is a branch of the field of psychology that studies human well-being, develops theoretical models to explain it, and devises and tests strategies to improve it. Its founding scholars define it as

> a science of positive subjective experience, positive individual traits, and positive institutions [that] promises to improve quality of life and prevent the pathologies that arise when life is barren and meaningless. [6, p. 279]

Among the different types of research conducted in this branch, Positive Psychology Interventions (PPIs) are a widespread outcome. PPIs are scientific tools and strategies focusing on supporting well-being and positive cognitions and emotions [7]. Examples of PPIs are protocols and guidelines for therapy and coaching, workshops for self-knowledge, meditation, gratitude, or for better communication, etc. This can be seen as a form of design, in the sense that there is an effort to develop something to bring a situation from A to B.

Positive design, is a field of design that does just that, using research on well-being to inform its methodologies and aims [8]. Specifically, it intends to design for opportunities—rather than for problems—and to purposefully support people's well-being, focusing on embodying some of these proposals (e.g., gratitude, self-knowledge) in concrete interventions (products, digital environments, games, spaces, interactions, etc.).

The example described in this paper, which used positive design frameworks and tools to develop PPIs, is a particularly interesting case because it shows how these fields closely intersect and can be mutually supportive. Currently we observe psychology informing design, but not the other way around—which justifies the approach we present in this paper, as an exercise to illustrate the assumed benefits of design thinking for psychology (and potentially other fields).

1.2 Teaching with Design Mechanisms

Human-centered design is a methodological approach originating from the design field. It focuses on human perspectives throughout the design process, including psychological and emotional factors [8]. This approach, applied to human-focused fields like psychology, can potentially help practitioners gain a more granular understanding of people and their contexts. In fact, in psychology, some design methodologies are already tacitly used. For example, in positive psychology (a branch of psychology that focuses on the study of well-being, devising and testing interventions to improve it, as described above), this is possible to observe. In master and doctoral theses from this branch, we can trace similar methodologies than those used in design [e.g., 9]. For example, the problem-based approach, an approach that uses "problems as the stimulus and focus for (...) activity" [10, p. 2]. This approach has also been used in design, and overlaps in many aspects with a project-based approach and design thinking [11].

Qualitative design methods [12, 13] such as the VIP (Vision in Product Design [14])—which encourages an examination of what the designer wants people to experience, or *Contextmapping* [15]—which encourages people (users) to manifest their understanding of their own context, present interesting possibilities for other fields. They can be used for team collaboration, co-creation, stakeholder involvement, and assessment of interventions in their context, which are research activities we find widely reported in the research output of positive psychology [16].

Another mechanism from design pedagogy that can be interesting for other disciplines is the studio format. "The design studio is (...) an educational setting where students fundamentally learn by practicing under the supervision of a teacher" [5, p. 22]. It deals with ill-defined problems and relies on a close exchange with the teacher, building on horizontal and reciprocal knowledge transfer, as opposed to a classical top-down teacher-student model. A design studio approach can bring these interactions (between students, and between students and teacher) to a next level of sharing and working together. Combined with a human-centered approach—the nuanced and empathic understanding of people and human perspectives around problems, the studio setting presents a promising educational setup.

We propose, therefore, that making these mechanisms explicit in an educational setting outside of design allows us to grasp their real value for the field. Using design tools is also a possible strategy to improve other educational settings.

Design tools are instruments that assist in the design process. They can be defined as compact formats (cards, booklets, canvases, digital guides, etc.) with data in the form of text and image, often with game elements. In addition, used in the design process these can be inspirational, informational, methodological, or a combination. They can provide systemic strategic support for a project or specific aid for part of it, as "powerful resources [that] intrinsically seem to reinforce [designers'] capabilities and capacities" [17, p. 607]. A well-known example of a design tool with a card set format is IDEO's Method Cards [18].

For this study, the design tools that were developed and distributed were six canvases to use in a class exercise. They were based on previously developed design methods, and adapted for the class context (see the Materials section for a description). Twenty-four students participated in an Applied Positive Psychology class, integrated in an executive master study in Applied Positive Psychology. The class was part of a module comprising of a theoretical class and a practical one, focusing on positive psychology interventions (i.e., interventions to improve people's well-being). The practical class was setup as an eight-hour workshop, with a one-hour lunch break (4+1+4 h). This allowed for an intensive studio session, going through the whole design process in a condensed period. The class was divided into four parts:

1. a slide presentation with an introduction to the topic and a description of the exercises;
2. the exploration of a user or group for whom to create an intervention;
3. the exploration of a situation for which to design; and
4. the sketching and detailing of an intervention to improve well-being.

1.3 Materials

In this section, we describe in detail the materials used in the class, and explain the design methods these were based on, as well as the aim of using these particular methods.

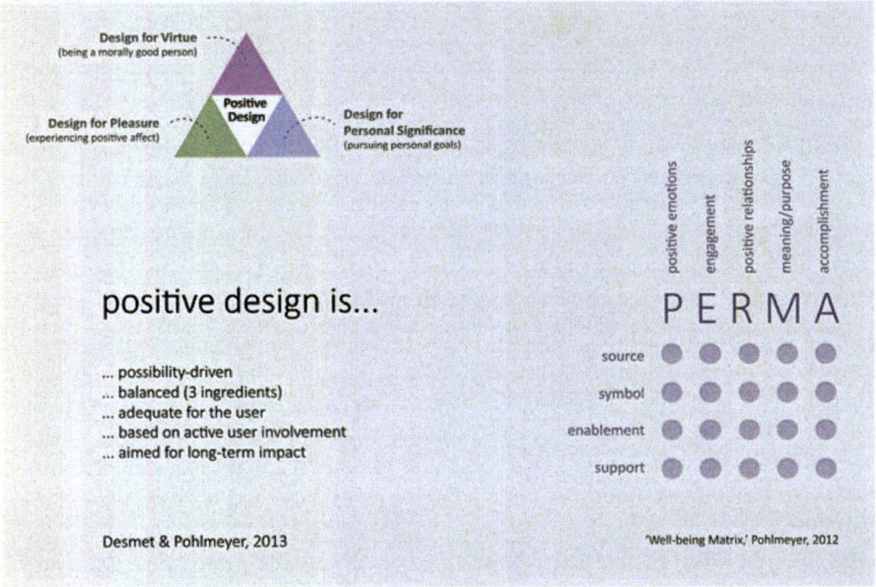

Fig. 1 Presentation slide about the positive design framework, the positive design ingredients (*Source* [8]) and the Well-being Matrix (*Source* [19])

Introduction Presentation. The class began with an introduction about positive design, based on the work of Desmet and Pohlmeyer [8]. In it we presented the Positive Design Framework, the ingredients of positive design, and the Well-being Matrix [19], as a basis from which students could work (Fig. 1). The latter provides a matrix of possible combinations between types of embodiment and well-being determinants, to generate ideas for interventions.

In the presentation, we went through the canvases one by one, explaining how to apply them to existing projects, to create new ideas, or to use the students' own person/context to go through the whole process and understand it from the user research until the intervention stage.

Design Canvases. Six canvases with design methods were developed to introduce in the class exercise: the first two canvases focused on the user, the third focused on mapping the situation/context to identify intervention opportunities, the fourth focused on the exploration of an effect for the intervention, followed by the exploration of possible intervention formats, and the last focused on the intervention detailing. The canvases were based on well-established design methods such as the *persona* method [20] or the Well-being Matrix [19].

Canvas 1: Persona. The first part of the workshop focused on understanding the user. For this, students were asked to either consider a familiar group, to imagine a group they would like to work with, or to consider themselves as a user, in order to go through the steps and understand the procedure.

The first canvas was entitled "Canvas about the Person (*Persona*)," and was based on the well-established *persona* method [20], originating from the marketing and consumer research fields.

> The personas approach proposes focusing on specific or canonical users. The principle is therefore to design a product adapted to different types of people, usually about a few dozen, representing typical consumers. From this viewpoint, the notion of a persona draws on its etymology: The actor's mask, each character playing a particular role during the performance of the play. [*ibidem*, p. 39]

The canvas comprised of six spaces to fill information about the potential user of the intervention (Fig. 2). Specifically, in the first space it asked the identification of the user (name, age, physical appearance, limitations, and talents). In the next space it asked how the person is, providing six sliders with two contrasting personality traits in each (extrovert-introvert, rational-intuitive, cautious-adventurous, imaginative-conventional, skeptical-naïve, and a blank pair to fill in). The next space was entitled "tasks" and asked questions about the person's profession and what they do in their spare time.

On the bottom row, the first left most space, asked about the person's context. Specifically, it posed five questions: where the person lives; where they come from; what their family looks like; the school level they achieved; and the groups they are a part of. The middle space asked about capabilities, namely the level of technology the person uses and the devices with which they interact. The final space, at the

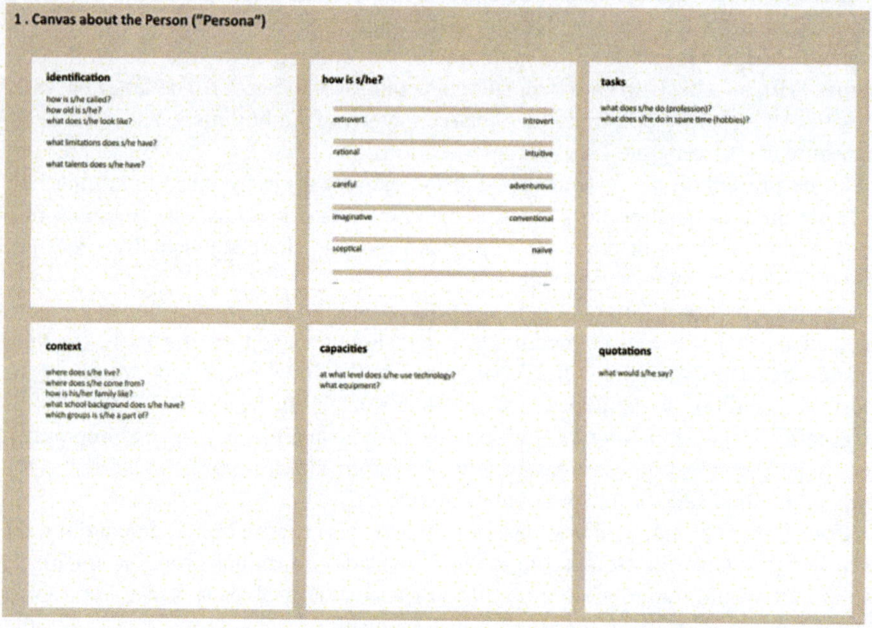

Fig. 2 Canvas 1 focused on general information about the target group/person (original printed size was A3)

bottom right side, asked for quotations of the person (these could be imagined even if the student was focusing on a real target group or person).

Canvas 2: Empathizing with the user. The second canvas, called "Empathy Canvas," also focused on understanding the user, however in a more profound way, based on the method of *empathy mapping* [21, 22].

> An empathy map is a simple, easy-to-digest visual that captures knowledge about a user's behaviors and attitudes. (…). The mapping process can help synthesize research observations and reveal deeper insights about a user's needs. (…) It can help guide the construction of personas or serve as a bridge between personas and concept deliverables. [21, para 5–7]

This canvas aimed to go deeper into the user's reality, their experiences, and their perception of the world around them (Fig. 3). It was organized in five spaces, radiating from a central space in the shape of a human head. The first space, at the top left side, began with the question: "WHO are we empathizing with?" (intentional emphasis on *who*). This was followed by three more specific questions, namely about who the person we want to understand is; what situation they are in; and what their role is in that situation.

Going in a clockwise motion, the second space begins with the question: "What does that person need to DO?" (intentional emphasis on *do*). Four additional specific questions followed, about what they need to do differently; the tasks they want or need to accomplish; the decisions they need to make; and the way they know they have succeeded.

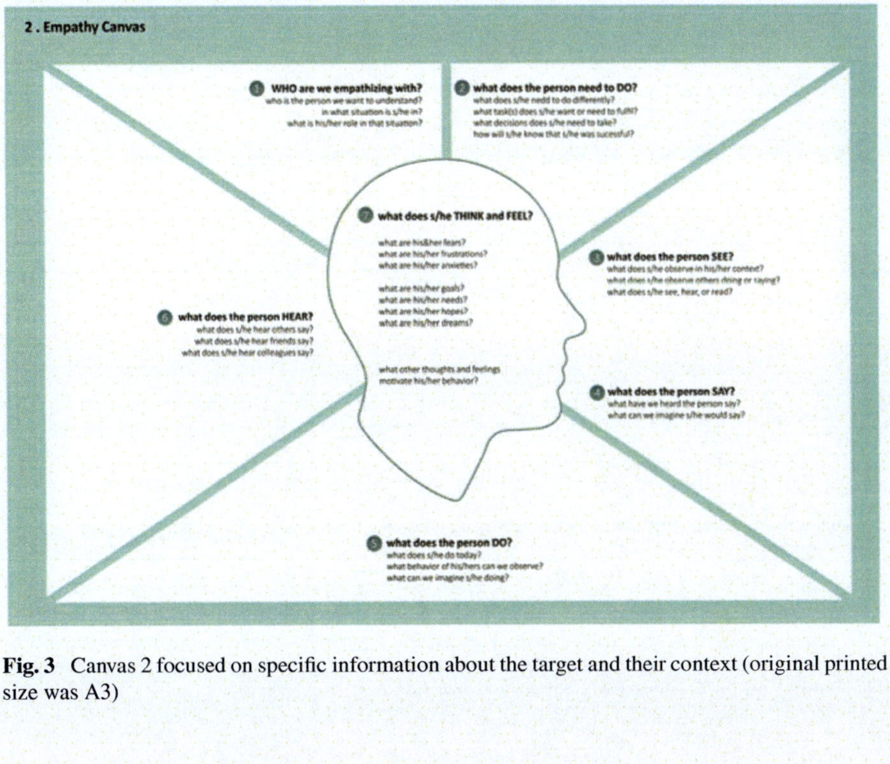

Fig. 3 Canvas 2 focused on specific information about the target and their context (original printed size was A3)

On the right most side of the canvas, we find a wider space with two heading questions and some follow up questions for each. The first question was: "What does the person SEE?" (intentional emphasis on *see*). This was followed by sub-questions about what the person observes in their context; what they observe others doing or saying; and what they see, hear, or read. The second question was: "What does the person SAY?" (intentional emphasis on *say*). This was followed by sub-questions about what we (designer/student) have heard the person say; and what we can imagine they would say.

Canvas 3: Mapping the journey. The third canvas, called "Journey Canvas," used the journey mapping method to "visualiz[e] the process that a person goes through in order to accomplish a goal" [23, para 1].

> (…) journey maps (also known as "experience maps" or "customer experience maps") (…) add a third dimension to traditional personas by focusing on a diachronic outline of a user's experience (…) over time. As the name suggests, journey maps provide a graphic visualization or a map of a customer's or user's experience (…). [24, pp. 10–11]

The canvas was divided into four parts (Fig. 4). On top, it asked for a description of the current scenario a person is going through, as well as for a description of the goals and ambitions for the person/situation. The middle section, the largest space, was destined for the journey visualization, in which students were asked to schematize the path and highlight the most important moments or key steps. Bellow, a smaller

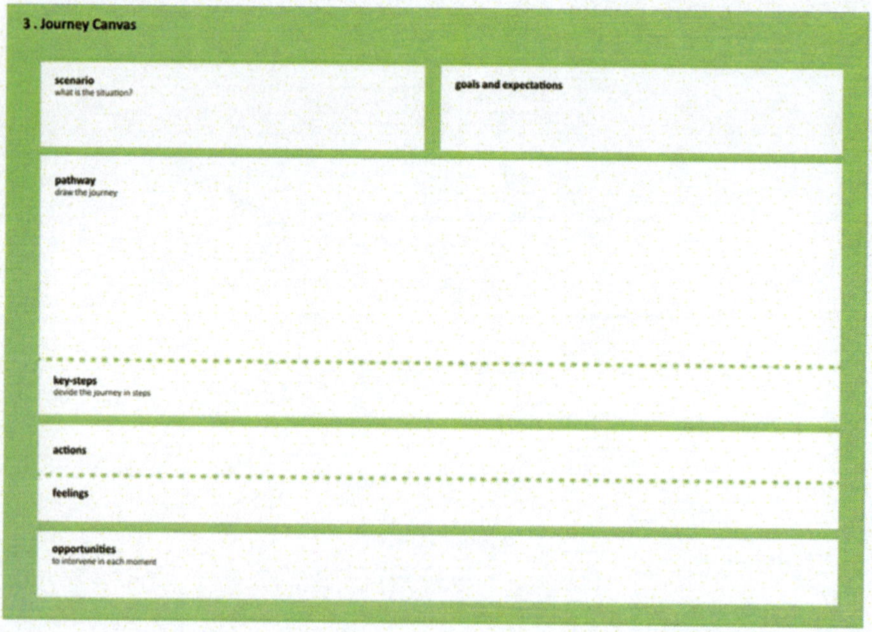

Fig. 4 Canvas 3 focused on a journey or situation the target user goes through to identify opportunities for intervention (original printed size was A3)

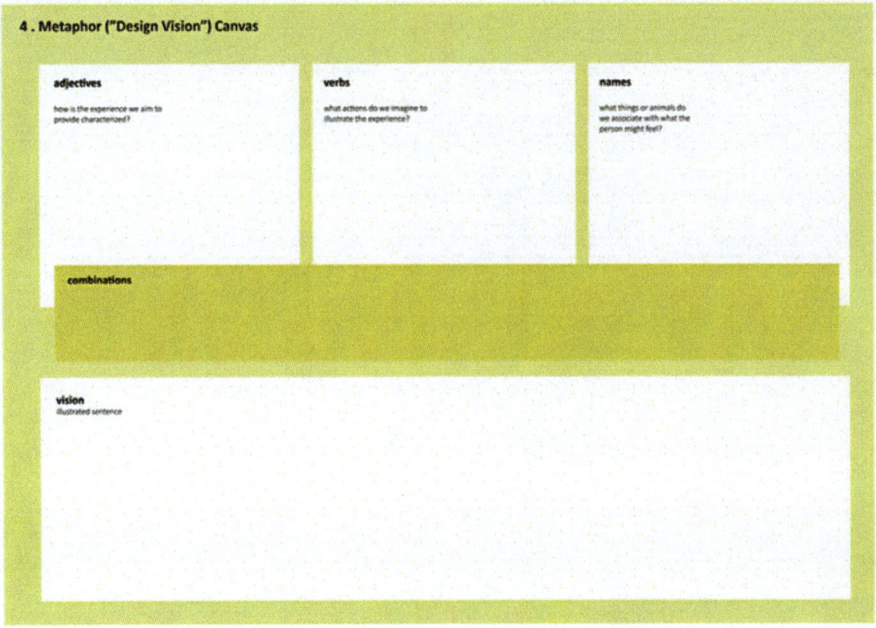

Fig. 5 Canvas 4 focused on envisioning an effect for the intervention through a metaphor (original printed size was A3)

space asked for the actions and feelings/emotions that were experienced in each of those moments. And finally, the last space in the bottom of the canvas asked students to identify opportunities they found throughout the ups and downs of the journey.

Canvas 4: Creating a design vision. The fourth canvas, called "Metaphor ('Design Vision,') Canvas" focused on the effect of the intervention (Fig. 5). It was loosely based on the Design Vision approach used at the Industrial Design Engineering Faculty of the Delft University of Technology [14, 25]: "A vision in the context of product design provides us with a personal, inspiring image of a new future situation created by a designer or a group of designers and/or other professionals [25, para 1]."

However, we aimed to used it more in the metaphorical sense, that is, to find an image that could illustrate an effect or a feeling, for the intervention. Metaphors build associations between different concepts, using the attributes of one to understand or represent the other [26]. Examples of this can be feeling as light as a feather as a result of using the intervention, or feeling as happy as a dog getting a treat, or even feeling like part of something bigger such as how a football cheering squad feels at the moment their team scores.

Canvas 5: Identifying design opportunities. The fifth canvas, called "Opportunity Matrix Canvas" used the Well-Being Matrix, proposed by Anna Pohlmeyer [19] to focus on finding opportunities for the intervention. The matrix crosses the PERMA model—the determinants of well-being as defined by Martin Seligman [27]: positive

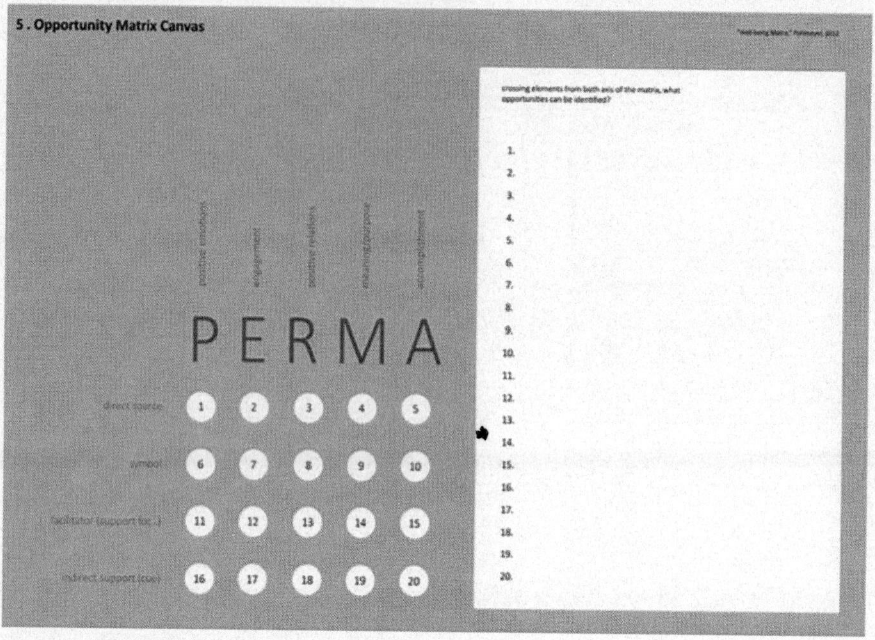

Fig. 6 Canvas 5 focused on identifying possible formats for the intervention (original printed size was A3; *Source* [19])

emotions, engagement, positive relations, meaning, and achievement—with possible embodiment categories—direct source, symbol, facilitator, indirect support.

The canvas is composed of the Well-being Matrix on the left side, and a list of the possible combinations on the right side (Fig. 6).

Canvas 6: The idea. The sixth and final canvas, called "Intervention Canvas" was the simplest out of all six, containing only one large space, which asked for a sketch of the intervention and its interactions, step-by-step (Fig. 7). During the class introduction, students were encouraged to use drawings, non-linear text, and diagrams as much as possible over linear text to explain their solution.

2 Method

To assess the appropriateness and usefulness of design tools in psychology education, a questionnaire was given to students to fill in a few days after the class. The questionnaire was send via social networking sites (*WhatsApp* mobile and desktop application) and via e-mail.

Fig. 7 Canvas 6 focused on detailing the intervention through drawings or diagrams (original printed size was A3)

2.1 Questionnaire

The questionnaire included four open-ended questions about the limitations students faced in the class and the results they obtained; three yes/no questions about whether they found the class material appropriate for the given educational context; and several Likert scale-based questions (ranging from 1 to 5) to address the usefulness, clarity and potential of the canvases.

The questionnaire began with an informed consent stating: "The collected data aims to improve class materials and to inform the study of the use of design tools in psychology education. Any collected data is anonymous."

The specific questions were:

- I accept information gathering: Yes/No
- Demographic data: age, occupation
- Do you believe you understood the content of the class? Yes/No
- What limitations did you encounter in understanding the content of the class?
  - Options: material, slides, teacher, language, other
- Specify other issues you encountered.
- Did you learn new content about psychology? Yes/No

- Rate the class materials (canvases and presentation) using a scale from 1 (not at all) to 5 (excellent) on each separate element:
  - in general; on clarity; on usefulness; on potential
- Rate each canvas using a scale from 1 (not at all) to 5 (excellent) on each separate element:
  - Canvas 1—persona canvas: on clarity; on usefulness; on potential
  - Canvas 2—empathy canvas: on clarity; on usefulness; on potential
  - Canvas 3—journey canvas: on clarity; on usefulness; on potential
  - Canvas 4—metaphor/vision canvas: on clarity; on usefulness; on potential
  - Canvas 5—opportunity matrix canvas: on clarity; on usefulness; on potential
  - Canvas 6—intervention canvas: on clarity; on usefulness; on potential
- What impact did these tools have in understanding the process of designing interventions?
- What impact do you envision for using these tools in your professional practice?
- Do you consider design tools to be adequate for psychology education? Yes/No.

3 Results

From a sample of 24 students, 70% responded to the online questionnaire, a few days following the class. All participants were female, with an average age of 41 (ranging between 24 and 59). Nearly half (47%) were psychologists or coaches.

All students (100%) reported having learned new content, and nearly all (94%) having understood the class content. Language and presentation (29%) and program organization (not related to content = 35%) were mentioned as the hindering factors to content comprehension.

Regarding the appropriateness of using design-based tools in psychology education, all students (100%) responded positively.

Students considered the provided material (presentation + canvases) to be clear or very clear (4 or 5 out of 5 = 83%), to be useful of very useful (4 or 5 out of 5 = 89%), and to have a high or very high potential (5 or 4 out of 5 = 89%), as seen in Table 1.

Table 1 Assessment of provided material (presentation + canvases) on a Likert scale (1–5) in percentages

	1-Not at all	2-Not so much (%)	3-Could be better (%)	4-Good (%)	5-Very good (%)	Average (points)
Clarity	–	6	11	33	50	4.2
Usefulness	–	–	11	33	56	4.4
Potential	–	–	11	22	67	4.5

Table 2 Assessment of individual canvases (average rating) on a Likert scale (1–5) in points

	Canvas 1	Canvas 2	Canvas 3	Canvas 4	Canvas 5	Canvas 6
Clarity	4.5	4.3	4.2	4.0	3.3	4.8
Usefulness	4.7	4.6	4.5	4.3	4.1	4.6
Potential	4.8	4.6	4.4	4.3	4.4	4.6

Focusing on each canvas, we asked students to score them individually in three categories: clarity, usefulness, and potential (Table 2). The best score in terms of clarity was given to Canvas 6 (the intervention canvas, see Fig. 7), which was the simplest out of all six, with only the request to detail the intervention step-by-step. The lowest score on clarity was given to Canvas 5 (the opportunity matrix canvas, see Fig. 6), which was considered the most difficult to understand.

The best score in terms of usefulness was given to Canvas 1 (the persona canvas, see Fig. 2), which was often mentioned, even after the class project was complete, as the most novel and perspective-changing canvas. We observed this particularly when students presented the projects and reflected on the experience. The lowest score on usefulness was given to Canvas 5 (the opportunity matrix canvas, see Fig. 6), because, as mentioned above, it was considered the most difficult to understand and apply.

Regarding potential—for idea generation and use in their coaching practice, for example—the lowest score was given to Canvas 4 (metaphor/design vision canvas, see Fig. 5). This canvas was also somewhat difficult to understand—it had the aim of illustrating the *effect* of the intervention, through a metaphor (e.g., "feeling light as a feather"). However, we observed that when used successfully, students were satisfied with its contribution to the project. The highest score in the potential category was Canvas 1 (the persona canvas, see Fig. 2), which, as highlighted above, was mentioned as the strongest overall canvas, and the most perspective-changing.

Following the numeric rating, we asked some open questions about the impact and adequacy of the tools that were presented. On the impact of the presented tools in understanding the process of designing interventions, students mentioned:

- "It is a great framework to create PPIs in a structured and targeted way."
- "[The tools] helped me to focus on my user and to understand them in a much deeper way than if I created ideas in my head without an analysis. So the impact was big."
- "The canvases are very useful to organize ideas and allow the link between problem and resolution to become much clearer."
- "To clarify the creative process with the potential client, to understand the user better, and to reach results."
- "It improves the diagnostic and the intervention methodology."

On impact considering their professional practice, students mentioned:

- "Because I work with people development within organizations, I will be able to use it in my interventions."

- It will shed a lot of light, especially to think and plan new projects and marketing strategies. The empathy mapping (Canvas 2, the empathy canvas, see Fig. 3) is exceptional to amplify the view on the target audience."
- "It can be used in coaching sessions."
- "It can be very useful in the citizenship perspective."
- "I believe that it will be useful to organize my thinking and minimize the action-reaction type of behavior that I tend to have."
- High impact, I intend to use it professionally to assist my business plan development."
- "It is highly adequate for my professional reality."

4 Discussion

4.1 *Insights from the Results*

Results indicate that design tools can be useful to understand people and groups, and to create an intervention to improve their lives, in the context of positive psychology. In addition, design tools were considered as having a great potential to turn theoretical content actionable, and expedite the process of generating positive psychology interventions.

This leads to two types of insight: first, that design methods and tools are potentially self-explanatory and accessible to be used by non-designers (which had been verified in previous research conducted with children [28], elderly [29], etc.). Second, that the design studio format is applicable in other educational contexts, to convey and apply non-design knowledge—it streamlines interactions, facilitates discussion, and promotes horizontal knowledge transfer.

The study reported in this paper is not about design knowledge or a contribution to the design field, *per se*. Rather, it is about the tools and methods of design, and their ability to streamline (educational) processes in other fields. Just as design (the field of knowledge) often relies on mechanisms from the social sciences, humanities, etc., to improve its own processes and agenda, so it is possible to use the mechanisms of design to advance other domains of expertise. Whereas design often resorts to methods and tools from other disciplines—e.g., anthropology, sociology, psychology [30, 31]—the opposite is not so common. With this exercise, we aimed to explore the possibility of using mechanisms from design to bring value to the educational settings of other fields.

4.2 *Implications for the Future of Education*

We propose that design tools and educational modalities (like workshops, or studio classes) are appropriate to introduce into other disciplines, while simultaneously considering their specific needs and aims—like a *glocalized* approach to education: general and globalized, yet specific and localized. We call this *designerly* approach to education *Education through Design.*

The designation *Education through Design* is not necessarily new. It has been used previously to describe ways to improve design education [32]. Our proposal, however, is a parallel to Frayling's [32] concept of Research through Design, meaning designerly ways of teaching. The concept of Research through Design (RtD) proposes using the distinctive perspective of design to understand and intervene in diverse settings outside the field of design, mapping contexts, people, opportunities, problems, and solutions. Similarly, design pedagogy offers a unique set of mechanisms—including the use of design-based tools, methods, and the studio classroom setting—that can be valuable for other fields [33]. This is particularly the case in fields that already use design-like approaches in education, as the one illustrated in this paper.

This proposal seems to be of particular relevance in the context of the future of education, from two perspectives. Firstly, from a convergence tendency perspective, as observed by the streamlining of higher education at a European level—for example, in the ongoing discussion on the Bologna process [e.g., 35].

> While European Union member states do not show the specific desire to delegate control in matters of education, there is a progressive tendency convergence in terms of resulting degrees (e.g., ECTS system, mobility programs, the Bologna process). Since the member states are included in an intergovernmental and supranational system, and share common legislation in many policy areas, ideas about how to teach might be valuable to test, propose, and replicate, to create a globalized direction towards better education. [36, p. 601]

Secondly, this design-based approach to education might potentially be useful to reach twenty-first century education goals and skills [e.g., 37], further contributing to the debate of quality in education. The Knowledge Age brings with it both possibilities and barriers. We must address both to provide a more adequate, competitive, and modernized education. The proposal presented here can be a potential pathway for this, however it needs further work, both conceptually and in terms of validation.

4.3 Limitations and Future Research

The class reported in this paper was carried out in the context of a two-class module, a theoretical class and a practical, applied one. Due to administrative issues, the practical class—the one reported here—was given first. Consequently, students did not have a sound theoretical basis prior to the practical class. This led to some difficulties in the understanding and application of concepts, which had to be bypassed by ongoing explanations throughout the exercise. This influenced the assessment of the class quality and of the respective material, rendering it (seemingly) less useful—as indicated in some open-ended questions (35% of questionnaire respondents considered this issue as hindering content comprehension).

Furthermore, the current proposal is focused on results from a small sample and a single exploration. A larger sample, with several groups, would provide a more accurate picture of how these tools and mechanisms affect the education of fields outside of design. It also reports results from the field of psychology and specifically from the branch of positive psychology. As such, more research can focus on other fields, and even other branches of psychology, to understand the proposal's value.

For the setup of the class exercise, we opted to use a studio-based modality. It is relevant to note that not all scholars focusing on design education agree that studio-based learning is worth pursuing, mentioning that there is little foundation to support its value. In fact, some research argues that this modality can be inappropriate as it is often directly based on the master-apprentice model of learning, which in turn is often not a promising way to teach because it does not value the intellectual development of the student [35].

The Research through Design (RtD) approach, proposed by Freyling [32], focuses on designerly ways of conducting research with knowledge (and for new knowledge) outside of design. RtD, thus, was our starting point to propose an Education through Design approach (EtD). Analogously, it would correspond to the use of designerly ways of teaching and learning that are applied in other fields. Future research can develop this concept further, carrying out experiments to attest its value.

References

1. Berk S (2016) Designing for the future of education requires design education. Art Educ 69(6):16–20
2. Cross N (1982) Designerly ways of knowing. Des Stud 3(4):221–227
3. Schön D (1983) The reflective practitioner: how professionals think in action. Basic Books, New York
4. Lawson B (2005) How designers think: the design process demystified, 4th edn. Architectural Press, Oxford
5. Ferreira J (2018) Design conversations: an exploratory study of teacher and student interaction in the design studio (doctoral dissertation). Delft University of Technology, The Netherlands
6. Seligman ME, Csikszentmihalyi M (2014) Positive psychology: an introduction. In: Csikszentmihalyi M (ed) Flow and the foundations of positive psychology. The collected works of Mihaly Csikszentmihalyi, pp 279–298. Springer, Dordrecht
7. Chowdhury R (2019) Online: https://positivepsychology.com/positive-psychology-interventions/. Last accessed 30 Jan 2020
8. Desmet PMA, Pohlmeyer AE (2013) Positive design: an introduction to design for subjective well-being. Int J Des 7(3):5–19
9. Weiss LA (2016) Direction: happiness. Improving well-being of vulnerable groups (doctoral dissertation). The Netherlands, University of Twente
10. Boud D, Feletti GI (1997) The challenge of problem-based learning. Routledge, London
11. Stokholm M (2014) Problem based learning versus design thinking in team based project work. In: Bohemia E et al (eds) DS 78: Proceedings of the 16th international conference on engineering and product design education (E&PDE14), Design education and human technology relations. The Design Society, Glasgow, pp 268–274
12. Sanders E, Stappers PJ (2013) Convivial toolbox: generative research for the front end of design. BIS Publishers, Amsterdam
13. van Boeijen DJ, van der Schoor R, Zijlstra J (eds) (2013) Delft design guide: design strategies and methods. BIS Publishers, Amsterdam

14. Hekkert P, van Dijk M (2016) VIP vision in design: a guidebook for innovators. BIS Publishers, Amsterdam
15. Sleeswijk Visser F, Stappers PJ, van der Lugt R, Sanders E (2005) Contextmapping: experiences from practice. CoDesign 1(2):119–149
16. Schuling R, Huijbers M, Jansen H, Metzemaekers R, Van Den Brink E, Koster F, Van Ravesteijn H, Speckens A (2018) The co-creation and feasibility of a compassion training as a follow-up to mindfulness-based cognitive therapy in patients with recurrent depression. Mindfulness 9(2):412–422
17. Lutters E, Van Houten FJ, Bernard A, Mermoz E, Schutte CS (2014) Tools and techniques for product design. CIRP Ann 63(2):607–630
18. IDEO (2003) Method Cards. https://www.ideo.com/post/method-cards. Last accessed 30 Jan 2020
19. Pohlmeyer AE (2012) Design for happiness. Interfaces 92:8–11
20. Brangier E, Bornet C (2011) Persona: a method to produce representations focused on consumers' needs. In: Karwowski W, Soares MM, Stanton NA (eds) Human factors and ergonomics in consumer product design—methods and techniques. CRC Press, Boca Raton, pp 37–61
21. Brown JL (2018) Empathy mapping: a guide to getting inside a user's head. https://www.uxbooth.com/articles/empathy-mapping-a-guide-to-getting-inside-a-users-head/. Last accessed 30 Jan 2020
22. Mortensen D (2020) Stage 1 in the design thinking process: empathise with your users. https://www.interaction-design.org/literature/article/stage-1-in-the-design-thinking-process-empathise-with-your-users. Last accessed 30 Jan 2020
23. Gibbons S (2018) Journey mapping 101. https://www.nngroup.com/articles/journey-mapping-101/. Last accessed 30 Jan 2020
24. Howard T (2014) Journey mapping: a brief overview. Commun Des Quart 2(3):10–13
25. Olieman AM (2011) Design vision. http://wikid.io.tudelft.nl/WikID/index.php/Design_vision. Last accessed 30 Jan 2020
26. Cila N, Hekkert P, Visch V (2014) Source selection in product metaphor generation: the effects of salience and relatedness. Int J Des 8(1):15–28
27. Seligman M (2018) PERMA and the building blocks of well-being. J Posit Psychol 13(4):333–335
28. Van Doorn F, Gielen M, Stappers PJ (2014) Children as co-researchers: more than just a roleplay. In: IDC'14: Proceedings of the 2014 conference on interaction design and children. Association for Computing Machinery, New York, pp 237–240
29. Demirbilek O, Demirkan H (2004) Universal product design involving elderly users: a participatory design model. Appl Ergon 35(4):361–370
30. Almquist J, Lupton J (2010) Affording meaning: design-oriented research from the humanities and social sciences. Des Issues 26(1):3–14
31. Gunn W, Otto T, Smith RC (eds) (2013) Design anthropology: theory and practice. Bloomsbury Academic, London
32. Niederhelman M (2001) Education through design. Des Issues 17(3):83–87
33. Frayling C (1994) Research in art and design. Royal College of Art Research Papers, vol 1(1), pp 1–5
34. Giaccardi E, Stappers PJ (2017) Research through design. In: Soegaard M, Dam RF (eds) The Encyclopedia of human-computer interaction, 2nd edn. Interaction Design Foundation (2017). https://www.interaction-design.org/literature/book/the-encyclopedia-of-human-computer-interaction-2nd-ed/research-through-design. Last accessed 30 Jan 2020
35. Wächter B (2014) The Bologna process: developments and prospects, 39(3):265–273
36. Casais M, Christiaans H, Almendra R (2012) Sustainability curricula in design education. In: Buck L, et al (eds) DS 74: Proceedings of the 14th international conference on engineering & product design education (E&PDE12) design education for future wellbeing. The Design Society, Glasgow, pp 599–604

37. Bellanca JA, Brandt RS (2010) 21st century skills: rethinking how students learn. Solution Tree Press, Bloomington
38. Souleles N (2013) The evolution of art and design pedagogies in England: influences of the past, challenges for the future. Int J Art Des Educ 32(2):243–255

Design for Culture

Design for Culture

Helena Souto, Gabriele Oropallo, and Helena Barbosa

Abstract In an era of continuous scientific and technological progress, which gives rise to a contemporaneity invaded by communications and messages presented in an intangible way, Design asserts itself as the discipline which understands the physical properties of artefacts as well as the importance of intangible culture, with its semantic and symbolic values. Culture allows imaginary spaces, endowed with evocative power, to which Design methods are indebted; in that sense, the main focus of Design Culture is not only in the internal project, but in an ambition that transcend it. This chapter titled "Design for culture" presents contributions on issues that pursue the role of memory and heritage in Design Culture and methods, establishing the threads between the material and the intangible that have enabled the development of new scientific approaches and research in the multidisciplinary field of Design.

Keywords Design history · Design practice · Material and immaterial culture · Memory · Gender studies

1 Introduction

The essays collected in this chapter show that culture and design share a space of reified ambition. Design for culture extends the enquiry into the intention embedded

H. Souto (✉)
UNIDCOM/IADE - Unidade de Investigação em Design e Comunicação, IADE - Faculdade de Design, Tecnologia e Comunicação, Universidade Europeia, Av. D. Carlos I, 4, 1200-649 Lisboa, Portugal
e-mail: helena.souto@universidadeeuropeia.pt

G. Oropallo
School of Art, Architecture and Design, London Metropolitan University, London, England

H. Barbosa
Departamento de Comunicação e Arte, ID+ Instituto de Investigação em Design Media e Cultura, Universidade de Aveiro, Aveiro, Portugal
e-mail: helenab@ua.pt

E. Duarte and C. Rosa (eds.), *Developments in Design Research and Practice*, Springer Series in Design and Innovation 17,
https://doi.org/10.1007/978-3-030-86596-2_24

in the artefacts and communications' programme to the broader web of interactions in which they move.

Over the last three decades, design practice has experienced a transition from a service culture to one of self service [3]. The resulting fragmentation of its remit into a series of highly localized spheres of action now makes the practice of design both successfully universal and semantically slippery. Every design discipline taxonomy is inevitably contextual and temporary.

The semantic and symbolic values with which design is invested fluctuate while artefacts experience their life cycles. In the process, these things of material or immaterial design are actively inscribed with meaning, but they are also unwittingly imbued with records in a forensic fashion. As research on design trains its lens on the territory that lies between one thing and another, as well as between things and people, this very in-between space reveals itself as a continuous environment of information waiting to be decoded.

2 The Rehearsals: A Critical Presentation

The first essay of the chapter is "Anti-Amnesia: Developing a Collaborative E-learning and Digital Archive Platform towards Contributing to the Preservation and Revitalization of Handicraft Industries". The essay examines the development of a digital platform that aims to document Portuguese craft industries. The project has a twofold purpose. On the one hand, it should create a living archive of material and immaterial knowledge. On the other, the archive should be organized as a repository that can be directly used as an online teaching platform, featuring documentation in the form of text, audio, and moving and still images.

The experience of "Anti-Amnesia" highlights the role of memory as both a horizon and method in heritage preservation. The materials thus far collected include examples from Central and Northern Portugal. The platform employs different approaches all aiming to unpack the knowledge accumulated by generations of craftsmen and women and make it easily available. The project implies an integrally participative method. The platform users contribute to the larger process of heritage preservation when they activate and mobilize the traditional knowledge made available by the repository.

For the continuation of the project, the concern of the authors was not restricted exclusively to putting the fruit of their research on this platform. It was rather to create ties and synergies between the 'holders' of this culture, in conjunction with potential stakeholders, taking into account the reality and specificity of each protagonist within a given territory. The latter aspect is particularly emphasized by the authors of "Anti-Amnesia", who see their project as facilitating a meshwork of connection between traditional stewards of the material and immaterial culture reified in Portuguese crafts, and their potential interpreters.

At the same time, they highlighted the importance of a documentary survey encompassing approaches related to local, personal narratives and traditional practices, whose results and the creation of various archives, along with pedagogical concerns, would lead to the setting up of networks and cooperative work to transfer knowledge. Bearing these issues in mind, the platform would have to correspond to a specific design programme, and provide an effective means of establishing links between people and facts: this underlying cultural heritage and a whole patrimony of knowledge was to be shared and explored by the community from both a singular and plural viewpoint. As part of the process of better understanding how such a digital support would methodologically function, the authors chose to carry out an analysis of several digital platforms, in order to learn about what was available and understand their advantages and constraints, so as to present a viable solution for "Anti-Amnesia", which would be consistent with the objectives outlined. In doing so, it was possible to retain the necessary content and those considered important to facilitate the necessary learning to be outlined in this project which, beyond the situations identified above, sought to use a methodology focused on Design Thinking and User Centred Design. Consequently, although the research is ongoing, it is understood that there is a concern in this project, in several themes, to contribute towards both its enhancement and a comprehensive view of a project in which the whole and the parts are fundamental to its success.

Cemeteries are places of memory with particularly interesting cultural and heritage features. They are witnesses of history, socio-cultural diversity and different aesthetic paths. Besides its principal function as a necropolis, the cemetery has become a unique and extremely complex place: a space where the collective memory is often connected to particular ideologies.

In a work that remains a standard reference, the historian Philippe Ariès [1] highlighted the evolution of mentalities in the face of death in Western Christianity, which have presided over the construction of cemeteries as we know them today, with pathways designed for visits. A city within a city, the cemetery reflects the urban planning: it has an old core, extensions, pathways like streets, simple or monumental constructions and architectural styles.

The essay entitled "Letters to Eternity: Typefaces of the Prazeres Cemetery" by Gonçalo Falcão focuses on the Cemetery of Prazeres (literally meaning 'Pleasures') in Lisbon, Portugal, but introduces a different approach which is important for graphic design studies. It is "exclusively concerned with the letters used to make inscriptions, and, therefore, specifically typographic."

Graves can be 'loquacious', and display quality information. They talk about themselves, about us, yesterday and today. They also partly reproduce the socio-economic order—in fact, we can easily see the emergence of class distinctions between the various funerary monuments within the cemetery.

Falcão's analysis is supported through "exploratory research" during three long visits to the cemetery accompanied by the respective photographic survey. "Around 1000 photographs of the letters that seemed most relevant were collected", which allows us to see that the graphics of a text "are never casual facts, since they contribute to the meaning of the text they convey."

In the cemeteries of Lisbon, some of the earliest epigraphed monuments (*jazigos*) refer only to the place where they were built [4] and, in general, the builder's epigraph was a form of advertising. This kind of promotion was highly effective and led to the generalization of epigraphs in the 1860s (*Idem*).

Falcão notes how "decorative/commercial fonts are present and sometimes mashed up in several styles and weights, in a strange typographic epitaph potpourri". This potpourri reveals the fragility of Portuguese visual culture, a major problem that has been a constant in Portuguese graphic design history.

In the case study presented, we perceive that, at the end of the nineteenth century, epigraphs began to include more information, leading to growing typographical eclecticism with formal inconsistencies and construction errors.

Many of this new information reflects the value of the family whose polysemy denotes "a visual translation of the spirit of the 'family patriarch' who builds patrimony, while the wife builds matrimony" thus reproducing the socio-economic order of the living.

This is one of the avenues that the author leaves for future research. Falcão also assertively considers that the cemeteries of Lisbon and Oporto are privileged places for interesting lines of research, not only for graphic design, but from the viewpoint of different approaches to Portuguese design history.

In another area, related to visual memory and the importance of its hermeneutics, the article by Mariana de Almeida and Helena Barbosa entitled "Port Wine Visual Communication: Traces of Posters from the Past in the Current Urban Environment" focuses on an interpretation that connects past and present through the visual rhetoric present in posters. In this analysis, the authors highlight the arguments and graphic strategies related to what is generally considered the best-known and most iconic Portuguese product, both nationally and internationally. To do this, the approach focuses on the analysis of communicational features of three Port wine brands in terms of graphic design, noting strategic convergences, both diachronically and synchronically. In parallel, they explore other media beyond posters, but where, nevertheless, the design as a whole is still a 'space' for cultural mediation between the respective Port wine companies and the public sphere. In this sense, the authors establish relationships between the exclusivity of a product characteristic of a geographical area in Portugal, together with 'its path' from its origin to the respective cellars, and also the sale of Port wine and its graphic products, spread out in various locations at the national level. This article also highlights situations of belonging and the uniqueness of this territory, through the direct association of geographical areas, mediated by iconographic representations that refer to the 'experience' of the product; together with a whole argumentation that is not only part of this imagined world, but also of a reality that is present in cities in various forms, in terms of supports and types of artefacts. All the material culture generated around this product is characterised by its denotative sense, but the visual communication seeks to exalt a complete connotative component. It is mainly in this connotative context that the article explains and presents examples of urban spaces that are transformed, present through the inclusion of these types of rhetoric. It also explores the way different brands have used and (re)use approaches to get close to the public, showing, on the one hand, few graphic changes over time. This is true of the Sandeman brand, with the use of the male image "The Don", or in the case of Porto Cruz, which has a female figure with some similarities to the "The Don". These show the importance of symbols and of an entire visual strategy that has been consolidated over time, eventually becoming references, visible in the different reinterpretations conveyed by the female figure in the "Porto Cruz" brand. In the case of the Ramos Pinto company, the current representations have kept "The Kiss" intact, first presented in the 1929 poster, making it perhaps the most iconic image of this company. Consequently, it can be seen that these three brands ended up crystallising and perpetuating in the present images of the past, in order to reach and involve the coverage of memories and emotions that, just like Port wine, have been embodied through the constant exploration of values associated with culture, emotions and traditions. Indeed, if the past, time and history are essential ingredients for the creation of a good Port wine, the same happens with the brands that visually exalt these characteristics in their contemporaneity.

Another issue that arises in this chapter concerns an urgent debate which places design alongside the problem of gender equity. The lack of awareness regarding the representational character of women in various contemporary fields is part of a larger absence of women [2], among those responsible for these representations with regards to the 'asymmetrical' nature of the world.

Society remains imbued by pervasive masculinity through deeply embedded epistemological and institutional phenomena. The question of equality and recognition of women has been limited not on account of men's ill will or women's level of trust, but due to the nature of institutional structures and the vision of reality they impose.

For this debate, we highlight the contribution of the ongoing research thesis conducted by Antonia Sophia Hinz under the supervision of Flávio Almeida and Anabela Couto, entitled "Representation Between Waves of Change, A Visual Analysis of the Advertisement of Female Surfers", which focuses on how these women athletes are displayed in various mainstream media that reflect issues of gender, sexuality and objectification.

The circulation and exchange of experiences regarding women surfers have revealed "that gender related power relations in surfing remain contradicting and impugned". The authors point out the situation experienced in 2016 by Silvana Lima, "ranked one of the top ten women on the Australian Surfing Championship (ASP) World Tour", who was refused a sponsorship deal because she did not fit the requirements of the "heterosexy surfergirl" category (according to Georgina Roy's study as referenced in the essay).

With the fundamental aim of introducing the issue of online visual communication, the article analyses various visual advertisements aimed at female surfers, with a special focus on an advertising video released in 2013 by "the first brand in the industry to design specifically and only for women". Based on detailed observation and description of the content, despite its intention to advertise a surfing contest, the authors point out that, in fact, "0% of the video time showed surfing, while 46% of the time was spent focusing on an anonymous woman's barely covered backside". The supposedly "anonymous woman was later identified as Stephanie

Gilmore, seven times world champion surfer". In our era of continuous scientific and technological advancement, while being bound by daily digital communications, voices were immediately and positively raised on Facebook and "critics claimed the video was sexist and soft porn".

Here is an inference that this ongoing study has already observed and involves other research goals related to the "'sexism, which is indicated by the APA Dictionary of Psychology to be discriminatory and prejudicial beliefs and practices directed against one of the two sexes, usually women (…) associated with acceptance of sex-role stereotypes".

History that crosses sports, geographies, genders and races is being rewritten in an increasingly broad register that condemns power relations, brings women athletes out of the shadows and highlights successes in gender parity. This can only encourage "a radical dialogue regarding the role that visual representation plays in addressing this societal issue of gender portrayal in sports, particularly in surfing" as this study is already establishing.

3 Final Considerations

After presenting these essays embodying the vast field of design research, we return to our subject through the prism of its interaction with society throughout design history and culture.

The economic models of each era, with their corresponding social attitudes, are embodied in their design. The industrial age, by spurring the transition from the artisanal artefact to the mass-produced object, generated as much fear as hope. It was in reaction to these feelings that the major currents of the late nineteenth and early twentieth centuries were formed, from the *Arts and Crafts* movement (with its desire to safeguard artisanal know-how), to the *Deustscher Werkbund* modernism of Peter Behrens and his disciple, Walter Gropius, the founder of the Bauhaus school.

Following the end of World War II, the gradual dissolution of Universalist values, combined with a period of strong economic growth, led to the confrontation between the Germanic modernism of the *Hochschule für Gestaltung Ulm* and the American *Styling*, which made design an ally of mass capitalism [5]. The last fifty years of the twentieth century were marked by both manifestations of commitment rather than excessively commercial behaviour.

Designers pursue the resolution of a given problem by understanding that artefacts accumulate, from project to consumption, meanings and values. The circuit of culture destabilizes the presumption of fixity; moreover history teaches us that the world is composed of change, but that only by understanding the chain of memory and creating from it is it possible to innovate.

References

1. Ariès P (1975) Essais sur l'histoire de la mort en Occident. Du Moyen Âge à nos jours. Editions du Seuil, Paris
2. Buckley C (1986) Made in patriarchy: toward a feminist analysis of women and design. Des Issues 3(2):3–14
3. Dreher A et al (2018) Apples and dragon fruits: the determinants of aid and other forms of state financing from China to Africa. Int Stud Quart 62(1):182–194
4. Queiroz JF (2000) Canteiros de Lisboa: construtores do cemitério romântico. Olisipo 13 (Julho/Dezembro):55–70
5. Woodham JM (1997) Twentieth-century design. Oxford University Press, Oxford

Port Wine Visual Communication: Traces of Posters from the Past in the Current Urban Environment

Mariana Almeida and Helena Barbosa

Abstract Port wine is one of the most famous national ambassadors in the world. Its origin is local, but its reputation is global and benefits from exceptional connotations. Although based on the attributes of wine, the visual commercial discourses play an unavoidable role in the creation of a symbolic framework that reverberates in the collective imagination both within and beyond borders. Since the XIX century Port wine posters have been a tangible part of this phenomenon; they represent a printed visual heritage, where the reality of the product is blended with commercial argumentation. The impetus of graphic communication, as a business auxiliary operational tool, has been developing alongside the sector itself, adapting to the context of action and in a perspective of approaching the consumer market—of the drink and, more recently, tourism. Based on a database of Port wine posters conceived within the scope of a master investigation and on some selected photographic records made during the course of the research, the article presents an observation exercise that highlights recognizable visual references in the 'traditional posters' of the past converted into more recent communication media that can be identified as 'posters of today', found in Oporto and its surroundings. The main objective is to understand how the Port wine communication transports and transforms its visual heritage into other physical media, with focus on the Sandeman, Porto Cruz and Ramos Pinto brands. Following this path, the article aims to contribute to the recognition of particular forms of expression in both traditional and present-day communication of Port wine.

Keywords Port wine · Posters · Visual communication

M. Almeida (✉) · H. Barbosa
Departamento de Comunicação e Arte, ID+ Instituto de Investigação em Design Media e Cultura, Universidade de Aveiro, Aveiro, Portugal
e-mail: mga@ua.pt

H. Barbosa
e-mail: helenab@ua.pt

E. Duarte and C. Rosa (eds.), *Developments in Design Research and Practice*, Springer Series in Design and Innovation 17,
https://doi.org/10.1007/978-3-030-86596-2_25

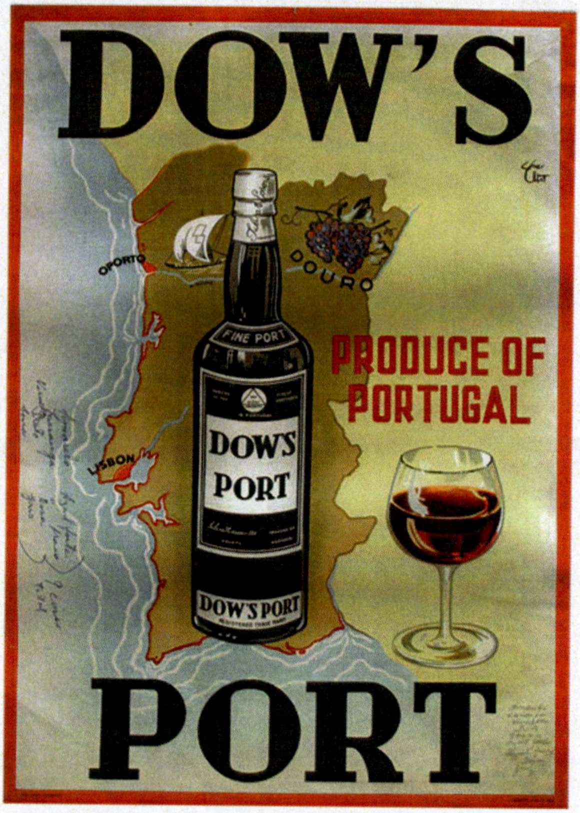

Fig. 1 Poster, Dow's. *Produce of Portugal*. Cesar Abbott. 1962. Lithography. Empreza do Bolhão, Porto. Packigrafica collection/Mariana Almeida digital collection [MA-dc]

1 Introduction

Vinho do Porto.
Vinho de Portugal.
E vai à nossa!
À nossa beira-mar…
(Carlos Paião, Vinho do Porto, 1983)

It is well known that Port wine is a Portuguese national emblem: Port wine is a "produce of Portugal" (Fig. 1) and Portugal is "the wine country[1]" (Fig. 2). The

[1] Since the 'Estado Novo' dictatorial regime (1933–1974)—in a moment of national exaltation of the most intrinsic characteristics of the country, Portugal was announced as "The country of wine". An example of this is the poster conveyed by the Wine National Council (Junta Nacional do Vinho). It represents 'Portugal wine districts' all around the territory—mainland and islands—proclaiming Portugal as a wine country in its entirety [2]. This idea has not ceased to be conveyed since it was introduced into the official tourist propaganda that was produced, particularly between the 1930s and 1960s.

Fig. 2 Poster, Junta Nacional do Vinho. *Portugal—Wine districts*. Mário Costa. 1958 (reprinting). Off-set, Artistas Reunidos. Nationwide distribution. Helena Barbosa digital collection

recognition of Portugal's winemaking vocation and the exceptionality of Port wine make both synonyms of each other and worthy of worldwide praise. This product reinforced throughout time its psychological performance and placement as 'ambassador of Portugal' in the world.

The same wine has been traded for centuries, however, always evolving. "*Fino*, *generoso*, de *embarque*, the *Feitoria*… *Port* at last" [5, p. 7]. The author lists the names by which Port wine was and is known today. It thus points out that the contemporary product has an extensive history and is the result of a combination of constant tense efforts between maintenance and processing, in order to continue to satisfy consumers who are attentive to tradition and others who are thirsty for novelties.

Fig. 3 Poster, Port Wine Institute. *Portwein Export*. Unknown author. c. 1937. Lithography. [s.l.]. MA-dc. Subtitle: *Port wine expansion since the seventeenth century (…)*

2 From a Local Origin to a 'Global' Reputation

Over time, Port wine has been promoted insistently by the different brands, without neglecting the strategic interventions of the State, as an excellent export product with high quality, income potential and traditional renown, linked to the territory. This local product conquered the national market and the markets abroad (Fig. 3) and even achieved a global reputation. It became an *ex-libris* of a city, of a region and even referring to the national identity of the whole country.[2]

It is a unique product with a shared genesis, whose preparation and ageing location does not coincide with the production area or with its nominal origin. The wine is produced in the Douro Demarcated Region (Fig. 4), and ages in Vila Nova de Gaia, but it was Oporto, a historically dynamic town with a commercial vocation, that baptized the certified wine (Fig. 5).

The city-based merchants made the wine known to the world, while modeling it to satisfy what was initially a mainly British market. Concerning this relationship with Oporto, Guichard wrote that, since then, "Port wine donates to the city (…) disseminates from it—and for it—an extraordinarily valuable image (…) the wine and the city have united their destinies in a way now inseparable to the world" [9, p. 29] (Fig. 6).

It is true that, from the middle of the seventeenth century until the Second World War (1939–1945), Port wine was the main provider of the Portuguese trade balance

[2] Along with the Barcelos rooster.

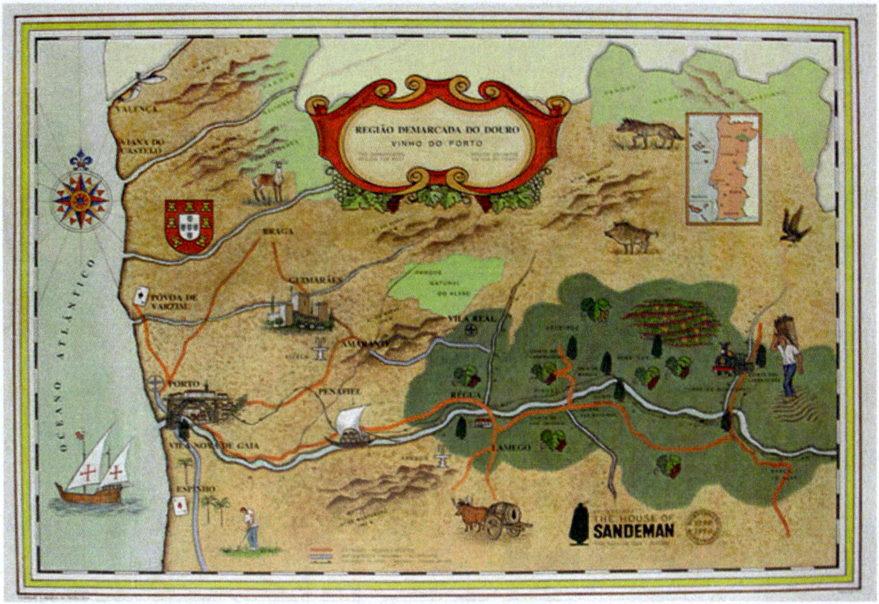

Fig. 4 Poster, Sandeman. *Port wine and Douro region.* Vinício Cruz. 1990. Off-set, Litografia União, Vila Nova de Gaia. Sogrape Historical Archive/MA-dc

and so one of the most dynamic sectors of the Portuguese economy (Fig. 3). However, today, with the diversification of the industrialization and tertiarization of the national economy, the preponderance of the Port wine commercial performance has relativized. Thus, Port wine's symbolic strength translates into a psychological impact that goes far beyond its objectivity as bottled merchandise. As Guichard also said "the simple beverage became an emblem of civilization; a food product became a cultural pillar" [8, p. 39].

Port wine no longer has the preponderance of a dominant sector by the number of jobs in the industry or the sales volume but maintains a symbolic dominance in the city reinforced by homonymy.[3] Its role has been reconverted and its potential widely capitalized in the current context, which seeks to meet the demand for gastronomic and touristic products. The versatility of Port wine is revealed, once again, in the variety of offer that is organized around it.

[3] In Portuguese, 'Porto' (city) and 'Porto' (wine) are called by the same word. In English, the terms are still very close.

Fig. 5 Poster, Grémio dos Exportadores de Vinho do Porto. *Oporto the city that gave name to Port wine.* Unknown author. 1953. Off-set, Empreza do Bolhão, Porto. Packigrafica collection/MA-dc

3 Port Wine Visual Communication and Oporto Urban Environment

Portugal (in general and the main cities in particular, such as Lisbon and Oporto) has registered a growth in tourism, due to an investment in self promotion and the strategic development of its potential, and benefits from the international acclaim in different platforms. Corroborating the dynamics seen in the tourism sector in 2019, the country received the Best European Tourist Destination award for the third consecutive time at the World Travel Awards, an event that delivers the tourism awards equivalent to the cinema Oscars [6].

The North region and Oporto have also been widely awarded and recognized for their multipurpose tourism offer [12], in which the products associated or associable with Port wine occupy a prominent place as one of the main attractions of the area.

There is a wide range of Port wine offerings: from hotels and restaurants of excellence, to Douro river tours and cruises—at the river mouth or from there to the vineyard valley—as well as the opportunities offered by Oporto and Gaia: the

Fig. 6 Sketch. *Port/Oporto. The city [Clérigos tower] in a bottle.* Unknown author. Ink on paper. ETP, Porto. Packigrafica collection/MA-dc

local shops and restaurants, the riverside area, the cellars (Fig. 7). A multifaceted product, Port wine adapts to various target audiences. It is popularly accessible and democratized, accompanies exquisite luxury or rises to utmost sophistication in specialized wine tourism programs.

Its presence in the urban landscape is unsurpassable and contributes to urban dynamics that are currently lived in Oporto and in Vila Nova de Gaia. It is possible to find references to Port wine on the building walls, in the traditional shop windows and the supermarket doors (Fig. 8), on the ground and fluvial transportation (buses, tuk-tuks, tourist trains and *rabelo* boats), in the terraces, in the urban furniture (Fig. 9), throughout the riverside landscape and surroundings.

These are examples of commercial discourses orchestrated by graphic design—and today more supported by comprehensive marketing strategies—that have been following the development of Port wine and increasing its promotion promising new experiences (Figs. 7 and 9), or the same experience but with a new label image (Fig. 8). Visual rhetoric plays a relevant role in the creation and maintenance of a symbolic framework that pervades and renews itself in the collective imagination.

Fig. 7 Picture. Sandeman cellars. *Hostel & Suites + Restaurant: 'a new way to experience Porto'*. Vila Nova de Gaia. July 2017. Mariana Almeida personal archive [MA-pa]

Graphic resources, as auxiliary operational tools for companies, were adapted to the context and developed with a view to approaching the consumer market. As a tangible part of this phenomenon, the posters—design artefacts and cultural interface objects—have played, at least since the nineteenth century, an important role in the creation and dissemination of exalting messages about Port wine, conveying commercial narratives.

An interpretative study was carried out on the narratives of Port wine posters, based on a collection of "traditional posters" dating from the end of the nineteenth century to the beginning of the twenty-first century.

It was found that, as printed graphic material, the posters embody a significant part of the visual heritage related to this product, where the reality of wine is blended with the artificiality of commercial argumentation. Thus, they are repositories of references and a semantic source of their own [1].

It should be noted that, according to Barbosa, a 'traditional poster' is one that "usually does not raise so many doubts as to its classification as an object; it is that poster whose definition is intuitively recognized and is associated with the paper medium". Therefore, it is justified that the Port wine posters analyzed—printed in various formats, on paper or similar, of static exhibition in public space (indoors and outdoors)—were understood as 'traditional posters'.

Despite the taxonomic proposal adopted denouncing a chronological character through the designation of each typology, Barbosa emphasizes "the existence of

Fig. 8 Picture. Supermarket entrance. Matosinhos. December 2018. MA-pa

these 'various types of posters' simultaneously" (2011, p. 137), and the very delimitation of typologies does not aim to deny the association with other artefacts identified by semantically close terms of poster, nor eliminate "the coexistence that their simultaneity fortunately expressed" (2011, p. 137).

From the paradigm changes from technological and advertising formulas evolution, results what is considered the 'poster of today'. In this context, 'poster of today' is one that results from "constant metamorphoses, which have gone beyond its 'traditional' nature and implied significant changes" (2011, p. 140).

It is these significant changes that underlie the different typologies. Therefore, Moles made this synthesis of the poster's trajectory, referring to striking aspects of its evolution: "in general, the poster that came out of the edict, then of the black and white, then activated and illuminated, for reasons of attraction, will tend to become a place where the fixed image and the animated image are intermingled (…)" [11, pp. 248–250].

Fig. 9 Picture. Calém cellars MUPI. Porto. July 2017. MA-pa

In terms of chronology, Barbosa considers that the 'poster of today' does not refer to the end of the last century or to the beginning of the present, but "it emerged during the twentieth century, and distinguished itself (…)" (2011, p. 141). Due to the need to adapt the poster to new circumstances, this type is characterized by the combination of "technologically available hypotheses, which allow large formats, to be illuminated, in motion, to be three-dimensional, interactive" (2011, p. 140).

It is in this border area that this article is focused, between the Port wine 'traditional poster' and the 'posters of today'. Using the artefacts gathered in the database, conceived within the scope of the investigation, it is now proposed to carry out a comparative observation exercise. The aim is to highlight recognizable visual references in the 'traditional posters' of the past reconverted into more recent communication supports that can be identified as 'posters of today', found in Oporto and its surroundings, through the observation of informal photographic records made during the research.

Through the photographs taken, one can perceive the ability to adapt to 'new' situations that make the poster a metamorphosing object, capable of transforming itself as many times as necessary to effectively respond to the communication goals for the public sphere. The 'actuality' does not, therefore, dispense the mixture with the 'traditional', for what is offered to the posters exposed in the public spaces, the possibility of, in large scale and/or in movement, to be present, for instance, in façades, walls, fences, sliding doors, transport vehicles or in proper structures of urban furniture.

So, the goal is to understand how the Port wine communication transports and transforms its visual heritage into other physical advertisement materials that can be recognizable as posters considering the 'poster of today' taxonomic designation.

4 Sandeman, Porto Cruz and Ramos Pinto: The Power of Visual References

In the photographic sample, Sandeman, Porto Cruz and Ramos Pinto are the main brands represented, although some images related to brands such as Ferreira, Offley, Rainha Santa, Real Companhia Velha and Calém were recorded. Next, observing the photos, the appropriation (in whole or in part) of the visual references is highlighted through a comparative study between the database 'traditional posters' and their alteration into other visual materials.

Dating back to 1928, it is the poster by Scottish artist George Massiot Brown (1881-?) that would define and guide Sandeman's visual identity in the future. More than a recurring visual resource, this silhouette constitutes a true brand image (Fig. 10).

In this poster, on an abstract background, the artist drew a black masculine figure of which one can only perceive the silhouette, dressed in the Coimbra students black cape and the sombrero, popular in the Andalusia region. The high reference potential of the pieces of Portuguese and Spanish clothing together with the presence of the calyx, promotes the perennial association, succinctly supported by the text, between Sandeman and the Iberian wines it sells (Port and Xerry).

Only the contours of the character can be perceived, at first somewhat rigid. Then one notices the hand in which a glass of Port or Sherry stands out. The gesture contains just the sketch of what might be a shy toast, but still is a polite "greeting to all of us, consumers, appreciators, little by little educated in the cult of the beautiful and the good" Port wine [8, p. 148].

Fig. 10 Poster. Sandeman. *The Don*. George Massiot Brown. 1928. Lithography. Sogrape Historical Archive/MA-dc

In 1930, this image debuted in the press, even before being integrated into the company's labels and becoming one of the best-known logos of the wine and beverage industry in general. In the campaign for a new wine, the 'Dry Don', in 1935, the character acquires the name by which he is still known today—Don Sandeman.

Since 1935, this iconic silhouette of simple cut, quickly assumed unquestionable protagonism in the brand's communication, namely in the posters. It evokes an aura of mystery, underlined by the enigmatic black connotation, which stands out over the ochres and the yellow of the original background.

Both colours, black and yellow, became part of the brand's main chromatic palette of its visual identity (Fig. 11). The brevity of the text together with the systematization of colours and shapes are visible on posters and other communication pieces that show the Don by the river or even by the sea (Figs. 12 and 13).

Examples from the 1990s already reveal the intervention of digital tools, not only in the text font, but also in Don's own image, in which some suggestive lines of volume, shadow games and overlays begin to appear. The background loses 'texture' and is flattened out into a solid color area (Fig. 11).

The mural painted on a riverside façade is a faithful copy of this poster (although the painted name is highlighted by a yellow 'shadow') (Fig. 12). It is a painting executed in a highly visible place. Standing by Douro river, on the Oporto side, it pairs with the Don on the other side, in the Sandeman's cellar building (Fig. 7). Whether the observer is on Gaia's or Oporto's side, there will be a Don on a large scale in sight, a reminder that this is Port wine territory.

Fig. 11 Poster. Sandeman. *The Don.* Unknown author. c. Anos 90. Off-set. Sogrape Historical Archive/MA-dc

From known later artefacts, the figure has been further worked on and retouched, sometimes even acquiring a convivial character when it appears accompanied by a slender female silhouette (Fig. 17).

In the 2005 poster series the palette diversifies. The backgrounds—blue (Fig. 14), gold, bordeaux and green -, illuminated by a clipping light, acquire shades depending on the narrative in the version presented. The glass displayed on the blue poster is replaced in other versions by objects, directly or indirectly, linked to the enjoyment of wine and life, which reinforce Don's character as the protagonist of power and seduction games. The key-phrase is as suggestive as the images: "Famous for pleasure".

An advertising panel displayed by the escalators at Oporto Airport Arrivals has been occupied by Sandeman images for several years. A photographic record from 2013 presents an image that shares the concepts of Don Sandeman's association with Port wine and pleasure. The juxtaposition of the two images illuminated by a light projected towards the viewer suggests a parallelism between the vertical dark silhouette of the Don and a Port wine bottle surrounded by cheese—an exceptional accompaniment to this drink. The phrase "the pleasures of Porto" closes the idea (Fig. 15).

Since the 2005 posters, Don's silhouette acquires more apparent three-dimensionality through the movement given to the cape. However, the static and mysterious Don's character does not change significantly, since the face is never revealed. A 'humanization' process has been giving the character other contours and

Fig. 12 Picture. Sandeman painted wall. Porto by the river. 2012. MA-pa

volumes. Since its creation, mystery and sensuality remain concentrated on this icon, which has already survived far beyond its first public appearance.

The 2015 panel presents an image that departs from the previous images, maintaining the slogan. This poster is composed of several photographs, in which various Port wine spaces are combined. Visual rhetoric is based on a cumulative juxtaposition of meanings aimed at enhancing the product's origin. The image is an invitation to experience "the pleasure of Port wine" on the spot: in the Douro vineyard and in Gaia cellars. Notice, however, how Don's black silhouette also appears framed by a mandorla of golden light (Fig. 16).

In the most recent Sandeman panel, the composition focuses again on photography, with digital retouching and vector drawing to introduce the text and framing

Fig. 13 Picture. Sandeman wind shield. Agudela beach, Leça da Palmeira. August 2017. MA-pa

the product on a yellow background, referring to corporate colours. The photograph is staged; it looks like a scene of a black and white film (Fig. 18). In the picture Don no longer wears a cape but an overcoat and the sombrero is now a more conventional brimmed hat. The text seems to represent the speech of the suggestive female who addresses the Don when he enters the bar and catches the attention of those present. The interaction recalls Sakura Satoshi's posters, in which Don appears in the company of a woman in an evening dress with a very similar silhouette (Fig. 17). In these images, without any other support, it is up to the observer to imagine the dialogue through the visual clues at his disposal. The contours mystery intensifies the sensuality of the images.

As a central figure in the communication of the company's brand image, the characterization evolved, without—between rhetorical advances and setbacks—losing the characteristics that turned Don into an unmistakable icon.

Don Sandeman, a striking male character in Port wine communication, finds in The Woman in Black, of the Porto Cruz brand, a female parallel. This woman's figure concentrates references to the iconography of Portuguese folklore[4] and to the traditional figure of the *Fado* singer. The brand has made this character the protagonist of its marketing campaigns, especially for the French market. This aesthetics has been reinvented over time, reflecting on the posters.

[4] There are references to the black clothes of the women of Nazaré and to the accessories (such as the basket) of the traditional fish sellers *(varinas)*.

Fig. 14 Picture. Sandeman poster on the street. 2005 campaign. 2017. MA-pa

In the characterization of this Woman in Black, the dress and shawl that surround the body and hide the face stand out. Like Don's her face is never completely revealed: this way the respective mysterious auras are preserved.

Demonstrating a strategic vision and concern with conceptual coherence, the communication follows the motto "Pays où le noir est couleur", which defines Portugal as the country where black is colourful. The same slogan has been repeated for about 30 years, in compositions based on contrasts between the background and the figure, between the bright colours and the black of the silhouette.

There is, however, an evolutionary path on the posters. Several campaigns with different scenarios are known, which have been progressively tending towards the abstract. Variations of the same concept have been used, without drastically deviating from the original principles. This ensures the coordination of the brand image, expanding its symbolic potential.

In the Oporto, Woman in Black was photographed at various times on advertising media such as terrace tables (Figs. 21 and 22), tourist buses (Fig. 24) and *rabelo* boats (Fig. 25), referring to the posters.

Fig. 15 Picture. Sandeman advertisement. Arrivals, Sá Carneiro Airport, Porto. September 2013. MA-pa

In an approach that combines photography with vector drawing, terrace tables show the Woman in Black with a basket of small pumpkins balanced on her head, as can be seen also in Fig. 20. The poster produced is exclusively photographic, with a brand reference at the top of the image through a label detail. However, on square tables, the image that appears is digitally stylized, in a transition to backgrounds based on vector experiments.

The character's pose is replicated from the poster. The fringes of the shawl flutter behind the thin silhouette and her right hand is resting on the side of her protruding hip. The pattern of Portuguese tiles reference (Fig. 21) is not a debut in the brand's campaigns. In a 1996 poster, the Woman in Black is photographed in front of a blue and white panel with a basket of yellow flowers (Fig. 19).

Notice how the colour scheme in these cases emphasizes the contrast between black, blue, white and yellow. These are intense colours that the brand often uses in its compositions. The wine bottle presence is mandatory, usually introduced as a miniature superimposed on the image.

More recently, a campaign (c. 2010) features a variation of the original sentence. The poster series, with a different colour range corresponding to each variety of Port—white, tawny (Fig. 23) and rosé—has the slogan "Toutes les couleurs du Porto"—"All colours of Port".

Fig. 16 Picture. Sandeman advertisement. Arrivals, Sá Carneiro Airport, Porto. September 2015. MA-pa

Slightly modified in the layout of its elements, the poster image was printed in vinyl and applied to the rear of a bus (Fig. 24). Note the placement of the glass and the bottle, as well as the displacement of the brand with the rearranged lettering. Thus, the image travelled the Oporto streets on tourist routes. In this way it gained visibility, not only for scale reasons, but also for the possibilities of travelling in the public space, on privileged routes.

In turn, the transformation of ancient *rabelo* boats into a kind of floating advertsign outdoor is impossible to ignore. In this situation, the poster's image is adapted to the horizontal and more compact format (Fig. 25). Therefore, the female silhouette is cut at the waist and the bottle/cup appears on the bottom line, resulting in a more concentrated composition, with the brand occupying the upper right corner.

As shown in Fig. 25, it is increasingly common to place billboards on the cabin of boats. It was noticed that the boats could be capitalized on tourism and their potential

Fig. 17 Poster. Sandeman. *As rich as your imagination.* Sakura Satoshi. c. 1990. Off-set. Sogrape Historical Archive / MA-dc

monetized through the visibility of their surfaces and the attraction they naturally raise. This strategy also represents a functional reconversion of what is a vestige of a craft closely related to the Douro tradition.

The boats, parked in the Gaia riverside area or going up and down the river on tours through the bridges area, make up a landscape, perfect to be spread around the world as an element of Port wine identity. The *rabelo* boat is a Douro *ex-libris*. A retired collaborator of Douro prosperity has risen above its operational status, incorporating itself into the riverside landscape and, like the Douro River itself, into the image vocabulary of Port wine.

In Portugal, since the 1980s, new urban furniture was introduced for the presentation of posters in public spaces, which allowed for movement (when motorised) and backlighting of posters (Barbosa, 2011, p. 445). These are, for example, MUPI (the French acronym for *Mobilier Urbain Pour Information*) adopted in Portugal with the same acronym, in direct translation. The posters are thus displayed in conditions that safeguard their outdoor public exposure, night visibility and the prolonged integrity of the object (Barbosa 2011, p. 141). The name of the media is confused with the name of the graphic piece. So, the posters with the appropriate format for placement on this equipment can be also called MUPI (Fig. 9 and 26).

Fig. 18 Picture. Sandeman advertisement. Arrivals, Sá Carneiro Airport, Porto. December 2018. MA-pa

Porto Cruz campaigns have been appearing annually in MUPIs, especially at Christmas time, when Port wine is frequently offered as a gift. In recent years, since 2013, an image has been repeated (Fig. 26). The composition of this photographic image does not change significantly compared to others of the same brand. The conceptual line followed predominates in the essence of the image and the slogan returns 'to the original'. The female figure, in her pose appears in a 'non-place', framed by a scenery of coloured cloths. The huge purple flowers basket that she carries on her head refers curiously to the poster that shows a branch with flowers of the same colour falling on a yellow wall (Fig. 20).

In high visibility outdoor locations, in addition to the MUPIs, large-format paper or canvas prints are placed on specially designed metal structures, with a variable area of a few square metres (Fig. 28).

Fig. 19 Poster. Porto Cruz. *Pays où le noir est couleur.* Unknown author. 1996. Off-set. Porto Cruz digital collection/MA-dc

These posters are usually called billboards our 'outdoors'. More recently, giant bright screens have been implemented. They make the same space profitable by displaying several digital advertisements, with still or animated images, in a timed sequence (Fig. 27). In this case too, the adaptation to the horizontal format conditioned the image layout to a structure close to that observed in Fig. 25.

It should be noted that the most recent posters comply with the legal obligation to call for moderation and responsibility for consumption. This speech component was introduced under European legislation on alcoholic beverages. It is a discursive aspect that is added to the 'today's' Port wine commercial message.

Don and the Woman in Black are successful cases of coordinated communication strategies that focus on updating their visual discourse while maintaining a cohesive conceptual line around their main character. However, they seem to follow opposite paths: while Don humanizes himself, the Woman body tends towards the abstraction of a silhouette.

The public is not disappointed by the conscious and systematic campaigns updates: neither those who react better to the renewal of the advertising message

Fig. 20 Poster. Porto Cruz. *Mulher de Negro*. Unknown author. 2000. Off-set. Porto Cruz digital collection/MA-dc

nor those who favour stability and nostalgia for a past known and integrated into the collective imagination.

Another resounding success example is the image conveyed by Ramos Pinto: *The Kiss* (Fig. 29), by the French illustrator René Vincent (1879–1936). In this 1929 poster, "eroticism is limited to kissing, or rather, to preparing the kiss" [8, p. 58], which gave the work its name. On a black background, the gesture appears in anticipation on the two characters' lips but is forever interrupted by an interposed Port wine glass raised by a "playful cupid" [7].

In the image composition, woman and man converge simultaneously for each other: two profile heads "of a brilliantined young man with a sports collar and of a young woman with 'garçonne' haircut (1981, [p.s.]). It is the glass raised by the cupid that separates them, the wine glass that gets in the way as an obstacle to the kiss. The glass thus becomes a metaphor for the thirst for one's own passion, untouched and intoxicating. There is a latent expectation of mutual fruition, because the benefit of the wine pleasures is imagined to be shared by both.

Because "in advertising, suggestion is much more powerful than consummation, eroticism is infinitely more commercial than realism" [10, p. 25], this kiss—which

Fig. 21 Picture. Terrace table. Ribeira, Oporto. February 2012. MA-pa

Fig. 22 Picture. Terrace table. Ribeira, Oporto. February 2012. MA-pa

Fig. 23 Poster. Porto Cruz. *Toutes les couleurs du Porto.* Unknown author. 2010. Off-set. Porto Cruz digital collection/MA-dc

was never exchanged—perpetuates in the image. The narrative of this poster remains suspended, suggestive, fascinating in its simplicity, widely disseminated and used by the Brand in its communication (Figs. 32 and 34), in merchandising objects (Fig. 30) and advertising (Figs. 28, 31 and 33).

The small colour palette is sober but striking: black, white, red and yellow. Also, a compositional balance was achieved through the proportion and arrangement of the various elements. The economy of this poster's compositional resources makes its figures easily detachable for application in the various supports due to the flat neutral background on which they were originally created (Figs. 30 and 34).

The text of the poster does not reveal evidently persuasive content; it is brief and informative: it mentions the product 'Port' and mentions the Ramos Pinto brand. Figures 28 and 33 alone show the Ramos Pinto signet associated with this image, perhaps because it is, by itself, already a symbol of Ramos Pinto branding.

The outdoor (Fig. 28) is the piece that brings together Ramos Pinto identity elements: the signet with the cavalry motto "*in hoc signo vinces*"—"*for this sign you shall conquer*", the *Kiss* characters and yet a schematized representation of the company peculiar yellow headquarters building.

Fig. 24 Picture. Vinyl applied to the rear of a bus, Porto Cruz. Clérigos, Porto. October 2015. MA-pa

Fig. 25 Picture. rabelo boat advertising panel. Porto Cruz. Ribeira, Porto. December 2014. MA-pa

Fig. 26 Picture. Porto Cruz mupi. Matosinhos, Metro station. December 2014. MA-pa

The banner (Fig. 31) introduces a novelty to graphics. The horizontal band, quite narrow, features a linear digital gradient from bordeaux to matt black, innovating when it comes to the use of the original smooth black.

From left to right appears ‘the love triangle’, the brand’s name and the trilingual directions of how to get to the cellars. Another aspect that distinguishes this composition from the poster is the blur introduced around the characters contained in a silhouette that can be a bottle of wine (which by its outline looks more like table wine than Port[5]). The cupid, due to the proportions of the banner, is cut at the waist line, a feature that we have already seen to be used in adaptations of Porto Cruz posters to horizontal formats.

Even so, in the diversity of media in which this triad of characters is represented, the integrity of the original drawing is preserved, although digital retouching is evident in softening features, while the text undergoes frequent changes.

Focusing on the same image, Ramos Pinto has been launching an annual competition: *The Kiss Contest*. Outside its Gaia’s headquarters a large panel was placed with the image of the poster *The Kiss*, in which the faces of the characters appear

[5] Barata et al. [4].

Fig. 27 Picture. Luminous display with Porto Cruz advertising. Highway Oporto-Lisbon. February 2017. MA-pa

Fig. 28 Picture. Ramos Pinto outdoor. Infante street, Porto. January 2012. MA-pa

Fig. 29 Poster. Ramos Pinto. *The Kiss*. René Vincent. 1929. Litografia. Vercasson, Paris. Douro Museum collection/MA-dc

cut out (Fig. 33). Through social networks, visitors are invited to place themselves behind the panel, have themselves photographed, publish the result on the company's official page and obtain the highest number of 'likes', so that, annually, a prize is awarded to the winners.

Through this panel, the man and woman can be anyone standing behind it assuming their place to share this Port wine kiss. This initiative promotes interactivity between visitors, the brand and its most famous image, leading to the dissemination of that interaction results.

The Kiss, and other posters by Ramos Pinto, exemplify how "old posters (...) continue stubbornly to sell the image of Port wine" [10, p. 17]. Regarding this company communication, the authors consider that "(...) because of an unusual commercial strategy that one hundred years later manages to keep up to date, the firm Ramos Pinto has not had much need to renew itself in this area" (2001, pp. 16–17).

The same happens to posters created by Jean d'Ylen (1886–1938) for Sandeman at the end of the 1920s, still reproduced and sold as souvenirs in the Cellar shop (Fig. 35), or to other Rainha Santa and Amadeu posters, available in notebook format at gift shops (Fig. 36). The reuse of all these posters, even of already extinct brands, provokes a graphic revival that coexists with manifestations of the more contemporary image of the product.

Fig. 30 Picture. Merchandising and shopping bags. Ramos Pinto cellars shop, Gaia. December 2018. MA-pa

5 Final Remarks

In the wine itself resides an organoleptic memory that is articulated between fidelity to a successful past and an impetus of auspicious reinvention for internal and external enjoyment and dissemination.

As the results in advertising are not perennial, the commercial poster has been following and obeying the requirements of each time. As part of its communication exercise to the public sphere, the poster remains in transformation: not only because visual rhetoric codes are changing, but also because of the change in the technological and cultural factors involved.

The communication by the Port wine poster demonstrated, in the course of the studied spectrum, a general trend, regardless of the brands, towards a product reality sublimation, producing a kind of timeless myth that insists on qualities as arguments: exclusive origin, authenticity, subtlety or exuberance, elegance, charm, temptation, abundance, distinction, uniqueness, refinement, to give just a few examples.

A symbol—or a set of them—condenses and conveys a favourable message about the product and its context or simply reminds of its existence, giving the set of

Fig. 31 Picture. Outdoor banner with directions to Ramos Pinto cellars. Ribeira, Gaia. February 2012. MA-pa

artefacts a sense of belonging to a specific brand sphere and, in a broad sense, to the significant Port wine universe.

By exploring posters' visual heritage and their transformation into other objects functionally close to the 'traditional poster', it was found that visual references relating to each brand are identifiable over time and remain in display in the public space. In the three selected cases, all detectable in the street, narrative lines adopted consistently over time were noted.

Several communication objects were identified that can be associated in close complementarity with the purpose of achieving good commercial results. The market is competitive and aims to sell more than just the drink: it is the whole Port wine universe that is increasingly available to be enjoyed by the curious onlookers and by the loyal connoisseurs.

This exercise showed that Port wine communication does not easily dissociate itself from first half of the twentieth century images, although adaptations are being made to the 'classic' images and new ones are being implanted in the collective imagination that Port wine continues to inspire.

Perhaps because it is a product closely related to 'time' and 'history', the perpetuation of the 'timeless tradition' reveals a need to insist on the images of success guaranteed by its past. Thus, some brands strategically seek to carry an image/narrative of the past, reintroducing into the public space an interpreted version attending to the present

Fig. 32 Picture. Information sign. Ramos Pinto cellars entrance, Gaia. December 2018. MA-pa

needs, without wasting the status that these references have already acquired. This is the value of the memory that Port wine distils between tradition and innovation.

Fig. 33 *The kiss* panel board. Ramos Pinto cellars entrance, Gaia. December 2018. MA-pa

Fig. 34 Picture. Ramos Pinto brand flag. Ribeira, Gaia. December 2018. MA-pa

Fig. 35 Picture. Poster reproduction for sale at the Sandeman's cellar store. Vila Nova de Gaia. February 2012. MA-pa

Fig. 36 Picture. Port wine posters reproduced in notebooks. Souvenir shop. July 2017. MA-pa

Acknowledgements This work is financed by national funds through the FCT—Fundação para a Ciência e a Tecnologia, I.P., in the ambit of the grant PD/BD/150509/2019 and the project UIDB/04057/2020.

References

1. Almeida M, Barbosa H (2017) Hermenêuticas do cartaz do vinho do Porto do final do século XIX ao início do século XXI (Master dissertation, Faculdade de Belas Artes da Universidade do Porto)
2. Almeida M, Barbosa H (2019) Authority and Austerity: in uence ows between Estado Novo and Port wine in posters from 1920's to 1960's. Paper session presented at the 4T Design and Design History Association (Biannual Symposium), Izmir, Turquia
3. Barbosa H (2011) Uma história do design do cartaz português do século XVII ao século XX (Doctoral thesis, Universidade de Aveiro)
4. Barata M, Branco V, Providência F (2009) Identidade do vinho do Porto, pela tradição da sua embalagem (Master thesis, Universidade de Aveiro)
5. Barreto A (1990) Prefácio. In: Martins CA (ed) Memória do vinho do Porto. Instituto de Ciências Sociais da Universidade de Lisboa, Lisboa, Portugal, pp 5–19
6. Curvelo P (2019) Portugal eleito melhor destino turístico europeu pelo terceiro ano. Retrieved from https://www.jornaldenegocios.pt/empresas/turismo---lazer/detalhe/portugal-eleito-melhor-destino-turistico-europeu-pelo-terceiro-ano
7. França JA (1981) Ramos Pinto 1880–1980. Ramos Pinto, Vila Nova de Gaia, Portugal
8. Guichard F (2001) Rótulos e Cartazes no vinho do Porto. Ed. Inapa, Lisboa, Portugal
9. Guichard F (2003) Os espaços do vinho do Porto. In: Guichard F, Pereira GM, Guimaraens D, Peixoto F (eds) O vinho do Porto. Instituto dos Vinhos do Douro e Porto, Porto, Portugal, pp 9–35
10. Guimarães G, Correia AF (2001) Vinho de Adão, uvas de Eva: a mulher na arte publicitária. Adriano Ramos Pinto, Vinhos S.A., Vila Nova de Gaia, Portugal
11. Moles A (2005) O cartaz (1st edn: 1969). Editora Perspectiva, São Paulo, Brasil
12. Regina C (2019) Porto e norte de Portugal distinguidos com quarto prémios nos World Travel Awards. Retrieved from https://anoticia.pt/2019/06/10/porto-e-norte-de-portugal-distinguido-com-quatro-premios-nos-world-travel-awards/.

Representation Between Waves of Change: A Visual Analysis of the Advertisement of Female Surfers

Antonia Sophia Hinz, Flávio Almeida, and Anabela Couto

Abstract This paper is concerned with the way in which females who surf are displayed in mainstream media channels and are experiencing issues of gender, sexuality and objectification. In particular, the article is based on the analyses of an advertisement video released by ROXY, the first brand in the industry to design specifically and only for women, in 2013. The video was chosen due to the remarkable number of responses it caused throughout the online female surfing communities and other media channels. The analysis was established with knowledge from visual culture, image analysis and feminist issues and is based on qualitative aspects taking into consideration the technical and visual grammar of moving images. The methods used are based on specialized literature on film and image analysis. It was found that the video, which advertises a surf competition for women by depicting Stephanie Gilmore, seven times world surfing champion, showed strong heteronormative bias in the visual narrative construction as well as in the depiction of the surfer. Considering the cultural background of surfing, this result is not surprising yet highly relevant. Women make up about half of the clientele at surf schools and are a particularly pronounced factor in the growth of the sport, also concerning consumption and visibility worldwide these days. However, since the original "surf- hype" swept across the United States in the 1950s and 1960s, the culture has been shaped by heteronormative and sexist narratives of gender and sexuality, with the strong image of the male surfer underlining the patriarchal American masculinity. By the 1980s, occurrences of sexism began to develop into a widespread cultural stance against women in the surf. By denying female surfers subjectivities through sexual objectification, and the 'symbolic annihilation' of women as active surfers, women were positioned as "objects" and "subjects" within the surfing culture. The topic has since remained unsettled and ambivalent, complicated by the trajectories of modern rationality, postmodernity and commodification. The topic remains relevant as an example of a new "lifestyle" sport, which will soon be part of the Olympic games and possibly continue grow exponentially. This long paper demonstrates, that gender related power relations in surfing remain contradicting and impugned. It therefore

A. S. Hinz (✉) · F. Almeida · A. Couto
IADE - Faculdade de Design, Tecnologia e Comunicação, Universidade Europeia, Av. Dom Carlos I, 4, 1200-649 Lisboa, Portugal

E. Duarte and C. Rosa (eds.), *Developments in Design Research and Practice*, Springer Series in Design and Innovation 17,
https://doi.org/10.1007/978-3-030-86596-2_26

encourages a radical dialogue regarding the role that visual representation plays in addressing this societal issue of gender portrayal in sports, particularly in surfing, and demands a powerful and potentially feminist way of re-interpreting the female surfer.

Keywords Surfing · Feminism · Visual representation · Visual culture · Sexism

1 Introduction

Houston, 1973. Twenty-nine-year-old Billie Jean King, sitting on a plushy pink palanquin, is carried into the Astrodome towards what is nowadays commonly referred to as "the battle of the sexes". A historic tennis match between her and Bobby Riggs, which became a symbolic milestone in gender equality in sports. When Billie Jean King won the match, she not only proved to an audience of approximately ninety million people worldwide, that women are able to play tennis just as good as men, but she also achieved a breakthrough for equal pay and media coverage for female athletes.

Fast forward to 2019 and one might assume that during the past 46 years the issue of gender inequality and heterosexism in sports should be resolved. However, when Georgina Roy from the University of Brighton investigated "Feminism in New Sporting Spaces: Gender, Subjectivity and the Female Surfer in Britain" in 2013, she found that "surfing still has strong links to standards of white heteronormativity, and the heterosexy image of the "surfer girl" is a central aspect of surfing's commercial mainstream" [15]. In 2016, a female surfer who did not fit Roys description of the "heterosexy surfergirl" was refused sponsorship for not being attractive enough. Silvana Lima, ranked one of the top ten women on the Australian Surfing Championship (ASP) World Tour, stated the she is ""not a babe… (but) a surfer" and argued that women were "excluded" and "disposable" to sponsors [3]. As a response, brand strategist Rob Frankel reassured Lima that "in the era of Photoshop and in the hands of skilled art directors, (she) has the basics to be more than pretty enough" [10]. This example illustrates, how the surf culture still limits and objectifies women today, despite women making up about half of the clientele at surf schools [5] and are a particularly pronounced factor in the growth of the sport, also concerning consumption and visibility across forms of popular culture worldwide [7].

On the other hand, several surf brands started female targeting sub-brands and the World Surf League (WSL), organizer of the World Surf Championships, celebrated equal price money for men and women in 2019 [22]. Silvana Lima eventually did get sponsorship, and in 2018 became the world's first female athlete to be sponsored by a crypto currency [8]. Other successful professional female surfers such as Carissa Moore and Tyler Wright who also do not fit in the previously mentioned stereotypical image of the heterosexy surfer girl, have received full support by the industry and top of the World Tour according to the interview with an anonymous marketing executive of a large surf brand, which was published in STAB magazine in 2016 [17]. However, despite this acclaimed "progression" of the sport, the topic remains unsettled and ambivalent, complicated by the trajectories of modern rationality, postmodernity and commodification [15], which provokes interest in the current situation of visual online communication of relevant surf brands.

Ultimately, the advertisement of female surfers is discussed in this article as an example of a new "lifestyle" sport, a category of sports which has exponentially grown since the 1990s [6]. The sport will be included in the 2020 Summer Olympics and its surrounding lifestyle image is used across several marketing campaigns of other industries. Therefore, the topic of gender inequality, sexism and heteronormative objectification of female athletes in surfing is considered highly relevant and has been further investigated throughout this article.

2 Methodology

To clarify the direction of this article and to choose the most appropriate research strategy, data collection and analysis techniques, Saunders et al. suggests formulating the starting point of the research by deciding on a research topic and objectives through a clearly defined research question. Additionally, it needs to be considered whether the time frame, financial recourses and necessary research skills are available. Finally, the topic also has to fit the specifications and meet the standards set by the examining institution (2016).

Therefore, the methodology states the research objective, central research question and sub questions as well as the hypothesis, theoretical framework and limitations for this work.

Following these directions, this article's main objective is to introduce the problematic of visual online communication of relevant surf brands displaying female athletes by briefly analyzing key features of visual advertisements of surf brands aimed at a female audience.

With this main objective in mind and considering the actual state of sexism in contemporary society and research done previously, the key features of visual advertisements of surf brands aimed at a female audience are questioned by this research. According to the introduction, the hypothesis is that the advertisement of female surfers is currently still influenced by a heteronormative and sexist narrative.

To gather the information necessary for this matter, the following theoretical framework has been applied. Firstly, relevant surf brands have been defined through a literature review. Secondly, to analyze a previous advertisement for female surfers in detail, the DAIJ model by Edmund Feldman has been used. The specific video has been chosen due to its popularity and the impact it caused on the surfing community. To investigate whether a heteronormative and sexist narrative was apparent, both of these abstract concepts have been operationalized and measurement entities were introduced. Finally, three visual examples of relevant surf brands' online communication have been selected due to their relevance to briefly consider the strategies of surf brands in terms of the current visual advertising online. The examples have been selected according to them being the main landing pages of the relevant surf brands for the "equipment" section, hence a first impression and indicator of the brands current visual advertising online. Due to the limited length of this paper and the academic expectations set by the institution, only a very limited sample of visual online communication by surf brands has been considered. This amplifies the possibility of bias and an unrepresentative impression of visual advertisements of surf brands aimed at a female audience.

3 Relevant Surf Brands

In 2016, Andrew T. Warren, researcher and lecturer at the University of Wollongong in Australia, identified the three biggest surf brands to be Billabong, Quicksilver and Rip Curl [20]. He came to this conclusion through analyzing the companies' revenues, asset write-downs and earnings while investigating subcultural enterprises, brand value, and limits to financialized growth. All "big three" brands were established in the 1970s during which the surf hype swept across California and Australia. The founders were avid surfers themselves, which helped promoting their brands within the subculture and also across it's boarders. Their gear and clothing represented laid-back, counter cultural values which equally appealed to surfers and non-surfers at the time [20]. However, as the companies expanded rapidly, they began to estrange themselves from the rebellious youth culture which were their core consumers. Since 2013, the brands attempt to reconnect with their origins through several marketing initiatives, sponsorships and reference to their original roots [20]. Meanwhile, the Californian brand Volcom entered the market in 1991 and grew rapidly through the slogan "Youth Against Establishment" in 2001. After being acquired for $607 million by French luxury group Kering S.A. in 2011, Volcom quickly outperformed the "big three" with a sales growth of 19% during the first half of 2013 [20]. However, as Volcom is mainly focusing on the fashion industry and does not produce a considerable amount of surf gear in form of wetsuits, boards or other equipment, the brand has been disregarded for this research, which leaves the relevant surf companies investigated to be Billabong, Quicksilver and Rip Curl.

4 Video Analysis

A starting point for this research was a video released by Quiksilvers female sub-brand ROXY in 2013, which advertised the professional female surfing contest "Roxy Biarritz Pro". To understand the impact of the video, the DAIJ model by Edmund Feldman has been applied. The model is based on a detailed observation and description of the content, before interpreting the motivation and evaluating relevant variables for the construction of a critical judgment.

To better understand the advertising, a detailed description of the script and the visual narrative created for the event has been laid out as follows:

Capturing the first rays of sunlight, the camera pans slowly across the shore of Biarritz in France. Waves crash softly on the golden colored sand while the beach promenade seems to be still asleep at such an early hour. The scenery is accompanied by the sound of a song by Flume called "Sleepless", which plays as the only audio throughout the entire video.

In the next scene, the body of a woman only covered by a pair of silk teal knickers appears laying in between soft white bedsheets. On the wooden bedside table in the foreground an alarm seems to wake her up. The camera shows the woman from another angle, a close up looking over her right shoulder, catching a few wisps of her long blonde hair through which she runs her fingers while laying on her belly and angling her legs in slow motion. In the next shot, she reaches over to turn the alarm off, lifting the phone displaying the HTC logo on its rear panel. Her nails are perfectly manicured, and the viewer can see pink and light blue handknitted bracelets on her wrist. Again, the camera jumps to look over her shoulder, showing the woman's backside and blonde hair, before moving to an angle from the bottom of the bed showing her rolling around between the sheets. Her tanned skin stands out against the white, her sleek, shaved legs gleam in the morning sun. The next shot shows her from behind, sitting on the long edge of the bed facing a window. She stretches, runs her fingers through her hair again. On the left side next to the bed stands a bouquet of pink roses and a white candle on another bedside table, in front of a red Fender Squire electric guitar which leans in the corner next to the window. On the other side of the room, the viewer can recognize a DHD surfboard, the DX2 model with a black Creatures of leisure tail pad. Jump shot, the woman is now standing facing the window, putting on a transparent white man's shirt against the light. She walks over to a black desk, before a closeup of a laptops touchscreen shows the woman briefly checking her email before a low shot shows her dropping the white shirt and entering the shower. Her right leg shows a tan line from a leash, a rope which connects a surfer to the board. The next shot shows water raining from a large shower in slow motion, before another shot shows the drops landing on the woman's neck and right shoulder. Jumps shot to her leaving the building through a large glass door, towards a black SUV. She is shown from her shoulders down, dressed in dark blue jeans shorts and a black and white striped top. The SUV appears to be from SLAVI, a French car rental service, as the sticker on its side indicates. A close up of the surfboards side shows the DHD logo, the woman runs her right hand tantalizingly slow over the rail. The still anonymous woman drives the car, the camera looks over her shoulder towards the front screen, showing her blonde hair and the edge of a black pair of sunglasses. The next scene shows the bottom half of her slender legs, exiting the car barefoot displaying her pink toe nails. With the surfboard under her arm, the woman walks towards the calm, waveless ocean. The camera follows her with a closeup of her backside in the dark blue jeans shorts. Suddenly, her hips are shown from the front, walking through a slid in a large pink banner with "ROXY" printed on it. Then the camera switches to her backside again, the banner closes like theatrical curtain. The woman's body is now shown in a pink and black bikini, she walks past the camera throwing her striped shirt on the lens. The next shot shows her upper body dressed in the pink bikini while she is waxing her DHD board in front of the pink

ROXY banner. Shown from behind, the woman then pulls a pink lycra over her head. When the camera switches to her front, the viewer can see her nipples through the thin fabric. A large ROXY logo is displayed on the lycra's front and back. When she pulls it further down, the brands "SOSH" and "Biarritz" appear. Afterwards, the camera dives up from underwater, as the unidentified woman paddles past the lens on her board towards Biaritz' famous rock formations. Her bikini bottoms ride up between her legs, as the sun reflects from her bare wet skin. The end of the video is marked by the logo of the "ASP Women's World Surfing Championship" appearing above her, right over "ROXY PRO 2013" and the dates of the event. All sponsors, "Surface", "Sosh", "Biarritz", "Melty", "Jeep", "Region Aquitaine" and "Virgin Radio" appear on the screen surrounding the surfer.

Overall, the video appears to be rather harmonious through the soft and warm lightning and music. However, its main intention, which is to advertise a surf contest, seems to be not met, which becomes apparent when the video is timed. The overall video showed 0% of the time surfing, therefore 46% of the time an anonymous woman's barely covered backside [19]. The anonymous woman was later identified as Stephanie Gilmore, seven times world champion surfer. Critics claimed the video to be "sexist" and "soft porn" in Facebook comments [13]. One user wrote: "This is disgusting. Just another ad for men. Where are the women surfing? Why aren't pro female surfers good enough to promote on their skill alone? Never seen a male surf comp promoted like this. Poor Form Roxy. Redo it!" (Alice McClintock 2013). After several of such accusations, the company, which claims in their mission statement to "exist to empower women" [14], responded: "(...) we are disappointed by recent mischaracterizations of the ROXY brand. (...) At Roxy, we will never stop celebrating female athletes. That's our brand promise, and we will continue to strive to live up to that goal." [11].

In order to explore whether Roxy has been misjudged or the video actually showed exemplary how a key feature of visual advertisements of surf brands aimed at a female audience are made with a heteronormative and sexist narrative, the two abstract concepts of "heteronormativity" and "sexism" have to be operationalized. Firstly, the indicator for heteronormativity as coined by social theorist Michael Warner is "that male–female differences and gender roles are the natural and immutable essentials in normal human relations" [1]. Therefore, the variable for this analysis is whether Quiksilver made a different video for the male surfing event in the same year. The measurement entity for this difference is the time during which the athlete in the video is shown surfing. As previously mentioned, the ROXY advertisement shows for 1.46 min 0% of Stephanie Gilmore surfing, therefore 46% of the time her barely covered backside. The video released by Quiksilver in the same year, advertising the male "Quiksilver Pro France", shows during the overall 1.54 min 51% of the time male athletes surfing, while showing 0% of close ups of their backsides [18]. Further remarkable differences are that the woman in the Roxy clip has been shown without revealing her face and therefor identity, while in the Quiksilver video several famous surfers can be easily identified. According to the company, the identity of Stephanie Gilmore was meant to be a surprise, and guesses were supposed to be made under the hashtag #WhoAmIJustGuess as indicated in the video description. However, this is cannot be considered a reason to explore the erogenous zones of

Table 1 Comparison Roxy/Apartadox

ROXY (Image Source: https://www.youtube.com/watch?v=GCji6TiJjbE)	APARTADOX (Image Source: https://www.apartadox.com/Acompanhante-de-Luxo-Maia-A-4640.html)

	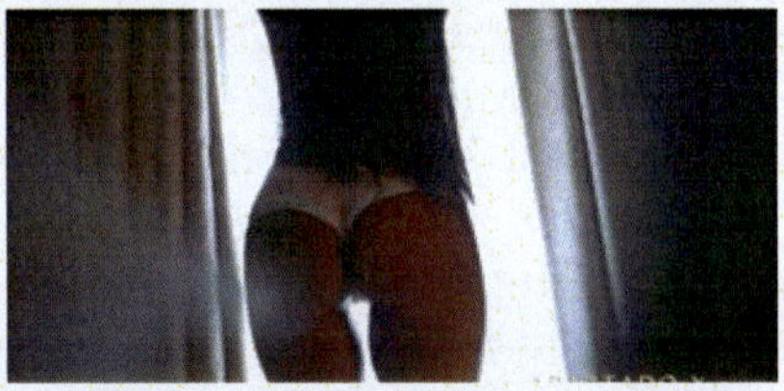
	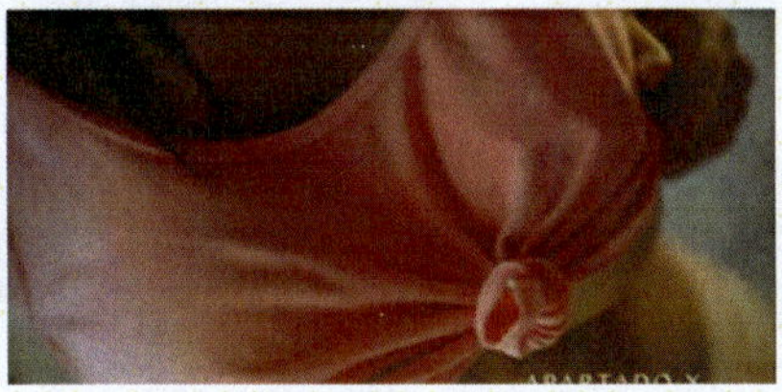

Gilmores body in the way described above while showing her waking up, checking emails and showering instead of surfing. It would have been easily possible to show her in action, surfing waves, without revealing Gilmores identity. Another difference between the advertising videos is, that Roxy does not show the topic of surfing from the beginning, only in the end Gilmore is shown in the ocean, while Quiksilver shows surfing in the first scene already. This clear ascription of differences between the advertising for a male versus female surf contest confirm the assumption of gender roles through a heteronormative narrative as suggested by Warner. The second abstract concept which this research aims to investigate is sexism, which is indicated by APA Dictionary of Psychology to be "discriminatory and prejudicial beliefs and practices directed against one of the two sexes, usually women (…) associated with acceptance of sex-role stereotypes" [2]. The latter has been further defined by Dr. Gail Dines, a sociologist who argues that we live in a hypersexualized culture which becomes particularly apparent in today's pornography, since it feeds and amplifies our gender stereotypes remarkably [9]. Therefore, the variable for "sexism" in the case of the Roxy video, is measurable through it's comparability to pornography. Table 1 contrasts on the left side the ROXY advertisement to a video from the escort service website APARTADOX on the right side.

The similarities in lighting, setting and even body positions in both videos are surprising. Both women are shown anonymously, and the videos both focus on erogenous zones of their bodies. The difference of one being a professional athlete and the other being an escort service only becomes clear towards the end of both clips, when the female surfer paddles out into the ocean and the escort service undresses in front of the camera. For APARTADOX, this is a matter of protecting the identity of the escort and to sell her body. In the Roxy video, the identity is also supposed to be kept a secret, but that does not justify the body exploitation in a softporn style. Especially since the Roxy video has the intention to advertise professional sports and is not labelled "soft porn", such visual portrayals of women are problematic, because according to Betterton [4] they "make up a discourse on what it means to be feminine in our culture (and) for women in particular, the images are impossible to ignore". Kate Millet (1970) supports this argument, as she wrote that "sex-role stereotyping ensures the social control of women (...) because they keep women from social activity", or in this case, from surfing. Maggie Humm agrees in her book "The Dictionary of Feminist Theory" (1995) that gender stereotypes displayed in visual language "limit the activities of women" significantly. However, some might argue that we live in postfeminist times, implying that such arguments are no longer needed as women have achieved legal equality [12]. But through the video analysis above, it becomes apparent that such criticism is indeed valid, because the Roxy video has clearly been made with a heteronormative and sexist narrative. However, since the video has been released in 2013 and received negative comments as mentioned, one might wonder whether the industry has reacted and changed their approach.

5 Visual Communication

To briefly analyze whether the criticism which followed the described advertisement above had an impact on the surf brands' visual online communication, three examples in form of screenshots from the relevant websites have been considered. The content has been selected according to being the main landing pages of the relevant surf brands either immediately as in the case of Billabong or for the "equipment" section of Quiksilver/Roxy and Rip Curl as a first impression and indicator of the brands current visual language online regarding the sport rather than fashion content. Due to length limitation of this paper, only the most significant aspects have been highlighted through red circles in the images below and are discussed in this section (Fig. 1).

On the wetsuit section of the ROXY website 2019, one might get the impression that wetsuits are used for walking on the beach. There is no obvious connection to wave riding, except the pink/light blue/white colored surfboard, which is carried through calm, ankle deep water with no waves in sight. Again, the company decided to not show world champion Stephanie Gilmore surfing, but instead modeling on the beach. The spring suit she is wearing in the picture in the upper right corner has a zip in the front which can only be found in female wetsuits. It supposedly helps to get in and out of the suit, however due to its placement in the front, often cold water gets

Fig. 1 Online communication ROXY, 2019. Retrieved from www.roxy.com/wetsuits

in and the zipper catches wax from the top of the surfboard while paddling, hence it must be placed there for optical reasons only. Another surprising element is the companies' suggestion to consider buying a beach bag before looking at wetsuits, as seen in the bottom left corner (Fig. 2).

When looking at the male counterpart of Roxy's website, there again appears to be a remarkable gender difference. On Quiksilvers landing page for wetsuits, a surfer is performing a difficult maneuver on a wave, wearing one of the wetsuits. There is a clear connection between high performance surfing and Quiksilvers gear.

Fig. 2 Online communication Quiksilver, 2019. Retrieved from www.quiksilver.com/wetsuits

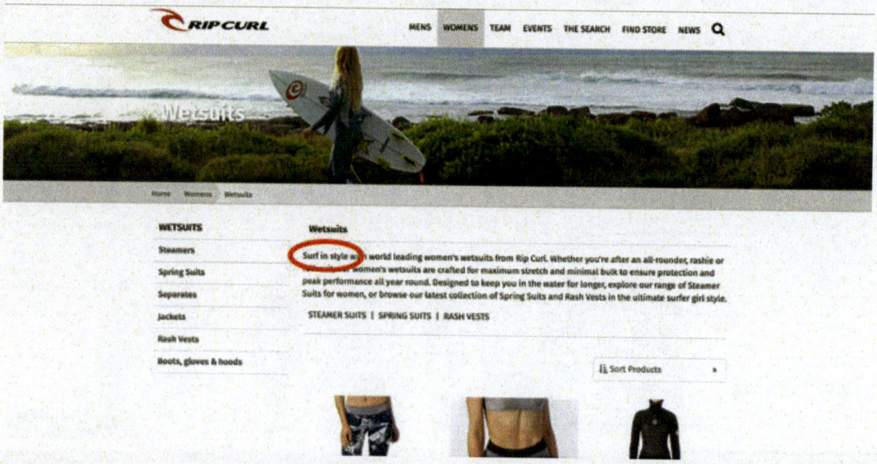

Fig. 3 Online communication Rip Curl Women, 2019. Retrieved from: www.ripcurl.com/women/wetsuits

On Rip Curls landing page for the same equipment sold to a female audience, a woman, most likely professional surfer Alana Blanchard, is not surfing in her light blue wetsuit either. Instead, she is carrying a surfboard while looking at the waves on her right side. She cannot be clearly identified, since the picture is not showing her face, again leaving the female surfer anonymous. The text beneath the image highlights the importance of "surf(ing) in style", while the shopping suggestions underneath show leggings and what appears to be a sports bra next to one full wetsuit (Fig. 3).

Meanwhile, Rip Curls male version of the website suggests to "stay ahead of the pack" with the new technology of their wetsuits. A special heat generating material is displayed, indicating the purpose of wetsuits: to keep the surfer warm in colder waters. In this case, there is no imagery connection to surfing, however there again is a link to high performance through Rip Curls gear. Beneath the text are three suggestions of full wetsuits to buy for the viewer (Fig. 4).

Finally, the last example is Billabongs landing page from 2017, which sums up what has been found on the other websites. On the left side, the site visitor can choose the men's section by clicking on an image with a surfer performing an "air", a difficult maneuver where the athlete is jumping meters above the wave. Opposite on the right side lays a woman on the beach, the image is not showing her face, but only her body instead. She is wearing a fashionable bikini which is unsuitable for surfing, since it would not hold up when performing the sport (Fig. 5).

The company has since changed the landing page, since it received a considerable amount of negative feedback from the surfing community much alike the Roxy video [21].

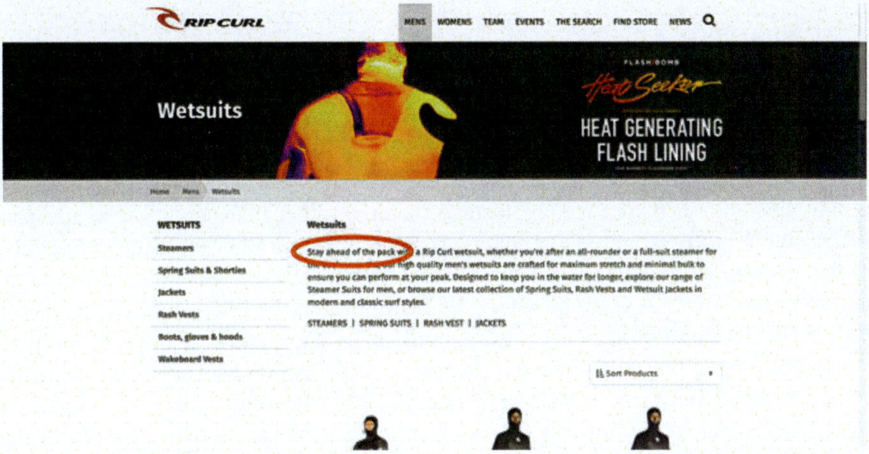

Fig. 4 Online communication Rip Curl Men, 2019. Retrieved from: www.ripcurl.com/mens/wetsuits

Fig. 5 Landing Page Billabong, 2017. Retrieved from www.metro.co.uk/2017/08/20/billabong

According to all examples above, the focus of female surfers lays on fashion, often stereotypically involving floral prints and pink shades, while the websites aimed at a male audience show a strong emphasis on technical details, high performance and the sport of surfing itself. Such fundamental differences indicate, that despite acclaimed progress in the industry, the current issue of sexism and a heteronormative narrative of the sport lies deeper than equal price money and sponsorship contracts.

6 Discussion

The impressions from this article support the hypothesis which claimed that the advertisement of female surfers is currently still influenced by a heteronormative and sexist narrative. The key features of visual advertisements of surf brands aimed at a female audience appear to be fashion inspired rather than surf sport related and differ significantly from male counterparts which focus rather on technical details, high performance and the sport itself. However, due to the limited length of this paper and the academic expectations set by the institution, only a very limited sample of visual online communication by surf brands has been considered. This amplifies the possibility of bias and an unrepresentative impression of visual advertisements of surf brands aimed at a female audience. Further research needs to be conducted in order to quantify findings and thoroughly analyze the topic. Nonetheless, a first impression and indicator of the surf brands current visual advertising online has been established and so far confirmed the hypothesis, that the advertisement of female surfers is currently still influenced by a heteronormative and sexist narrative.

Slow progress is being made as discussed, concerning sponsorship and price money additional to several new surf brands for females appearing on the market. It needs to be further investigated, how such advertising and acclaimed progress is perceived by and influences the females it is targeting.

The topic remains relevant as an example of a new "lifestyle" sport, which will soon be part of the Olympic games and possibly continue grow exponentially.

This long paper demonstrates, that gender related power relations in surfing remain contradicting and impugned. It therefore encourages a radical dialogue regarding the role that visual representation plays in addressing this societal issue of gender portrayal in sports, particularly in surfing, and demands a powerful and potentially feminist way of re-interpreting the female surfer.

References

1. APA Dictionary of Psychology (2019) Retrieved from https://dictionary.apa.org/heteronormativity
2. APA Dictionary of Psychology (2019). Retrieved from https://dictionary.apa.org/sexism
3. BBC (2016) The Surfer Who Wouldn't Take No for an Answer
4. Betterton R (1987) Feminism, femininity and representation. In Looking On. Pandora Press, p 1
5. Booth D (2001) From Bikinis to Boardshorts: Wahines and the Paradoxes of Surfing Culture. University of Otago, Dunedin, New Zealand
6. Booth D (2004) Surfing: from one (cultural) extreme to another. Understanding lifestyle sports: consumption, identity and difference. Routledge, London
7. Comer K (2010) Surfer girls in the new world order. Duke University Press, Durham
8. Currie JP (2018) Surfer Silvana Lima becomes world's first female athlete sponsored by cryptocurrency. Retrieved from The Inertia. https://www.theinertia.com/business-media/surfer-silvana-lima-becomes-worlds-first-female-athlete-sponsored-by-cryptocurrency/
9. Dines DG (2015) Growing up in a pornified culture (TEDxNavesink, Interviewer)

10. Falzone D (2016) Pro Surfer Silvana Lima says she can't get sponsors because she doesn't have model good looks. Fox News
11. Haddad S (2013) Cooler. Retrieved from https://coolerlifestyle.com/news/roxy-release-statement-to-draw-line-under-the-roxy-pro-biarritz-shower-scene-trailer.html
12. Humm M (1995) The dictionary of feminist theory. Prentice Hall
13. Levy M (2013) The Sydney Morning Herald. Retrieved from https://www.smh.com.au/business/companies/roxy-hits-out-at-backlash-as-controversial-advert-makes-waves-20130708-2pls1.html
14. ROXY (2019) roxy.com. Retrieved from https://www.roxy.com/customer-service-corporate-information-about-us.html
15. Roy G (2013) Feminism in new sporting spaces: gender, subjectivity and the female surfer in Britain. The University of Brighton, Brighton
16. Saunders M, Lewis P, Thornhill A (2016) Research methods for business students. Pearsons Education Limited, Essex
17. Smith J (2016) The painful truth of Silvana Lima's Sponsorship Struggle. STAB
18. Teaser QP-O (2013) Youtube. Retrieved from https://www.youtube.com/watch?v=ceTtZL2q3Sk
19. Teaser RP (2013) Youtube. Retrieved from https://www.youtube.com/watch?v=GCji6TiJjbE
20. Warren A, Gibson C (2017) Subcultural enterprises, brand value, and limits to financialized growth: the rise and fall of corporate surfing brands. Geoforum 86:177–187
21. Willis A (2017) Metro. Retrieved from Billabong ran these two surf adverts and people got angry. Read more https://metro.co.uk/2017/08/20/billabong-ran-these-two-surf-adverts-and-people-got-angry-6863285/?ito=cbshare Twitter: https://twitter.com/MetroUK | Facebook: https://www.facebook.com/M: https://metro.co.uk/2017/08/20/billabong-ran-these-two-surf-adverts-and-people-got-angry-6863285/
22. World Surf League (2019) Equal price money celebrated at the Surf Ranch. worldsurfleague.com.

Letters to Eternity: Typefaces of the Prazeres Cemetery

Gonçalo Falcão

Abstract The Cemetery of Prazeres (freely translated to 'Pleasures') is the second largest in the Portuguese capital city, Lisbon. It was built after a cholera epidemic in 1833, and it is almost exclusively made up of private deposits [1]. It serves Lisbon's west side population, where aristocratic residential districts were located in the nineteenth century (it remains today a very expensive area to live in). For this reason, it became the graveyard of the city's most prominent families, whom, with their particular taste and financial means, ended up bestowing a certain monumentality to the site. The cult of death is particular in Portugal, where some investment is made in all kinds of ceremonial procedures, including 'the last address'—privately owned small stone houses inside the cemetery, whereby families put their dead to rest: *jazigos*. Almost half of the cemetery is occupied by these small architectures for death, the family *jazigos* (it is said to be the biggest *jazigo* cemetery in Europe with more than 7000 of these monuments). The idea of the cemetery of the elite extended past the city's geographical surroundings and became the place to bury the "great names" of Portuguese culture until present days (e.g., writers, singers, painters, actors). Within its walls, Catholic and the Masonic imagery compete (free masons were a important group in the organization of the regicide in 1910), along with tombs without specific group affiliation. Messages to immortalize people and their ideas and work are cut into stone or engraved in bronze with carefully designed texts and typefaces. These letters in the *jazigos* and tombstones contribute to the understanding of ideas and tastes, but also convictions and beliefs of these people. These aspects are made evident in the rich and varied symbolism that adorns these funerary places. Shapes along with text are carefully planned in the construction of everlasting memory. The relevance of these structures, the types and shapes that adorn them, the texts they immortalize, as well as the stories about the people who erected them can be a relevant source of information. Specifically, it can inform about the dynamics of communication and its respective symbolism within society; for example, in terms of wealth, power, prominence, affiliation, etc. Furthermore, it can have repercussions

G. Falcão (✉)
Faculdade de Arquitetura. CIAUD, Rua Sá Nogueira, Polo Universitário, Universidade de Lisboa, Alto da Ajuda, 1349-055 Lisboa, Portugal
e-mail: gfalcao@fa.ulisboa.pt

E. Duarte and C. Rosa (eds.), *Developments in Design Research and Practice*, Springer Series in Design and Innovation 17,
https://doi.org/10.1007/978-3-030-86596-2_27

for heritage studies, tourism (namely, Dark Tourism, and design and art research. This paper focuses on these letters to eternity, the texts and shapes they produce, and what and how they memorialize. It reports a photography safari, through which examples were collected, and their subsequent analysis. The research focuses on the way memory is engraved into letters, the types chosen to do so, and the content and format used for it. Results are discussed in terms of the constructed taxonomies and future studies.

Keywords Typography symbolism · Design for loss · Dark tourism · Collective memory

1 Introduction

The theme "Lost in (G)localization" suggests an innuendo between "local" and global". From this perspective, this study proposes an exploratory approach of a local phenomenon that has a global context: death. It is inevitably global, and at the same time, unique in the way Lisbon cemeteries display messages for and about deceased people. It also deals with a specific kind of monuments, the *jazigos*– small stone houses that wealthy families build inside public cemeteries, in order to honour their departed loved ones. While this practice is not exclusively Portuguese, it does have some particularities.[1]

This study is centred on *Cemitério dos Prazeres*,[2] in Lisbon. It differs from previous research done about cemeteries, and specifically on the Prazeres' cemetery, because it is not epigraphic, architectural, anthropological, genealogical, nor iconological. It is exclusively concerned with the letters used to make inscriptions, and, therefore, specifically typographic. It hopes to contribute in a productive and meaningful way to the field of graphic design and other related fields, opening territories for further investigation. It acknowledges that letters are a way to graph a text, and that the choice of a letter and the way texts are composed are never casual facts, since they contribute to the meaning of the text they convey. In Lisbon's cemetery, writings are often short, informative, suffered, or poignant (or all of the above), resting on heavy architecture—costly constructions planned as a statement of social power but also of character—one that takes care and respects his departed loved ones.

[1] Even between Lisbon's *jazigos* and Oporto's *jazigos* there are differences. In Lisbon, visitors are not supposed to view the inside, which is kept private, while Oporto's *jazigos* tend to be more open to visitors so they can see the inside.

[2] The word *prazeres* in Portuguese translates to "pleasures" in English, so the name of the cemetery is freely translated to English as'Pleasures Cemetery.' It is not an irony. The Cemetery of "Prazeres was installed on the grounds of a farm from which it was named *campo dos prazeres* (something like *field of pleasures*). The name was kept for geographic reference (but also because it was found suitable).

1.1 Stone Death

"Burials and cemeteries from antiquity, indeed from all periods of history, reveal a great deal about culture" [1]. The Prazeres cemetery is the second largest in the Portuguese capital city, Lisbon, and it is almost exclusively made up of private deposits [2]. The reduction of burials inside churches (*ad Sanctus*) or in its surroundings, for sanitary reasons, was a nineteenth century Enlightenment hygienist cause that, in Portugal, was lead by doctor and diplomat Ribeiro Sanches [3]. The establishment of civil cemeteries as a public service in Portugal was ruled in 1835, and primarily aimed at responding to an emergency (an 1833 morbus cholera epidemic). Moreover, it was also due to the fact that "Parish graveyards could no longer cope with the demands of the industrial cities" [1]. In 1834, the liberal party took power and the cemetery became a symbol of progress (some important liberals were exiled since 1828 and might have had contact with the Parisian Père Lachaise Cemetery [1804], as well as others public cemeteries).

It serves Lisbon's west side population, where aristocratic residential districts were located in the nineteenth century (it remains today a very expensive area to live in). For this reason it became the graveyard of some of the city's most prominent families, whom, with their tastes and financial means, ended up bestowing a certain monumentality to the site. The cult of death is particular in Portugal, where sometimes sumptuous investment is made in all kinds of ceremonial procedures, including these privately owned small stone houses (sometimes literally) inside the cemetery, whereby families put their dead to rest: the *jazigos*.

The vast majority of the space of the Prazeres cemetery is occupied by these small architectures: the family *jazigos*. It is said to be the biggest *jazigo* cemetery in Europe with more than 7000 of these units, some quite monumental; the biggest one belongs to the Duke of Palmela's family, has the capacity to house 200 coffins on three floors, with a private church (including a Canova sculpture, the only one from this author in Portugal). The entrance to this *jazigo* is done through a long corridor where the employees of the family are buried, making it a private cemetery inside a public one.

The idea of a cemetery for an elite surpassed the city's geographical surroundings and this graveyard became the place to bury the "great names" of Portuguese culture until present days (e.g., writers, singers, painters, actors, polititians) [4]. Within its walls, the Catholic Church and the Masonry imagery compete (free masons were a important group in the organization of the regicide in 1910), along with tombs without any specific group affiliation. Messages to immortalize people and their ideas and work are cut into stone or engraved in bronze with designed texts. These letters in the buildings (*jazigos*) and stones help us get a sense of the ideas, tastes and a context of peoples and times. These aspects are made evident in the rich and varied symbolism that adorns these funerary places. Shapes along with text are carefully planned in the construction of everlasting memory. The relevance of these structures, the types and shapes that adorn them, the texts they immortalize, as well as the stories about the people who erected them can be a relevant source of information about the dynamics of communication and its respective symbolism within society. For example, it can inform about wealth, power, prominence, affiliation, etc.; and can have repercussions for heritage studies, tourism (namely, Dark Tourism [5]), Death and Design, [6–8]).

As mentioned, this study focuses exclusively on the letters to eternity, the texts and shapes they produce, the way these are written and what they can memorialize, besides their main function. The funerary inscription is a small part of the memorial; most of the effort is devoted to the design of the monument, its architecture and materials. Letters are, most of the times, left to an informational function. This is a particularity that has to be noted and that might be different from other funerary lettering practices [1].

Studying only the letters, detaching them from the building, will help us raise some questions and open research topics that might not be evident if we would pay attention to the monuments as a whole. These tombs are 'great chronicler of taste throughout the world' [9].

2 The Epigraphy of Death

Megalithic societies are anepigraphic, and so, most markings in funerary monuments in pre-writing societies are impossible to be fully contextualized in terms of their meaning. We start understanding ancient cultures from the moment we can read their ideas, and most of these recordings are obtained from the inscriptions and drawings in their monuments for death. It is not very controversial to mark the emergence of the epigraphy of death in Classical Greece and Ancient Roman periods. 'The creation of tomb-cult became a feature of the post-classical Greek world' [1] and the Roman letters carved in stone are still a major reference for type anatomy until present days.[3]

2.1 Public Cemeteries and the Contemporary World

The Prazeres cemetery was designed much like a city, with avenues and streets, emanating from a central point in which the church plays a dominant role. Each *jazigo* belongs to a family and can house several coffins from different household members, throughout several years. It is described in common language as the "the last address," and this idea is taken sometimes almost literally in the design of these "houses" where the deposit of the dead is made above ground. In these places, relatives keep on "living" together in a post-domestic way, representing an eternal harmony of families (as cruel, ironic or both as it might be sometimes). In most of the cases observed, their design answers to this fact by writing on the top of the building

[3] The Trajan typeface, designed from the letters at the basis of the Trajan Column in Rome is the preferred typeface for Hollywood movie posters (more than 400 movie posters in 20 years use Trajan as the main typeface, according to Yves Peters's FontFeed blog), although, in recent years, the typeface Gotham is gaining territory (another font inspired by street signs).

the name of the family patriarch or of the family member that payed the housing, and by having a decoration in the front made of equal stone panels that serve as a background to each of the next entombment inscription. They can be seen almost as mailboxes or door bells that have the name of the family member that stays inside, and so a new inscription is made on the outside for each of the family member that are placed inside, in horizontal shelves in vertical alignment (usually 6 or 8).

Some dates can be found in these inscriptions or in the cemetery registers. Further research on this issue would demand careful dating, since most of the monuments do not show any foundational dating, and they can change tenants. *Jazigos* are not eternal… they can be bought in municipality auctions as families abandon them or can no longer pay the cost of owning and maintaining them.[4] They represent a big investment in terms of ground, architecture, building (crafting the stone and bronze), and maintenance. The last auction had starting prices of 10,000 Euros for the smaller units. After it is bought, restoration or rebuilding has to be payed and conservation has to be done. With all this money involved, one could expect to have an equivalent investment on lettering and text, but a first visit to the cemetery leaves a different impression: very seldom letters and text are treated with care. Very rarely there is a distinction between the people who design and those who produce: we found only one example where the signature refers "drawn and produced by" (desenharam e executaram). There is at least one case in which the *jazigo* was made in Paris, and imported to the cemetery in Lisbon.

3 Method

The work presented in the current paper can be classified as exploratory research, and it was pursued through (an adapted version) the "touchstone tour method" [10]. It is exploratory because it is "focused on gaining a solid knowledge base of the design territory and existing artefacts and forging an empathic sense of the people targeted by the design work" [10]. The "touchstone tour method" (or "guided tour), "is a contextual, empathic method that efficiently immerses the designer in a participant's world, to understand how he or she organizes information and systems through the use of space and cognitive artefacts" [10]. "Touchstone tours" are documented with video or photos as these "provide an excellent record that can be easily sorted and annotated and sent to others for discussion and analysis (…). The outcome of touchstone tours may suggest general design implications, but it is largely an exploratory method for designers to establish baseline familiarity with a territory in early phase research" [10]. Three organized photographic expeditions and ten walkthrough visits were done in preparation for this paper.

[4] We even found one *jazigo* being sold directly in an online second hand sales website, although we could not understand whether the offer was real ot not, and how would this selling process would be consummated in bureaucratic terms.

For the research reported in this paper, I started by visiting the cemetery regularly, not as a tourist but in search of information, alone and focused on the purpose. The first trip was equivalent to a first scanning. The subsequent, more informed visits were conducted with more directed eyes in search of specific elements (nevertheless, some new ones always appear). Scanning visits were a way of realizing and identifying the space, occurrences, similarities, finding patterns, etc. Recording visits (3) documented all the things that presented themselves as important to my eyes and regarded them afterwards carefully, in isolation, selecting among these the typographic evidences. Similarly, it was also not scientific (recording a specific kind of predetermined information), but rather it was the mood of a designer aiming to make a visual program or a book for the cemetery, looking for visual information for his work (recording what might be important and what impressed and amazed). Most of the time while seeing, we are thinking about it, registering, and making a database of species in our mind. A certain pleasure stems from looking at letters—it is associated with a professional defect that one cannot avoid after years of practice, the excitement from the design of letterforms.

Three visual expeditions were done specifically for this research, each averaging more than 2.5 h. The first visit covered the left part of the cemetery's corridors in a serpentine walk through. The second visit focused on the right part. The third visit covered the cemetery's central axis. Around 1000 photographs were collected of the letters that seemed more relevant. In the third visit, a sense of purpose and aim was more present, so it was possible to make more informed choices of things that should be photographed (although surprises never stop), while in the first visit the registration was more unintentionally done.

While visual expeditions are not yet formally classified as a method and remain a loosely described procedure, there is clearly a task to be done to enlarge the visual research methodologies In Memoriam.

4 Letters in the Prazeres Cemetery

What was most evident since the first visit to the cemetery was the amount of "slipups," that is, visual mistakes, intentional or not, that are noticeable. Several inscriptions present problems of letterspacing, of composition, and even of typography choice (at this stage, it becomes increasingly evident in more recent inscriptions). This is not surprising, since Portugal never had a strong visual culture, especially concerning letters. It is common to see, even today, architects making basic typographic inconsideration in excellent architecture. If it is difficult to have in mind today all the difficulties of carving stone in the 1800s, so it is easy to understand that recent *jazigos*, done after the year 2000—for a figure like Mário Soares, for example—should not be graphed in Geometric 706 typeface black condensed[5] or in a handwritten "comic" font. It should also be noted that even in the 19th or early

[5] Designed by Wilhelm Pischner (1904–1989).

twentieth century, decorative/commercial fonts are present and sometimes mashed up in several styles and weights, in a strange epitaph typographic potpourri. Although these older engravings have to be read without a deeper historical insight, dating them with precision and comparing them with other typographic practices renders it impossible to miss these strange "Frankenstein" types in tombs were a certain ceremony and sobriety is required (Fig. 1).

Capital letters slanted left are also visible. This is not a common practice in books or magazines of these periods, so it is important to point them in funerary practices. Slanting left and right is also noticeable, and it is an idea used by several builders/architects. As far as this first study was able to find, it is not an "in house" style of one of the resident builders/engravers (Fig. 2).

Mixing serif with san serif, or with slab serif types is also present in solutions that do not consider the respective visual "weights," thereby producing black blocks (contrast) that point out certain text portions over others. While this could have been done intentionally, to a certain extent, in some cases it does not make sense: why should dates or articles stand out over names? (Fig. 3).

From the early 19th until the early twentieth century, it is common to see some letters that have a lot of decoration. A kind of a Baroque attitude towards letters, were

Fig. 1 Three examples of peculiar typographic choices and composition mashed up of different styles and weights

Fig. 2 Slanted left capital letters

Fig. 3 Mixing serif, slab serif and san serif

Fig. 4 Decorative lettering

decorative elements, torsions, and effects are overimposed (as in Raphel Bordallo Pinheiro's lettering).[6] We find this kind of elements also in a context where we would imagine they could be seen as excessive or too light. Letters with terminals in the shape of a fish tail, letters with bubbles (maybe lachrymals…?) coming out, letters without interiors, rich shading, and 3D trompe l'oeil (today associated with circus lettering)—there is a rich thesaurus of special effects over letters that make any graphic designer happy. This might also contribute to dismantle a common belief about Romanticism as dark and macabre, but rather seeing the Romantic Era bringing new and captivating views on death (of which the institution of the cemetery, with its trees, is in itself a good example) (Fig. 4).

The use of compressed type is very common, mostly because the space available for writing is usually scarce, and names in Portugal are long (five or more names is common). Most epitaphs are merely informative; very seldom, we encounter a text that adds a flair of poetry to the information about the deceased. We tried to look for printers to see whether their *jazigos* would show some differences; having found just one and, considering that it was not possible to conclude anything substantial given

[6] These kind of decoration over letters (it is usually the letter with decoration added on it) was still visible in 1980 in national roads, were commerce would use some bizarrely over ornamented letters.

Fig. 5 Gregório Fernandes funerary monument: small letterspacing problem in the inscription

Fig. 6 There are a lot of letterspacing problems in a great number of funerary inscriptions. Even in cases where it would seem simple

the unique sample, it pointed at a special care with the typeface choice and with composition. José Gregório Fernandes was the national printing press director; he has a distinguished *jazigo* standing in the crossroad of the cemetery streets. Modern Cassandre type letters with an Art Nouveau feeling were used and well designed; however, even him was not free of a small letterspacing problem in his tomb[7] (Figs. 5 and 6).

[7] Gregório Fernandes published the book "*A Exposição de Leipzig*" [The Leipzig Exhibition] in 1910, in which some risks in composition and type choice were taken which shows interest for a renovation in printing styles having in mind what he saw in Leipzig.

Most contemporary *jazigos* (and there are some outstanding examples of interesting solutions) do not use any kind of letters.

4.1 Iconography

The study of iconography in the cemetery merits a field of research in itself, and was already initiated by several authors (e.g. [11]). A brief mention of some key elements is done here, to acknowledge that letters are very frequently accompanied by small symbols, which sometimes should be considered along with the text as they build a second line of discourse. Flowers are extremely frequent when putting thoughts into stone. Flowers named saudade (artichoke, whose name in Portuguese means something like "longing"), and *perpétua* (globe amaranth, whose name in Portuguese means something like "perpetual") are present allover. The language of the flowers is currently not so easily understood; however, at the time, it was fluently unspoken.

Another highly recurrent symbol is the hourglass with wings (life passes by flying), present also in the entrance of the cemetery facing outward. Symbols of profession like the military, or other affiliations like the masonry, are also common. Other symbolic repetitions include the Alfa and Omega (beginning and end), the ship (death ship), Caduceus (Hermes/Mercury), the skull, the owl, the cross (resurrection), broken/burned tree, inverted torches (end of life), and scissors (cuts the thread of life).

4.2 A Small but Curious Detail

The word Família [family] appears frequently in the front of the *jazigos*. The scheme is repetitive; either "Familia XPTO" [Family surname], or the name of the "owner" (could be the founder, buyer, commissioner, it is uncertain), with an added "e Família" (and family). This seems to be the way to mark a position, given that the owners' name always appears bigger, and the rest ("e família") appears smaller. This is a visual translation of the spirit of the "family patriarch," who builds patrimony while the wife builds matrimony. He provides housing, food, clothing, and a social status in life, and the *jazigo* has the same spirit in death, gathering the family under a roof preserving the social "living" life.

A curious fact concerning the word *Família*, is that it has an accent mark (i.e., í), but it is frequently written without it in stone. Only with the 1911 Spelling Reform did the systematic use of accent marks become ruled. According to linguist Rita Marquilhas [12], "the graphic accentuation debate was, from [the late sixteenth century] until the nineteenth century grammaticists, completely peaceful. The use of a graphic accent was optional and its usage was usually chosen according to each specific context, with the exception of avoiding any ambiguity of meaning."[8]

Thus, adding an accent mark in the first "i" of the word *família* was optional. "In fact, if you consult the 'Thesouro da Lingua Portugueza' (1871), by Frei Domingos Vieira, you will see that 'family' [*familia/família*] is written without an acute accent mark, as happens with the word 'language' [*lingua/língua*], with this same spelling without an accent mark, in the title itself" [13]. And regarding the *jazigos* built after 1911? We should consider at least three explanations. The place in the cemetery is expensive and was a sign of social status. "New money" individuals like to buy lineage, and thus not using the accent mark could be a way to simulate an older family tradition. A second explanation could be that, both before and after the 1911 Reform, words written in capital letters could omit accent marks intentionally, due to a typographical association with the French usage.[9] Finally, a third reason could be the great resistance to the new spelling established with the 1911 Reformation, resulting in the accent mark's absence.

5 In Memoriam/Conclusion

The research reported in this paper is a first attempt to look at letters in Portuguese cemeteries, and particularly in *jazigos*, and put them in a design history context, a territory that has already been explored in other fields by several authors, namely the work of Paula André [13, 14] and others. Cemitério dos Prazeres is called an "open air museum" because of its rich architectural, urbanistic, sculptural, and heraldic resources. It needs a more consistent look from a designer's point of view, and this paper is only a first aerial view that tries to open future doors for investigation and assure that there is enough material to be looked at from the point of view of design (graphic and other). This is one of many works that desacralize cemeteries and search for their social, cultural, and pedagogical value, with a specific point of view.

Further studies can be done focusing on the work of certain builders/architects and their use of letters. Moreover, future research can compare letters used in certain periods and establish relationships between cemetery letters and other letters from the same period (book covers, magazines, newspapers, packaging). In addition, future studies can focus on overly decorated letters in a broader context. Comparison between Lisbon, Oporto, and other cities from the perspective of this kind of funerary monuments (e.g., Paris), would be an interesting research avenue. Other topics of potential interest are: the anatomy of nineteenth century engraved typefaces; a systematization of decorative typefaces in funerary context (the appearance

[8] A questão da acentuação tornou-se, a partir [de finais do século XVI], e até aos gramáticos oitocentista, completamente pacífica. Instituiu-se o uso de, no caso de cada contexto específico, se poder optar entre a marcação do acento e o silêncio em relação a ele. A única regra a respeitar-se-ia a de evitar ao máximo ambiguidade de sentido.

[9] https://fr.wikipedia.org/wiki/Usage_des_majuscules_en_fran%C3%A7ais#Historique.

of lachrymal, for example); type as a signifier [15]; the establishment of lettering normativity (architectures are scrutinized by the cemetery direction but letters are left to the builders/client view); comparison between noble families without noble titles; study of letters in free-masons *jazigos*; and others.

Although they are carved in stone for eternity, these letters are erased or overimposed in *jazigos* that are transacted and restored by other users. New lettering and architectures substitute what once seemed perpetual; but the design of letters lives on, looking much more interminable and timeless, even without having any king of materiality. They are just shapes and drawings, but when we see a typeface that was carved in Trajan Column[10] in 113 still being used, more than 1900 years later, to sell blockbusters films like "Titanic," we cannot avoid thinking about Hunter S. Thompson's words in "The Proud Highway: Saga of a Desperate Southern Gentleman" (1955–1967): "Wow! What a Ride!".

References

1. Oliver GJ (ed) (2000) The epigraphy of death: Studies in the history and society of Greece and Rome cover (1a). Liverpool University Press, Liverpool
2. Cemitério dos Prazeres (2011) [website História de Portugal] (2011, December 15) Retrieved November 11, 2019, from http://www.historiadeportugal.info/cemiterio-dos-prazeres/
3. Neves F (2004, Setembro 2) Dos monturos aos cemitérios públicos
4. Solon Sales e Silva J (2018) O Cemitério Revisitado. Baraúna, São Paulo
5. Stone P (2006) A dark tourism spectrum: Towards a typology of death and macabre related tourist sites, attractions and exhibitions. Interdisciplinary Int. J. 52(2):145–160
6. Death & Design (2014) Retrieved November 11, 2019, from http://deathanddesign.com/
7. Massimi M, Baecker RM (2011) Dealing with death in design: developing systems for the bereaved. In: CHI '11 Proceedings of the SIGCHI Conference on Human Factors in Computing Systems. 2011, May 7–12, Vancouver, Canada, pp. 1001–1010
8. Pallister J (2019) Reinventing death for the twenty-first century. Retrieved November 11, 2019, from https://www.designcouncil.org.uk/news-opinion/reinventing-death-twenty-first-century-0
9. Strang J (1831) Necropolis Glasguensis with observations on ancient and modern
10. Martin B, Hanington BM (2012) Universal methods of design: 100 ways to research complex problems, develop innovative ideas, and design effective solutions, Digital. Rockport Publishers, Beverly, MA
11. André P (1999) Os cemitérios de Lisboa no século XIX: pensar e construir o novo palco da memória. Lisboa (Master Thesis, Universidade Nova de Lisboa). Obtido de Biblioteca Nacional
12. Castro I et al (1987) A Demanda da Ortografia Portuguesa. Lisboa, Sá Da Costa
13. Ciberdúvidas da Língua Portuguesa (2019, Novembro 23) [Ciberdúvidas] Resposta à sua pergunta

[10] Trajan is a typeface designed in 1989 by Carol Twombly, based on the letterforms (*capitalis monumentalis)* carved at the abasement of the Trajan Column in Rome. Trajan is an all-capitals typeface that was used extensively in Hollywood posters whenever an epic film was produced. It became so popular that, from epic films, it migrated to other genres and become a kind of a "standard" for movie posters. This trend was noted by Graphic designer Yves Peters while analyzing more than 100 posters per month during 2006–2018.

14. André P (2006) Modos de pensar e construir os cemitérios públicos oitocentistas em Lisboa: o caso do Cemitério dos Prazeres, Edições Colibri/IHA—FCSH/UNL, Lisboa
15. Noble I, Bestley R (2005) Visual research: An introduction to research methodologies in graphic design. AVA, Lausanne

Anti-Amnesia: Developing a Collaborative e-learning and Digital Archive Platform Towards Contributing to the Preservation and Revitalization of Handicrafts Industries

Nuno Martins, Heitor Alvelos, Susana Barreto, Abhishek Chatterjeed, Eliana Penedos-Santiago, Mariana Quintela, and Cláudia Lima

Abstract The study presented in this paper consists of the development of a platform for digital archiving, collaborative and e-learning, within the framework of the craft industries. This work is part of Anti-Amnesia, a project which aims to investigate Design as an agent for the regeneration and reinvention, narratives and materials, of disappearing Portuguese cultures and manufacturing techniques. In general, this project aims to: develop methodologies, through digital means and tools, that contribute to the preservation of the heritage and cultural heritage of the craft industries; contribute to the sustainability of the craft heritage, through the revitalization of the tradition of these industries, ensuring a continuous transmission of skills, techniques and craft knowledge to future generations; and promote research and research in this area of Portuguese manufacturing. Specifically, in this work to create the platform, we seek, through digital technology, to make information more accessible and universal, facilitating access to knowledge. Through the platform, and with the support of educational tools and services, we will seek to promote the learning of Portuguese manufacturing techniques, in order to attract young people to recover older practices. Based on this strategy, it is intended to value tradition, preserve it, and transmit it to future generations.

Keywords e-learning and digital archive · Cultural heritage · Craft industries · Education · Art and design

N. Martins (✉)
Research Institute for Design, Media and Culture, School of Design, Polytechnic Institute of Cavado and Ave, Barcelos, Portugal

H. Alvelos · S. Barreto · A. Chatterjeed · E. Penedos-Santiago
ID+/University of Porto, Porto, Portugal

M. Quintela
Polytechnic Institute of Cavado and Ave (IPCA), Barcelos, Portugal

C. Lima
ID+/Universidade Lusófona, Porto, Portugal

E. Duarte and C. Rosa (eds.), *Developments in Design Research and Practice*, Springer Series in Design and Innovation 17,
https://doi.org/10.1007/978-3-030-86596-2_28

1 Introduction

This study is part of the project Anti-Amnesia based at the Research Institute for Design, Media and Culture (ID+). This work consists of the development of a platform (for desktop and mobile formats) for archiving, collaboration and e-learning, dedicated to the Portuguese craft industries.

The platform consists of a collaborative model, with priority on archives and digital bookshops, of documents, testimonies and historical facts. Through this digital space we intend to make a strong contribution to the preservation of the cultural heritage of these craft industries.

Another important objective of the platform is to stimulate the learning of the manual techniques used in these craft industries and to provide online e-learning courses, in video format, so that the knowledge is passed on to all potential stakeholders, namely future generations.

This technological solution aims to recognize the protagonists of these industries and enhance their work by providing the possibility for artisans and designers to exchange ideas, collaborate, make partnerships, and give rise to new creative models of products, thereby consolidating their participation in the market.

This platform, aimed at researchers, designers, crafts communities and other interested entities pertaining to the area in question, aims to translate into a cooperative infrastructure of support, through the creation and display of digital content, in the form of text, graphics, image, audio, and video.

The digitalization and online accessibility of cultural material makes it possible, in addition to the possibility of preservation, to promote wider access to the material through the public domain, for own use and reuse for commercial and non-commercial purposes.

2 Project Anti-Amnesia

The design research project Anti-Amnesia is based on growing evidence that traditional small-scale manufacturing cultures in the Northern and Central regions of Portugal are facing critical viability challenges [1–4] in the wake of new global commercial and industrial realities. These industries typically operate on older manufacturing paradigms that rely comprehensively on competencies that are specifically manual in nature and often exist within human-centric subsystems.

The move to a digital future has not been a straightforward process of evolution for the associated communities in lieu of their primary vocation being established on hand-operated application and skills garnered through innumerable years of dedication to their respective crafts. Therefore, the ensuing practice and identity balance—for example, being known as a weaver or a shoemaker—does not allow for a smooth transition from manual to digital without compromising on factors pertaining to the authenticity of practice; specialized knowledge built upon generations of individual and community involvement; implicit local/regional cultural identities; definitive customs and practices; and critically, means of sustenance in case of absence of conditions to undertake the required re-calibration [5, 6].

Under these circumstances, the project considers four examples of traditional industrial practices: artisanal weaving; azulejos tilework; shoemaking; and letterpress printing.

The project Anti-Amnesia respectively observes the various connotations of the ensuing interrelationship between traditional crafts and digital technology and comprehends that while it may not be its position to counter shifting manufacturing paradigms, it can make meaningful contributions towards the continuation of specialized knowledge through design. Its corresponding strategy thus pays close attention to advancing the public and academic inscription of particular traditional practices, detailing their industrial and cultural legacies and surrounding social narratives; conducting on-ground ethnographic documentation to recover and archive informal, contingent, and tacit streams of knowledge in addition to related instrumentalities [7]; involving practitioners in impact-driven pedagogic activities with students of design that target critical issues concerning traditional manufacturing, such as communication and product innovation; building a collaborative network between the said industries, along with institutional and business partners towards potentiating knowledge and know-how exchange and examining future avenues of application.

In accordance, the project recognizes digital technology not particularly as a hindrance to the sustainment of traditional practices, but as a complementary factor that can help establish conditions for incremental innovation and extend capacities for involved individuals and communities in times of change.

Similarly, the project utilizes digital technology for activities pertaining to audiovisual documentation, archiving, design, and dissemination, thereby indicating an additional manner in which elements of the ongoing technological proliferation can be leveraged on the basis of intent and approach. In doing so, it is keen on understanding and acknowledging the surrounding human and social sensitivities, which it considers as its prerogative towards assuming the role of a moderator within the scheme of things.

The project, thereby, works in close contact with practitioners belonging to the said contexts in order to consensually explore and develop ways to calibrate the eventuating state of affairs into a viable and mutually constructive ecosystem. Its corresponding line of intervention operates on the inference that an approach centered on design and identity can play a decisive role in the dignified recovery and sustainability of the said contexts of production.

In this scenario, digital media, through the development of an online platform, can make an important contribution to the preservation of artisan wisdom and ensure its passage to newer generations of makers. We believe that these digital tools—due to the greater ease of access, interaction, and potential collaborative networking—may be more appealing to younger generations in current circumstances than the traditional means of knowledge transfer with respect to craft industries.

3 Cultural Heritage in Artisanal Industries

The Cultural Heritage embodied in traditional handicrafts is an integral part of any nation and reflects the culture and tradition of a region.

Cultural Heritage is explained in UNESCO documents as "our legacy of the past, what we live today and what will be passed on to future generations"[1] [8].

There is a wealth of traditional Portuguese industries and crafts that face fundamental challenges in their recognition and longevity, largely due to new models of production, changes in consumption, and a public absence of associated narratives.

The handicraft industries in Portugal are diverse, and closely linked to popular culture, particularly in rural areas, however, due to the ongoing exodus of younger generations who are migrating to urban centres of Portugal and the rest of Europe for better livelihood, the traditional industrial sectors are facing a rapid decline.

During the twentieth century, the strong growth of industrialised production—cheaper and more profitable than craftsmanship—has led to the rise of consumerism and the gradual decline of the craft business. This depression has also resulted in a declining interest of young people towards these professional activities and, consequently, the risk of various traditional techniques and skills disappearing arises.

Digital media can be an important response to this problem, through its contribution to the preservation and pro-motion of the traditions, techniques and cultural and artistic skills prevalent in traditional craft industries.

By implementing technological measures that preserve crafts, new activities and employment opportunities are encouraged, especially in local and rural communities, thus contributing to their sustainable development [9, 10].

4 Analysis of Reference Digital Platforms

This chapter looks first at examples of digital archiving platforms that are intended to contribute towards preserving cultural heritage; e-learning platforms; and, examples of digital platforms that are dedicated to revitalising craft industries and acting as a reference for this study.

4.1 Europeana

Europeana is a European Union (EU) archive platform recognised as a digital cultural heritage, containing digitised cultural material from collections of libraries, archives, museums, galleries and audiovisual collections across Europe.

[1] UNESCO. Cultural Heritage. Accessed on August 8, 2019, at: http://www.unesco.org/new/en/santiago/culture/cultural-heritage/

It provides access to approximately 30 million cultural objects, with more than 3000 institutions from across Europe contributing to its archive and collection, notably including the Rijksmuseum in Amsterdam, the British Library, and the Louvre Museum.

Funded by the EU, Europeana is a digital service infrastructure that provides Europe's cultural data, tools and services, widely reusable and freely accessible, enabling the public to get to know and explore Europe's cultural and scientific heritage from prehistoric to contemporary times.

4.2 UNESDOC

UNESDOC, like the previous example, is an online public access digital bookstore used to research and access books, articles, documents and electronic materials produced by UNESCO.

UNESDOC serves as a repository for knowledge of high-quality information on UNESCO's activities in the areas of education, natural sciences, social and human sciences, culture and communication, containing more than 350,000 titles, dating back to the year 1945.

4.3 NEMECH, New Media for Cultural Heritage

NEMECH is a Competence Center established by the Tuscany region and the University of Florence, which through its connection with research centers, organizations and institutions, promotes the transfer of know-how and re-search in laboratories, providing access to information to all interested parties.

This Center develops technological solutions that promote the preservation of cultural heritage. NEMECH develops projects and builds innovative prototypes, using 3D modeling, computer vision, computer graphics, multimedia presentation, natural interactivity and mobile applications for cultural heritage. It also provides courses, workshops and conferences related to digital technologies.

4.4 Lynda

The Lynda website, founded in 1995 by Lynda Weinman, a multimedia teacher, is an online library of books and classes. The success of this website led to the interest and purchase of this platform by LinkedIn in 2015. Currently, Lynda.com has been renamed to LinkedIn Learning and each user is required to create an account on LinkedIn to access its content.

This website provides online courses in various areas of knowledge, working through videos and interaction with teachers (always online). The online courses have a subsiption-based payment scheme after the first month of free use.

4.5 Khan Academy

Khan Academy is a non-profit educational organization established in 2008. The purpose of this academy is to offer an online toolkit to help students learn a variety of school subjects (math, science, history, etc.).

The academy, through its platform, provides short lessons in video format, and also offers complementary practical exercises and materials for educators. The site and its contents are mainly available in English, with certain contents available in different languages.

4.6 Made In

Made In, within the artisanal industries, is a research, design and heritage initiative that encourages collaboration and exchange of knowledge between traditional artisans and contemporary designers.

This project, aims to promote European Artisanal Heritage and contemporary design, through various activities: research and mapping of crafts; public conferences and expert seminars; design and craft residencies; workshops; publications and exhibitions; and the development of an online platform, all of which are intended to facilitate the transmission of knowledge and provide professional development opportunities for artisans, designers and connected professionals.

The aim of the project is to revitalize the tradition of craftsmanship and educate designers about material and immaterial heritage, thus creating an authentic and more sustainable image of contemporary design. It also aims to create a platform that enables the exchange of knowledge, constructive dialogue and, finally, new collaborative practices between artisans and designers.

5 Platform Development

For the development of the platform, the identified methodology is based on Design Thinking and User Center Design.

The Design Thinking model comprises five phases as proposed by the Institute of Design in Stanford [11], namely: empathy, for a better understanding of the problem; the definition, where the data collected in the first phase are analyzed; the idea to develop, in which one can begin to create sketches, according to the needs of the user; the prototype, in which some versions of the product are produced in order to understand how it works; and the test, to verify the usability of the application.

In "The Design of Everyday Things", Donald A. Norman [12], cognitive science researcher and creator of the term User Center Design, describes design based on user needs.

For the development of the platform, the Agile UX concept was adapted, which is a management process that allows the consumer to be involved during the whole process [13].

Through the Agile UX concept, the product is designed to ensure the best usability for the consumer and for this it becomes necessary to define the personas (profiles of users idealized as the target audience), a term created by Alan Cooper [14].

When starting the development of the platform, a workflow or mind map has been required to be determined, which illustrates the entire navigation scheme of the platform and contains a detailed description of all its features.

Therefore, the creation of wireframes (Fig. 1), a visual guide with the structure of the platform, allows to define the information architecture and the interface design in low fidelity through an initial pre-finalized version of the project.

In order to identify flaws or inconsistencies in the design, a prototype will correspondingly be created to examine all the functionality and interactivity of the platform and perform usability tests.

Accordingly, the "Sprint Design"[2] strategy by Jake Knapp [15] is used, which is based on a cycle of five phases (Fig. 2), spread over five working days, to idealize, create sketches, develop and test.

This continuous process stimulates productivity and enables the immediate obtaining of feedback at each stage, until determining a suitable solution, with an efficient flow [16, 17].

The completion of the project scheduled for April 2020, will entail the development of a high-fidelity prototype in accordance with the principles of UX Design [18, 19].

[2] Flowchart of the Sprint Design method. Avaiable in: https://uxplanet.org/whats-a-design-sprint-and-why-is-it-important-f7b826651e09. Accessed on: 03–11-2019.

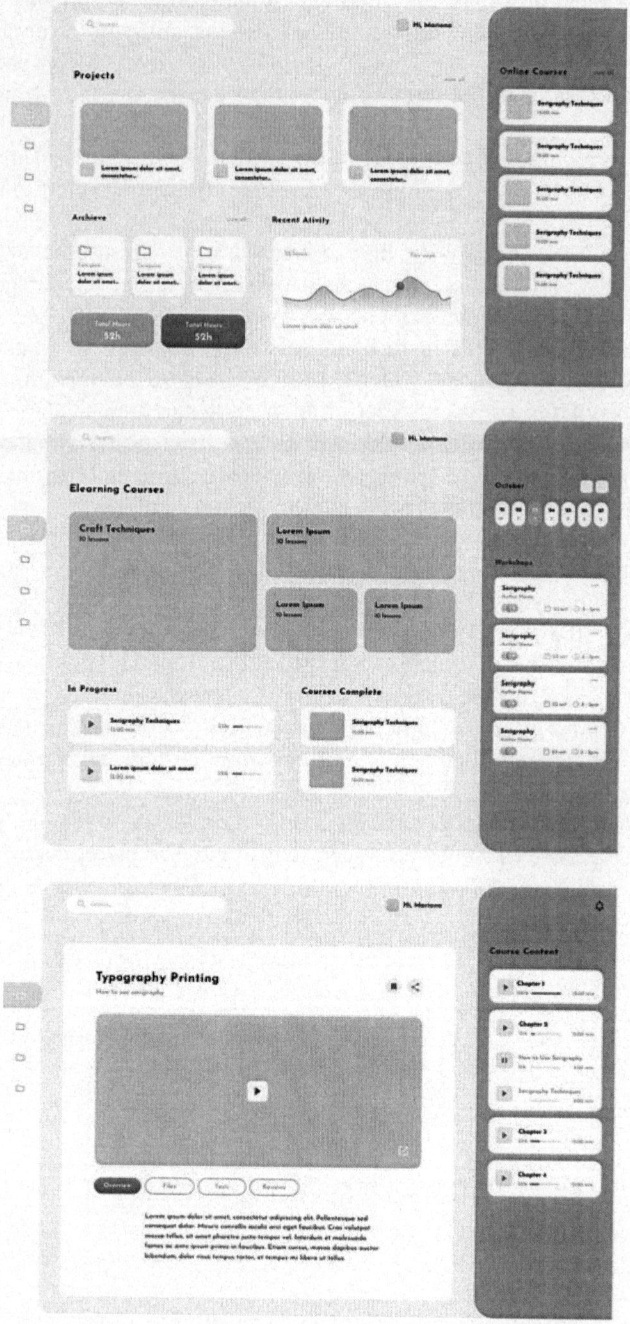

Fig. 1 First wireframes developed. From top to bottom: Home page; courses available; and course detail page

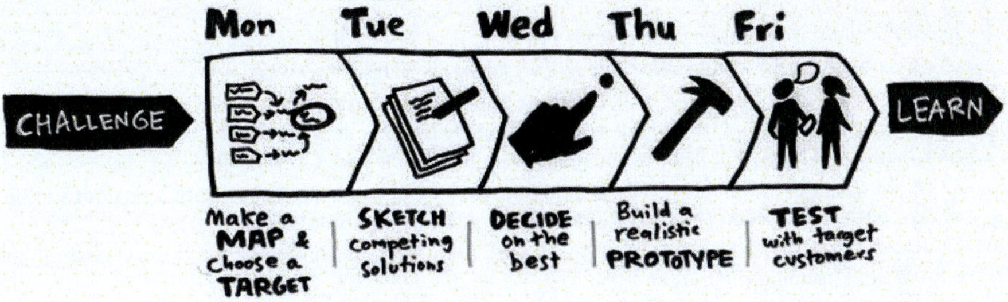

Fig. 2 Flowchart of the sprint design method

Acknowledgements The research is being developed within the framework of the project "Anti-Amnesia: design research as an agent for narrative and material regeneration and reinvention of vanishing Portuguese manufacturing cultures and techniques", co-financed by Portugal 2020, along-side the European Regional Development Fund and the Founda-tion for Science and Technology, Portugal (ID+/Unexpected Media Lab: POCI-01-0145-FEDER-029022; 2018-2020).

References

1. Alvelos et al, Anti-Amnesia. http://endlessend.up.pt/antiamnesia. Last accessed 19 November 2019
2. Gomes A (2016) Design, tradition and craft: The case of Almalaguês. Universidade do Porto, Presentation
3. Martins N, Brandão D, Alvelos H, Silva S (2020) E-Marketplace as a tool for the revitalization of Portuguese Craft Industry: the design process in the development of an online platform. Future Internet 12(11):195. https://doi.org/10.3390/fi12110195
4. Albino C (2017) À procura de práticas sábias—Design e Artesanato na significação dos territórios. Presentation, Universidade de Aveiro
5. Alvelos H, Barreto S, Chatterjee A, Gomes AJ, Almeida PC (2019) Anti-Amnesia: the viability of millenary weaving in a World of acceleration. In: European Conference on Arts and Humanities. IAFOR, Brighton
6. Alvelos H, Almeida PC, Barreto S, Monteiro R (2017) Anti-Amnesia: Design research as an agent for narrative and material regeneration and reinvention of vanishing Portuguese manufacturing cultures and techniques. Ciência 2017. FCT/MCTES, Lisbon, July. http://www.encontrociencia.pt/files/A4_41530_s4_2_Heitor%20Alvelos_%20Pedro%20Carvalho%20de%20Almeida_%20Susana%20Barreto_%20Rui%20Monteiro_Co_@.pdf
7. Fleck J, Tierney M (1992) The management of expertise. University of Edinburgh, Programme on Information and Communication Technologies, Edinburgh
8. UNESCO, Concept of digital Heritage. Retrieved from: https://en.unesco.org/themes/information-preservation/digital-heritage/concept-digital-heritage
9. Interreg Europe (2018). Digital solutions in the field of cultural heritage. Retrieved from: https://www.interregeurope.eu/fileadmin/user_upload/plp_uploads/policy_briefs/2018-08-06_Policy_brief__on_digital_technologies.pdf. Last accessed 3 October 2019
10. Yang Y, Shafi M, Song X, Yang R (2018) Preservation of cultural heritage embodied in traditional crafts in the developing countries. A case study of Pakistani handicraft industry. Retrieved from: https://www.mdpi.com/2071-1050/10/5/1336

11. Dam R, Siang T, Personas, A simple introduction, interaction design foundation. Retrieved from: https://www.interaction-design.org/literature/article/personas-why-and-how-you-should-use-them#targetText=Personas%20are%20fictional%20characters%2C%20which,%2C%20experiences%2C%20behaviours%20and%20goals, last accessed 2019/09/17.
12. Norman DA (1988) The design of everyday things. MIT Press, Cambridge
13. Korkishko I, Agile UX vs. lean UX, medium. Retrieved from: https://medium.com/@Iren.Korkishko/agile-ux-vs-lean-ux-e901ea9d0cbb. Last accessed 10 September 2019
14. Cooper A, Reimann R, Cronin D (2007) About face 3: The essentials of interaction design, 3rd edn. Wiley, Indianapolis
15. Knapp J, Braden K, Zeratsky J (2016) Sprint: How To solve big problems and test new ideas in just five days. Transworld Publishers, London
16. Benyon D, User centered design, The encyclopedia of human-computer interaction, 2nd edn. Retrieved from: https://www.interaction-design.org/literature/topics/user-centered-design. Last accessed 3 October 2019
17. LIU D (2017). Agile UX vs lean UX: don't force yourself to choose, designer, medium. Retrieved from: https://uxplanet.org/agile-ux-vs-lean-ux-dont-force-yourself-to-choose-designer-61f8d60f4f7a. Last accessed 21 October 2019
18. Martins N, Alvelos H, Chatterjee A, Calado I, Quintela M (2020) Multimedia as mediator of knowledge between older generations and present-day students of art and design. In : The 4th international conference on education and multimedia technology (ICEMT 2020), 19–22 July 2020, Kyoto, Japan. ACM, New York, NY, USA. https://doi.org/10.1145/3416797.3416827
19. Martins N, Brandão D, Pereira L, Raposo D, Neves J, Silva J (2021) Craft Academy: design of an e-learning solution for the handicrafts' industries preservation. In: Proceedings of 16th Iberian conference on information systems and technologies (CISTI), Chaves, Portugal, 23–26 June 2021. IEEE. https://doi.org/10.23919/CISTI52073.2021.9476456

Design for Society

Design for Society

Sara Gancho, Isabel Farinha, Robin Teigland, Paula Trigueiros, Teresa Franqueira, and Nuno Sá Leal

Abstract This chapter presents the "Design for Society" concept and provides a backdrop for the papers included as chapters in this section of the book. We suggest that the Design for Society concept can be viewed as an umbrella for a range of concepts related to how we can leverage the principles of design research and practice to create holistic solutions to societal challenges. Design can drive societal innovation, and the collection of papers we have compiled below focus on how design can enable society to become more equitable and inclusive. Our hope is that this chapter and the subsequent chapters might inspire design practitioners and researchers to learn from and build on each other's ideas - as opposed to reinventing the wheel or even making the same mistake, so that we may move forward in tackling our societal challenges and achieving the UN SDGs.

S. Gancho (✉) · I. Farinha · N. S. Leal
IADE - Faculdade de Design, Tecnologia e Comunicação, Universidade Europeia, UNIDCOM/IADE - Unidade de Investigação em Design e Comunicação, Av. D. Carlos I, 4, 1200-649 Lisboa, Portugal
e-mail: sara.gancho@universidadeeuropeia.pt

I. Farinha
e-mail: isabel.farinha@universidadeeuropeia.pt

R. Teigland
Division of Entrepreneurship & Strategy, Department of Technology Management and Economics, Chalmers University of Technology, SE-412 96, Gothenburg, Sweden
e-mail: robin.teigland@chalmers.se

P. Trigueiros
Lab2PT—Laboratório de Paisagem, Património e Território e Escola de Arquitetura, Arte E Design da Universidade Do Minho, Braga, Portugal
e-mail: paula.trigueiros@arquitetura.uminho.pt

T. Franqueira
Universidade de Aveiro, Coordenadora Internacional DESIS Network Unidade de Investigação: ID+, Aveiro, Portugal
e-mail: teresa.franqueira@ua.pt

E. Duarte and C. Rosa (eds.), *Developments in Design Research and Practice*, Springer Series in Design and Innovation 17,
https://doi.org/10.1007/978-3-030-86596-2_29

Keywords Design for society · Environment · Sustainability · Design for change · Design for all · Societal challenges

Since the turn of the century, new societal challenges have continued to arise, making it all the more clear that social, economic, and environmental issues cannot be dealt with separately, as was described in the Brundtland Report published by the UN World Commission on Environment and Development (also known as Our Common Future) [1]. In this section, we present a series of papers that highlight the emerging role that Design for Society has in enabling a holistic approach to tackling these challenges.

Today we live in a "Designed World", a human-made artificial world in which design has been involved in every aspect of creating our living environments [2]. However, design as a creative act is intertwined with the act of destruction as everything created requires something else to be changed [2]. Some argue that the societal challenges we are facing today can be seen as a result of "user-centred" design in which the focus was merely on fulfilling short-term human-centred desires without taking into consideration the long-term destructive effects on the environment and society [2]. Nigel Whiteley in his book, *Design for Society* (1993), pointed out the problems of consumer-oriented design in which the design process generally did not go beyond fulfilling the immediate client's requirements, thereby ignoring the greater environmental impact, which in the long run led to environmental unsustainability. Looking forward, the question arises as to how we instead can leverage the principles of design research and practice to create solutions to not only help "fix" the problem but also to regenerate the planet and society through "destroying" harmful behaviours and practices.

Already today there are signals that we are moving in this direction where the term "Design for Society" takes on a broader meaning and could serve as an umbrella for a range of concepts. For example, design for inclusion strives to create equitable products and services by embracing the opportunities and challenges presented by a diverse population. Human-centered thinking plays a central role in developing and delivering solutions that are not only accessible but that also provide enriching and engaging experiences for as many people as possible, regardless of age, gender and (dis)ability. Further, concepts such as design for social innovation and design for sustainability walk hand in hand. Design for sustainability tends to have a more technical approach to the intervention and construction of the artificial world while design for social innovation tends to have a more "liquid" and enabling intervention in the way people organize and are called to act on behalf of a common well-being. Several strengths of design for social innovation projects, such as understanding user experiences, rapid prototyping, visualisation and systematic approach, have been illustrated by Mulgan [3] [4]. Within design for sustainability, these same tools can be used to design regenerate and repair actions to occur in tandem. Today numerous designers are working to diminish the ecological footprint of products and services throughout their lifecycle by designing for recycle, reuse and repair instead of producing new "future" waste while also striving to reduce consumption excesses through designing new circular and sharing economy models. Other areas include supporting and designing suitable processes to assist self-organized citizens in their own efforts to save and regenerate our common home.

Yet it is important to remember that design has several levels. For example, the Danish Design Ladder or the Design Management staircase [5, 6] describes design as having three levels: (1) at an operational level, design is for achieving a specific task, (2) at the functional level, design is applied as a process or discipline; and (3) at a strategic level, design is holistically used for strategic intent [4]. We view Design for Society as incorporating all three levels with a particular focus on the strategic level in which a systems approach should be used to analyze potential interactions among actors, and even with existing and emerging technologies, and their effects on the environment and society. This analysis should include not only the "here-and-now", but it should also be intertwined with scenario thinking, in which an analysis of how potential possible futures might play out is conducted to better understand the ramifications of one's design efforts today.

The above is a challenging task, and as has been noted by several authors, design, and in particular design for social innovation, is wrought with complications, such as the use of expensive methods and tools, lack of awareness of conditions or limitations beyond the project, and lack of resources, especially by entrepreneurs designing new solutions [4]. Another considerable challenge is the difficulty to learn from others' experiences working with design for society projects. It is with this challenge in mind that we have assembled this series of papers. Our hope is that these papers might inspire design practitioners and researchers to learn from and build on each other's ideas—as opposed to reinventing the wheel or even making the same mistake, thereby wasting valuable resources. In this manner, we hope to move forward in fulfilling the UN's Agenda 2030 goals. Thus, the collection of papers we have assembled here as chapters includes a number of topics related to design research for societal issues, the environment, and sustainability, and they present us with challenges, ideas and some solutions. Design can drive societal innovation, and this collection of papers demonstrates just that. Whether addressing globalization issues, implementing social change, finding alternative means of transportation, bringing innovation to traditional organizations, making use of global sustainable practices and circular business models, allowing access to play for all or facing the challenges of smart mobility, design is the key. These papers focus on redesigning society and making it more equitable and inclusive, and they give us a new perspective on what design can bring to society in terms of new ways of living.

Appendix: Chapters

Addressing Glocalization Challenges Through Design-Driven Innovation Approaches

The *Addressing Glocalization Challenges through Design-driven Innovation Approaches* suggests a framework for design-driven glocalization, making use of seven case studies from all over the world, cross comparing them to other design-driven approaches. Globalization has brought the opportunity for companies to be present in international markets. However, Globalization strategies, adaptation strategies and even Localization strategies although dynamic, have created continuous tensions and are of considerable consequence for corporate strategy decision making and processes.

The option for a Global strategy or Glocalization has its challenge as well but it is of relevance to the UN 2030 sustainable development goals. Design driven glocalization can lead to the development of sustainable cities and communities, it can preserve local cultural heritage, maintain cultural atmosphere of the local economies and even reduce environmental impact. It can also build work opportunities and economic growth.

It can also result in the support of local craftsman from rural areas, it can help build resilient infrastructure and innovation. Allowing for the investment in new technologies that focus on sustainability, responsible consumption and production. It can also allow for the rethinking of the fast street fashion system that is highly based on a product's fast lifecycle and planned obsolescence.

Inclusive Design as Promoter of Social Transformations: Understanding Androgyny in Contemporary Society

This chapter contributes to deepen the knowledge about the diversity of which contemporary society is constituted. This is a challenge for the design to constantly renew itself and meet people's expectations. Minorities and people who stray from some common 'norm' or stereotype are often the target of prejudice and discrimination—problems that Design can help to alleviate. In this article some questions about gender identity and proposals about genderless / androgyny are addressed.

This paper addresses the role of Design in the fight against stereotypes and prejudices related to gender issues and the way it influences design knowledge and consumer choices. This can directly contribute to SDG No. 5 'Achieve gender equality and empower all women and girls' and, in general, to SDG No. 16—promote peaceful and inclusive societies for sustainable development, provide access to justice for all and build effective, accountable and inclusive institutions at all levels'.

Dynamic Ride Sharing As An Alternative Transportation Mode For Commuting Among Metu Campus And Eryaman

The paper presents us with a study of TAG app (let's ride in a single car) and is very interesting for Design for society. The work clearly explained, and the methods are very well described. The paper will allow other research to structure a similar research because the process is all documented and detailed. The authors were able to present the findings in a very consistently way.

Dynamic ride sharing is an alternative way of commuting to individual car use with the utilization of improving available seat capacity for new riders through which more sustainable trips are ensured among drivers and riders who have similar itineraries in closer time slots. The main objective of the study was to acquire in depth knowledge about the phases of dynamic ride sharing practices, experiences of individuals commuting via ride sharing, their concerns and acquisitions from ride sharing, and to determine potential solutions for improving the overall ride sharing practice which would make it an effective and preferable way of alternative transportation in a sustainable manner. The study revealed user's underlying motivations towards ride sharing, their concerns, needs and desires which would help generate better approaches to performing ride sharing practice, in terms of both optimized and secured ways for better experiences in the light of fresh insights gathered in specific themes.

This chapter highlights a lot of challenges for society. Mode choice decisions for travelling depend on many different issues such as convenience, comfort, safety, time considerations, accessibility and cost, yet these decisions tend to be affected by social, economic and environmental impacts of commuters' acts and behaviours. It seems that there has been substantial interest in promoting sustainable transportation alternatives due to climate change, aging infrastructure, and raising green culture. Population-based changes in individuals' behaviours, attitudes and knowledge are important to accomplish wide-spread adoption of alternative transportation modes. Moreover, for the improvement in the quality of life of campus residents, some measurements should be implemented through sustainable transportation which will return in many social, economic and environmental benefits. In this respect, dynamic ride sharing could be one of the most prominent solutions for METU Campus where hitchhiking culture is truly favoured among all its members.

There are several economic, environmental and social benefits of dynamic ride sharing. In general terms and considering identifiable objectives for the sustainable development for UN, as current transportation models are not both environmentally and socially sustainable, road transportation is one of the most challenging issues of the new world. In one respect, car pollution directly causes global climate change as being one of the major sources of greenhouse emissions. In another respect, traffic congestions have a great influence on the reduction of people's quality of life worldwide. In this respect, many researchers from diverse disciplines such as transportation, economics, and behavioural, social and environmental psychology have described dynamic ride sharing as an effective solution to the inefficiency of current transportation models. Moreover, circumstances such as finite oil reserves, rising gas prices, traffic congestions and related environmental concerns have given a direction to people to use their personal automobiles more wisely with an increased interest in services such as dynamic ride sharing. Further-more, ride sharing allows participants having automobiles to share travel expenses while offering those without cars an enhanced mobility. In overall ride sharing, in other terms the joint travel of two or more persons in a single car has initiated a common way of sharing also the costs and benefits together of a shared private car.

Castelo Branco Embroidery—Tradition and Innovation

The chapter *Castelo Branco Embroidery—Tradition and Innovation* touches on both sustainability of materials and of tradition. It deals with the cultural heritage of Castelo Branco, a region of the Portuguese country as well as it touches on the area of sustainability.

This study presents a process of developing garments with the application of Castelo Branco Embroidery. It describes three different phases: bibliographical revision Embroidery of Castelo Branco, the brief that was present within a Design Challenge to fashion designers, and the products resultant from the proposed brief. The authors describe deeply the materials and types of stiches, colours, motifs and symbology of the Castelo Branco Embroidery and the reflections on the clothes are very clear.

Each product tells a story. To know this technology and resort to the acquisition of the resulting products presenting the Castelo Branco embroidery and history highlights challenges for society, as they try at a social level, to involve people who are not active in society and they are given this important role in making embroidery. Still in terms of social sustainability, it foresees an increase in the supply of jobs. It is important as it deals with the cultural heritage of a region of the country, at the same time it touches on the area of sustainability.

Playponics In India—Local Hydroponics Playground Gardens Utilising Kinaesthetic Learning To Promote Global Sustainable Practices.

The paper, *Playponics in India—Local Hydroponics Playground Gardens Utilising Kinaesthetic Learning to Promote Global Sustainable Practices*, is in this chapter because it presents questions of identifiable objectives for the sustainable development for UM. The goals identifiable objectives for the sustainable development for UM are number two, End hunger, achieve food security and improved nutrition and promote sustainable agriculture number four, ensure inclusive and equitable quality education and promote lifelong learning opportunities for all, and.

This is an important paper because this research and development project set out to investigate possible means by which future populations might have or build closer links with environmental Is-sues. The work has integrated the diverse fields of sustainability and STEM education, global products and local production. It aims to bring these aspects together in holistically considered learning environments where these elements can co-exist in engaging ways.

The paper concluded that multi-modal research is essential when we seeking to design learning systems that aim to have holistic benefits. The study has demonstrated degrees of 'validity' to the original proposition and strongly indicates high levels of in principle acceptance of the developing concept. Where the aim is to instil these understandings in children so that they may carry sustainability knowledge with them into adulthood, and where we propose integrating learning modes, mechanisms and topics as a means of achieving that, the breadth of our research needs to be as equally all-inclusive. Designers of learning systems must undertake research into each, often (on the face of it) diverse topic areas as holistically as possible, and it is this breadth of research that informs designs appropriateness for progression.

Preventing Single Use of Plastic Packaging. Design Strategies for Circular Business Models: Refill, Reuse and Recycle

Due to the changes that the circular economy approach is demanding, product design, rather than being destined for disposal, must integrate resources that should maintain their utility and value and flow back into the cycle. *Preventing single use of plastic packaging. Design strategies for circular business models: refill, reuse and recycle* shows the importance of integrating circular economy principles in an early stage of the design process. Because once the product specifications are being made, only minor changes are usually possible. Ultimately, from general design parameters, innovation in products must emerge taking into consideration human aspirations and worries—answering to a more participative, more environmental-conscient and more demanding end-user.

Design, as a discipline-oriented to study product usage and services experience become key to support business to close the loops on their business models—engaging more participative users/consumers. The acceptance that the activities of man and society have a negative impact on the planet's resources has been accepted for the first time by the user, the consumers. Nevertheless, there are huge challenges for society, citizens, business and policymakers, such as, the creation of disruptive innovation on business models for global economies, both at a manufacturing and consuming level.

Even though the content of this chapter aims to impact the sustainable development goal 8, 9 and 11, its main theme is on the 12—Responsible Consumption and

Production. Aligned not only with the Climate Action—United Nations Sustainable Development but also with the Paris Agreement. Designers, innovators, and decision-makers in businesses, aim to become active actors on the necessary shift from an industry relying on fossil resources. The study on preventing how to prevent the single use of plastic is motivated by the assumption that the anti-plastic sentiment is distracting us from the net environmental benefits of plastic and identify how design can be a source of innovation for using plastics as a highly circular resource. Accordingly, to this study, the selling drivers should prioritize services that profits from the flow of resources over time, continually reuse products and materials and using renewable energy.

Changing The Game: Social Engagement and Cultural Adaptation of Young Refugees Through Playful Design

The paper *Changing the Game: Social Engagement and Cultural Adaptation of Young Refugees through Playful Design* is aimed at understanding how design can have a positive influence on the cultural adaptation of migrant refugee children who had to flee their countries because of war, climate change and/or political or religious persecution.

Promote peaceful and inclusive societies for sustainable development, provide access to justice for all and build effective, accountable and inclusive institutions at all levels, this paper addresses one of the main themes of present days, the problem of forced migration, commonly referred to as seeking refuge. The target user of this study is exiled children who, in most cases, arrive at their destination alone and with few or no basic survival conditions.

The target user of this study is the exiled children who, in most cases, arrive at their destination alone and with few or no basic conditions of survival. In addition, it presents to worrying situations, as more than 69 million people worldwide have had to leave their countries of origin due to wars, climate change and/or political or religious persecution.

It is a practical example of how Design can make a difference for Society, in these case for migrant refugee children's lives, by trying to include them and educate them. These children are especially vulnerable as their basic conditions are far from optimal.

It aimed to develop an innovative solution through gaming and playing, to learning the customs and culture of the new country, preventing the users to be in dangerous situations to which they are vulnerable, also, to provide a different way of learning the new no native language. The main goal of the game is to materialize the experience of a good welcoming moment, through skills and knowledge construction.

Regarding the UN 2030 sustainable development goals this solution can promote the well-being of children, provide quality education and reduce inequalities. It is also based on a circular economy model which promotes climate change, responsible consumption and production and sustainable growth.

Using Focus Groups to Design for Sustainable Behaviour: User-Oriented Challenges of Smart Mobility

The relevance of this article is related to the fact that "Urban mobility gained importance in the context of smart cities, considered one of its main pillars, and one aspect with a stronger impact on its sustainability." This impact depends to a large extent on the change in citizens' behaviour regarding their daily choices and decisions.

Mobility is an important requirement for social development and a core part of everyday life. Future smart mobility systems offer so much convenience and solutions for user's frustrations. However, "sharing" options which are highly important for sustainability, are not appealing to users regardless of possibilities. Plus, improving the quality of transportation detract users from walking, which is not only non-motorized, but also vital for human health. Therefore, some good intentions may also turn into undesirable. Major challenge is understanding the user's mind and changing them. If we run more design for behaviour studies like this and may explore user problems effectively, we can introduce more innovative and beneficial solutions for society.

This is the theme and contribution of the work presented in this article—which proposes to proceed with: a"step intended to identify the main obstacles that can prevent users from adopting more sustainable behaviours in what regards their urban mobility."

By contributing to change behaviours concerning the use of public/ sustainable transport options, this work can illustrate the role of Design for ODS 11—Make cities and human settlements including, safe, resilient and sustainable, namely considering the role of transport in sustainable development.

In truth, "Sustainable cities and communities" is one of the global objectives of sustainable development. By 2030, intention is to provide access to safe, affordable, accessible and sustainable transport systems. Plus, reducing the negative impact of cities, including by paying attention to air quality. This study is exploring challenges that come from user demands and frustrations related to current and future mobility modes including sustainable and unsustainable ones. Benefiting from the results, future work is motivating users to prefer and be satisfied with non-motorized mobility. It also provides data for other researchers to find possible solutions for wants and needs. In this sense, it serves the objectives for the sustainable development for UN, by revealing challenges of sustainable and smart mobility for possible solutions.

Conclusions

Design, as a project method that is transversal to all areas of knowledge and directed to the human being, leads to diversity being permanently present in everything that surrounds the designer and that is part of his projects.

On the other hand, sustainability must be appreciated according to the society for which it is designed. This means that a project completed in a defined society, according to accepted principles of sustainability, may not behave as well when moved to a different culture, that is, a different society.

As such, there is a strong need to respect basic principles that govern the environment, the economy, society and human diversity as to be able to concretize their projects, upon the request that diversity requires.

All these problems can be solved through a creative, sustainable and interdisciplinary approach. There is an increasingly call for change, coming from millions of individuals across the world.

A need for a paradigm shift towards a new economy his now requested globally. The pandemic we are living through, at this moment, puts the consumer society before a new way of looking at consumption, pointing it towards an announced end.

Design is able to question, stimulate and attempt to change the collective consciousness, in order to help us go into a new direction, a better future perhaps?

Design for society offers a systemic and holistic design ethos which integrates the frameworks and approaches of the leading design thinkers we have gathered here in these chapters.

To be able to redesign society in a way that is more intentional, equitable and inclusive, the main goal being for it to be more successful than traditional methods.

A new social and sustainable way of living, producing and consuming in a sustainable way, using concepts like circular economy and upcycling, projecting for diversity without no fear of changes of paradigms.

This is what we can read in these articles that approach not only different societies, different ways of living and different forms of seeing the results design can give to the human being, working in the direction of inclusivity and respect for the world that is the home of all life.

References

1. Oh S (2017) From an Ecodesign guide to a sustainable design guide: complementing social aspects of sustainable product design guidelines. Archives Design Res 30(2):47–64
2. Jiang Y, Jachna TJ, Dong H (2020) Game for complete care: A means of connecting 'user-centered design'with sustainability. Sustainability 12(24):10555
3. Mulgan G (2014) Design in public and social innovation, what works and what could work better. Retrieved from http://www.nesta.org.uk/sites/default/files/design_in_public_and_social_innovation.pdf
4. Pérez D, Hands D, McKeever E (2017) Design for Society: Analysis of the adoption of Design practices by early-stage social entrepreneurs. Des J 20(sup1):S3020–S3034

5. Best K, Kootstra G, Murphy D (2010) Design management and business in Europe: A closer look. Design Management Rev 21(2):26–35
6. Danish National Agency for Enterprise and Housing (2003) The economic effects of design. København. Retrieved from https://erhvervsstyrelsen.dk/economic-effects-design-2004.
7. Blokdyk G (2018) Human-centered design A Complete Guide. Starcooks, London
8. Boylston S (2019) Designing with society: a capabilities approach to design, systems thinking and social innovation. Routledge
9. Human Centered Design (2011) Second International Conference, HCD 2011, Held as Part of HCI International 2011, Orlando, FL, USA, July 9–14, Proceedings
10. Neo M (2021) Design sustentável. Novas Edições Acadêmicas
11. Whiteley N (1993) Design for society. Reaktion Books, London, UK

Castelo Branco Embroidery—Tradition and Innovation

Alexandra Cruchinho, Ana Sofia Marcelo, Fernando Raposo, and Paula Peres

Abstract Castelo Branco Embroidery and Innovation are the two major focuses of this research—by identifying the focus the following questions are asked—Is it possible to develop garments with the application of Castelo Branco Embroidery? Is it possible to apply Castelo Branco Embroidery on accessories and footwear? It is around these questions that the objectives began to be outlined. The Castelo Branco Embroidery is an identifying element of the culture of a region in the interior of the country and presents itself as a reference of the Albicastrense cultural heritage and was traditionally used mostly in bedspreads and panels. The richness of this heritage element lies essentially in the raw material used—silk and linen—but also in the vast diversity of elements and motifs that together construct a narrative that may vary according to the composition chosen. This richness is also represented in the vast palette of colours and the diversity of stitches that are used. The methodology consisted, in a first phase, in the research and bibliographical revision about the theme—Embroidery of Castelo Branco. This was followed by the preparation of a briefing to present the challenge to the Fashion Designers invited to participate in the project. At the same time, a national competition was held which resulted in new proposals or ideas that came to fruition. The products resulting from the initiative should reflect the use of a cultural element and the tradition of a region in a contemporary approach where the work of fashion designers is presented as a differentiating element, important for adding value and sustainability. Embroidery as a traditional element is associated with innovation through the creation of new proposals.

Keywords Sustainability · Cultural heritage · Embroidery · Tradition · Innovation

---

A. Cruchinho (✉) · A. S. Marcelo · F. Raposo
CICANT, CIPEC, Universidade Lusófona De Humanidades e Tecnologias, Lisboa, Portugal
e-mail: p2668@ulusofona.pt

P. Peres
CIPEC, Escola Superior de Artes Aplicadas do Instituto Politécnico de Castelo Branco, Castelo Branco, Portugal

E. Duarte and C. Rosa (eds.), *Developments in Design Research and Practice*,
Springer Series in Design and Innovation 17,
https://doi.org/10.1007/978-3-030-86596-2_30

1 Introduction

The Castelo Branco Embroidery is a hand-made embroidery that has as its privileged products bedspreads or bed covers embroidered with twisted silk on linenabric, which, in Portuguese territory, are highlighted as "the richest, most beautiful, most complex work" [11, p. 15], " the masterpiece of the art of embroidery (...), the great manifestation of women's ingenuity and sensitivity" [8, p. 16).

This embroidery owes its name to the city of Castelo Branco because it was in this locality that its execution resumed in the early twentieth century [9, p. 3; 10, p. 9].

At the origin, these sumptuous textile products of high patrimonial and cultural value can be considered hybrid products, the result of the intermingling of both oriental (China, India, Persia, ...) and European influences.

The decorative/aesthetic grammar and technique employed—above all the Castelo Branco stitch—are, according to Pinto [10], the distinctive features of this embroidery. Other studies add to these characteristics the presence (intentional and visible) of a design or debux on the fabric [1, p. 245] and the use of linen and silk as raw materials [5, p. 11; Annex to Order (extract) No. 15559/2016, of December 27].

The lack of information and written documentation refers the dating of these embroideries to the motifs used [e.g. 7, 9, 10, 12]. Thus, the most recent studies lead to the assumption that the production and marketing of products recognized as Castelo Branco Embroidery date back to at least the eighteenth century [e.g. 7, 10, 12]. Some scholars, such as Chaves [3], Arruda [1] have raised the hypothesis of the seventeenth century being the beginning of this activity.

In the twenty-first century, Castelo Branco Embroidery was consolidated as a "living cultural heritage, breaking with the atavisms and practical blockades that came from the Estado Novo period" [7, p. 158]. The publication in 2007 of the booklet Embroidery of Castelo Branco—Specification Booklet, by the Institute of Museums and Conservation/Francisco Tavares Proença Júnior Museum, marked the district of Castelo Branco as the production area of this embroidery.

In terms of certification of this embroidery, the Attachment to the Order (extract) no. 15559/2016, of December 27, contains three major categories, defined from the examples in the Francisco Tavares Proença Júnior Museum: classic embroidery, those that faithfully reproduce the embroidery produced in the seventeenth, eighteenth and nineteenth centuries; embroidery of classic recreation, those that are inspired by the old embroidery and introduce some alterations to them; embroidery of contemporary creation, those that pave the way for innovation, whether deeper in terms of design, structure and colour (when proposed by artists or art and design professionals) different and with diversified functions [5, p. 97].

2 The Castelo Branco Embroidery and Innovation

Creativity is now embraced within national and regional policies and is a central concern of policy makers as a factor in economic and social development. It should be noted that already in the "Lisbon Agenda" in March 2000, Europe is committed to becoming the most competitive and dynamic knowledge economy in the world.

As noted in the Macroeconomic Study Developing a Creative Industries Cluster in the Northern Region, "Creativity is now recognized as a key economic and social driving force in generating wealth and employment and sustainable development, incorporating technological change and promoting innovation. competitiveness and enhancing the competitiveness of cities, regions and countries" [12, p. 14].

Conscious of the importance that endogenous resources and their specificities represent for the affirmation of the identity of the territories, making them more competitive, ADRACES—Association for the Development of the Southern Centre Border, Francisco Tavares Proença Júnior Museum, the Polytechnic Institute of Castelo Branco and the Castelo Branco City Council, carried out the "Ex-Libris" project—Reconverting/Adapting/Certifying the Castelo Branco Embroidery, as part of the EQUAL Community Initiative, funded by the European Social Fund.

The project, which took place during 2005 and 2008, had as its primary objective "to ensure the preservation of the Castelo Branco Embroidery in relation to its genuineness, authenticity, aesthetic and technical quality" [13, p. 6).

This process began about 6 years ago, with the creation of the Castelo Branco Embroidery Workshop-School, an entity under the supervision of the City Council, with the aim of guaranteeing the continuity and quality of this Embroidery production.

At the same time, in June 2017, the City Council inaugurated the Castelo Branco Embroidery Interpretation Centre, which became the nerve centre for the production, marketing, promotion and study of this unique form of artistic expression in the country and one of ex-libris of the City and County, identity card of the territory.

In August 2017, Castelo Branco Embroidery was noted with a positive opinion by the Consultative Commission for the Certification of Traditional Handicraft Productions on the application for registration of the traditional production "Embroidery of Castelo Branco" in the National Register of Certified Traditional Handicraft Productions.

It is in this context that, in 2017, the City Council proposed to hold the 1st Contest in the area of fashion design with the application of Castelo Branco Embroidery on clothing, footwear and accessories.

If Castelo Branco Embroidery has always been associated with bed covers and gradually has also found its place in panels, tray cloths, integration in decorative elements and others, in this project, we tried to find in the author's clothing a integration of the Castelo Branco Embroidery.

The Innovation Project at the Castelo Branco Embroidery started in 2016 with a challenge launched by two National Fashion Designers, Luís Buchinho and Alexandra Moura. In 2017 the Storytailors and Katty Xiomara were invited and in 2018 Pedro Pedro and Hugo Costa. The project already has three collaborative editions of Designers and two of the National Competition.

Designers should know about Embroidery and its applications and develop their proposal for a piece made with Embroidery to be executed by professional Embroiderers and then presented to an audience on a runway show.

Each proposal results from the sensitivity of each Designer and represents their interpretation of the various elements that characterize this ancient technique.

In this process of innovation in the Castelo Branco Embroidery, on the one hand, the Embroidery gains the own interpretation of several national designers who present their proposals for innovation and reinterpretation of embroidery. On the other hand, young designers competing in the national competition in both categories (clothing and footwear and accessories) also made their approaches and interpretations to embroidery.

The result of the creations with the application of embroidery is presented at a Fashion event (Castelo Branco Moda) created with the purpose of enhancing and promoting this traditional element of the Albicastrense region.

The Designers needed to have contact with those who consider themselves the main defining and defining elements of the Castelo Branco Embroidery. Thus, raw materials, stitches, colours, motifs and symbology are the target of their analysis, study and reinterpretation in contemporary pieces.

2.1 Raw Material

Flax and silk are the two textile fibres used, almost exclusively since its origin, in Castelo Branco Embroidery. The privileged use of these two natural raw materials—linen, in taffeta fabric, and silk, in twisted thread—is one of the aspects that contributes to the enhancement of this embroidery. In the case of historical embroidery, some examples of Castelo Branco bedspreads use only one of these fibres either in fabric or in embroidery [10, p. 13; 5, p. 15].

The fabric that is currently used in the Castelo Branco Embroidery, "for the sake of economic survival of the embroidery itself, results from the manual weaving of woven cotton web" [5, p. 16].

Linen fabric was, and still is, usually made up of several cloths to the desired width [10, p. 13) and has shades ranging from darker shades (brown/raw) to lighter shades (bleached).), and may also be tinted blue or brown [10, p. 13; BMI/Francisco Tavares Proença Júnior Museum, 2007, p. 15). Thus, in the historical bedspreads there is a predominate use of finer linen cloths and darker shades (raw/dark) than those used today [5, pp. 15–16].

Silk, produced in China about 5000 years ago, and extracted from the cocoon of the insect Bombyx mori was one of the most precious goods exchanged between East and West, and it was not until the fifth and sixth centuries that it began to be produce abroad; In Portugal the oldest records of its production date back to 1253 [5, p. 18].

In Castelo Branco, APPACDM (Portuguese Association of Parents and Friends of the Mentally Disabled Citizen) promoted the recovery of silk production; its annual production already amounts to about 15 kg of both extra quality silk and wild silk [5, p. 18].

The distorted silk thread used in Castelo Branco Embroidery is in either natural or dyed tone. In Portugal at least in the eighteenth century there were treatises on the art of dyeing fibres that included specific recipes for silk [10, p. 15). Some of the products made from the second half of the twentieth century resort to viscose.

2.2 Stiches

About 50 stitches are currently used by the Embroiderers of the Regional Embroidery Workshop-School, which are present in the collection of quilts at the Francisco Tavares Proença Júnior Museum. The Castelo Branco stitch, which stands out for its widespread use in the Castelo Branco Embroidery products, has been referred to in certain historical periods as broad stitch, loose stitch, Oriental stitch, Hungarian stitch, Bologna stitch [15, p. 170]; It has as its peculiarities: to allow to save raw material by the fact that only the right of the motif is filled with the silk thread; enable technical variations (Fig. 1), namely the variant in the orientation of the silk thread (made vertically, horizontally or following the shape of the pattern/motif), variant in the orientation of the locks perpendicular to the cast, following the orientation of the pattern/motif), zigzag), colour variant (use of two contrasting colours in equal stripes) [5, p. 29].

In the oldest historical bedspreads, only three to five stitches were used, namely the Castelo Branco stitch (Fig. 1), the flower stem stitch, the rooster-foot point, the simple criss-cross stitch and the stitch cast in a grid (a net of squares) [5, p. 87].

Fig. 1 Variants of the Castelo Branco stitch. *Source* IMC/Museu Francisco Tavares Proença Júnior [5, p. 29]

Currently, the application of the stitches has no strict rules except for the application of the Castelo Branco stitch to fill the tree trunks and the remaining stitches are dependent on the embroiderer's creativity and "it depends on everyone else around it and if the work done demands more or less exuberance more or less simple, of greater or lesser technical richness" [5, p. 28]. In the Technical Specification Booklet of this embroidery the possibility of introducing new stitches is left open "in the face of challenges posed by new designs, or discoveries of old pieces with hitherto unknown stitches" [5, p. 28].

2.3 Colors/Chrome Palette

Colour is one of the elements that contributes to the aesthetic quality of a piece of this embroidery. The strong colours of the silk thread applied to the harmoniously coupled Historical Bedspreads nowadays have pastel shades that "result from poor colour fixation and the much use that has been made of them" [9, p. 3]. There are currently 49 colours certified in the Castelo Branco Embroidery Technical Specifications that correspond to the colours used in the Francisco Tavares Proença Júnior Museum, at the time of publication of the referred brochure (2007).

The application of colour to the pieces of Castelo Branco Embroidery must follow certain principles so as not to neglect the aesthetic quality: (1) the colours to be used in the reproductions must be the same as the piece from which they are inspired; (2) the blue and yellow colour matching should preferably be used in the tile compositions; (3) the use of black and dark brown colours should be restricted to the representation of hair, shoes, eyes, clothing buttons and small embroidery notes; (4) pastels and gradients are not applied to historic bedspreads and their use should be avoided; (5) monochrome is an aspect that should be explored; (6) In the embroidery of contemporary creation, colour changes are allowed, provided that these are presented by professionals in the fields of arts and design [5, p. 92].

2.4 MOTIFS

The diversity of motifs is one of the defining elements of the richness of Castelo Branco Embroidery; The motifs used in this hand embroidery are the result of influences that extend into space and time, namely "seventeenth and eighteenth century European design and engraving, Portuguese tiles, Indian textiles and Chinese textiles and porcelain" [5, p. 81]. Arruda, when referring to the decorative grammar of Castelo Branco Embroidery, points out the similarities with Chinese and Indian oriental embroidery, with painted cloths originating in India, with oriental decorative arts, namely rugs, porcelain or earthenware [2, pp. 241–242).

They are currently grouped into five typologies: plant motifs, anthropomorphic motifs, zoomorphic motifs, mythological motifs of specific symbology and inanimate motifs [5, p. 81].

The group of plant motifs, also called phytomorphic, includes plants, fruits, leaves and flowers, generically represented in a very stylized way which, in some cases, makes their identification difficult and leads researchers to diverse conclusions [5, p. 82]. The motifs that present the greatest diversity of stylization are carnations, pomegranates and leaves, and the ones most often represented in Castelo Branco Embroidery are carnations, tree trunks, ivy, pomegranates, peonies, roses, artichokes, forget-me-nots, marigolds and acorns. [5, p. 82].

Hybrid objects include plant embroidery, whether from the medieval repertoire, with strong religious symbolism, such as lilies, thick roses, clovers, carnation (carnation with small flowers), tendrils, grapes and other small fruits from the east like pineapple, pomegranate, lotus which, symbolizing the renewal of life, could be assimilated by Christianity as symbols of eternal life [15, p. 213).

Vila Tejero [15] argues that the plant motifs are "the majority of times (…) the protagonists of the decorations, on the ground by the beauty of its flower elements, also because tree trunks and branches are used to organize and articulate the rest of the decorative elements" (p. 212).

In the group of anthropomorphic motifs, representation of human beings, stylization is also visible; Both female and male elements are represented mainly in the Historical bedspreads, so their design must be respected in the reproductions of these elements [5, p. 82]. The male element rarely appears isolated, but in some cases its representation is made with a flower in hand or mounted on horseback; the feminine element, when isolated, is represented above all in the centre [5, p. 83]. In pairs, there are male/female and also female/female representations [5, p. 83].

In the group of zoomorphic motifs there is also a huge stylization. Uncommon representations of horses, dogs, reptiles and winged animals appear [5, p. 83]. The bird is undoubtedly the most represented zoomorphic motif and with many variations in its design. Its representations, although very stylized, appears to be adjectives of domestic and exotic [5, p. 83].

In the group of mythological motifs and specific symbology, the only element considered is the two-headed eagle. This motif, similar to the motifs included in the previous groups, is also represented in a stylized way and sometimes "with the heart pierced by an arrow, constituting the most bizarre form of representation" of this element [5, p. 84].

In the group of inanimate motifs there are representations of various elements such as flower centre pieces/vases, palm leaves, shells, ribbons and bows, as well as other elements "which resemble the Gothic and Baroque architectural ornaments" [5, p. 84].

2.5 Symbology

Ormonde [8] states that embroidery is more studied as artisanal production, considered a subsidiary activity because it is performed mainly by women. However, he defends the study of another component that addresses "embroidery as a bearer and producer of meanings and values and thus participates in the symbolic sphere" (pp. 10–11).

A symbol/motif element that is associated with this embroidery is the tree, "symbol of life, in perpetual evolution, ascending to the sky (...), grows, loses its leaves and recovers them and therefore regenerates: dies and is born again and again" [5, p, 93; 10, p. 45]. For Vila Tejero [15] the tree of life is the symbol of all cultures and all religions—the symbol par excellence.

The bird is also one of the symbols/motifs widely used in this context and is considered synonymous with message; It also serves as a symbol for the relationship between heaven and earth [5, p. 93]. The rooster, one of the most frequently used birds, "symbolizes pride, courage and vigilance" [5, p. 94].

The carnation deserves prominence as a symbol/motif often used in these embroidery; "Of Turkish origin (...) in the Portuguese and later tapestries of Castelo Branco, seems to respond to Indo-Mogul models and the most accepted symbolism, if it corresponds to the happiness of the country" [15, p. 213), "with living and pure love: (...) the white carnation symbolizes an even more exacerbated passion than that of the red carnation" [5, p. 94].

The flower generically symbolizes "the paradisiacal state and feminine beauty; the opening of the flower bud represents the creation and energy of the sun "and can be associated with youth and vitality as well as fragility and transience due to its ephemerality [5, p. 94].

The apple is used as symbol/motif in many ways; "Sometimes it is the fruit of the Tree of Life, sometimes the Tree of Science of good and evil: unifying knowledge that grants immortality, or separating knowledge, which causes the fall" [5, p. 94].

The man (human figure) is also represented in the embroidery of Castelo Branco and in the Bible is considered the work of God, made in his image; thus "the body is considered a microcosm, a replica of the structure of the cosmos" [5, p. 95].

The pomegranate, which appears in this embroidery with many variations, is in Christianity the symbol of divine perfection. In classical civilizations it was, on the one hand, associated with the return to life in Greece, on the other hand worn by brides in the crowns with which they adorned their heads in Rome [5, p. 94].

The serpent can be used with double meaning, "evil (... or ...) regenerating power" [5, p. 94]. The vase, or flower centre piece, symbolizes a secret force, it is the reservoir of life [5, p, 94].

As embroidery with secular history, the Castelo Branco Embroidery Technical Specification Notebook recalls that "symbolic language adapts, loses, recovers and metamorphoses" [5, p. 92].

Regarding the embroidery of Castelo Branco, Vila Tejero [15] recalls that "it has been written that it is like a cared for garden of symbols" but prefers to say that "it is inhabited by elements of the collective imagination of a people who have inherited and assumed with that which has formed part of its great history" [15, p. 164]. However Vila Tejero also points out that, "undoubtedly, the embroidery of Castelo Branco is much more than a technical and formal description. It is a small universe that (…) always surprises us" (p. 198) because each quilt " seems to be endowed with its own life and effort to show us its uniqueness, and it is very rare that it cannot" (p. 198).

3 Castelo Branco Embroidery Applied to Clothing

The methodology for the implementation of this project focused, initially, and after identifying the Designers to develop the proposal, bringing them to Castelo Branco to collect all the information necessary for the development of their proposal. The Designers were invited to visit the Francisco Tavares Proença Júnior Museum, where they could find some copies of the Castelo Branco Embroidery Bedspreads and familiarize themselves with the various motifs and different compositions.

The Silk Museum was another of the spaces chosen to make the visit. In this space, designers were able to learn a little about the history and origin of silk, production method, characteristics and applications. This was followed by a visit to the space where embroiderers are the main characters, the Castelo Branco Embroidery Interpretation Centre. Here designers have a closer contact with the Castelo Branco Embroidery, knowing stitches, types of linen used in bedspreads, motifs, colours, etc.

After the visit to gather information, we sought to answer the main question of this research by the development of the creative process by the Designers. Designers should take into account the characterizing and defining elements of embroidery, their application, type of audience and market that it is aimed at and should be aware of the costs inherent in the materialization of the pieces.

The development and organization of the various stages of the creative process differs according to the methodology of each Designer. Some Designers seek to "take apart" all research and place fabric as the first aspect that helps define the theme and is often the basis of the collection. Others, however, immediately identify the theme or concept from the decoding of all research and only then interpret all elements for their visual narrative such as fabrics, colours, shapes, silhouettes, etc. [6].

Only after the interpretation of all elements of the research does the Designer meet the conditions to start the design of the collection, however, the whole process begins with the development of several possible ideas to arrive at a final idea that will be the basis of all. the work. "Creating a collection requires developing a wide range of related ideas to produce pieces that work not just as individual models, but fit into one theme" [4, p. 99].

4 Different Approaches—Results Analysis

Each artist's interpretation of the Castelo Branco Embroidery is very distinct and diverse and reveals very different focuses of interest and aesthetics. All proposed silhouettes are clear indiosyncrasies of each Designer.

Alexandra Moura (Fig. 2) interprets Embroidery in all the elements that characterize it, in terms of raw material, motifs, symbolism and stitches. The search for a narrative very close and associated with a traditional quilt (Adam and Eve) joins the choice of linen as the basis for embroidery, with stitches representative of this technique, and the making of the piece.

Luís Buchinho (Fig. 3) supports his interpretation using a wide and diverse combination of stitches. The basic raw material for the production of the piece is the one chosen for his collection and the colours are those identified in the colour palette he proposes for this season. The Designer creates a visual pattern by combining stitches and colours.

Fig. 2 Coat by Alexandra Moura

Fig. 3 Dress by Luís Buchinho

The Storytailors duo (Fig. 4) uses the linen to create the proposed piece, a multi-purpose piece and ways to use it. The Castelo Branco stitch is the only stitch used to embroider a motif that marks the symbolism of the Castelo Branco Embroidery—the Tree of Life. A stylized tree of life proposal, representative of a designer's own interpretation that maintains some elements that characterize it, namely the hills at the base of the tree.

Katty Xiomara (Fig. 5) uses the base for linen embroidery, matching the colour palette to the one she proposes in her collection but found in the colour palette for the silk thread. The chosen stitches reveal a diverse panoply of nets and stitches and the motifs are those from her collection.

The identification and association with embroidery is respected in the raw material, the colour palette and the stitches used, where the Castelo Branco stitch has strong representation.

Pedro Pedro (Fig. 6) proposes a skirt with a silhouette taken from the collection he proposes this season. Although it uses silk for Embroidery, the base is not linen but in the fabric chosen by the Designer and that integrates the proposal of his collection,

Fig. 4 Piece by StoryTailors

as well as the chromatic palette adopted for embroidery. Embroidered motifs result from the stylization of flowers. The stitches, especially that of Castelo Branco are the most representative of the Embroidery.

Designer Hugo Costa (Fig. 7) presents a men's piece proposal where the base is his own dyed linen. The raw material (silk and linen) is respected and used by the Designer, but with its interpretation and manipulation. The motifs are taken from Embroidery and the stitches are used predominantly from the Castelo Branco stitch, but the author's very own colour palette.

5 Embroidering Promotion and Value

The current impact of the fashion industry on the European and world economy requires new communication strategies. It is in this context that it is imperative to

Fig. 5 Dress by Katty Xiomara

define an integrated communication strategy that allows the embroidery of Castelo Branco to be designed not only nationally, but especially internationally. An example of this are the different application projects of the Castelo Branco Embroidery, the result of collaboration with Designers Luís Buchinho, Alexandra Moura, the Storytailors duo, Katty Xiomara, Pedro Pedro and Hugo Costa.

One of the ways the Designer can present the final product of their creative process is through the runway presentation. When it comes to a fashion show from the creator's collection, the concept is also respected and reinforced in order to lead the audience to the emotional and creative universe of the Fashion Designer. "Today, the fashion show is a show that is an integral part of the fashion industry. It gives the brand an authentic opportunity to express their identity, design and vision for the next season" [6, p. 40].

Among the planned communication actions, the highlight was the presentation by Luís Buchinho, as early as September 2016, of the dress proposed in the project and

Fig. 6 Skirt by Pedro Pedro

two other embroidered dresses later to be part of the Designer collection in Paris—Paris Fashion Week. This presentation took place again in Porto, in October, at the Designer fashion show at Portugal Fashion.

Alexandra Moura, in September 2017, presented her own piece at the Portuguese Embassy in London. This presentation took place under the London Fashion Week initiative, and was supported by the Portugal Fashion Association and the National Association of Young Entrepreneurs.

In conjunction with this action, the Designers present their collection on show and in the showroom space. Thus, in a space whère potential buyers, international press and other fashion industry professionals intersect, and as a result of an integrated communication strategy, which includes other actions, the Castelo Branco Embroidery gains international prominence. in an area, fashion, environment privileged for the value of this icon of the Portuguese cultural heritage.

Highlight for other dissemination initiatives, including the traveling exhibition of Castelo Branco Bedspreads, from the eighteenth and nineteenth centuries, and was the first exhibition held at the Lisbon Museum, in partnership with the Lisbon

Fig. 7 Suit by Hugo Costa

Municipality, between August and October 2015. This exhibition also passed by Guimarães and, more recently, in Óbidos.

At the same time, a series of other initiatives were taking place and, in 2017, a large project by Portuguese architect and artist Cristina Rodrigues, which brings to Manchester Cathedral a permanent work composed of seven large panels in the Castelo Branco Embroidery, in honour of Queen Elizabeth II. This work was opened in September 2017, on the same weekend that in London Alexandra Moura's piece was presented at the Portuguese Embassy.

6 Conclusions

As Ormonde [8] states: "The work of the designer plays an increasingly important role in the creation of consumer objects, including the recovery projects of these traditions. This role or intervention may become problematic when it comes to preserving national identity emblems, but it does respond to a contemporary tendency to blend tradition with innovation. In a way, it is with the category of objects that results, in which the old and the new, the local and the global, in which territorial relations are blurred and in which the cosmopolitanism made up of pluralities of "locations" is blended.", And from the practices associated with it, that today new identities and new circles of consumption of embroidery are being built, whether this is a piece for the masses or an antiquity" (p. 25).

The challenge of applying the Castelo Branco Embroidery in the design of garments led to the search for knowledge of the history and production method of this cultural product. Key elements resulting from research such as colour, shapes, silhouettes, textures, details and details will determine the choices and influence all the aesthetics and narrative to develop.

The Designer seeks his inspiration through research and reveals his own sensitivity and visual aesthetics in the development of unique pieces that unify the initial concept as a whole.

You can reinterpret the Castelo Branco Embroidery by giving it a new approach associated with Product Design. As a result, its diffusion is shown on the one hand to reinforce the identity and culture of a region, on the other hand it can boost the development of production for these products by increasing their demand.

References

1. Adere-Certifica (2017) Manual de Certificação—Bordado de Castelo Branco. Adere-Certifica
2. Arruda L (2007) Bordado de Castelo Branco. Estética e desenho. In: IMC/Museu de Francisco Tavares Proença Júnior e ADRACES (Coord.). *Colchas de Castelo Branco—Percursos Por Terra e Mar*. "Ex-Libris", Castelo Branco, pp 240–290
3. Chaves L (1980) Colchas de Castelo Branco. In: Cardoso R, Chaves L, Moura M, O Bordado e as Colchas de Castelo Branco. Museu de Francisco Tavares Proença Júnior, Castelo Branco, pp 14–20
4. Fernández A, Roig G (2007) Draw for fashion designers, Editorial Estampa, Lisboa, ISBN 978-972-33-2388-7
5. IMC/Museu Francisco Tavares Proença Júnior (Coord.) (2007) *Bordado de Castelo Branco - Caderno de Especificações Técnicas*. "Ex-Libris", Castelo Branco, Parceria
6. Matharu G (2011) What is the fashion design?, Bookman, Porto Alegre. ISBN 978-85-7780-853-3
7. Melo D (2008) A 'Cidade-Símbolo' de Castelo Branco e o Bordar da Identidade Beirã (séc. XX)" In: IMC/Museu de Francisco Tavares Proença Júnior e ADRACES (Coord.). Colchas de Castelo Branco - Percursos Por Terra e Mar. "Ex-Libris", Castelo Branco, pp 65–89
8. Ormonde H (2008) Bordar no quotidiano e no imaginário. In IMC/Museu de Francisco Tavares Proença Júnior e ADRACES (Coord.). Colchas de Castelo Branco - Percursos Por Terra e Mar. "Ex-Libris", Castelo Branco, pp 8–31
9. Pinto C (1992) Bordado de Castelo Branco. Catálogo de desenhos. Colchas—I. Lisboa: Secretariado de Estado da Cultura, Instituto Português de Museus e Museu de Francisco Tavares Proença Júnior
10. Pinto C (1993) Colchas de Castelo Branco. Fundo Vip e Instituto Português de Museus, Lisboa
11. Pinto C (2004) O Bordado de Castelo Branco. In: Pinto, H. (Org.). Artes Tradicionais de Portugal. Fundação Calouste Gulbenkian, Lisboa, pp 23–37

12. Pires A (2008) Bordado de Castelo Branco, elementos para uma geografia. In: IMC/Museu de Francisco Tavares Proença Júnior e ADRACES (Coord.). Colchas de Castelo Branco - Percursos Por Terra e Mar. "Ex-Libris", Castelo Branco, pp 292–397
13. Rechena A (2007) Introdução—Colchas de Castelo Branco—Percursos por Terra e Mar. In IMC/Museu de Francisco Tavares Proença Júnior e ADRACES (Coord.). Castelo Branco: Parceria "Ex-Libris". 6.
14. Teixeira M (2000). Moda Portuguesa Tradições e Contradições. In Gaspar, F. (coord.). Contradições—Moda Portuguesa. Coimbra: PPART. 11–16
15. Vila Tejero M (2008) La universalidad del bordado através de los Bordado de Castelo Branco. In IMC/Museu de Francisco Tavares Proença Júnior e ADRACES (Coord.). Colchas de Castelo Branco—Percursos Por Terra e Mar. Castelo Branco: "Ex-Libris", pp 162–239

Playponics in India—Local Hydroponics Playground Gardens Utilising Kinaesthetic Learning to Promote Global Sustainable Practices

Avika Sood, Heath Reed, and Andy Stanton

Abstract This research and development project set out to investigate possible means by which future populations might have or build closer links with environmental issues. The work has integrated the diverse fields of sustainability and STEM education, global products and local production. It aims to bring these aspects together in holistically considered learning environments where these elements can coexist in engaging ways. This paper highlights the intended benefits to students of engagement in these systems such as cognitive development through practical or 'kinaesthetic' learning. The outcomes intent is that students who take part also learn how to implement basic mechanical knowledge that, according to school curriculums, are currently part of textbook education alone, "…most Indian classrooms remain dominated by rote-learning" [Brinkmann S (2015) Learner-centred education reforms in India: The missing piece of teachers' beliefs]. Our intended forms of learning aim to impart a spirit of innovation and experimentation, whilst being surrounded by elements of nature. In this work we explore how we learn about and 'instil' senses of responsibility and empathy, community and co-operation in future populations towards greater environmental sustainability. We report and discuss various strands of the research to date including methods to identify and develop the concept. We conclude that evidence to date strongly indicates high levels of acceptance of this approach and report on our progress, findings, and our anticipated next steps. This project is being funded through Global Challenge Research Fund (GCRF) and the Research England, Expanding Excellence England (E3). The work is collaboration between Designers from Lab4Living at Sheffield Hallam University in the UK, businesses in Delhi, and has been delivered jointly by Playponics design research teams in the UK and India.

Keywords Sustainability · STEM education · Kinaesthetic learning · Play · Hydroponics

A. Sood (✉)
Ativa Design, 1/31 Sadar Bazaar, Delhi Cantt, New Delhi 110010, India

H. Reed · A. Stanton
Sheffield Hallam University, 153 Arundel St, Sheffield S1 2NU, Sheffield, UK

E. Duarte and C. Rosa (eds.), *Developments in Design Research and Practice*, Springer Series in Design and Innovation 17,
https://doi.org/10.1007/978-3-030-86596-2_31

1 Introduction

This study involves apparently unconnected topics; physical play, environmental sustainability and early years STEM education delivery. The research has led the team to propose a new kind of sustainability education concept, a physical play park for schools that 'captures' children's play energy and uses it to help sustain edible crops. Although this applied research is at early to intermediate stages, this disclosure discusses and positions our thinking, rational, research themes and methods when designing in these 'spaces'. From practicing designers' perspectives, we consider what kinds of learning mechanisms can be capitalised upon when designing educational experiences, and how these diverse headline topics might be holistically considered and integrated into new learning experiences. We explore the role of physical play in learning about STEM and design education, with the aim of increasing awareness in future populations of environmental sustainability issues. We focus on and discus four key aspects of the study, namely issues about Sustainability, Education and STEM education, the provision of globally relevant products that support sustainability and locally sourced production. Finally, we discuss the concept and conclude.

2 Sustainability

Our focus in this study is early year's education and relationships future generations will have with our world, how we consume, and the balance between consumption and living in more sustainable ways. With India's middle classes on the rise (data indicates the group is set to grow rapidly, from around 80 million today to 580 million people by 2025), is it sensible that these developing societies 'ape' the consumption patterns the so called 'developed' countries have? Excessive consumption, pollution, natural resource depletion and a disconnect from nature have been the warning cries of environmentalists for many years. This project specifically explores ways children might engage in learning about those issues, how we may instil knowledge about sustainable living, through education which in turn may help us to reconnect with and value natural systems. '...today's global imperatives—to eradicate poverty and improve wellbeing, while restoring the Earth's balance—form a single agenda, and that the most effective means of achieving it is education [1].

A motivation for undertaking this work is, in a pragmatic sense, 'how' we achieve this learning and potential behaviour change as 'Alongside any form of promotion of learning about issues such as climate change, ...there needs to be consideration of how children learn about (it)' [2]. Our research takes 'why do it' very much for granted, and focuses on 'how'.

This is perhaps most adequately captured in the World Economic Forums article 'Why education is the key to sustainable development' which includes that '...education can bring about a fundamental shift in how we think, act, and discharge our responsibilities toward one another and the planet. After all, while financial incentives, targeted policies, and technological innovation are needed to catalyse new ways of producing and consuming, they cannot reshape people's value systems so that they willingly uphold and advance the principles of sustainable development. Schools, however, can nurture a new generation of environmentally savvy citizens to support the transition to a prosperous and sustainable future' [1]. 'Reshaping people's value systems' is of particular interest and importance to our work.

3 STEM Education

Reports from various media channels confirm that the condition of government schools in India is poor; 'The Union government may have made right to education a fundamental right by bringing into force the Right To Education (RTE) Act of 2009, but government schools are lagging far behind in providing quality education…' [3]. Further, a significant proportion of the population cannot afford private education. It is referred to as a commodity that is not affordable for all; 'Being a publically provided private good, Education needs a greater focus on Accessibility, Equity and Quality…' (ESAG-2018.pdf, 2018) [4].

Many schools lack basic amenities; in short transformation of education in India, and around the world, is needed. Keeping in mind sustainability education problems faced by these students we have researched, designed and propose school playgrounds that harness 'play energy', to help power that change. In one embodiment, we proposed that playground equipment is used to facilitate hydroponic crop growth and is constructed and deployed using locally sourced materials like bamboo, readily available at economically viable rates in the country.

'Too often, as architects, engineers and owners, we get caught up in sustainability codes and guidelines and neglect to consider what might be most logical–and effective- -for a specific building, client or locale' [5].

Working with educationalists and schools in the Delhi region, designers established broad frameworks of understanding and developed their knowledge of different types of education provision in India. This included lay understanding of government and privately funded school provision, the quite diverse nature of infrastructural situations and a sense of school staff opinion and willingness to engage in sustainability education curricula.

A finding of school visits and subsequent investigations was that there is opinion within teaching staff that traditional pedagogical methods need to change, or at least require augmentation, as the very nature of children's experiences, expectations, and their inherent capabilities have changed. These aspects were cited as being due to increasingly consumerist behavior and a change towards faster paces of family life, but predominantly due to technological advances. This view was reinforced by literature in terms of changing pedagogy in the context of sustainable lifestyles, in that '…today's youngsters are born into an India that readily embraces consumerism with all its accompanying trappings. Besides being surfeited by a plethora of toys and gadgets, the average middle-class child is tethered to a global network that tends to homogenise childhood. As childhood has undergone tumultuous shifts, the dynamics of classroom life have been altered' [6], and that consequently, 'Teachers cannot rely on plain chalk-and-talk anymore but have to keep pace with a generation raised on a multimedia diet' [6].

Alongside methods used to teach children, who are reported as 'having increasingly short attention spans', notions were expressed that the curricula itself can be 'out of date' and 'bland'. '... syllabi for various subjects continue to remain dreary and uninspiring. Students continue to cram information to score marks without engaging actively and meaningfully with content' [6]. It was also found that although embraced by some schools, and evidenced in some state/authority curricula, educational syllabuses seldom have a specific focus on sustainability issues delivered through experiential learning. '...even after Environmental Education was introduced as a compulsory subject, we have not managed to cultivate a green conscience in children.' [6].

Reportedly, current '...awareness of environmental issues comes not from direct engagement with the environment itself but from a more passive and indirect understanding of these issues. Direct interaction with the natural environment appears to be increasingly absent in children's lives and this new phenomena gives rise to concern because such experiences are essential in developing children's knowledge and understanding of the world' [1]. Although somewhat anecdotal, discussions proved highly valuable in terms of building understanding in designers of current contexts, which served to reinforce designs premises.

The design team (not being educationalist) needed to build understanding of what kinds of learning mechanisms, or 'modes' exist, and which ones may engage learners to greater or lesser extents. Discussions with educationalists led the team to understand degrees to which current provision relies on 'traditional' repetitive, information transfer, or 'rote-learning'. Investigation confirmed that "...most Indian classrooms remain dominated by rote-learning" [7], or at least that was reported to be the case up to 2015. The team investigated a range of teaching modes. By way of summary, we adopt educational theorist Neil Fleming's VARK model in explanation. 'VARK is an acronym that refers to four types of learning styles: Visual, Auditory, Reading/Writing Preference, and Kinesthetic.' (The VARK Modalities [8]) A Guide to learning Preferences).

Design researchers came to understand this model and that it upholds that there are favoured ways learners absorb and hold onto knowledge. The designers came to appreciate that the Playponics proposal may be more in line with the Kinaesthetic learning modality because, by definition, it refers to "perceptual preference related to the use of experience and practice (simulated or real). Although such an experience may invoke other modalities, the key is that people who prefer this mode are connected to reality, either through concrete personal experiences, examples, practice or simulation" [9].

Therefore, the principles and practices of kinesthetic learning are of particular interest to the design team, as it is 'a learning style in which learning takes place by the students carrying out physical activities...', and that, 'The key is the reality or concrete nature of the example. If it can be grasped, held, tasted, or felt it will probably be included. People with this as a strong preference learn from the experience of doing something and they value their own background of experiences and less so, the experiences of others' [8]. It is not suggested that other cited modalities will be void in this example, as '…such an experience may invoke other modalities'.

The team have invested in research to build knowledge about educational mechanisms that ultimately are the purpose of this work. However, a wide range of other technical, maintenance and facilitation research was required. This includes investigation of physical formats of play equipment and appropriate energy capture systems integration, building knowledge about what is preferred, safe and practical. A pilot workshop with Indian children aged 11 years was undertaken where the kids were asked what kinds of play format they thought was 'fun'.

Based on feedback from local schools the design intent now includes the development of materials intended for the education of staff and students in regard to maintenance of play systems, the hydroponic (and/or horticultural), and back-up systems. Back-up systems were agreed to be essential to ensure crop survival out of school hours. We further established that it would be feasible, in principle, to integrate these backup and maintenance regime into the taught programmes.

Through these engagements the team further recognised that, in the spirit of sustainability, proposed systems needed to be designed with respect to local materials fabrication, manufacture, installation and maintenance. Being trained and experienced product designers (and capable makers), a phase of research activity focused on local resource provision and experimental fabrication. This was undertaken to ensure proposals could be locally made, both in terms of materials, available skills and tools.

Our focus is to make each setup effective and efficient using sustainable materials in different parts of the country. This will help the children in identifying sustainable materials and practices. The crops produced by the setup can be used in the school kitchen, sold in the market to generate funds for the school/community or to maintain hydroponics playgrounds themselves. Following successful test runs, educational content will be developed and distributed to students and faculty members of the schools to inform curricula.

Another focus of this study is to make the setup as simple as possible so it is easy to maintain for the students, and become part of their early stage STEM education. The project aims to reach out to as many under privileged schools as possible following our first prototype installation.

4 The Concept

At the emerging concept's heart is learning through physical play enabled by bespoke designed playground equipment. This equipment is augmented with systems that in turn enable crop production. In this scheme, physical play energy expended by children is made tangible, harnessed, stored and used to help facilitate crop growth. In this way, from an educational perspective, we aim to help address shortfalls in current 'information transfer only' (or 'chalk-and-talk') type teaching such that 'Rather than just being informed about the environment and the wider world, children will be supported in both understanding and experiencing (it)' [10].

Amongst the many business infrastructure, economy, materials and fabrication facilitation challenges around how such systems will be implemented in India, the team's focus was very much on how Playponics system would manifest. This diverse research set led the team to further refine and develop the Playponics system (see Fig. 1). In this most recent concept we bring together and integrate diverse subject matter. For example, in technical facilitation we can demonstrate that children riding a seesaw [A below] fitted with pumps (engineering) [B], can effectively transfer water around the play learn system. As water is pumped it is retained by a 'header tank' (energy storage) [C] that in turn gravity feeds (physics) [D] nutrient rich water through a hydroponic crop growing frame (horticulture) [E].

The evolved installation design requires levels of maintenance that include mechanical, electromechanical and hydroponic systems management. The original concept ethos was that these processes include the child and school community. In discussions with educators on the ground this 'maintenance' was not deemed overly complex. It was felt that integrating specific topics into taught curricula may be problematic, but that in time these could be overcome.

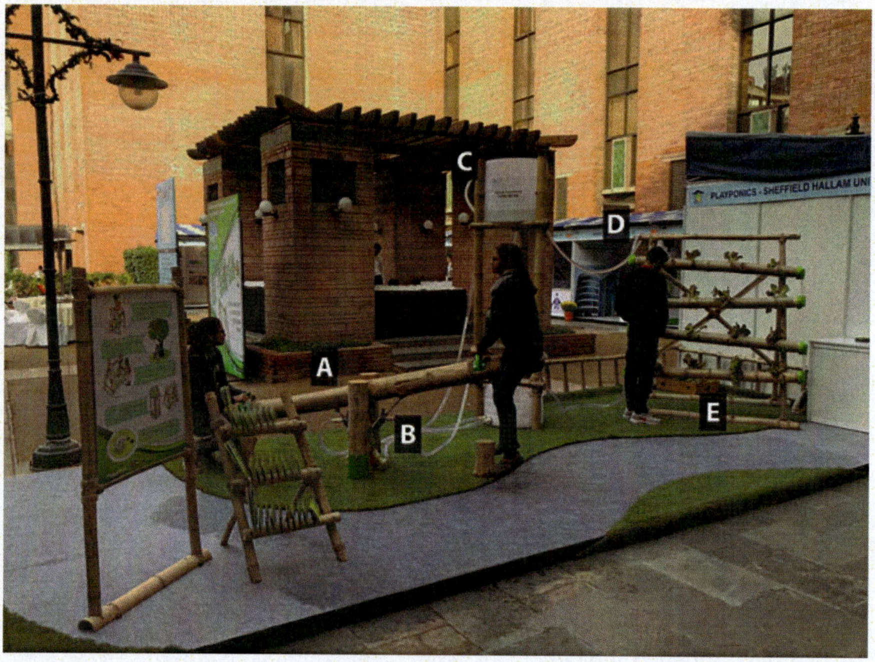

Fig. 1 The resulting 'Playponics' set up on display at the Griha Summit, Delhi, December, 2019

5 Next Steps

Our applied research has been planned in two distinct phases. Firstly, we identify known elements impacting upon the design of proposed educational play systems that include technical, teaching community, horticultural, and at a natural environmental level. This first level research informs physical design of systems we implement and test. A mixed methods approach is used in this first phase including desk based, interview and experiential research. Secondly, we plan to install and monitor systems arising from first stage research and attempt to measure their education benefits and impact. At the time of writing (December 2019) we have not commenced second stage research. The following outlines some key aspects of our enquiry to date.

To date we have developed these designs and the rationales for their implementation to proof of concept levels. Our current challenge is to refine these designs, install working systems and gather the pedagogical evidence to a stage that enables 'buy in' from schools. One school in India has stated that the proposal fits well with their ethos and ambitions for their pedagogy.

6 Discussion

We explore the position that engagement in this play 'ecosystem' by children (aged 6–12 years) will show benefits in developing primary stage understanding of STEM and Design subjects alongside broader lessons about cause, effect and cooperation. The term 'ecosystem', most usually used to define 'a biological community of interacting organisms and their physical environment' (Oxford Dictionaries definition) is used literally here as we explore integration of child learning through play, with building and embedding of early stage knowledge associated with horticulture.

In respect to learning about sustainability issues, education is key to our population's health, wealth and security. Although a simple idea, the benefits of this approach to play and sustainability technology education could be multifaceted. As a result of our research, we now consider the benefits in three primary ways.

i. Benefits derived as a result of taking part in physical play, both cognitive and in terms of a child's healthy, physical development.
ii. In the child's developing mind, the building of understanding about the relationships between physical effort made and a crop's subsistence (knowledge building around biological systems, cause and effect, ecology and symbiosis).
iii. Design and STEM educational benefits derived from very real and tangible, physical interactions with the mechanisms and systems that enable energy capture, storage, transmission and utilisation, through the kinesthetic learning modality.

Added to these primary benefits for the child are potential wider social lessons, about community (teacher, parent, student, school) engagement, co-operation, nutrition and the bigger topics of future sustainable living. We further plan to test if the dependency of the crops on the use of the playground equipment actually encourages the children to play more.

The key anticipated benefits of the original concept were those associated with regular Physical Activity (PA) in promoting children's health. These, the team found, are well known and understood. Further though, PA is associated with improvements in children's intellectual development, 'PA has a positive influence on cognition as well as brain structure and function' [11]. However, for many schools, it can be hard to put programs into practice that bring about such holistic advantages. On presentation to teaching staff it was agreed the concept may be an appropriate 'platform' to enable multiple benefits.

The concepts were based on multiple lay understandings across a range of specialisms (education, hydroponics, physical activity, etc.) and as such initially, arguably, held a simplistic premise. Nonetheless, the concept acted as a catalyst to help identify, probe and validate aspects of the proposal across a range of more complex topics, from local cultural contexts to enabling technologies. Their subsequent development and manifestation, in the form of sketches and working prototypes sparked the imagination of the designers and collaborators. The concept's iterative development has resulted in what we now term a 'play grow ecosystem', and these approaches continue to gather momentum.

7 Conclusions

This phase of research concludes that multi-modal research is essential when seeking to design learning systems that aim to have holistic benefits. The study has demonstrated degrees of 'validity' to the original proposition and strongly indicates high levels of in principle acceptance of the developing concept. Where our aim is to instill these understandings in children so that they may carry sustainability knowledge with them into adulthood, and where we propose integrating learning modes, mechanisms and topics as a means of achieving that, the breadth of our research needs to be as equally all-inclusive. Designers of learning systems must undertake research into each, often (on the face of it) diverse topic areas as holistically as possible, and it is this breadth of research that informs designs appropriateness for progression.

Acknowledgements We would like to thank GCRF and E3 funding programs for their support, Vahani Scholars-Tanya Chaudhary, Mansi Chauhan and Pradeep Kumar for their involvement during the GRIHA summit. Also, children and schools in the Delhi region too numerous to mention personally.

References

1. Bokova F (2015) World economic forum
2. Bourn, Hunt, Blum, Lawson, [2016] Primary Education for Global Learning and Sustainability
3. Government schools imparting poor quality education (n.d.). Retrieved October 21, 2019
4. https://www.education.gov.in/sites/upload_files/mhrd/files/statistics-new/ESAG-2018.pdf
5. AP, LBCAL (2009) Site-specific sustainable strategies embodied in Lisbon project, July 21
6. Sankaranarayanan (2011) The Hindu
7. Brinkmann (2015) Learner-centred education reforms in India Learning Styles (2019) Teach
8. The VARK Modalities (2019) A guide to learning preferences
9. Make a Difference Fleming, Mills (1992)
10. Rickinson (2006) RSPB
11. Donnelly (2016) Physical activity, fitness, cognitive function, and academic achievement in children
12. Brinkmann S (2015) Learner-centred education reforms in India: The missing piece of teachers' beliefs

User-Oriented Challenges of Smart Mobility: An Analysis of Focus Group to Identify User Behaviour

Sevgi Gaye Ayanoglu, Madalena Pereira, and Emília Duarte

Abstract Mobility in big cities has been pointed as the cause of growing unsatisfaction of citizens with a direct effect in their quality of life and overall sustainability of these places. Recently, due to new technological developments, cities are becoming "smart" and smart mobility has growing importance. However, little is known about smart mobility effectiveness and how it may positively affect sustainability in cities daily life. This study aims to understand the potential users' behavioural intentions about their mobility choices and to identify which are the biggest perceived obstacles that may prevent them to take more sustainable mobility decisions. Six focus groups were run with students and young adults that were living in different districts of Lisbon intended to discover their expectations, problems and needs regarding mobility. A card-sorting task was used to measure their preference and satisfaction levels about a set of options. Fifteen topics of major demand and nineteen problems were identified. Results also confirm both positive and negative impacts on preference and satisfaction levels of users on current city life and smart possibilities of mobility. These results provide a useful basis for the ideation phase, as part of a design process that aims to find solutions able to motivate users towards more sustainable choices.

Keywords Sustainable behaviour · Smart mobility · Human-centred design · Focus group · Card sorting

S. G. Ayanoglu (✉) · M. Pereira · E. Duarte
UNIDCOM/IADE - Unidade de Investigação em Design e Comunicação, Lisbon, Portugal

M. Pereira
e-mail: mmrp@ubi.pt

E. Duarte
e-mail: emilia.duarte@universidadeeuropeia.pt

S. G. Ayanoglu · E. Duarte
IADE - Faculdade de Design, Tecnologia e Comunicação, Universidade Europeia, Lisbon, Portugal

M. Pereira
FibEnTech, UBI—Universidade da Beira Interior, Covilhã, Portugal

E. Duarte and C. Rosa (eds.), *Developments in Design Research and Practice*, Springer Series in Design and Innovation 17,
https://doi.org/10.1007/978-3-030-86596-2_32

1 Introduction

This paper presents findings from a user research phase, run as part of a doctoral study aiming to promote sustainable behaviours in everyday life in large cities. In particular, we are interested in exploring the potential of fashionable wearables for promoting sustainable mobility behaviours, in the context of smart cities.

This study follows a Research Through Design approach [9]. Thus, the findings gathered from this phase are expected to work as a keystone to back the development of alternative designs that can be considered adequate solutions for the problem of current low adoption of sustainable mobility behaviours in large cities. As mobility choices have an impact on citizens' behaviours and vice versa, this study is also about how to design for behaviour change.

The transition to sustainability introduces new ideas and behaviours as well as ambiguity [15]. Although people are optimistic about sustainability and want to live in a way that treats ecosystems we depend on with care and respect, most often they found themselves engaged on unsustainable daily behaviours that have negative environmental impacts. The reasons for this are diverse. In some cases, they are uncertain about how to behave sustainably and/or how to implement it in their everyday life. In other cases, they might either need social proof that convinces them the effort serves the purpose or they just need some help to change some social norms. For example, many unsustainable daily actions, such as driving alone, still are well accepted by the community while alternative options (e.g., share the ride or use public transports) are seen as signals of lower status and avoided due to their negative connotation [15, 22].

Despite people's rational minds know that change is necessary, we know that human behaviour is not always drive by reason [15]. Also, when confronted with rational arguments against or questioning their actions, people are very good at finding strategies to avoid uncomfortable feelings and find excuses without changing beliefs or actions [8]. Previous studies on the common excuses that allow people to feel justified not behaving desirably (e.g., quit smoking) [8], perform a pro-environmental behaviour [16] suggest that the argument used is that the "desirable" action is not appealing to the associative system. Thus, regarding the topic we are exploring, it is legitimate to assume that people want to be sustainable, however, they think their current life situation does not allow them to behave accordingly. Also, they may not be either able to see an immediate consequence of their behaviour change, sometimes aggravated by witnessing other people behaving in an opposite direction, giving them the idea that their effort was useless. Altogether, these arguments give support to a "valid" excuse, turning their unsustainable actions "reasonable" at their point of view.

Many of these "reasonable" behaviours are currently taking place in cities, with strong impacts on the environment and the community itself, which is foreseen to aggravate in a near future due to the increasing pressure for urbanization, as it is expected that nearly 60% of people will live in urban areas by 2030 [26]. The need for urbanization is addressed as "hope for a better life" [6] and "hope to gain access" to necessities of life, knowledge, other people, and some other opportunities [7].

However, people will need to deal with much more complexity in urban life such as overpopulation, pollution, poor infrastructure, depletion of sources, and so forth.

To deal with these issues, modern cities are increasingly becoming "smart" thanks to networks of Internet of Things (IoT), requiring new and distinctive codes [23] that take into account not only the information technology capabilities and limitations but also sustainable issues. Urban mobility gained importance in the context of smart cities, considered one of its main pillars, and one aspect with a stronger impact on its sustainability. According to [11] approach, smart mobility is one of the six characteristics that emphasize the accessibility of citizens, together with sustainable and innovative transportation systems and efficient infrastructure of Internet and Communication Technologies. Likewise, [23] addresses the importance of smart mobility as it directly and locally affects the quality of air and the quality of life in terms of generating congestion, pollution, and preventing freedom of accessibility. [13] emphasize the smartness of mobility and addresses smart mobility as a tool to achieve sustainable cities.

[23] defines smart mobility as:

> The ability to guarantee a good availability of public, innovative and sustainable transportation services; the support of low environmental impact transportation means such as bikes or pedestrian routes; ruling the access to historical centres; the adoption of advanced solutions for the mobility management through info-mobility, managing the mobility of individuals within the city and towards the neighbouring areas.

In a recent study, [14] proposes the term "smart urban mobility", defining it as "connectivity in towns and cities that is affordable, effective, attractive and sustainable". In this definition, "connectivity" acknowledges that the physical mobility of people and goods is only one means of providing access but also considers that motor vehicles are not the only option to allow such mobility. Being "affordable" and "effective" involves understanding the differing users' needs and abilities in cognitive, physical, and financial terms about connectivity. The definition also claims concerns related to being "attractive" for everyone meaning, for example, that the mobility system must meet the needs of individuals as citizens, urban dwellers, or business owners. Finally, the achievement of all these requirements must have a long-term basis that emphasizes the "sustainable" determinant of the term.

Whereas behaviour change is quite hard to achieve, mostly due to the huge diversity of individual and situational variables, we agree that if "one cannot design behaviour, one can influence people's preconditions for acting with technology by (re-)designing artefacts; designing for behaviour" [24]. In this sense, motivated by a desire to design successful solutions for smart mobility that can produce a positive impact in the promotion of sustainable behaviours, and convinced that such an endeavour requires a Human-Centred Design approach, we plan and conducted a series of analysis intended to facilitate the understanding of the needs, expectations and problems of the potential users of smart cities. In this context, in this paper, we present the results of a step intended to identify the main obstacles that can prevent users from adopting more sustainable behaviours in what regards their urban mobility. Knowing the demands and problems of the users, which can generate the obstacles to adopting a given behaviour, would provide us with a strong basis for the ideation of solutions, as well as could raise topics that could be later addressed in the development of the solutions. Thus, through the accomplishment of focus group sessions, incorporating a card sorting procedure, we gathered qualitative and quantitative data about the users' demands, problems, preferences and satisfaction levels, as well as their expectations regarding the potential impact of some given smart solutions on their future mobility choices. Focus groups are suited for idea generation as well as discovering problems, challenges, frustrations, likes, and dislikes among users and card sorting is often used to inform or guide the development of information [1].

2 Methodology

To better understand the demands and problems of users, their preferences and satisfaction levels, as well as the perceived impact of smart solutions into users' mobility choices, we organised this research into four main phases:

- **Information:** In this phase, we accomplished a literature review to understand the state of the art about smart mobility systems, human behaviour, design for behaviour change and sustainability.
- **Planning and preparation:** In this phase, we determined the types of data to be collected and the criteria to organise it according to the objectives of the research. Also, we prepared all the materials and procedures for the focus group and card sorting sessions. This phase also included a pilot test, which offered us the chance to do the necessary modifications/improvements before running the final sessions.
- **Data Collection:** In this phase, we conducted six focus group sessions with a convenience sample composed by students and young adults that are currently living in different districts of Lisbon. The sessions were audio-recorded and the transcripts were subject to content analysis to assess the topics that gathered more agreement, support or disagreement between the participants. Notes taken by the observer/moderator during the sessions were also considered.
- **Data Analysis and discussion:** There is no standard procedure for the evaluation of data collected in focus groups [10] however, the interaction between the participants is highlighted by various authors [10, 17, 18]. Thus, we conducted a tape-based analysis to both content and interaction data, which was collected in the form of verbal (words, sentences) and non-verbal (body language and facial expressions) expressions. The tape-based analysis is wherein the researcher listens to the tape of the focus group and then creates an abbreviated transcript [18]. We opted by this type of analysis because it allows us to focus on the research question and only transcribe the portions that assist in a better understanding of the phenomenon of interest [18]. A content analysis performed with MAX Qualitative Data Analysis (MAXqda) software, provided us with data about the frequency that some keywords were referred by the participants. Heatmaps were created with the data gathered from the card sorting, which are a visual representation

showing the most selected (hot areas) and less selected (cold areas) quadrants in which the cards were positioned by the participants, after and before the debate.

2.1 Focus Groups

Focus group is defined as "a technique involving the use of in-depth group interviews in which participants are selected because they are purposive, although not necessarily representative, sampling of a specific population, this group being 'focused' on a given topic" [12, 20] and widely used in social sciences for developing hypothesis in exploratory phases of research projects [2]. Compared to other methods, the interactive and synchronous group discussion format allows participants to discuss, agree, or dissent with each other's ideas and to elaborate on the opinions they have already mentioned [17]. Therefore, it is assumed that this method has a high potential to enhance our understanding of user behaviour. The method also enables the collection of different types of data, as well as testing underlying assumptions and research questions [4, 10].

Demographics

A convenience sample comprising 29 volunteers participated, which were distributed by 6 independent focus groups (5 groups with 5 participants each and 1 group with just 4 participants 4 due to a last-minute impediment from one participant).

Table 1 shows the sample demographics. The age of the participants ranged from 18 to 31 + years old. The majority of the participants were female (82.8%), with ages comprised between 21 and 23 years old (41.4%), single (62.1%), and college students. Although the majority of participants were Portuguese (48.3%), 10 different nationalities participated, allowing the comparison between these participants' former experiences in different cities and Lisbon.

Instruments, materials and procedure

The focus groups sessions were structured into 3 phases and were undertaken both inside the IADE campus and in other locations in Lisbon, Portugal. Initial demographic data were collected and open-ended questions targeting content and characteristics of the mobility system, pillars of the mobility system, their experiences in Lisbon, and comparison with other cities they lived before, were asked.

In the second phase, a mixed card sorting was conducted, both in the context of current and future mobility systems. Participants were given 10 cards representing "Modes of Mobility", which were designed on purpose for this study. They were also told they could add more if wanted. The cards depicted current and planned modes of transportation, examples gathered from smart mobility projects, reports, and other sources covering both motorized and non-motorized, public and individual means of transport [3, 19, 21, 27]. The 10 options shown were: (1) Walking, (2) Bike-sharing, (3) Public Transport, (4) Multi-modal Transportation, (5) Ride-sharing, (6)

Table 1 Participant demographic profiles distribution

Value	F	%	Value	F	%
Gender			*Civil status*		
Female	24	82.8	Married	11	37.9
Male	5	17.2	Single	18	62.1
Age			*Nationalities*		
18–20 years old	6	20.7	Portuguese	14	48.3
21–23 years old	12	41.4	Brazilian	6	20.7
24–26 years old	3	10.3	Spanish	2	6.9
27–29 years old	3	10.3	Turkish	1	3.4
30–32 years old	3	10.3	Salvadorian	1	3.4
33 years old and older	2	6.9	Russian	1	3.4
Professional status			French	1	3.4
Student	21	72.4	Dutch	1	3.4
Employed	5	17.2	Mozambican	1	3.4
Student and employed	3	10.3	Romanian	1	3.4
Total ($\sum$)	29	100	*Total* ($\sum$)	29	100

F: Frequency/%: Percentage

Autonomous Vehicles, (7) Scooter-sharing, (8) Car-sharing, (9) On-demand transport, (10) Personal cars. We asked the participants to sort the cards on a chart divided into 4 quadrants by 2 axes representing the level of preference and satisfaction. In a first moment, they were told to sort the cards having in mind their current daily life and as a group, then they watched the video of future mobility scenarios [5] that includes all the modes of mobility and they were asked to replace the cards if needed.

In the third and last phase, we gave them 9 quotes that define random citizen approaches and common excuses that were gathered from previous studies on behaviour change [8, 15, 25, 28] and we asked for their interpretation about that.

3 Results and Discussion

3.1 Demands and Problems of Users

The contents collected during all three phases of focus groups were analysed for the identification of Demands and Problems related to mobility systems. As Table 2 shows, 15 topics were obtained for the demands and 19 for the problems.

Table 2 Demands and problems of users—topic frequency

Topics: Demands from mobility	F	%	Topics: Problems of mobility	F	%
Convenient price	58	13.36	Security/Trust	68	12.08
Frequent	57	13.13	Time	65	11.55
Secure	49	11.29	Demand additional preparation	56	9.95
Enough space	37	8.53	Cost	46	8.17
Fast	34	7.83	People Density/No space	42	7.46
Punctual	30	6.91	Frequency	35	6.22
Convenient routes	29	6.68	Facilities	34	6.04
Comfortable	26	5.99	Confusing	26	4.62
Easily reachable	26	5.99	Lack of option	25	4.44
Reliable information	22	5.07	Comfort	24	4.26
Clear information	20	4.61	Digital app	24	4.26
Offer efficient facilities	18	4.15	Lack of route	23	4.09
Sustainable	14	3.23	Personal space	20	3.55
Maintained/Repaired	10	2.30	Personal excuses	19	3.37
Clean/Hygienic	4	0.92	Traffic/Parking	17	3.02
			Maintenance/Regulations	11	1.95
			Weather conditions	11	1.95
			Lack of information	8	1.42
			Hygiene	6	1.07
			Sustainability	3	0.53
Total ($\sum$)	434	100	Total ($\sum$)	563	100

F: Frequency/%: Percentage

Demands:

The demands were identified mostly from the analysis made to the answers to questions such as: "What do you want from a mobility system?" and "What does the best mobility system must have?". Additionally, we examined the whole conversations and interactions, looking for complaints and the reasons why the participants claim they prefer a specific mobility mode, disregarding others. We gathered the keywords and phrases.

The most referred demands were: "convenient price", "frequency", "safety/security", "space" and "speed". In short, these results suggest that users demand "cheap" or "free" transportation, with a frequent schedule, in which "people should be able to sit" or, at least, one that is less crowded so that they don't have to "crush or push people". They would like "to go from point A to B in the fastest time possible".

Problems:

Regarding the problems, safety, security and reliability were the most mentioned issues. According to the participants' opinions, the mobility systems are not safe as should be, encompassing both robustness and reliability of the mechanical components and the associated technology, also embracing the payment process and the quality of the maintenance. About security, the participants declared they do not always trust people, referring, in many cases, to the driver or the other people co-sharing the vehicle with them. Other less frequently mentioned topics were the poor quality of the information provided on how to use the transport system. Participants state they need clear and reliable information that can avoid/reduce difficulties about understanding the routes and timetables. They also say that they often doubt the information provided is updated and/or is accurate/valid. The "Sustainability" topic was also raised during the focus groups, however, the percentage of times was quite low, this suggests sustainability was not a hot concern or demand.

The statements below reflect the general perspective of users that mostly refer safety/security issues on using/sharing a given transport mode:

- "For me, everything that you have to share with people is not an option. I come from a city that is not safe, especially for a woman. I know that rape and harassment are not common here, but still, I would never take these options (sharing) in any circumstances."
- "I am kind of afraid of car sharing, I can crush a car that's not mine."
- "The only reason why I won't choose autonomous vehicles that I'm scared that something might go wrong. So public transportation is my choice, I mean, there's also a possibility that I'll crash there, but at least, I'll have somebody to follow."
- "As a woman, I wouldn't prefer it (ride-sharing)."
- "I don't feel safe in Uber. Sometimes because of the driver and sometimes because of the payment method. They take me from a long distance."
- "I don't want a stranger in my car."
- "Drivers don't like bikes on the road, they hate it."

Statements related to "time" concerns were also frequently indicated. Associated with this issue we could find some concerns with the lack of confidence in apps that, according to them, provide little credible information about timetables. In cases where the user needs to reach somewhere quickly, personal cars or on-demand transport, which were less sustainable options, were chosen immediately. The statements below illustrate the participants' time concerns.

- "I use it every day. The apps are accurate mostly but sometimes it just doesn't work. Or it says five minutes. You wait, and then the bus appears, you realize that thirty minutes passed."
- "I work in Carnaxide and to get there is terrible. It's 15 minutes by car but it's 1,5 hours by bus. If my car breaks down, I'll cry."
- "I don't want to wait 40 minutes; I'm not going to be here (bus stop) forever."
- "Big issue about walking is the time that it takes."
- "You need to download (the app for scooter), put your card number, then discover how the thing works. It takes time."

- "If I really need to go somewhere quicker, I think I would call for Uber."

Users claimed that some modes of mobility which "demand additional preparation" before the usage, cause demotivation. The need for validating tickets, waiting on queues, the necessity of choosing appropriate apparel, additional payment, downloading different brand's digital applications were the mostly stated complaints:

- "Coming to work on a bike you have to wear very comfortable clothing like sneakers, shorts, t-shirts. Sometimes you have meetings where you have to dress up so you'd have to have a place where you can put your clothes."
- "Sometimes I just want a green line, without using transport cards, validation on gates or buses. It would be so good."
- "I go to university with my computer and lots of other stuff, I can't ride a bike with them."
- "I already pay for public transport, why am I going to pay for bikes or scooters additionally?"
- "You don't have a helmet and you can be in the middle of the cars so it's dangerous. No one is walking with helmets."
- "I don't want to change my clothes, to carry other clothes, and change it. It is so hard (for biking)."
- "Too many brands (scooter) confuses. I don't have the time to actually research and stuff."
- "If you don't have a (transport) card, you can pay to the driver but there must be a system that you should be able to pay with your credit card directly inside."

The complaints regarding the "cost" of mobility were frequent, as the price of the rides was the primary concern of users. However, the cost was not as relevant as the other issues raised by the participants. The results suggest that if the choice was secure, fast, and did not require any extra action, the price could be considered acceptable or could be negotiated. The statements below illustrate the participants' concerns with price/cost.

- "It'll be great (autonomous vehicles) but it also depends on the price. If they had some kind of systems like the public transport that you pay monthly and a lot cheaper, I would definitely use them."
- "Sometimes it (scooter) can get more expensive than picking on-demand transport."
- "I think public transportation should be almost free or free. Transportation is a right for everyone."
- "It doesn't make sense to me that there are zones and the price is changing. I can't afford to live in the centre and I have to pay more than the people who can afford to live in the city centre. With one ticket you should go everywhere."

Another issue that was mentioned very frequently was the poor availability of public transports given the high demand, especially at the rush hours, which results in overcrowding, causing discomfort and affecting people's choices.

- "Rush hours are very crowded. I have to stand and I am small, so I have to travel with armpits around me."
- "It's so full that you cannot even get inside (bus)."
- "The anxiety when you need to get out of public transport, but you cannot because of people. It is too crowded so you have to crash or push people."
- "There was one time I almost faint."
- "Bus is not always great because it is crowded and uncomfortable, there is no place to sit, you just stand up and it shakes a lot."

Other assorted problems, which could also be subject of improvement, were the number of alternative options available and their convenience, namely in terms of the routes available that not always cover the entire city or require a complex combination of alternate transports to get to certain destinations. Some participants also found the information systems (e.g., maps, timetables, signage) confusing and not as inclusive as they should be. Overall, "Hygiene" and "sustainability" topics were the least mentioned problems.

3.2 *Preference and Satisfaction Levels About the Modes of Mobility*

To analyse the results gathered in the card-sorting phase, we used a heat map, which provided a visual representation of the users' preferences and satisfaction level. The heat map was created having in consideration the number of cards positioned by the participants in the diverse regions of a map drawn by us, which was divided into four quadrants, employing two-axis representing preference and satisfaction.

We created heat maps per each participant and group, in two distinct moments: 1st considering the current experience with the transport systems and 2nd anticipating future experiences of use after watching the smart mobility scenario video. Table 3 shows heat maps representing the overall placement of cards.

Overall, the heat maps suggest the users were extremely satisfied and moderately prefer the "On-demand Transport". They seem to prefer primarily the "Public Transports" and, secondarily, the "Multi-modal transportation systems". However, they were moderately satisfied with these options. The "Walking" mode attained a high preference level and was participants were slightly more satisfied with it than with the other options.

The "Sharing" modes were the least satisfying and the least preferable. Users were not satisfied with these options and barely preferred "Ride-sharing" and "Scooter-sharing". This trend gets worse when looking at "Car-sharing" and "Bike-sharing", which were the least preferred options, with poor satisfaction levels. The placement of the cards of "Autonomous Vehicles" and "Personal Car" was very divergent and we could find a clear trend there. Without more data, we can't state the reason for this result, but we can speculate that the indecision may be due to contradictory

Table 3 The heat maps of overall card placement

Walking

0	0	0	2	2	18
0	0	0	2	0	0
0	0	2	2	4	0
0	0	3	3	4	0
0	0	2	4	0	0
0	0	0	0	0	0

Bike-sharing

0	0	0	0	0	0
0	0	8	0	8	0
1	1	0	2	0	0
1	1	5	3	8	0
0	0	7	3	0	0
8	0	2	2	8	0

Public Transport

8	0	0	0	6	6
0	0	8	1	8	4
0	0	5	4	0	1
0	0	0	0	0	0
0	0	0	0	0	0
0	0	0	0	0	0

Multi-modal Transportation

0	0	4	6	0	8
0	0	4	2	0	8
0	0	6	2	0	0
0	0	8	0	0	0
1	6	0	0	0	0
8	5	0	0	0	0

Scooter-sharing

0	0	0	0	0	0
0	0	0	0	7	0
0	0	1	1	1	0
0	0	1	1	0	0
0	0	5	0	0	0
8	8	14	7	0	0

Ride-sharing

0	0	0	0	0	0
0	0	2	6	0	4
0	0	0	0	0	0
0	0	8	0	0	0
2	12	0	0	0	0
13	2	0	0	0	0

Autonomous Vehicles

0	0	0	0	8	0
0	8	0	0	8	0
0	0	2	4	0	2
0	0	5	1	8	2
0	0	0	0	1	0
0	0	0	0	4	0

Car-sharing

0	0	0	0	0	0
0	0	0	8	8	0
0	0	0	3	0	0
0	0	3	5	0	0
0	0	6	2	4	0
0	0	0	0	10	0

On-demand Transport

0	0	0	4	4	4
0	0	0	4	0	0
0	0	0	0	9	0
0	0	8	0	11	8
0	0	0	0	4	0
0	0	0	0	0	0

Personal Cars

0	0	2	6	0	0
0	0	0	0	0	0
0	4	0	0	8	0
8	0	0	0	0	0
2	0	0	4	0	0
2	0	2	6	1	3

Colour scale/Placement Times

VALUES
20
16
12
8
4
0

Chart axis details

Mostly preferred Least satisfied	Mostly preferred Mostly satisfied
Least preferred Least satisfied	Least preferred Mostly satisfied

opinions about personal cars and, in the case of autonomous vehicles, to the absence of previous experiences with this type of vehicles.

3.3 *The Impact of Smart Solutions on Expectations About the Mobility System*

As said before, we requested participants to revise their cards placement after watching a video about smart mobility solutions. The intention was to show them a glimpse of a future scenario of mobility within smart cities and collect their anticipated user-experience.

Results reveal no big differences in participants' preferences and satisfaction on most of the sharing possibilities: "Scooter-sharing", "Car-sharing", and "Ride-sharing", with exception of "Bike-Sharing", which we could observe a positive shifted. Without more data, we cannot say with certainty whether people continue to have doubts about these modes of mobility, even when they are presented in apparently perfect working condition (i.e., without most of the problems previously identified) or if the scenario presented was not credible enough to affect their assessment. Nevertheless, "On-demand Transport" and "Personal Car" preference levels got lower, which could be considered a positive change in favouring sustainability. Plus, users were more satisfied and willing to adopt "Public" and "Multi-Modal" transportation modes. Overall, this suggests, at least to a certain degree, that people are sensitive to changes regarding their mobility decisions if the system can offer a good user experience all over the journey. However, we also observed that "Walking" became the least preferred and least satisfactory option, losing terrain against less sustainable options.

4 Conclusion

This research is part of our efforts to understand users' behaviours and the obstacles that can prevent them to be engaged with sustainable mobility modes.

Understanding uses' minds represent a major challenge when designing a future smart and sustainable mobility system. The solutions that are better judged (i.e., better user experience) might have better odds to became adopted, no matter their sustainability degree. By identifying the users' concerns and demands, their biggest complaints and the problems they judge as more relevant, which can hinder their daily displacements, this research provides informative data for designers and other professionals in different fields to develop future mobility solutions.

The present findings reveal some of the users' most relevant demands and problems about mobility modes, which need to be dealt with appropriately when designing future mobility systems. The results also show that a future smart city, at least one that fits the scenario showed, can have both positive and negative impacts on users' mobility choices. Options which are currently being suggested as good choices for sustainability, such as "sharing" mobility modes, were not appealing to users regardless of future smart possibilities. Also, improving the overall quality of transportation negatively affected the attractiveness of "Walking", which is the most advocate mode for non-motorized sustainable mobility.

Even though "sustainability" topics were mentioned as demands for mobility and asserted as a problem that must be solved, the frequency it was brought to the conversation was very low compared to other topics. This also suggests that sustainability in mobility is not yet the main concern for people. Thus, researchers, designers, and experts in related fields concerned with sustainability cannot count with much support from users currently. Which highlights the need to run more design for behaviour studies like this one, in an attempt to change the current preferences and expectations for paving a better path for introducing innovative solutions with success.

Mobility is an important requirement for social and economic development and today's culture of mobility is unsustainable [19]. Even though smart mobility systems, smart cities, and IoT possibilities create various beneficial opportunities for our society, the power of being able to choose to use these opportunities hold by people. Benefiting from these results, future work should concentrate on enhancing users preference and satisfaction levels of people about non-motorized mobility by finding possible solutions to their wants and needs.

References

1. Baxter, K., Courage, C., & Caine, K. (2015). *Understanding your users. A practical Guide to user Research Methods* (2nd ed.): Morgan Kaufmann.
2. Bloor M, Frankland J, Thomas M, Robson K (2002) Focus groups in social research. Sage Publications, London, UK
3. Chow J (2018) Informed urban transport systems. Classic and emerging mobility methods toward smart cities, Elsevier
4. Cyr J (2015) The pitfalls and promise of focus groups as a data collection method. Sociol Methods Res 45(2):231–259. https://doi.org/10.1177/0049124115570065
5. Deloitte University Press (Producer) (2016) The future of mobility: ben's journey. Deloitte insights. Retrieved from https://www2.deloitte.com/us/en/insights/multimedia/videos/roadmap-for-future-of-urban-mobility.html
6. Eremia M, Toma L, Sanduleac M (2017) The smart city concept in the 21st century. Procedia Eng 181:12–19. https://doi.org/10.1016/j.proeng.2017.02.357
7. Etezadzadeh C (2016) Smart city—future city? smart city 2.0 as a livable city and future market. https://doi.org/10.1007/978-3-658-11017-8
8. Festinger L (1985) A theory of cognitive dissonance, 2nd edn. Stanford University Press, Stanford, California, USA
9. Frankel L, Racine M (2010) The complex field of research: for design, through design, and about design, July 2010. Paper presented at the Proceedings of the Design Research Society (DRS) international conference. http://www.drs2010.umontreal.ca/data/PDF/043.pdf
10. Gailing L, Naumann M (2018) Using focus groups to study energy transitions: researching or producing new social realities? Energy Res Soc Sci 45:355–362. https://doi.org/10.1016/j.erss.2018.07.004
11. Giffinger R, Fertner C, Kramar H, Kalasek R, Pichler N, Meijers E (2007) Smart cities: ranking of European medium-sized cities. In V. U. o. T. Centre of Regional Science (ed)
12. Guest G, Namey E, Taylor J, Eley N, McKenna K (2017) Comparing focus groups and individual interviews: findings from a randomized study. Int J Soc Res Methodol 20(6):693–708. https://doi.org/10.1080/13645579.2017.1281601
13. Hessel V (2015) Smart mobility—a tool to achieve sustainable cities. http://www.vt.bgu.tum.de/fileadmin/w00bnf/www/VKA/2014_15/150212_Smart_Mobility_v5_TUM.pdf
14. Lyons G (2018) Getting smart about urban mobility—aligning the paradigms of smart and sustainable. Transp Res Part A: Policy Practice 115:4–14. https://doi.org/10.1016/j.tra.2016.12.001
15. Manning C (2009) The psychology of sustainable behavior. Tips for empowering people to take environmentally positive action, pp. 34. Retrieved from https://www.pca.state.mn.us/sites/default/files/p-ee1-01.pdf
16. Manolas E (2015) Promoting pro-environmental behavior: overcoming Barriers. Aegean J Environ Sci 1:13–21

17. Nili A, Tate M, Johnstone D (2017) A framework and approach for analysis of focus group data in information systems research. Commun Assoc Inf Syst 40:1–21. https://doi.org/10.17705/1CAIS.04001
18. Onwuegbuzie AJ, Dickinson WB, Leech NL (2009) A qualitative framework for collecting and analyzing data in focus group research. Int J Qual Methods 8(3):1–21. https://doi.org/10.1177/160940690900800301
19. Perschon J (2012) Sustainable mobility. Recommendations for future-proof transport strategies. (36). https://www.eurist.info/wp-content/uploads/SEF_Paper_English-1.pdf
20. Rabiee F (2007) Focus-group interview and data analysis. Proc Nutrition Soc 63(4):655–660. https://doi.org/10.1079/PNS2004399
21. Reis L (2017) The challenge of future mobility. https://www.anacom.pt/streaming/1LuisReis.pdf?contentId=1423665&field=ATTACHED_FILE
22. Sadalla EK, Krull JL (1995) Self-presentational barriers to resource conservation. Environ Behav 27(3):328–353. https://doi.org/10.1177/0013916595273004
23. Sanseverino RR (2014) Competitive Urban Models. In R. S. E., S. R., V. V., Z. G. (eds) Smart rules for smart cities. SxI—Springer for Innovation/SxI—Springer per l'Innovazione, vol 12, Springer, Cham
24. Selvefors A, Strömberg H, Renström S (2016) What a designer can change: a proposal for a categorisation of artefact-related aspects. Paper presented at the proceedings of DRS 2016, design research society 50th anniversary conference, Brighton, UK, 27–30 June 2016
25. Solomon MR, Bamossy GJ, Askegaard ST, Hogg MK (eds) (2016) Consumer behaviour: a European perspective, 6th ed., Pearson
26. United Nations (2016) The World's Cities in 2016 Retrieved from https://www.un.org/en/development/desa/population/publications/pdf/urbanization/the_worlds_cities_in_2016_data_booklet.pdf
27. Viechnicki P, Fishman T, Eggers WD (2015) Smart mobility: reducing congestion and fostering faster, greener, and cheaper transportation options. https://www2.deloitte.com/content/dam/insights/us/articles/smart-mobility-trends/DUP_1027_Smart-Mobility_MASTER1.pdf
28. Wendel S (2014) Designing for behavior change: applying psychology and behavioral economics, 1st Ed, O'Reilly Media.

Preventing Single-Use of Plastic Packaging. Design Strategies for Circular Business Models: Refill, Reuse and Recycle

Ana Espada, Isabel Farinha, and Carlos A. M. Duarte

Abstract The analysis of the plastic packaging value chain is fundamental to understand how to avoid the “end-of-life”, excess of recycles plastic, the logistics and new services among the supply chain, supported by the new technologies (e.g. IoT, AI, big data). Recognizing Earth as a limited source of resources and energy and look at the waste and pollution as a potential and not as a defeat, are the foundations of the circular economy philosophy. Product design in particular, as well as business in general, are feeling an increasing pressure on moving from a linear to a circular approach to help reduce our global sustainability pressures [5]. Designers, innovators, and decision-makers in businesses, aim to become active actors on the necessary shift from an industry relying on fossil resources (Bakker, C. (2019). Products That Last: Product Design for Circular Business Models. Laurence King Publishing.). To give insights in the current product design and circular business model strategies able to promote the shift from a linear to a circular economy, preliminary interviews are presented as the first confrontation to a collapsible plastic bottle. The goal is to develop knowledge to support the framework to a more detailed research.

Keywords Product Design · Circular Economy · Plastic Packaging

1 Introduction

Moving from a linear economy to a circular economy have felt the combined effects of changes in design, rather than being destined for disposal, materials should maintain their utility and value and flow back into the cycle.

The transition to a business, which creates monetary and environmental benefits, has paved the way for the development of new business models focused on increasing resource efficiency, ensuring that the value of products, materials and waste is maintained.

A. Espada (✉) · I. Farinha · C. A. M. Duarte
UNIDCOM/IADE - Unidade de Investigação em Design e Comunicação, IADE - Faculdade de Design, Tecnologia e Comunicação, Universidade Europeia, Lisbon, Portugal

E. Duarte and C. Rosa (eds.), *Developments in Design Research and Practice*, Springer Series in Design and Innovation 17,
https://doi.org/10.1007/978-3-030-86596-2_33

The relationship between design and actual success of business has been often addressed, nevertheless, the overall objective of this study is to explore design strategies related with the circular economy, where product-design is a source of innovation to business models, impacting on a system level, the plastic-packaging value chain.

This research explicitly addresses the problematic of the financial and environmental costs in transportation, storage and collection of blow-moulding plastic bottles, produced in a centralized business model.

Design strategies for circular business models may urge the plastic-bottles manufacturers to answer to their ownership obligations and shift from linear to a circular economy. This research focus on:

1. Avoid single-use of plastic bottles through refill and reuse.
2. Can design be a source of change to prevent single-use of plastic bottles?

It is important to understand the process of design for sustainability to find whatever implications may occur and what does centralized production means in the present design approach to circular business models.

1.1 Motivation and Relevance

Integrating circular economy principles in an early stage of the design process is important because once the product specifications are being made, only minor changes are usually possible. Ultimately, from general design parameters, innovation in products has to emerge taking into consideration human aspirations and worries – answering to a more participative, more environmental-conscient and more demanding end-user.

The circular business models are being advocated but yet they are not widely practiced. Today's world lives beyond large changes at topics such as industrial symbiosis, zero waste goals, non-energy industrial materials and biowaste, that influence all activity sectors, moreover the people day-by-day lives. The plastic manufacturers are one of the business stakeholders that are actually suffering large oscillations of environmental and societal paradigm.

"The anti-plastic sentiment is distracting us from the net environmental benefits of plastic relative to alternative materials when properly recycled" [23]. Replacing plastic with alternative materials such as paper and cardboard, glass, steel, aluminium, textiles, rubber, cork, results in a significant net negative environment impact [7, 23].

The plastic-packaging industry model is linear and dates from the Industrial Revolution. Plastic Packaging global economy develops around the model of consumption "take-make-use-dispose", but various social, economic and environmental factors make it no longer sustainable.

2 Brief Literature Review

Recognizing Earth as a limited source of resources and energy and look at the waste and pollution as a potential and not as a defeat, are the foundations of the circular economy philosophy.

Product design in particular, as well as business in general, are feeling an increasing pressure on moving from a linear to a circular approach to help reduce our global sustainability pressures [5]. Designers, innovators, and decision-makers in businesses, aim to become active actors on the necessary shift from an industry relying on fossil resources. The selling drivers are now centre on generate services that profits from the flow of resources over time, continually reuse products and materials and using renewable energy [4].

2.1 The Theoretical Foundation

The origins of industrial ecology date from 1972, a global environmental movement started at the United Nations Conference on the Human Environment in Stockholm. Almost 30 years after the Stockholm declaration, General Motors published the paper "Strategies for Manufacturing" where the environmental impacts of manufacturing were discussed and the resources reduction and the waste accumulation were speculated [9].

"The traditional model of industrial activity … should be transformed into a more integrated model: an industrial ecosystem. In such a system the consumption of energy and materials is optimized, waste generation is minimized, and the effluents of one process … serve as the raw material for another" [15].

It was in a publication in 1994, in the fields of industrial ecology, where the idea of industrial metabolism was first presented [3]. The goal of industrial ecology was to achieve a state close to optimal, where nature prevails the most. Andrews [1] The industrial metabolism integrated all the industrial process that transform materials and energy, plus the labour inherent to those transformations, into products and industrial waste. A systematic approach that "complete or nearly-complete internal cycling of materials" [3].

Systems approach

Circular design involves many disciplines. Among them, the technical, financial and environmental fields, of the anthropology, the sociology, the ethical philosophy and design established an analogy between industrial and natural ecosystems [9].

In 2009, the book "Materials and the Environment: Eco-Informed Materials Choice", written by the engineering professor Michael Ashby of Cambridge University, was a piece of evidence for the broad acceptance of the design contribution for the environmental strategies in business: eco-design [2].

In 2010, Jackson [9] definition of sustainability is one of the most succinct:

Sustainability is the art of living well, within the ecological limits of a finite planet.

Below, is presented a historical review of industrial ecology main concepts, based on the paper "Industrial Ecology's First Decade": [17].

- **Life-cycle assessment** (LCA) is the methodology that seeks to identify the environmental impacts of a product or process at each stage of its life cycle.
- **Design for environment** consider environmental factors such as minimizing energy requirements, decreasing manufacturing discards, choosing more sustainable materials.
- **Material flow analysis** (MFA) is the methodology for quantifying the stocks, flows, inputs, and losses of resource.
- **Socioeconomic metabolism**, the ultimate task of this field of study is to relate resource transitions to societal change and to prospects for and measurement of sustainability.
- **Urban Metabolism** quantified flows of human and animal food, glass, plastics, sewage, sulphur dioxide emissions and the analyses of intensification of food, water and material consumption per capita.
- **Industrial Symbiosis** is the organization of industrial organisms and their processes so that "the waste of one process is the material for another".
- **Circular economy** is an industrial system that is restorative or regenerative by intention and design. It replaces the 'end-of-life' concept with restoration, shifts towards the use of renewable energy, eliminates the use of toxic chemicals, which impair reuse, and aims for the elimination of waste through the superior design of materials, products, systems, and, within this, business models [10].

Today, the parallelism between the circular economy and design principles is fully assumed. In both disciplines, between the technical, economic and environmental fields; anthropology, sociology and ethical philosophy establish an analogy between industrial and natural ecosystems [9]. The acceptance that the activities of the man and society have a negative impact on the planet's resources has been accepted for the first time by the user - the basic premise of sustainability.

Figure 1 represents the main moments of the historical evolution of environmental philosophies applied to design [19].

The circular approach evolves the end-user more than ever, especially in the elimination of waste, however, user awareness and contribution in a wide range of product lifecycle phases, or even by maintaining a more direct relationship with the manufacturers of the products (online shopping), is setting a rising trend of

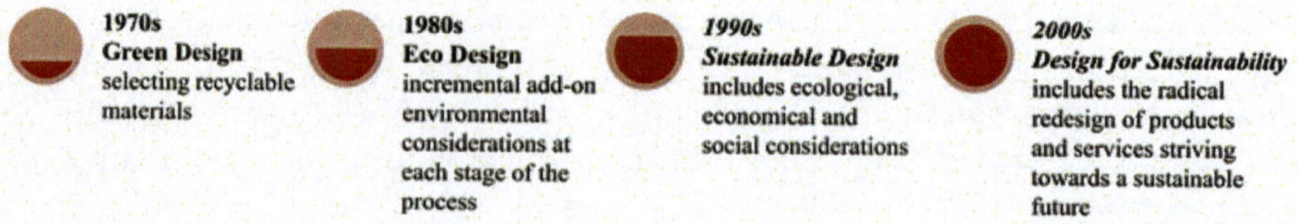

Fig. 1 Historical evolution of environmental philosophies applied to design [19]

direct participation on the shift to circular economy. Next section briefly explores the contemporaneous sentiment that has been rising in a modern consumptive but also highly aware society.

2.2 *Anti-Plastic Sentiment*

The confrontation with a continent size patch of garbage composed mainly of plastic floating in the Pacific Ocean produced the anti-plastic sentiment among our modern consumptive society. It highlighted, in fact, that much of the discarded plastic packaging is not actually reused or recycled, rather, it is disposed in landfills, ending-up in the ocean and other waterways, provoking large damage on Nature [8]. Environmental awareness has grown and raised around the world, an anti-plastic sentiment.

As can be seen in Fig. 2 the landfill disposal of solid plastic waste (SPW) in Europe has decreased by 38% between 2016 and 2014. In parallel, the SPW used for energy recovering and recycling increased 46% and 64%, respectively. An optimistic trend for the plastic industrial sector and for the consumers.

Moreover, despite the anti-plastic sentiment, plastic is also crucial as a circular recourse. Polymers melting temperature is about 300 °C, glass melting point is at 1600 °C. The energetic consumption is definitely outstanding. In the other hand, if we look at paper and cardboard as an alternative, the level of water consumption during and recycling is extremely high.

Those observations raise some issues about the fundaments behind the anti-plastic sentiment and the ethics on the communication being done. And the pollution of the air and the water scarcity may be a bigger problem to face [22].

To look at the circular economy as a simple closing loop "take-make-use-recycle" may not be sufficient. The single-use policy [14] demands 90% separate collection target for plastic bottles by 2029 (77% by 2025), as well as a target to incorporate

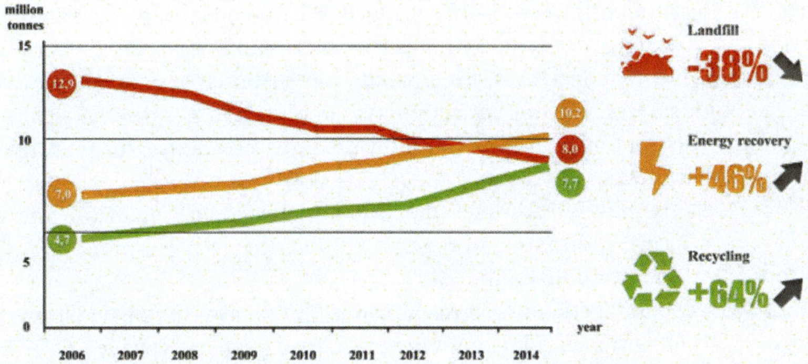

Fig. 2 Evolution of landfilling, recovery and recycling of SPW in EU [21]

30% of recycled plastic in all plastic bottles as from 2030 [14]. Furthermore, the annexe III of the European strategy for plastics set the target that 10 million tonnes of recycled plastics find their way into products in Europe in 2025 (4 million tonnes in 2016) [13].

The controversy is in increasing recycling as the main strategy to scale up the circularity of plastics, but the industry has no demand for all the recycled material [25] [22]. It is a capacity that nobody needs. Polyethylene terephthalate (PET), is already 100% recycled, and only in 10 years is expected to close the loop, since the stakeholders of the value chain need to adapt [25].

3 The Evidences of Plastic Paradigm

The main assumption that emerge from the literature review is that it will be required to understand the circular business models archetypes. It is crucial to generate guidelines that will draft the future recommendations to product design in general and in particular for the plastic industry. The transformation is needed to occur on a system and ecosystem level and not only on a product level, but the integration of circular economy principles in an early stage of the design process, with impact in all value chain, is fundamental to shift from a linear to a circular economy approach.

3.1 Gap in the Literature

The plastic-packaging industry model of consumption is linear and dates from the Industrial Revolution. Plastic Packaging global economy develops around the model of consumption "take-make-use-dispose", but various social, economic and environmental factors mean that it is no longer sustainable.

Centralized production and air transportation. The decentralization of the plastic packaging production (or in-hole production), by opposition to the centralized production, was a natural consequence of the regulation. However, that restriction is a demand for alimentary products. Meanwhile, the sustainable concepts have been generalized widespread, for market segments such as house-care and healthcare, the centralized production is still a common scenario nowadays for packaging of non-alimentary products.

This research explicitly addresses the problematic of the financial and environmental costs in transportation, storage and collection of blow-moulding plastic bottles, produced in a centralised model.

Design for long-last, refill and reuse, are huge challenges for plastic industry and for designers. To integrate circular economy principles in an early stage of the design process, because small changes made to existent products, serve a linear economy but it is no longer fitting the circular economy philosophy.

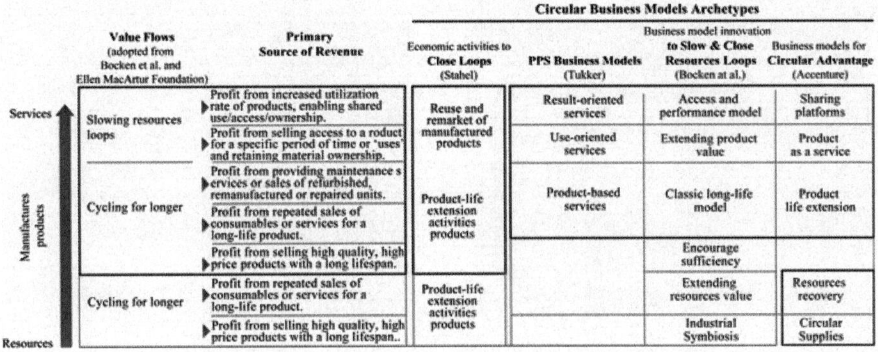

Fig. 3 Circular business model archetypes crossed with value flows

The plastic packaging value chain analysis is fundamental to understand: (i) how to end the "end-of-live" concept; (ii) how to avoid the excess of recycled plastic; (iii) the changes required in logistics and new services among the supply chain (e.g. take-back systems, remanufacturing technologies); (iv) how to use digitalization and new technologies (e.g. IoT, AI, big data) to support circularity.

The categorization of circular business model archetypes The chosen archetypes to the categorization of the circularity in the business models are based on the model presented in the Fig. 3, suggested by Moreno et al. [19]which crosses the value flows synthesized from Bocken [5] and Ellen MacArthur Foundation [10]. The choice was made to emphasize the manufacture of the product that integrates services in the value proposition (outlined in red, Fig. 3).

The methodology established consists of the analysis of three circular business models archetypes (refill, reuse and recycle), and position of the stockholders along the value-chain of a plastic-packaging, in a case study analysis.

3.2 *Selected Model for the Analysis*

In order to focus the analysis, the case study of WisePack is used. The principle of WisePack is the collapsibility of the packaging (Fig. 4) before the filling phase and after the use. Reducing transportation costs and the carbon footprint, as well as it facilitates its collection and reduce the contamination after usage, dropping the negatives impacts of the washing process [11, 12].

The research focuses on the activities or processes along with critical parts of the circular approach, to examine what are the requirements to upgrade efficiency and innovation capacity in the plastic packaging value chain through the design process. Promoting plastic packaging reliability, and duration through product-service systems (refill cycles, reuse and recycling).

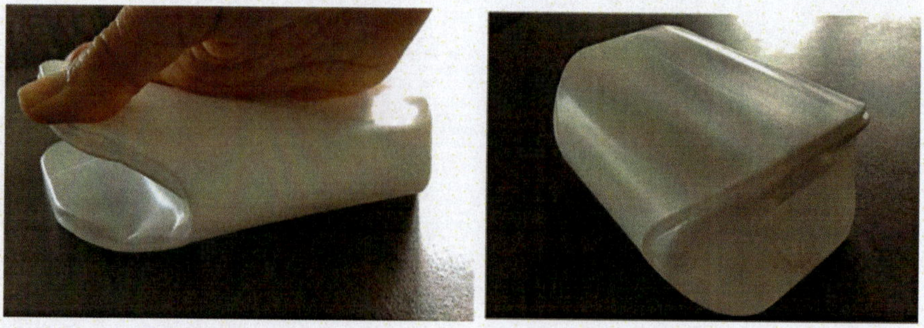

Fig. 4 WisePack functional prototype 2019

4 Research Structure and Methodology

This research problem aims to gain concrete, contextual, in-depth knowledge about the specific problem, such as WisePack that allows exploring the key characteristics and implications of the design in the context of circular business models (Fig. 4).

According to Sampieri, Collado and Lucio (2006), qualitative research gives data depth and a wide possibility of interpretation, as well as details and flexibility. The choice of different collection resources (Table 1) is due to the qualitative nature of the research and aim to raise information from different angles to understand the constraints and objectives among different stakeholders of the value chain.

The research structure is described below, and this article presents the findings from the preparation phase:

Preparation relating theory and problem delimitation: literature review and semi-structured interviews;
Exploratory research to uncovering new concepts and ideas that need to be incorporated: collective case study and value chain analysis;
Execution exploring the design case through prototypes to verify (or not) the established assumptions: design conceptualization/ user behaviour driven;

A collective case study was selected in order to improve knowledge about the universe that the problem belongs. Gomm et al. [16] It aims to collect data that provide greater familiarity with the problem, with a more refined view in other to generate design ideas or identify constraints.

The first phase consists of the value chain analysis, aiming to identify the key-stakeholders and their constraints or requests that implicated the product or service design (Fig. 5).

The second phase of the exploratory research comprises semi-structured interviews with scientific experts, industrial association, manufacturers, fillers, retailers and end-users. Selltiz [24] The semi-structured interviews have been chosen, in order to obtain subjective information. Boni and Quaresma [6] Accordingly to Guerra [18],

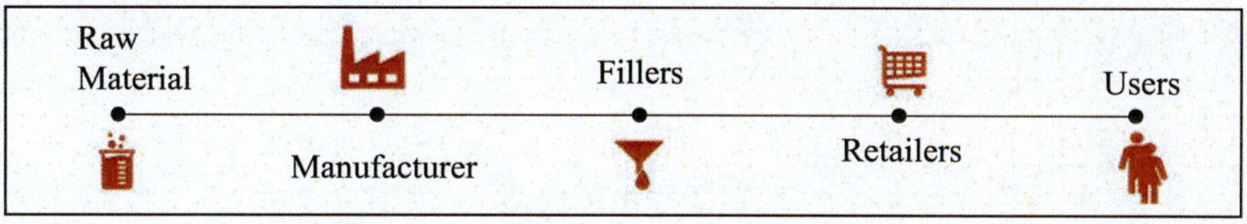

Fig. 5 The supply chain of the plastic packaging industry

"to protect sources" is important to "presupposing neutrality, confidentiality, clarity of ideas to be able to transmit them and generate valid results".

The interviews were not conducted mechanically. This approach allowed spontaneous follow-up questions to be asked in order to clarify the details of the subject. The interviews ranged from 60 to 90 min. An overview of the conducted interviews is shown (Table 1).

The *interview script* has been completed over time and the order of the questions were not necessarily followed to maintain a fluid speech.

Interventions were reduced to a minimum and the content and language were kept true to the interviewees' discourse [18] and translated into English whenever the interview was conducted in Portuguese. During the interviews, general aspects of the observed behaviours were collected and will be included in the next section.

4.1 Results of the Interviews and Value Chain Analyse

Three major actions raised from the interviews and are the main required actions to promote the shift from a linear to a circular business model in the plastic packaging industry specifically (and plastic industry in general):

(1) to increase collaboration of all value chain to develop waste-free design
(2) to increase communication towards design for long-last, refill and reuse;
(3) to reduce the dependency on fossil resources, introducing recycled resins.

In addition, a simplified life cycle assessment (LCA) emerged from the interviews, especially from the last interview mentioned in Table 1, the LCA researcher. This interview turned into a series of brief discussions about possible approaches for an LCA.

The extrapolation of the data extracted from the interviews and the researchers' ethnographic analysis, suggested that a detailed LCA should be set aside at this stage. Due to the recognition that the reduction in volume during transport and the prolongation of the circulation time of each package (Fig. 6), are advantage known and accepted by all the stakeholders.

However, when the object of analysis is a specific packaging for a specific product and context of use, for a specific transition, an LCA analysis is recommended.

The relevant information to retain from this exploratory phase of the research is in one hand, that the ethnographic research conducted used interviews and observation

Table 1 Overview of the conducted interviews

Type of organization	Design and Value Creation What is the first impression?	Process and Operations What are the key transformations?	Partners and network What are the key partners or channels?	Craving to innovate How does it promote innovation?	Perception and expectations What happened afterwards?	Reflection How do you think about it afterwards?
Manufacturer (centralized production) Portugal	.massive impact on final cost .environment is our priority	.possible to produce with the current machinery and materials	.other flexible need new processes/mach. .users are afraid of spilling chemicals (dislike flexible pack.)	.we need to differentiate, but only with realistic innovation	.we sell less .charge to license per each time the packaging is refilled?	.interesting, changes the geometry but not require significant investment on technology, material, machine or mold
Manufacturer (decentralized production) Portugal	.not interesting (...) reusable packaging goes against our business model	.hole-in-the-wall reduces shipping .weight reduction . incorporate PCR PET / HDPE	.work with key actors in the supply chain .incorporate recycled raw materials in customers products	. promote recycling . reduce the consumption of natural resources . help mold consumer's behavior	. implementation of deposit return systems .design and producing reusable packaging .reduce emissions and operational waste	.the shelf is very important, the contact with the end-user .prioritize recycling
Filler (production of goods) Belgium	. >3 liters avoid pet competition . >3 liters, the advantage of volume reduction	.identify big consumer problems and look for business and tech. partners	.work closely together along the plastic value chain (collection, transformation, marketing.	. disruptive approach that enables us to accelerate and elevate our R&D processes	.customer valorize environment less than product lower costs	.cooperation between stakeholders is key .worfc together along the plastic value chain

(continued)

Table 1 (continued)

Type of organization	Design and Value Creation What is the first impression?	Process and Operations What are the key transformations?	Partners and network What are the key partners or channels?	Craving to innovate How does it promote innovation?	Perception and expectations What happened afterwards?	Reflection How do you think about it afterwards?
Retailer Portugal	.want to reduce the waste at our stores close to zero .wooing on ecodesign, looking for the best solutions	.understanding the client's needs and preferences, followed by a moment of experimentation	.promoting a stronger relationship between producers and academia and research centers	.products made from the surplus of the retail chain (...) made from end-of-life food products from our stores	.reduce, recycle and reuse or packaging its own brand plastic packaging, anticipating 2025 Sustainable Dev Goals	5 strategic pillars: .Energy & Climate Chan. .Circular Economy .Responsible Sourcing .Responsible Offer .Awareness & Education
Recycling Park Belgium	.package arrive broken, deformed	.all brands are collected together	.unaware of how others (recycle chain) proceed	.no innovation	.just waste .negligible value	.taxation based on weight or volume
Industrial Federation + Academic Expert Belgium	.citizens, need to simplification (IF)	.take-back process also impacts environment .each case needs an LCA (AE)	.it is difficult to motivate the take back . monetary incentive, as in glass bottles return on supermarkets	.it could use new materials, with sensors monitoring quality (IF)	.need policy development (AE) . IoT technologies can support the product- service design	.companies are very open to circularity, but their focus is economic benefits, not environmental

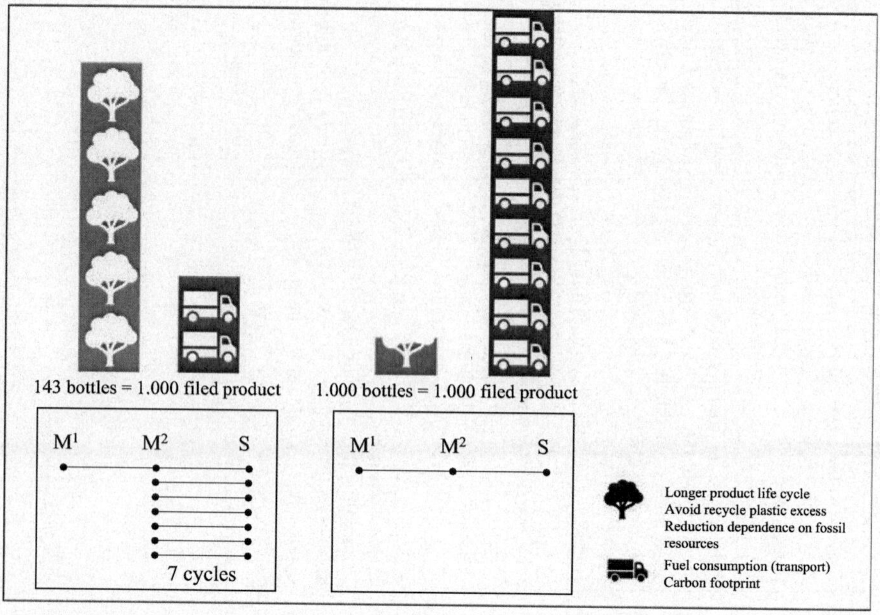

Fig. 6 Ethnographic life cycle assessment emerged from the interviews

as the preferential interaction means. And on the other hand, may have changed some of the assumptions made from the representants of the different stakeholders of the value chain.

In order to better exemplify those contradictory assumptions, an additional diagram is presented (Fig. 7). The position of the icon indicating the filler (e.g. detergent producer), gives high weight to the "standing-out in the shelf and brand trust" as a relevance to the sales volumes. And this assumption is mostly due to the user criteria in the moment of the purchase, but the end-user or buyer, prints more weight into their decision based on price and waste-free packaging design.

It is interesting to see that the position of the user's perception also favours environmental communication and the brand's popularity. Adopting a clear position in favour of cost reduction and waste reduction.

In turn, in line with the initial assumptions based on the literature review, packaging manufacturers place greater emphasis on reducing transport costs due to the popularity of their brand and environmental marketing. And retailers place greater importance on the brand's popularity, on environmental communication as a sales tool.

This cross-analysis will be addressed in further work. Namely, during the development of the case study and the scenarios description for circular business models. In addition, the analysis of the physical prototype during the interviews, stimulated a deeper understanding. This will also be validating the design requirements and

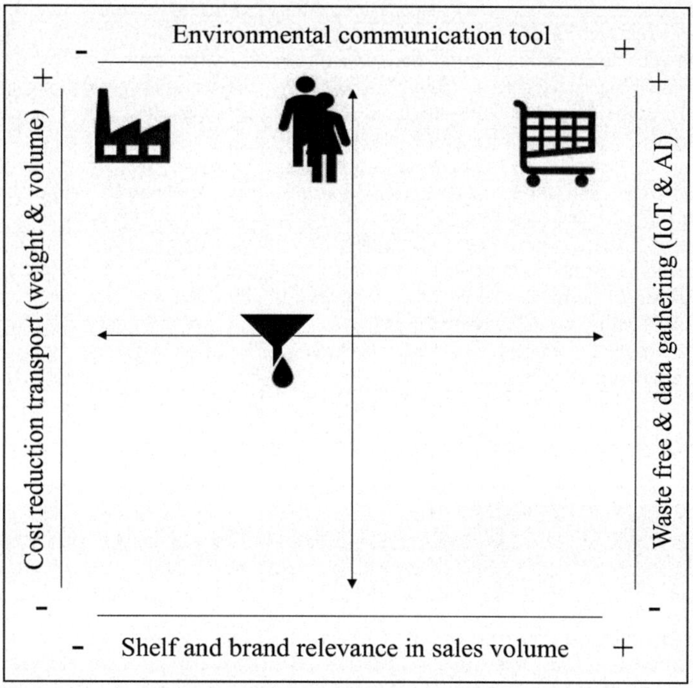

Fig. 7 Ethnographic assumptions emerged from the interviews

business points of view among the value chain, which will be fundamental to the delimitation of the research problem for a broader investigation.

References

1. Andrews D (2015) The circular economy, design thinking and education for sustainability. Local Econ 30(3):305–315. https://doi.org/10.1177/0269094215578226
2. Ashby MF (2012) Materials and the environment: eco-informed material choice. Elsevier
3. Ayers RU (1994) Industrial metabolism: theory and policy. 8
4. Bakker C (2019) Products that last: product design for circular business models. Laurence King Publishing
5. Bocken NM, De Pauw I, Bakker C, van der Grinten B (2016) Product design and business model strategies for a circular economy. J Ind Prod Eng 33(5):308–320
6. Boni V, Quaresma SJ (2005) Aprendendo a entrevistar: como fazer entrevistas em Ciências sociais. Em Tese 2(1):68–80
7. Chalmin P (2019) The history of plastics: from the Capitol to the Tarpeian Rock. Field actions science reports. J Field Actions Special Issue 19:6–11
8. Clapp J (2012) The rising tide against plastic waste: unpacking industry attempts to influence the debate. In Histories of the Dustheap: Waste, Material Cultures, Social Justice, pp 199–226
9. Clift R (2015) Taking stock of industrial ecology. Springer, Berlin Heidelberg

10. Ellen-MacArthur-Foundation (2015) Ellen-MacArthur-Foundation-towards-the-circular-economy: business rationale for an accelerated transition-vol.1.pdf
11. Espada A, Rosado L, Vargas L (2007) Método de redução e controlo ds tensões envolvidas na colapsibilidade de uma embalagem totalmente colapsável (Patent No. PT103553B). https://patents.google.com/patent/PT103553B/pt?q=colapsibilidade&oq=colapsibilidade
12. Espada A, Rosado L, Vargas L (2008) Collapsible package (WO Patent No. WO2008020776A1).https://patents.google.com/patent/WO2008020776A1/en?oq=WO2008020776A1
13. European Commission (16 Jan 2018a) A European strategy for plastics in a circular economy. https://eur-lex.europa.eu/legal-content/EN/TXT/HTML/?uri=CELEX:52018DC0028&from=EN
14. European Commission (28 May 2018b) Single-use plastics: new EU rules to reduce marine litter [Text]. European Commission—European Commission. https://ec.europa.eu/commission/presscorner/detail/en/IP_18_3927
15. Frosch RA, Gallopoulos NE (1989) Strategies for manufacturing. Sci Am 261(3):144–152. https://doi.org/10.1038/scientificamerican0989-144
16. Gomm R, Hammersley M, Foster P (2000) Case study method: key issues, key texts. SAGE
17. Graedel TE, Lifset RJ (2016) Industrial ecology's first decade. In Taking Stock of Industrial Ecology, pp. 3–20, Springer, Cham
18. Guerra IC (2006) Pesquisa qualitativa e análise de conteúdo: Sentidos e formas de uso. Lucerna
19. Moreno M, De los RC, Rowe Z, Charnley F (2016) A conceptual framework for circular design. Sustainability 8(9):937. https://doi.org/10.3390/su8090937
20. Plastics Europe (2015) Plastics—the facts 2015. An analysis of European plastics production, Demand and Recovery for 2015
21. PlasticsEurope, Ass. of P. M (n.d.) PlasticsEurope Annual Review 2017–2019. Retrieved 19 November 2019, from https://www.plasticseurope.org/en/resources/publications/498-plasticseurope-annual-review-2017-2018
22. Ragaert K (11 Jul 2019) How to make the Plastic Rehab happen—a cleaer view on today's challenges for sustainable plastics industry
23. Selke S (16 Dec 2018) Plastic is the most environmentally-friendly packaging material. ..Packaginginsights.Com/. https://pi.cnsmedia.com/a/v54lhSj7KzQ=
24. Selltiz C (1967) Metodos de pesquisa nas relações sociais. Herder : Ed. da USP
25. Wiesweg M (12 Jul 2019) Polymer market trends—outlook and sustainable impact

Social Engagement and Cultural Adaptation of Young Refugees Through Gaming and Playful Design

Vanessa Improta and Ana Margarida Ferreira

Abstract The document addresses one of the main themes of the present day, i.e., the problem of forced migration—commonly referred to as seeking refuge. The target user of this study is exiled children who, in most cases, arrive at their destination alone and with few or no primary survival conditions. UNHCR (2015) points out that more than 65 million people worldwide have had to leave their home countries for war, climate change and/or political or religious persecution. Within this number, 1 in 200 children worldwide are refugees, representing 52% of the total refugee population (UNICEF 2016). In this scenario, in addition to being victims of school dropouts due to migration, many children end up developing trauma, anxiety, and suicide attacks, resulting in less favourable conditions, sometimes living in shelters. The methodologies developed by Ellen MacArthur Foundation and IDEO—respectively, Design Thinking and Human-Centred Design—aimed to understand how design can positively influence the cultural adaptation of these children. For the best development of this research, the subject was studied through official data from the leading organizations operating within this panorama and the analysis of some projects that promote social inclusion. Finally, in response to the proposed issue, we defend the relevance of a playful, educational, sustainable and focused on circular economy design. Through it, the user could learn by playing the language and the customs of the new country. Besides, it could prevent the exposure of the users to dangerous situations—which they are vulnerable.

Keywords Inclusive design · Social innovation · Cultural adaptation · Young people · Refugees · Sustainability

V. Improta · A. M. Ferreira (✉)
IADE - Faculdade de Design, Tecnologia e Comunicação, Universidade Europeia,
UNIDCOM/IADE - Unidade de Investigação em Design e Comunicação, Lisboa, Portugal
e-mail: ana.margarida.ferreira@universidadeeuropeia.pt

E. Duarte and C. Rosa (eds.), *Developments in Design Research and Practice*,
Springer Series in Design and Innovation 17,
https://doi.org/10.1007/978-3-030-86596-2_34

1 Introduction

According to Pazmino [11], especially in the last decade, there is a need to create new solutions through real projects to solve social and environmental problems. For this, during the creative phase of a product or service development, it is necessary to prioritize those requirements while taking into consideration technical, economic, aesthetic and symbolic aspects.

The social aspect is also applied to design, targeting areas that the industry has shown few interests and providing solutions that can increase the quality of life. Thus, design can be rebuked along with ideas of solidarity and moral responsibility [11].

That said, this paper addresses one of the main themes of present days, i.e., the problem of forced migration, commonly referred to as seeking refuge. The target user of this study is exiled children who, in most cases, arrive at their destination alone and with few or no primary survival conditions.

According to the UNHCR Report (2018), more than 69 million people worldwide had to leave their countries of origin as a result of war, climate change and/or political or religious persecution. Although the United Nations High Commissioner for Refugees (UNHCR) [13] Data Report stated a drop in the number of refugees in Europe, a large percentage of people continue to risk their lives on alternative routes searching for safe territory. See Fig. 1.

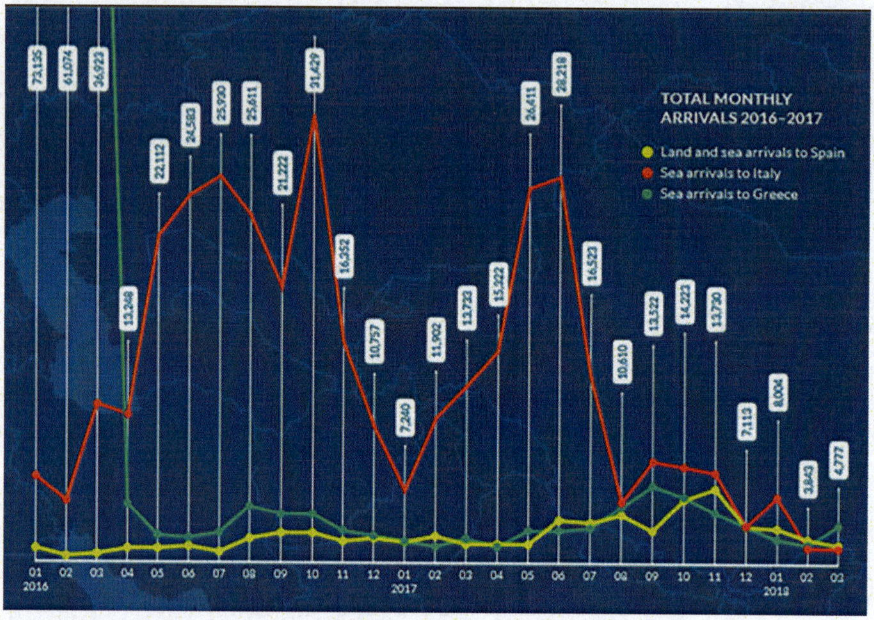

Fig. 1 Evolution of the number of refugees in Europe (Adapted from UNHCR [14])

9 out of 10 refugee children are estimated to be unaccompanied (UNICEF, 2016). The figures include the number of Syrian children arriving by sea in Europe, representing 92%. The European Commission describes this situation as the biggest humanitarian crisis since the time of World War II.

According to the UNICEF's report (2016), 1 in 200 children worldwide is refugees, representing 52% of the total refugee population. In this scenario, due to the time lost from migrating, the children are victims of at least 1-year delay in schooling. Besides, many children end up developing traumas, anxiety, psychological distress, and in worse cases: suicide. Often this results in less fortunate conditions, such as living in shelters.

> They know what hunger, physical suffering or fear are. They have marks that remain in the psyche and in the body, marks that appear in somatizations, in nightmares, in distrust, in the feeling of persecution. As various authors pointed out, these are events that place the child in a vulnerable situation due to a clinical mental health situation (Borges 2014) [10].

Based on the methodologies of design thinking, human-centred design and circular economy—developed by the Ellen MacArthur Foundation and IDEO—the researchers of this study intend to understand how design can positively influence the cultural adaptation of these children.

Brown [4] already mentions that design thinking is the set of principles that can be applied by many people to a wide variety of problems. The author also comments that it is a systematic approach that allows innovation and goes beyond the need to produce a product or service because it is assertive to the point of entering directly into the user's life and may even dictate certain future behaviours.

Thus, design thinkers are based on how to use the product, objects, and services that should be engaged. This enables pattern discoveries as well as the discovery of new ideas, converting problems into opportunities [4].

The phases of this study were based on the design thinking stages proposed by Baeck and Gremett [2], which are summarized in 5 stages: empathy, definition, idealization, prototype creation and testing. In addition, Fig. 2 addresses the main contributions of using this process. The attributes of the method are summarized as flexible, collaborative, constructive, inquisitive, empathic, holistic, iterative, non-judgmental and open-minded (Figs. 3, 4, 5, 6, 7 and 8).

According to Burdeck [5], until the late 1970s the design was done in a deductive and predictable manner. Only in the early 1980s, it began to be inductively considered. As a result, it became more important to search for answers in the user, for whom the product was being developed. The user desire and needs, as well as, the basis of their behaviour should be considered.

For IDEO [8], the process is called "human-centred" because the starting point is the people that the solution is developed for. The Human-Centred Design (HCD) process begins by analyzing the needs, desires, and behaviours of these people whose lives one wants to influence. We seek to hear and understand what they want and seek. Once the user's desire is identified, they begin to examine viable solutions.

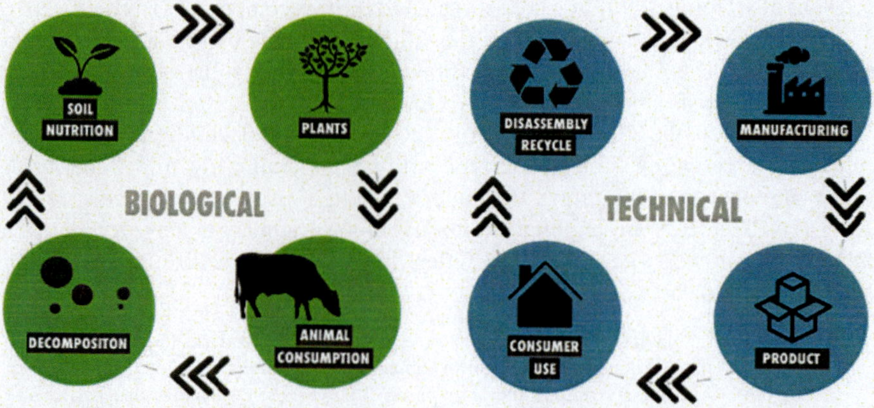

Fig. 2 Biological and technical cycle of the circular economy (Project Author)

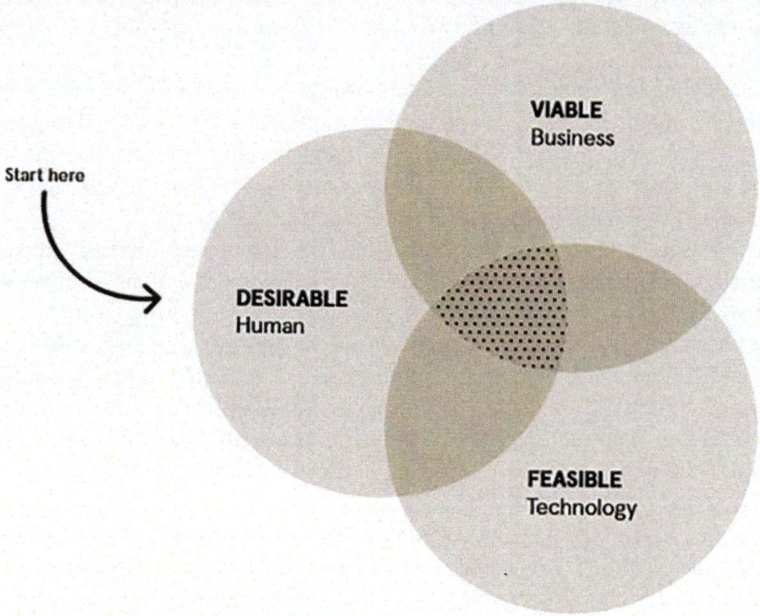

Fig. 3 HCD insertion lenses (IDEO 2017)

The Human-Centred Design process begins with a specific strategic challenge and continues through three main phases: listening, creating, and implementing. During the process the team will change from concrete to abstract thinking, identifying themes and opportunities, and then back to concrete with solutions and prototypes [8].

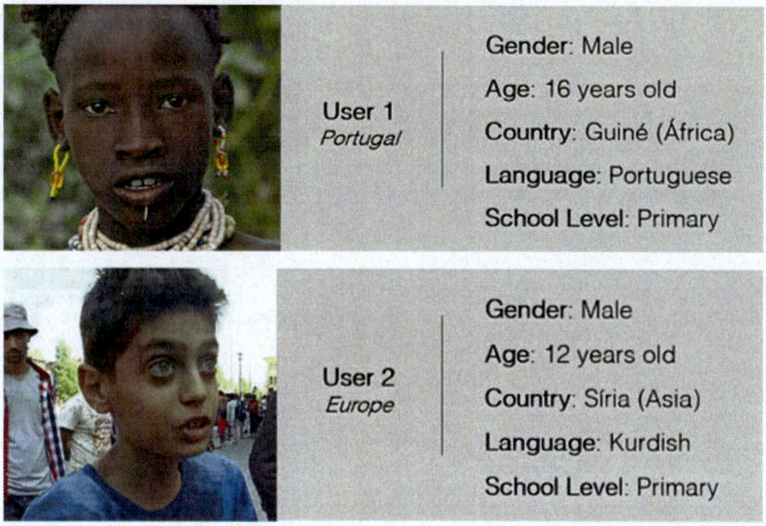

Fig. 4 User's profile (Project Author)

Fig. 5 TÎNI Card (Project Author)

Another methodology applied in this research was the circular economy model that has gained strength in society due to the possibilities of success while reducing dependence on primary resources and energy. The central idea of circular economy is that an open production system—where resources are extracted, used to make

Fig. 6 Action Cards (Project Author)

Fig. 7 Reward Notes (Project Author)

Fig. 8 Color Cylinder (Project Author)

products and become waste after consumption—should be replaced by systems where resources are reused, and energy is conserved [9].

Therefore, it is clear that achieving economic development—that combines prosperity with sustainability—requires moving from a short-term value-driven system focusing on the process to a long-term process-driven economy based on the value within a systemic framework (CNI, 2018).

In a circular economy, the product is designed to create minimal waste by allowing it to be easily repaired or its materials to be reused or updated. Value creation is built on the longevity of goods and new forms of consumption (Schulte, 2013). Thus, its integration into the product is still important in the initial phase since it is the period in which the specifications are created [3].

For the best development of this research, the subject was studied through official data obtained with the leading organizations operating within the refugee's context in Portugal. In addition, we analyzed cases of studies from projects that promote social inclusion, such as expressed below. Thus, some important data stand out:

- Origin of Refugees in Portugal (ACM, 2017, p.6): Síria (55%)
- Situation of Refugees in Portugal (ACM, 2017, p.5): Minors (24%)
- Gender of Refugees in Portugal (ACM, 2017, p.6): Male (63%)
- Age Range of Refugees in Portugal (ACM, 2017, p. 6): > 18 years old (35%)
- Refugee Children in the World (UNHCR, 2017, p.25): Children (52%)
- Refugee children attending school (UNICEF, 2018, p.15): Delayed school 1 year or more (58%)
- Refugee Children School Level (UNHCR, 2018, p. 15): Primary (61%)
- Situation of Refugee Children and Youth in the World (UNICEF, 2018, p.13): Alone (44%)
- Gender of Refugee Children / Youth in Portugal [7]: Male (84%)
- Origins of Refugee Children / Youth in Portugal [7]: Guiné (40%)
- Ages of Refugee Children / Youth in Portugal [7]: 15–17 years old (92%)

It was identified that the main focus of the product should be the cultural and school adaptation of refugees in the destination country, based on the information already highlighted in the research. Thus, two user profiles were created and adopted for this work, the first referring to the characteristics of a refugee in Portugal, and the second related to characteristic of a young and refugee child in the European Union.

The design conception began through brainstorming the main characteristics that the product, in this case, the game we call Tîni should have. Tîni is a reference to the word "welcome" in Portuguese, Kurdish, Creole. Thus, we developed the attributes that the game should work on:

- Learn and revise the local language of the destination country;
- Learn local habits;
- Interaction with other children/young people;
- Encourage emotional, financial and cultural independence;
- Encourage motor skills;
- Overcome cultural barriers;
- Create or give back to these children/young people dreams and hope in the future;
- Provide protagonism and dialogue to these children/young people;

- Present new possibilities;
- Positively influence their lives;
- Prevent risky situations;

We realized that it would be ideal for the user's profile to interact with something intuitive, simple and easy. In such a way that the most appropriate typology would be the set of paper and pen and card games. The summary of the subjects and their referred importance to the game approach are below:

Language Association	To add more domain and word revision of the language
Portuguese Habits and Customs	Learning habits of the population that may differ from the usual in another country
Educational and Financial Mathematics	Practical learning of the importance of math in their lives for greater financial independence, based on strategy, budgeting, the anticipation of expenses, etc
Refugee Protection Laws and Human Rights	In addition to learning about the subject, knowing what rights and laws protect them to feel more secure regardless of the country you are in
Art and Culture	To boost curiosity about the world and life
Ludic and Play activities	Stimulating creativity
Prevention of Risk Situations	To prevent and instruct them on how they should respond to such facts

Moreover, to be more interesting for the young audience and children we added, a set of playful and dynamic cards, as well as actions cards, like Pass the Turn, Reverse Order, Earn Points and Lose Points. Although severe reality issues are addressed, these cards soften the game, providing the positive and fun aspect proposed.

To encourage financial education, the points won and lost with the questions in the cards will be computed monetarily with bills. Our goal is to boost values such as game strategy, score calculation, budgeting, planning spences and profits. All the cards of the game, such as main cards and reward bills, were made with recycled paper. The reward bills were developed in the measures of 5.0 cm × 9.5 cm also on recycled paper, but in weight of 90 g. The design was inspired by the 5, 10 and 50-euro bills to promote the association with money.

To create the color cylinder we used a conventional wine's cork painted with varnish enamel paint and varnished. To use this cylinder, one must roll the cork horizontally and let it stop with one of the colors at the top. This will substitute the traditional dice.

Instructions: The game consists of cards and bills and can be played with 2 to 6 people. Each player starts the game with 100 assorted reward bills. Each player in their turn must roll the colorful cylinder, while the other player asks the question. Each card in the deck corresponds to a score so if the player gets the question right, earns the points and if gets the question wrong, lose the points and must return the points from their reserve to the game's bank. The player who asks the question must read in order: the theme, the score and then the question. Thus, the score— credited or subtracted from the player based on the correct or incorrect answer— corresponds to the value of the bills to be withdrawn or added to the game's bank. Wins the game the player with the highest savings or the player who most quickly reaches 1000 points.

For the last phase of the methodology— tests, evaluations and comments— we got in contact with groups that work with refugee children to test the game. Refugee organizations, groups and institutions were sought, but because the game has some sensitive subjects, the ethical council should approve it first, which would take at least 4 months.

Thus, we chose to do an informal test with refugee and non-refugee children. The test was carried out informally at a fair on May 25, 2019, with young refugees of 14 to 17 years old, originally from the Republic of Congo (1), Cameroon. (2), The Gambia (1) and Syria (1). After the test and interview with the group, we discovered the following information:

- Since some refugees from Cameroon had recently arrived and had not started attending Portuguese classes, but had already registered, we could attest that the language of the game was challenging;
- The colourful cylinder used in place of the traditional dice provoked curiosity among young people. All praised how it is innovative, practical, and easy to draw.
- The group did not show any difficulty in understanding the game and how to play it. The group perceived the game as intelligible and intuitive.
- They emphasized that for implementation elsewhere, subjects such as on Rights/Obligations/Regulations in the Service of Foreigners and Borders, Places and Geography of Portugal, Portuguese Habits and Customs, should be reviewed and adapted with their respective country's information.

Also, for the evaluation of the game we consulted professionals from the areas of pedagogy and psychology. In the area of pedagogy, we presented the game to the professor and doctor in education Maria Augusta Olivieri Sá Barreto. Maria Augusta is one of the most influential educators and researchers in the field of learning assessment and school planning. She works specifically in projects and schools concerning children and youth traumatized by social reality.

In the interview, the researcher highlighted the educational nature of the game and how the playful approach is important for learning since it relates to the child's reality. The schooling proposed is achieved through understanding the value of money, financial mathematics, and language association. This favours not only those who do not know the language by introducing it, but also favours those who are already familiar with the language by using the regional questions.

In the area of Psychology, the consulted professional Ana Cláudia Laviano, emphasizes the importance of the game regarding the prevention of dangerous situations since it provides awareness and knowledge about how the child should act in such experiences. She highlighted the sensitivity and care in the approach of the game and the welcoming aspect present in the exchange of dialogue between the participants. Even if there is no identification in the foreground, due to the way the question is approached, there is a later association that consequently brings about the change in the form of acting and perceiving the actions related to that content.

In the view of the specialist, the game is highly enriching in favouring interaction and the exchange of experiences between players concerning difficult subjects. The fact that the problems are recognized in-group makes the process very encouraging and welcoming, since it causes a sense of security and support. When the question is asked, the young person automatically makes the association, even if at first there is no awareness, as the game progresses, someone who knows the subject will identify it.

Concerning the issue of the game achieving the goal of providing to exiled youth and children the possibility of dreaming and hoping again: both professionals interviewed agree that it is a challenging but very important process in this scenario. As a result of a positive perspective of the future, one can change the situation in which one lives, causing a change of behaviour, which allows one to reach new possibilities.

Finally, concerning our proposed objectives, we designed a playful, educational product, supported on sustainability and circular economy pillars. We developed an innovative solution through gaming and playing as a learning strategy to achieve that.

2 Conclusion

The developed game offers a new way of learning the customs and culture of the new country, preventing the users from being in dangerous situations to which they are vulnerable and providing a different way of understanding the new native language. The game's primary goal is to materialize the experience of a good welcoming moment through skills and knowledge construction. The game should provide these vulnerable youth or children space and resources for a better adaptation process, integration and engagement in society. It also promotes the possibility of a playful moment of intercultural dialogues and sparks the ability to dream again.

Because the objective is to learn and revise the local language, we emphasize that the game should not be played in English, even though it is a universal language. The game can be used in Portuguese classes promoted by public institutions since the offer of these courses is expressive.

Adaptation potential allows its transformation to new social standards adopted in the country where it has to be applied. Although designed for Portugal, the game is universal. It can be adopted in other countries by changing the information on the cards about Country Geography, Habits and Customs and Rights/Obligations/System Regulations.

Last but not least, the present work is expected to open the way for further research and action for this significant forgotten portion of the refugee population. Several projects are developed in refuge initiatives, but most are for shelters and the migrant adult population. Supporting and welcoming projects for this underage audience can potentially save lives by promoting the ability of children and young people to dream again.

References

1. ACM.: Relatório De Avaliação De Política Portuguesa De Acolhimento De Pessoas Refugiadas (2017). Avaliable in : https://www.acm.gov.pt/documents/10181/27754/Relatorio_Acolhimento+Pessoas+Refugiadas_Dez.2017.pdf/d21546b3-7588-483d-92a3fa8185d61b5b [consulted on 2019, aphril 01]
2. Baeck, A., & Gremett, P.: Design thinking. *UX Best Practices How to Achieve More Impact with User Experience*. McGraw-Hill Osborne Media (2011).
3. Bocken NMP et al (2016) Product design and business model strategies for a circular economy. J Ind Prod Eng 1015:308–320. https://doi.org/10.1080/21681015.2016.1172124
4. Brown T (2010) Design Thinking: uma metodologia poderosa para decretar o fim das velhas ideias. Elsevier, Rio de Janeiro
5. Burdeck, B.: História, teoria e prática do design de produtos. Blücher, São Paulo (2010).
6. Confederação Nacional da Indústria – CNI.: Economia Circular: Oportunidades e Desafios Para a Indústria Brasileira. CNI, Brasília (2018).
7. CPR.: Relatório de Atividades (2017). Avaliable in: http://refugiados.net/1cpr/www/transparencia/CPR_Relatorio_2017_Web.pdf [consulted on 2019, aphril 03]
8. IDEO.: Field Guide to Human-Centered Design: Kit de ferramentas (2^0 edit). IDEO, São Francisco (2011)
9. Macarthur, E.: Growth within: a circular economy vision for a competitive europe. Ellen MacArthur Foundation, Cowes. Ilha de Wight (2015).
10. Mendes V.: Obrigadas a abandonar seus países, crianças refugiadas sofrem para se adaptar à nova realidade (2014). Available in: https://www.uai.com.br/app/noticia/saude/2014/06/01/noticias-saude,192309/obrigadas-a-abandonar-seus-paises-crancas-refugiadas-sofrem-para-se.shtml [consulted on 2019, January 03]
11. Pazmino, A.: Uma reflexão sobre Design Social, Eco Design e Design Sustentável. In I Simpósio Brasileiro de Design Sustentável, Curitiba (2007). Available in: http://naolab.nexodesign.com.br/wp-content/uploads/2012/03/PAZMINO2007-DSocial-EcoD-e-DSustentavel.pdf [consulted on 2018, November 01]
12. UNHCR. Global Report. (2015). Available in: http://www.unhcr.org/gr15/index.xml [consulted on 2018, December 06]
13. UNHCR. Global Trends (2017). Available in: https://www.unhcr.org/5b27be547#_ga=2.245472324.1191762606.1556739072-901197625.1536077594 [consulted on 2019, April 01]
14. UNHCR.: Global Trends (2017). Avaliable in: https://www.uhcr.org/5b27be547#_ga=2.245472324.1191762606.1556739072901197625.1536077594 [consulted on 2019, aphril 01]
15. UNHCR.: *Refugee children in Europe dream big in UNHCR project* (2018). Avaliable in: https://www.unhcr.org/news/latest/2018/1/5a5366a54/refugee-children-europe-dream-big unhcr-project.html [consulted on 2019, aphril 01]
16. UNICEF. Desarraigados: uma crisi creciente para los niños refugiados y migrantes. Nova York: UNICEF (2016).

Dynamic Ride Sharing as an Alternative Transportation Mode for Commuting Among METU Campus and Eryaman

Mehmet Erdi Özgürlük and Ismail Yavuz Paksoy

Abstract Dynamic ride sharing is an alternative way of commuting to individual car use with the utilization of improving available seat capacity for new riders through which more sustainable trips are ensured among drivers and riders who have similar itineraries in closer time slots. This paper describes a study in which dynamic ride sharing practice among METU (Middle East Technical University) Campus and Eryaman (district of Ankara) via both Facebook group and TAG (let's ride in a single car) app has been investigated with the participation of 12 users. The main objective of the study was to acquire in depth knowledge about the phases of dynamic ride sharing practices, experiences of individuals commuting via ride sharing, their concerns and acquisitions from ride sharing, and to determine potential solutions for improving the overall ride sharing practice which would make it an effective and preferable way of alternative transportation in a sustainable manner. The study revealed user's underlying motivations towards ride sharing, their concerns, needs and desires which would help generate better approaches to performing ride sharing practice, in terms of both optimized and secured ways for better experiences in the light of fresh insights gathered in specific themes.

Keywords Dynamic ride sharing · Sustainability · Behaviour change · Sustainable transportation · Campus commuting

1 Introduction

"Nowadays we are experiencing a shift of paradigm from ownership of goods to sharing goods and experiences" [1, p. 777]. Developments in technology have influenced the way people behave towards their actions through daily practices, including

M. E. Özgürlük (✉) · I. Y. Paksoy
Middle East Technical University, Ankara, Turkey
e-mail: ozgurluk@metu.edu.tr

I. Y. Paksoy
e-mail: ypaksoy@metu.edu.tr

E. Duarte and C. Rosa (eds.), *Developments in Design Research and Practice*,
Springer Series in Design and Innovation 17,
https://doi.org/10.1007/978-3-030-86596-2_35

sharing experiences and shifted them to another level. Uber, Lyft, Airbnb, Blablacar and similar services have given a direction to daily routine activities of people and sharing practices enabled the acquisition of a new meaning. In this respect, for ride sharing, Fogel and Nehmad [2] illustrate that people having profiles on social networking websites tend to take greater risks through their attitudes. Therefore, sharing a ride with people they have never met is not a big trust issue affecting their decisions. Inspired by those sharing practices of the era, this research explores ride sharing experiences of METU (Middle East Technical University) students commuting between Eryaman (district of Ankara) and METU Campus.

1.1 Dynamic Ride Sharing

All presence of mobile internet technology has provided, for those having similar itineraries and time schedules, new opportunities to share their rides on short notice [3]. According to Saranow [4] the demand for ride sharing services which meets the needs of people of common transport in similar routes and time slots has increased clearly in recent years. Specially, the spread of using Internet-enabled smartphones "allow people to offer and request trips whenever they want wherever they are, enabling dynamic, on-demand ride-sharing" [5, p.4]. In that sense, dynamic ride sharing differs from traditional car sharing or carpooling and vanpooling with its two distinct features [6]. In traditional ride sharing, the origin and destination points are fixed, yet through dynamic ride sharing one can simply match with random points and random times regarding the trip purpose. Another distinct feature of dynamic ride sharing is that trip arrangements are made very close to the trip time when a user needs immediate travel rather than arranging it from weeks ago.

There are several economic, environmental and social benefits of dynamic ride sharing. In general terms, as current transportation models are not both environmentally and socially sustainable, road transportation is one of the most challenging issues of the new world [7]. In one respect, car pollution directly causes global climate change as being one of the major sources of greenhouse emissions [8]. In another respect, traffic congestions have a great influence on the reduction of people's quality of life worldwide [9]. Considering that on average cars carry 1.6 passengers, only 25% of predicted amount of emissions are caused by people travelling and the rest is caused by empty seats [10]. In this respect, many researchers from diverse disciplines such as transportation [11], economics [12], and behavioural, social and environmental psychology [13] have described dynamic ride sharing as an effective solution to the inefficiency of current transportation models. Moreover, circumstances such as finite oil reserves, rising gas prices, traffic congestions and related environmental concerns have given a direction to people to use their personal automobiles more wisely with an increased interest in services such as dynamic ride sharing [14]. Further-more, Agatz [14] states that at an individual level, ride sharing allows participants having automobiles to share travel expenses while offering those without cars an enhanced mobility. In overall ride sharing, in other terms the joint travel of two or more persons in a single car has initiated a common way of sharing also the costs and benefits together of a shared private car [15].

1.2 Problem Statement

University campuses have high human concentration where many trips are being made daily. Travel to and from campuses is associated with diverse mode choices such as public transit, individual car use, ride sharing, walking, cycling, subway if available, and so on. Affecting the behaviour of commuters and guiding them to use more environmental-friendly modes is crucial for the improvement of sustainable transportation of the campus area and surrounding neighbourhoods. Limanond, Butsingkorn and Chermkhunthod [16] state that recently, the rising awareness on the issues of climate change and gasoline price crises have caused university planners to focus much more attention on the encouragement of alternative modes that are environmentally friendlier and less gasoline-dependent on campus. Transport planning needs to encourage using these alternative modes, while discouraging individual car use [17]. Mode choice decisions for travelling depend on many different issues such as convenience, comfort, safety, time considerations, accessibility and cost, yet these decisions tend to be affected by social, economic and environmental impacts of commuters' acts and behaviours. Fu et al. [18] state that there has been substantial interest in promoting sustainable transportation alternatives due to climate change, aging infrastructure, and raising *green* culture. Population-based changes in individuals' behaviours, attitudes and knowledge are important to accomplish wide-spread adoption of alternative transportation modes. Moreover, for the improvement in the quality of life of campus residents, some measurements should be implemented through sustainable transportation which will return in many social, economic and environmental benefits. In this respect, dynamic ride sharing could be one of the most prominent solutions for METU Campus where hitchhiking culture is truly favoured among all its members. Deciding on how to commute by either using the Facebook group or via TAG app can be another major means of transportation in campus level after some conflicts are resolved and the experience is made better with further improvements.

1.3 Aim of the Study

This paper presents a study conducted with the participation of both drivers and passengers to a specific diary study with the aim of gathering their concerns, needs, desires and acquisitions through dynamic ride sharing experiences. Commute experiences of different types of users, via ride sharing among METU Campus and Eryaman, are utilized to set a number of applicable criteria for better ride sharing experiences under diverse-themed positive and negative insights. This paper introduces the study which is constituted of pursuit of users' ride sharing experiences throughout a week and complementary stages for obtaining feedback on their acquisitions with their experiences. After gathering this feedback, the study finally discusses how the findings can turn into insights to be used for the further development of ride sharing experiences, under different themes, using affinity diagramming.

2 The Study

The study consists of three main phases. At the start of the study, mass data logs of the TAG app, an app used by commuters who are mostly university students to arrange ride sharing, are collected to determine the participants of the study. Furthermore, an experience chart is prepared, to reflect the experiences of participants regarding ride sharing. The second phase incorporates initial interviews with the participants, which also employs the participants to build an experience timeline using the chart created in the preparation phase. The third phase includes the conducting of the diary study. At the same time of the diary study, behaviours of participants are also monitored using the real-time device logs. Furthermore, post interviews are conducted to fill in the knowledge gaps in the prior part of the study.

2.1 Phase 1: Preparation

The first phase of preparation aimed to gather cumulative data regarding the ride sharing Facebook group titled "ODTÜ-Eryaman Araç Paylaşımı" (METU-Eryaman Ride Sharing) and TAG (Tek Araba Gidelim-*let's ride in a single car*) app to determine the profile of participants to conduct the study with. In addition, an experience chart was prepared in this phase to be employed during the first face-to-face encounter with participants, in the second phase. At the end of the preparation phase, 12 participants were determined to conduct the study with.

Collection of Mass Data Logs. Mass data logs from the developers of the TAG app were acquired in order to determine the sampling strategy for the study. The harvested data included general demographic data regarding the participants' gender, age, role (driver or rider), number of journeys with TAG app and frequency of application use. Sensitive data that may distinguish individual users were not included in these mass data logs.

Sampling. Quota sampling is a method of acquiring representative data from a group. Unlike random sampling, quota sampling requires each representative participant to be chosen out of a specific subgroup [19]. This study uses quota sampling technique to determine participants. The factors that are taken into consideration to achieve the quota sampling were demographic data such as gender and age, role of the user as the user can be either driver or passenger, frequency of application use and number of journeys made with TAG. Contemplating on the data harvested, which includes 340 users in total, 214 were passengers and 126 were drivers. Number of males and females were very close to each other. Considering that the study was conducted for a graduate course with limited time to complete the study, it was decided to conduct the study with 12 participants. As a result, the sampling was distributed as two female drivers, two male drivers, four female passengers and four male passengers.

Preparation of the Experience Chart. An experience chart was prepared for the participants of the field study to fill in their ride sharing timeline experience rating. With the tool, the aim was for participants to declare the steps of a general ride-sharing practice. For this reason, the A4 landscape formatted tool employs a timeline, where participants can add the significant steps of ride sharing. Along with a timeline, the tool also accommodates an area for participants where they can note the positive, neutral and negative experiences as they talk, corresponding to the steps declared. The positive, neutral and negative aspects of experiences are represented with emoji-like bullet points. There are also fields for distinctive characteristics of participants in addition to duration information of the interview.

2.2 *Phase 2: Initial Interviews and Experience Timeline*

The aim of the second phase was to get to know the participants and make them talk about how they plan their daily commutes, and how they make choices regarding their mode of transportation. Furthermore, they were encouraged to express their typical commute experiences explaining the positive and negative aspects of their experiences and problems they face during the commute. The interviews also aimed at extracting knowledge regarding the ethics, advantages and disadvantages of the ride sharing activity. TAG app, which accommodates ride sharing for both drivers and passengers, was also the topic of discussion, as one of the aims of the research was to understand how TAG works and how it could be improved.

To achieve all this, semi-structured interviews were conducted with each participant, considering the factors presented in Table 1. During the interviews, participants were asked to fill in the ride sharing timeline and experience rating tool (Fig. 1), as the tool clarifies the significant steps of the commuting and ride sharing experiences to sensitize participants to see their journeys from start to the end and talk about how they felt about a certain step or behaviour. An example of a ride sharing timeline & experience rating tool is presented in Fig. 2.

Table 1 Considerations for semi-structured interviews

Passenger	Driver
Planning the ride/method Deciding on method by available options	Deciding on the route Deciding on whether there should be additional passengers or not
Leaving home/campus	Heading to the car
Heading to bus stop/waiting point	Heading to gathering point/etc
Waiting for bus/car/etc	Waiting for the passenger/if available
Identifying bus/car Getting into the car	Identifying passengers who wants to join Letting passenger to get in
Social contact	Social contact
Quality of comfort/accompany	Quality of accompany/comfort
Timing	Timing
Traffic	Traffic
Driving (safe/sound)	Company (distracting/helping)
Reaching to the campus/home	Reaching destination
Drop off point	Drop off point
Final social contact for the ride comments on driving, declaring gratitude	Final social contact for the ride comments on accompany, declaring gratitude
Money transaction	Money transaction
Rating driver	Rating passenger

2.3 *Phase 3: Diary, Device Logs and Post Interview*

The third phase of the study employed a diary study to acquire knowledge regarding fresh commuting experiences of participants, device logs for the duration of diaries and a post interview to fill the missing gaps in the researchers' minds.

Diary. In the diary studies, reports on the events and experiences of participants' daily lives are journaled frequently by the participants themselves. The method allows participants to examine reported events and experiences in their natural, spontaneous context, providing information complementary to that obtainable by more traditional designs [20]. Diary study designs require careful consideration of the question(s), to increase effectiveness. With the assistance of technology, different media types (including photographs, videos, audio recordings, and location data) can be obtained as a response from the participants. Three types of diary studies are [20]: Signal contingent, event contingent, and interval contingent. The diary study was conducted with the 12 participants. When all interviews with the 12 participants were completed, they were asked to initiate their journaling on the app. The journaling process started at the same time for all participants, which took place for a week. Participants were asked to complete a short survey after each commute was completed. Questions mainly covered two areas: How their commute experience was, and how the usability

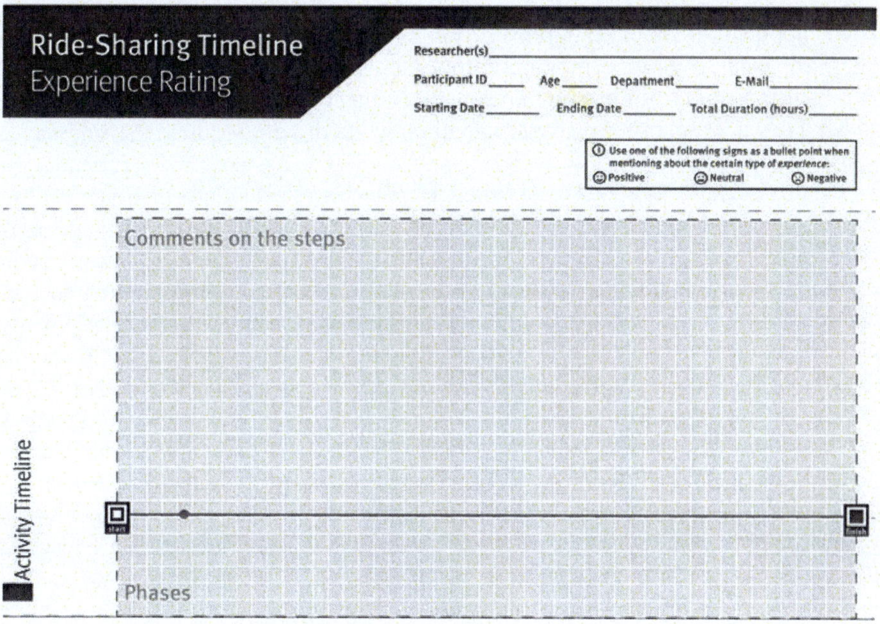

Fig. 1 Ride-sharing timeline and experience rating tool

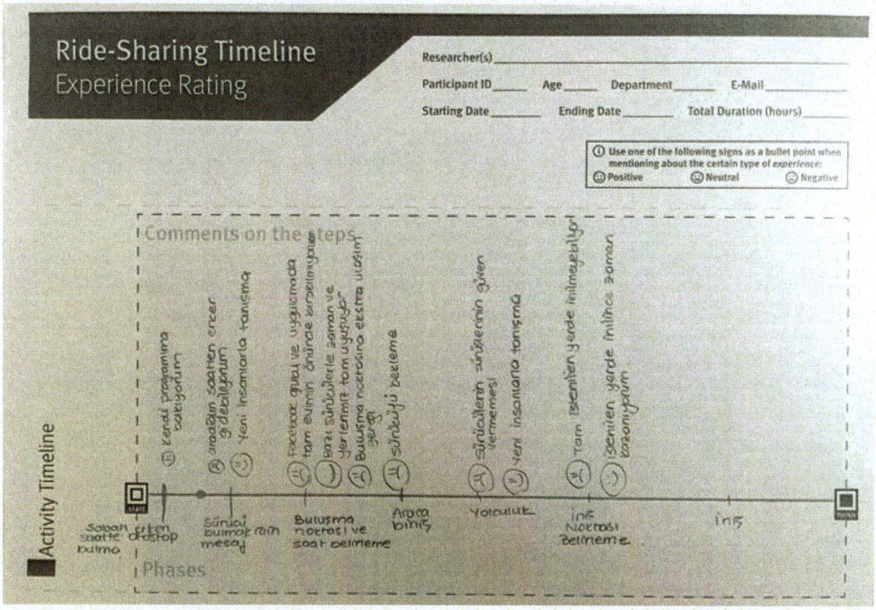

Fig. 2 Example of a filled ride-sharing timeline & experience rating tool

of the application could be improved. The study provided longitudinal data, which shows changes over time (commute to commute) for an individual and how the experience differs from person to person. Passengers were rewarded with free trips to METU or Eryaman for their participation, while drivers acquired their money as usual during the study.

Device Logs. Device logs were aimed at gathering relevant information regarding the TAG mobile application use. With the consents obtained from participants of the study, TAG app developers were asked to share real-time application screen recordings and quantitative application use data that revealed information such as application use time and duration, number of messages sent between participants, contents of the messages, location data with commute start and end time, UI action data that reveals information about number of taps and swipes on screen. The device logs were employed as a way to discover the usability of the mobile application. Furthermore, the logs also helped researchers to discover participants' motivations and behaviours regarding the factors that affect experiences independent from the app.

Post Interviews. Follow-up interviews was the last stage of the data collection procedure. According to the significant points based on the results of previous studies, this interview included refined, to the point questions for users, to extract more specific data for the study. All 12 participants were interviewed to learn more about the reasons behind their ratings in the diaries. As questions depended on participants' individual experiences, each interview progressed differently. This step also allowed researchers to gain more in-depth data regarding the better solutions to current experiences, as participants were more sensitized, enabling them to give suggestions for future ride sharing solutions.

3 Analysis and Findings

The collected data allocated significant amount of information since there were multiple steps of data collection including mass data logs, initial interviews, diaries, device logs and follow-up interviews. To translate all the data into meaningful insights, affinity diagramming with content analysis is employed. Furthermore, insights are presented in experience timeline of ride sharing, which are colour coded with their specific themes, source of information and number of mentions. Findings are presented in this paper under each specific theme and shared with the Facebook group members and TAG App developers.

3.1 Affinity Diagram

An affinity diagram is an analytical tool employed for managing numerous ideas into subgroups which builds up themes [21]. In the study, researchers first extracted

Fig. 3 Affinity diagramming with themes

insights from all the data collected, and wrote it down on post-it notes for classification. As the insights collected on the table, ride sharing related statements surfaced as TAG-positive, TAG-negative, General Experience-positive, General Experience-negative. Elicited insights were then grouped together based on their natural relationships, which created the themes. Analysing data via affinity diagramming produced six themes: social contact; trust, safety and comfort; timing; meeting and route; cost of ride and payment; TAG app-specific insights. Based on the content, negative insights were also marked with a sign, to signify them as a drawback/negative inference. Some of the insights were related to the phases of the ride sharing experience. Therefore, the insights were represented in a temporal fashion in the experience timeline (Fig. 3).

These insight themes (Fig. 4) were presented on a two-axis graph, in which the horizontal axis represented the experiences of ride sharing as a timeline of phases, and the vertical axis represented Facebook group-specific themes at the top, and TAG app-specific themes at the bottom. At the left end of the horizontal axis, themes independent of the phases of ride sharing were presented (Fig. 5).

3.2 Themes

Social Contact. Under the theme of social contact, it is discovered that sharing commute allows both drivers and passengers to gain new friendships. On the contrary,

Fig. 4 Distribution of mentions under related themes

Total 83 insights; positive 41% (34), negative 59% (49)			
social contact	%21,7 (18) over total	%50 (9) are positive	%50 (9) are negative
safety/trust/comfort	%15,7 (13) over total	%46,2 (6) are positive	%53,8 (7) are negative
timing	%21,7 (18) over total	%44,4 (8) are positive	%55,6 (10) are negative
meeting/route	%7,2 (6) over total	%33,3 (2) are positive	%66,7 (4) are negative
cost of ride/payment	%12 (10) over total	%50 (5) are positive	%50 (5) are negative
TAG App specific	%21,7 (18) over total	%22,2 (4) are positive	%77.7 (14) are negative

unnecessary and overwhelming social contact via Facebook group or TAG App, or on the road, disturbs both parties. In addition, there still are trust issues among the members of the commute, since a significant number of encounters happen for the first time. Meeting for the first time also make it hard to identify passengers to pick up, as mentioned by the drivers. It is also expressed by the drivers that they may not recognize some passengers although they met more than once with them. Furthermore, people who are coming from different backgrounds have instrumental idea sharing in the phase of journey, which is one of the topics of chit-chat during the journey. Moreover, popular culture is one of the topics of conversation, which makes long commutes more satisfactory. On the other hand, silence makes people nervous during the journeys. Lastly, drivers stated that they do not like to be treated as a taxi driver, as some passengers prefer to do other activities rather than having a conversation with the driver.

Trust, Safety and Comfort. As members of the METU have similar demographic background and have certain level of characteristic values, people tend to trust each other when they realize that they belong to the same community. Based on this reasoning, people also feel nervous when they encounter non-METU members. A few participants of the study also expressed that they make quick background checks for people they will meet for the first time. A significant number of riders feel that their home address should be kept private. On the side of drivers, it is expressed that riders should not leave trash on board. In addition, complying with traffic rules is important for both safety and comfort, however, most of the journeys did not have issues related to that, although a few riders reported that they feel uncomfortable when the car is driven in a hurry. Drivers also feel more responsible when they carry passengers.

Timing. In terms of duration of journeys, car sharing is more effective regarding time compared to mass transportation. In addition, door-to-door transportation is ensured via ride sharing. Thanks to ride sharing, passengers have more time to sleep due to the fact that journeys take half the time of mass transportation. In addition, being able to pass through security checks of the university gates and arriving at

Fig. 5 Insights presented on experience timeline

a closer point of their destination, not only lessens the travel time significantly but also makes the journey much more comfortable. Furthermore, being able to share rides at late nights, when there are rare bus rounds, makes ride sharing a significant contribution to the riders. In the negative side, when mutual responsibilities among riders and drivers are not met, inconveniences that take time occur for all members, which causes the journey to start later than expected.

Meeting and Route. When it comes to the meeting point and route to be followed, the proximity of driver's origin and pick-up points of riders is a significant factor for optimum journey planning. Thanks to the application and conventions developed in the Facebook group, members try their best to adapt to the journey plan. A few participants reported that drivers change their route to best serve passengers when they have time, which happens mostly coming back from the campus. However, especially when going to the campus, the app does not offer driver's route of daily commute to passengers. On the Facebook group, drivers must write down their routes manually so that passengers can know the nearest point to join the commute.

Cost of Ride and Payment. Most of the riders believe that ride sharing should be free or priced reasonably. For this reason, the Facebook group has much more activity due to being free and being established much earlier than the app. However, those who want to pay or get paid via the app suggested that there should be option to change the amount and type of payments, since the TAG app does not offer custom pricing for the commutes. In addition, the cost for the travel can be adjusted depending on the number of passengers, as the amount of cost per rider is fixed and independent from the number of passengers. Furthermore, as payment by credit card is the only option, participants expressed their complaints regarding not being able to pay in cash. Interestingly, as a few drivers stated that they do not want to get any profit from carrying students, riders want to make favours for drivers such as giving a gift, bringing coffee or hand-made pastry.

TAG App. As the app offers a promising starting point for a more sustainable transportation, there were still a lot to consider for satisfying members' expectations. Significant number of participants reported that they were not aware of the TAG app before the study. In addition to bringing solutions to the pain points mentioned under the other themes, the app can improve itself in terms of its user experience. Reportedly, transitions in the app is slow in addition to being hard to understand, which drops the motivations of users to continue using the app. Moreover, the app is not that favourable when compared to Facebook group, as the group has its own dynamics and is more informal by nature which makes users more comfortable. Furthermore, both drivers and passengers could be evaluated based on a rating system where there are different categories such as driving skills, friendliness, punctuality, respect to personal space, etc.

4 Conclusions

In this paper, we have presented a case study for ride sharing where it took place in Ankara for the commutes between METU and Eryaman. Facebook group was the main means of communication to arrange ride sharing practices at the time. In addition, TAG App was a newly emerged mobile application which specializes on the ride sharing practice with a payment support. We discussed users' concerns, needs, desires and motivations through their ride sharing experiences for further analysis of the acquisitions. With a three staged methodological approach, we revealed diverse insights under six specific themes regarding the ride sharing practice itself and tools to arrange and maintain it. The study showed that further improvements should be considered to develop better ride sharing experiences both for Facebook group and TAG App.

In order to achieve a broad grasp of the research aim, a variety of methods have been utilized throughout the data collection and analysis procedure. Empirical methods including experience chart, device logs and diary study is employed for data collection procedure. Initial and post interviews were conducted to gain knowledge about ride-sharing practices of students, in the context of Eryaman and METU. Device logs study provided valuable amount of data throughout the study. Thanks to the developers of TAG, which is a specialized tool for ride sharing, researchers were able to generate meaningful distribution of users, according to the provided data logs. Affinity diagram is the tool for analysis of the collected data, which gave a birth to the Experience Timeline chart above, that sums up the initial evaluation of the data. Themes, which were created with the help of affinity diagram, are formed to represent groups of issues/contributions.

The most significant limitation of the study was time, as the research was conducted under a graduate level course which was limited to 14 weeks. If there was adequate time for further research, the data would show more generalizable insights with in-depth characteristics with the participation of more commuters.

Acknowledgements As the research presented in this paper is conducted as part of the "ID730 Modelling User Experiences" course at METU, we would like to present our gratitude to the instructors Gülşen Töre Yargın, Sedef Süner and Aslı Günay. Furthermore, we would like to thank our supervisors Naz A. G. Z. Börekçi and Gülay Hasdoğan for their extended feedback and support in the writing of this paper.

References

1. Gargiulo E, Giannantonio R, Guercio E et al (2015) Dynamic ride sharing service: are users ready to adopt it ? Procedia Manuf 3:777–784
2. Fogel J, Nehmad E (2009) Computers in human behavior internet social network communities: risk taking, trust, and privacy concerns. Comput Human Behav 25:153–160. https://doi.org/10.1016/j.chb.2008.08.006

3. Agatz N, Erera AL, Savelsbergh MWP, Wang X (2011) Dynamic ride-sharing: a simulation study in metro Atlanta. Procedia—Soc Behav Sci 17:532–550. https://doi.org/10.1016/j.sbspro.2011.04.530
4. Saranow J (2019) Carpooling for grown-Ups—high gas prices, new services give ride-sharing a boost, Rating Your Fellow Rider
5. Agatz N, Erera A, Savelsbergh M, Wang X (2010) Sustainable passenger transportation: dynamic ride-sharing
6. Casey RF, Labell LN, Moniz L, Royal JW, Sheehan M, Sheehan T, Marks B et al. (2000) Advanced public transportation systems: the state of the art update 2000 (No. FTA-MA-26–7007–00–1). United States. Federal Transit Administration
7. Kleiner A, Nebel B, Ziparo VA (2011) A mechanism for dynamic ride sharing based on parallel auctions. In Twenty-second international joint conference on artificial intelligence
8. European Commission (2010) Energy European commission. Statistical pocketbook 2010, EU energy and transport in figures. Publications Office of the European Union
9. European Conference of Ministers of Transport (1999) European conference of ministers of transport. Traffic congestion in Europe. ECMT Round Tables. OECD Publishing
10. Jeanes P (2010) Can total car emissions be reduced enough by 2020?. Traffic engineering and control 51(1)
11. Morency C (2007) The ambivalence of ridesharing. Transportation (Amst) 34:239–253. https://doi.org/10.1007/s11116-006-9101-9
12. Brownstone D, Golob TF (1992) The effectiveness of ridesharing incentives. Discrete-choice models of commuting in Southern California. Reg Sci Urban Econ 22:5–24. https://doi.org/10.1016/0166-0462(92)90023-T
13. Nordlund AM, Garvill J (2003) Effects of values, problem awareness, and personal norm on willingness to reduce personal car use. J Environ Psychol 23:339–347. https://doi.org/10.1016/S0272-4944(03)00037-9
14. Agatz N, Erera A, Savelsbergh M, Wang X (2012) Optimization for dynamic ride-sharing: a review. Eur J Oper Res 223:295–303
15. Furuhata M, Dessouky M, Ordóñez F et al (2013) Ridesharing: the state-of-the-art and future directions. Transp Res Part B: Methodol 57:28–46
16. Limanond T, Butsingkorn T, Chermkhunthod C (2011) Travel behavior of university students who live on campus: a case study of a rural university in Asia. Transp Policy 18:163–171. https://doi.org/10.1016/j.tranpol.2010.07.006
17. Norzalwi N, Ismail A (2011) Public approach towards sustainable transportation in UKM's campus. Australian J Basic Appl Sci 5:1332–1337. https://doi.org/10.1007/s10640-009-9274-8
18. Fu TS, Mundorf N, Redding CA et al (2012) Promoting behavior change among campus commuters. 53rd Annual Transportation Research Forum, TRF 2012 2:994–1025
19. Moser CA (1952) Quota sampling. J Royal Statist Soc. Series A (General) A 115:411. https://doi.org/10.2307/2980740
20. Bolger N, Davis A, Rafaeli E (2003) Diary methods: capturing life as it is lived. Ann Rev Psychol 54:579–616
21. Plain C (2007) Build an affinity for KJ method. Qual Prog 40:88

Inclusive Design as Promoter of Social Transformations: Understanding Androgyny in Contemporary Society

Ana Catarina Carvalho Ferreira, João Carlos Monteiro Martins, and Maria Antonieta Vaz de Morais

Abstract This article results from a research work for the elaboration of a dissertation, within the master's degree course in Integrated Design of the Polytechnic Institute of Viana do Castelo—Portugal, developed from Androgyny studies and its influence on genderless product design. With the development of this research, supported by literature review and research works, survey and case studies, issues related to the identity of the individual were deeply studied and understood, namely those related to gender expression, as well as the interpretation of social and cultural action in Androgynous Identity formation. Bearing in mind this theoretical framework, it was possible to identify and analyse products designed under the influence of genderless concept and its definition. The few cases that fit within the scope of the study reveal a diversity of approaches that are considered to be more based on commercial arguments than on an honest concern with gender issues and their influence on the characteristics of the product. Therefore, it is thought that there is an opportunity to investigate more about the subject at least in order to highlight the theme framed by the purposes of Design.

Keywords Product design · Inclusive design · Androgyny · Identity · Social stigma

1 Introduction

With the approach of this theme, the aim is an investigation that contributes to a broader view of contemporary society. The idea is to encourage reflection on Androgyny, genderless products and an inclusive design process that can lead to

A. C. C. Ferreira (✉) · J. C. M. Martins · M. A. V. de Morais
Polytechnic Institute of Viana do Castelo, Viana do Castelo, Portugal

J. C. M. Martins
e-mail: joaomartins@estg.ipvc.pt

M. A. V. de Morais
e-mail: amorais@estg.ipvc.pt

E. Duarte and C. Rosa (eds.), *Developments in Design Research and Practice*, Springer Series in Design and Innovation 17,
https://doi.org/10.1007/978-3-030-86596-2_36

that typology. It is also intended to break with assumptions, trying to reflect upon the concept of androgyny, a topic that over time has always held a greater or lesser attention of society.

The lack of information on gender issues often results in products that could cause negative feelings and unnecessary exclusions. Thus, this topic was chosen to alert to a non-discriminatory view of the individual, seeking to respond to the need for greater integration within society, achieved through products respond to each person's daily needs and contribute to a more satisfying life.

For a better understanding of the topic in question, it is relevant to distinguish and frame some key concepts. According to the LGBTI[1] + Communication Manual [1], biological sex is inherent to the individual and identifies hormones, chromosomes and sexual organs characteristics. Sexual orientation is defined by the gender through which an individual is attracted on physical, emotional and spiritual level, considering his or her gender identity, gender expression or biological sex. Gender identity is characterized by the identification of the individual in relation to their gender, how they see themselves and what they think about themselves, which can be identified as men, women, transgender or cisgender. Finally, gender expression is classified by the set of physical and intellectual behaviours by which an individual externalizes his or her gender identity and the way they live and interact with others. Although society stipulates objects, ways of dressing and acting as male or female, today we have the most varied expressions that do not fit that binary gender pattern. They are ambiguous, neutral, multiple, partial expressions, among others.

2 Androgyny

Androgyny is a term whose etymology derives from the Latin androgy̌ nus, which derives from the Greek ἀνδρόγυνος, meaning andros (male) and gyní (female). Singer [2] defines androgyny as a being that contains two. For Robinson and Godbey [3], the androgynous individual is half formed by characteristics that are culturally defined as masculine and half formed by characteristics that are culturally considered feminine. Androgynous, therefore, is someone who has physical and behavioural characteristics of the female and male gender, making it hard to say to which gender he/she belongs to only by observing their physical and/or behavioural characteristics.

The concept of androgyny has emerged in virtually every culture from a mythological, legendary, philosophical or even religious point of view being thus possibly a millennial archetype. According to Singer [2] there are several items of evidence of androgynous tendencies in Western culture, present in people's moral as well as in society's habits or traditions.

[1] LGBTI + is an acronym for Lesbian, Gay, Bisexual, Shemale, Transsexual, Transgender and Intersex. The + symbol refers to the inclusion of other sexual orientations, identities and gender expressions.

The sexual revolution in the mid-1960s was also considered a milestone for the deconstruction of a social structure that accepted as legitimate only the male and female concept prevailing until then. At the time, questions were raised regarding social and sexual behaviours, enabling the ideal of androgyny, recognizing authentic individuals.

Singer [2] believes that individuals should abandon gender and gender stereotypes, as there will always be a difference between the male principle and the female principle, regardless of social definitions.

For Muraro and Boff [4], the androgynous being is classified as a heterosexual individual who coexists perfectly with the characteristics that conventionally belong to the opposite sex, referring to sensitivity, loss of fear and affection in man and creative intelligence in woman. "Only are androgynous those who are able to reunite the opposites within themselves: men and women, activity and passivity, mind and body" [4, p. 251]. It is necessary to emphasize that we cannot confuse androgyny with hermaphroditism, being hermaphrodite, or bisexuality. Júnior [5] believes that for a long time, the term androgyny was understood as synonymous of hermaphrodite, but this concept is already obsolete. According to Carvalho [6], the term hermaphrodite refers to a figure of sex whereas androgynous is related to gender and bisexual to sexuality. Nevertheless, according to Garber [7], the confusion between bisexuality and the androgynous or hermaphrodite terms is due to the meaning of the word bisexual, i.e. 'two sexes'. Nowadays, designations, among others, such as non-binary, genderfluid or genderqueer cover the topic of androgyny. Judith Butler [8], an important American philosopher in favour of the Androgyny concept, proposes queer theory as performative of gender acts that destabilize the categories related to body, sex, gender and sexuality, enabling approaches beyond binary gender concepts.

2.1 Androgyny in History and in Different Areas of Knowledge

The androgyny concept has gone through many interpretations throughout history and has been linked with various legends and myths. Androgynous characteristics have always existed, to a greater or lesser extent, and are present throughout the whole history of human reality and imagination. There are reports and studies related to Androgyny theme in Antiquity, in the Middle Ages, the Renaissance, during the Enlightenment and in the Victorian Era, among others.

In both the Fine Arts and Visual Arts, androgynous influences were found in works by artists such as Sandro Botticelli's (1445–1510) painting Virgin and Child with Five Angels (1483); Miguel Ângelo's (1475–1564) sculpture Sleeping Slave (1513–1516) or Claude Cahun's (1894–1954) photography Self-portrait (1920).

In Theatre, one of the most significant examples is the option to exclude women from stage, in favour of castrati inclusion. These boys, who had perfect singing voices, were subjected to castration in order to keep their high-pitched voice. Francesco Bernardi (1686–1758), for example, was a famous Italian castrato.

The film industry also plays an increasingly important role in disseminating information on gender minorities. The influence of cinema on addressing LGBTI + issues is undeniable. Nepomuceno [9, p. 2] states that "This generation of filmmakers stood out for making films with less sensational approaches on bodies, genders and sexualities' differences being instead more concerned about the complexification of ambiguous and transgressive subjectivities".

There are several films which have successfully approached gender matters, such as, among others, Marocco (1930) directed by Josef von Sternberg (1894–1969), in which the character played by Marlene Dietrich (1901–1992) appears in an erotic performance scene dressed in men's costumes and kissing another woman; Some Like It Hot (1959) by Billy Wilder (1906–2002), with Marylin Monroe's (1926–1962) participation, in which two musicians pretend to be transvestites to be part of a female band; The Adventures of Priscilla, Queen of the Desert (1994), a film directed by the Australian Stephan Elliott (1964-), which is about transsexuality and drag queens. More recently, several films by the Spanish Pedro Almodóvar (1949-), especially Tudo Sobre Mi Madre (1999), which features a transgender male character who tries to seduce men and women.

In the field of Fashion, and according to Roland Barthes [10], fashion is understood as a social phenomenon and as a manifestation of individuality, and any coverage used on our body is part of an organized, normative and recognized system by society. When we talk about fashion, we must go back to the significant changes that occurred during the fourteenth century, where the female silhouette began to differentiate itself from the male one. Later in the second half of the eighteenth century, between the 1770s and 1780s, in England, came out the macaronis, aristocratic men obsessed with fashion who dressed and spoke in an effeminate manner. By the end of the eighteenth and early nineteenth centuries, appeared the figure of the dandy, who is historically characterized as a man who paid attention to his silhouette and physical appearance, sometimes using false artifice and other extravagances to achieve the ambitious aesthetic perfection.[2] However, the dandy of the nineteenth century cannot be confused with the macarone or other young men with exacerbated customs. Riello [11, p. 64] states that while the eighteenth century macarone was fashionable by being excessive, the dandy stood for moderation. Moving into the twentieth century, fashion is somewhat ambiguous, where binary gender and female identity have undergone changes supported by fashion designers who were precursors in gender boundaries. In the crazy 1920s, young women took on a modern attitude. The desire for freedom was mainly associated with an intellectual elite and artists of this era laid the starting point for sexual and women's emancipation. Women could already express a new awareness about their body, thus emerging a new ideal of female beauty. The ideal of beauty associated with the rounded forms that prevailed over the past decade has been abandoned to make way for a slender woman with narrow hips, small chest, long legs and small head, an androgynous silhouette which put an end to women's

[2] Almeida Garrett (1799–1854) is considered by many to be one of the first Portuguese dandies.

curves and made them look like "skinny boys". For the first time, women's legs were seen as a sensual and an erotic part of their body in detriment of chest and hips. Their hair was cut very short which was nicknamed garçonne haircut. They used make-up and new methods to cover up the chest and female body shapes. The idea was to look thin at any price, even if they had to starve. The erotic ideal was therefore a certain androgyny. These advanced women were nicknamed Flappers [12]. When we talk about 1920s fashion, we must refer to Coco Chanel (1883–1971), known for her classic and timeless taste and for having felt the need to change the postwar women's wardrobe, motivated by her dislike for excessive ornaments and adornments.

Innovative in her creations, she brought many articles of men's clothing—some of which had been worn by women during the war years—into fashionable woman's wardrobe. Chanel turned blazers, cufflink shirts, cloaks, berets, and typically men's tailoring, in thick wool tweed, into women's fashion. The increasing acceptance of trousers by women was no longer considered eccentric but strictly of utilitarian use in sports and leisure. This innovation in women's outfit is largely due to the contribution of Chanel who, even before World War I, began wearing men's tight pants and shirts to go horse riding. Also, Yves Saint Laurent (1936–2008) determined the fashion of the 60's and 70's when he launched Le smoking in 1966, a women's tuxedo made up of pants, a coat and a gray silk blouse.

2.2 Androgyny Today

Contemporary androgyny presents itself today with increasingly evident characteristics, not only in individuals, but also in artists and other public celebrities, allowing a break in the dichotomy that promoted behaviors and categorizations with gender stereotypes.

In the LGBTI + Communication Manual [1] we can find the Genderqueer and Non-Binary Pride Flag. This flag which was created in 2010 by Marilyn Roxie and officially established in 2012 is defined by three colored bands with distinct meanings. The first band, lavender in color, represents androgyny and androgynous people; the second white band symbolizes gender neutrality, and finally the green band portrays identities that are defined beyond or without any reference to the binary gender system. According to the publication of Practical Androgyny [13], one of the representative symbols of Androgyny is the Necker Cube. This object, presented in 1832 by the Swiss crystallographer Louis Albert Necker (1786–1861), allow us an optical illusion: its design, though two-dimensional, can be visualized as a three-dimensional cube. This cube is often presented as a symbol of ambiguity and an illustration of the ability of the human brain to alternate between two states of perception.

Through an analysis of the search results for the term androgyny in Google Trends[3] [14] it is possible to conclude that, over the last twelve months, the worldwide demand

[3] Google Trends - Google tool that shows the most popular terms searched in a recent period.

for this word has remained constant, which allows us to infer the high level of interest on this topic, as showed by the following chart.

By filtering the search though different regions in the world, it is possible to conclude that the United States of America, Canada and Australia are the first three countries where the search on the web for the term androgyny was most significant. In the same survey, Portugal is the nineteenth country among the fifty-two countries where the research for the same term was evident.

Still within the internet, gender issues are also considered in social media. In 2015, Facebook started to allow free fill in the gender field without any specific descriptions [15, 16].

From the 14th July 2019 onwards, the United Kingdom has banned gender stereotypes in commercials. Advertising Standards Authority (ASA) made the decision based on evidences that suggest that stereotypes can "constrain the choices, aspirations and opportunities of children, youth and adults" [17].

In what comes to teaching, according to the publication of BBC News [18] at Egalia preschool in Stockholm, the teaching methodology involves a neutral and gender-free environment, from common spaces to toys. At this Swedish school, which opened in 2010, teachers avoid using pronouns like "he" or "she" and toys are made available to children without gender barriers so that everyone can play freely.

Talking about new perspectives and new generations, it is essential to mention Generation Z.[4] According to Parker, Graf and Igielnik [19], approximately one third of Gen Z knows someone who uses gender-neutral pronouns.

According to The Irregular Report [20], Generation Z "(...) has power over their identity - what pronouns they use, how they dress, who they love, and which bathroom they choose. And it is not a niche. It is not an isolated reality only common among urban locations or western countries. It's everywhere". The same study also adds that about 45% of Gen Z expect their gender identity to change two to three times and that 55% say that dress is extremely important in expressing their identity, adding that 62% of this generation think their pronoun doesn't fit their identity.

3 Identity and Consumption

Over the past decades, population growth and the development and spread of product sale companies have led to a significant increase in world consumer society. Thus, today, individuals need constant development and innovation in order to respond to their increasingly individualized needs.

Bauman [21] states that an individual's identity is always under constant construction, never reaching a total and complete state. For the author, the pace of "liquid modernity" has created an ever more fluid and ever-changing world.

---

[4] Generation Z refers to all people born between 1997 and 2012. Individuals born in this period are defined as digital natives, considered as the most tolerant and most diverse generation ever.

Hall [22] designates this event of displacement or decentralization of the individual, resulting in "identity crises", stating that "the old identities, which for so long stabilized the social world, are in decline, giving rise to new identities and fragmenting the individual seen as a unified subject" [22, p. 7].

According to Schweriner [23], consumption became part of people's lifestyle in the 1920s, when the economic and industrial "revolution" led to large-scale production with cheaper consumer goods,[5] which caused a significant change in people's identity. The constant search for transformations in media and technology has emphasized the rise of consumers' wants and expectations, leading to the emergence of less and less defined identities [24]. Therefore, Bauman states that the likelihood of total satisfaction is almost nil as there will always be new archetypes and new alternatives, leading to the consumer craving for more [25].

The element 'desire' presented itself as a factor of conflict in the standards of the past. As a result, the choice for more hybrid behaviors, thoughts and products on the gender issue prevailed, raising the issue of androgyny as a way of consumption. According to Baudrillard [26], "Men and women today increasingly end up meaninglessly in both registers (…). Both models are not descriptive; they regulate consumption" [26, p. 97].

3.1 Genderless Products

The examples presented are considered relevant to the correct perception of the study in question, elucidating the importance and up-to-dateness of the topic. Colour, shape or function are points of products that can play a prominent role when described under the theme of gendered products.

In 2019, the Wild Flowers brand introduced what it claims to be the first non-gendered sex toy designed for multi-body adaptation. This product, called Enby (NB), short for Non-Binary, is a vibrator for users of all genders. Designed with a saddle shape and open functionality, it is sold in black and purple and appears, according to the brand, by the need for a redesign of sex toys, allowing them to have soft shapes, leaving aside the traditional allusion to sexual organs. Beyond the intentionally abstract form, the choice of colours does not appear to be random though.

According to Hallock [27], blue is the favorite colour both for men and women whereas brown and orange are the colours with less preference. Beyond the blue colour, men tend to choose green and black, while women turn to purple and green. In American cosmetics brands "Dermatologica" or "Non Gender Specific", it is possible to conclude that the packaging was developed in neutral colours—black, white and gray—in order to promote a balance between male and female connotations.

[5] Usually known as the Fordist production system - developed by Henry Ford, in 1914, which refers to mass production and consumption systems.

In visual communication, there are differences and trends, for example in typography. According to the publication on the Visme website [28], typically feminine fonts, such as Atlanta, Denver or Chicago, tend to be softer, curved, fluid and rounded. In the case of male-associated fonts such as Phoenix, Jackson or Austin, they turn to straight lines, strong serifs, geometric spacing and thick lines. Also, according to Visme [28], currently, gender neutral sources are characterized by being 'classic', readable and safe for, for example, web uses. Some examples are Bookman, ITC Bauhaus, Open Sans or Helvetica.

In Fashion, Gucci creative director, Alessandro Michele, and Italian artist MP5 developed in 2019 the new identity of the Chime for Change campaign, called 'To Gather Together'. In the campaign image, it is possible to see yellow, black and white graphic elements, superimposed in silhouettes of human figures that hide their gender identification. With this, the brand intends to appeal to the global community's awareness of gender equality issues that are increasingly obvious in our society [29].

There are currently some specific brands for androgynous audiences, such as Tomboy Toes and Nik Kacy (footwear), Kirrin Finch, Peau de Loup, Androgynous Fox, Bindle and Keep (clothing), Luna Pads (intimate hygiene), Tomboy X (underwear).

In Interaction Design, "Q" is considered the first genderless artificial voice in the world. Created in 2019 through a partnership between Virtue, Copenhagen Pride, Equal AI, Koalition Interactive and Thirtysoundsgood, it aims to end gender biases in artificial intelligence systems by calling for inclusion and diverse representation in the technology sector. "Q" made its debut at South by Southwest 2019, the Austin, Texas music and technology festival. The Vice group argues that binary gender perception perpetuates with stereotypes that some sectors of today's society struggle to change and, in an interview, the creators said they were in negotiations to implement this neutral voice on brands like Amazon, Facebook, Google and Apple [30]. The authors further add that male voices are commonly used in more authoritarian sectors and female voices in service sectors. This groundbreaking neutral voice assistant was created based on a set of people voices who identified themselves as non-binary and has been approved by more than four and a half thousand people from different European countries [31]. Concerning Jewellery Design, in 2015 the Estonian researcher Darja Popolitova concluded her Master in Jewellery and Blacksmithing, taught at the Estonian Academy of Arts in Tallinn presenting her thesis entitled "Androgyny: Jewellery Beyond Gender". Popolitova developed an investigation in which she considered the concept of gender as a disaggregation of totality, that is, a deconstruction of the androgynous being, in order to create male and female duality. The main thought of his project was based on an amalgam concept, where natural materials and synthetic materials were fused, generating several pieces of neutral jewellery [32].

4 Questionnaire Survey—Materials and Methods

A questionnaire survey was carried out in order to validate the researched information and to understand the general constraints and needs of non-binary public. This is a sensitive topic that involves personal information, and therefore, to access an audience in person is not feasible. Thus, it was chosen to develop this work online. Therefore, the questionnaire with a total of ten questions was developed based on a convenience sampling method. It was carried out in the online network of communities based on people's interests, called Reddit, more specifically in admittedly non-binary groups.

4.1 Presentation and Discussion of Results

In total, two hundred and eight responses from respondents aged fourteen to sixty-eight were registered. It was possible to conclude a greater adherence of younger audience as 59% of the answers refer to people between fourteen and twenty-four years and 30% to people between twenty-five to thirty-five years old. Regarding the gender with which the respondents identified themselves, 53% were non-binary, 34% female and 13% male. Regarding the origin of the respondents, 63% said they were from the United States of America; 23% from European countries; 6% from Canada; 6% from Australia and 2% from South Africa, India, New Zealand and Hong Kong.

Based on the above percentages, it is possible to corroborate some of the information mentioned in item 2.2.

In a first generic analysis of the questionnaire, it was found that 51% of the respondents consider suffering from prejudice for being androgynous. When asked about the need to disseminate information about non-binary gender issues in schools and educational institutions, 90% agreed.

Regarding the existence of enough genderless products that meet the daily needs of androgynous people, 75% were dissatisfied with the existing offer. Similarly, 84% of the respondents were dissatisfied with the scarce supply of genderless brands offer currently available.

When asked about the genderless products they need the most daily, there was general dissatisfaction with the availability of clothes not only for common use, such as shirts, t-shirts or gender-free pants that could model for different forms of the human body, as well as with the clothes intended for pregnant women. The footwear sector was referred as limited in sizes available and always conditioned to female and male gender. The same goes for personal care products such as shampoos or deodorants. Intimate products used during menstruation have also been mentioned as being usually allusive to gender stereotypes, as well as underwear that needs to be more neutral. The respondents also expressed dissatisfaction with the limited supply of neutral jewellery as well as genderless cosmetic products.

Finally, they were asked about which they considered to be the best way to raise awareness and promote the acceptance of androgyny by society. In this topic, they alluded to representativeness in the media, as well as in characters from series and movies, considering the high number of people they can reach. They also suggested integrating comprehensive gender issues in the education sector to promote knowledge, respect and inclusion in schools.

5 Conclusions

Androgyny presents itself as an archetype widespread in different eras and areas of knowledge although, for certain periods of time, with a complex or still little explored symbology. Technological advances associated with new patterns of daily life and new social perspectives have allowed transformations of mentality and consequent changes in identity. Based on what has been analyzed and presented, it is acknowledged that this is a constantly developing theme and that the area of Design is not yet sufficiently explored in what gender issues is concerned. However, it is also possible to conclude that the reduction of gender stereotypes and the gradual choice for neutrality play an increasingly important role, as we can see in some successful Design cases that prove their need and feasibility, legitimizing the difference that this type of product may cause, promoting society reflection as a whole and individuals in particular. In an age marked by constant changes, it is necessary to break prejudice, promoting equality and social acceptance. Nowadays, we experience multiple, undoubtedly more complex and comprehensive behaviors that come from a growing number of people who do not identify with the binary gender system, seeking for freedom and striving for individuality.

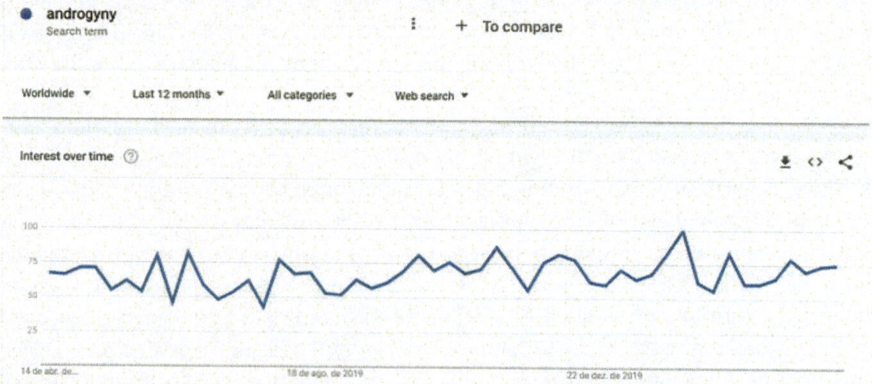

Fig. 1 Research chart of androgyny term in the last twelve months worldwide. Image taken from: https://trends.google.pt/trends/explore?q=androgyny. Access: April 10, 2020

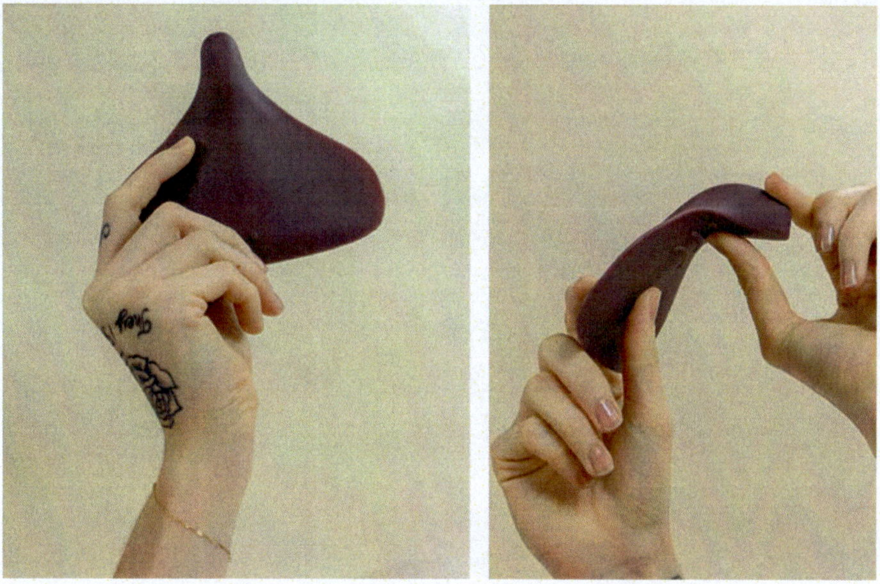

Fig. 2 Enby product designed by Wild Flowers, 2019. Image taken from: https://wildflowersex.com/products/enby. Access: August 10, 2019

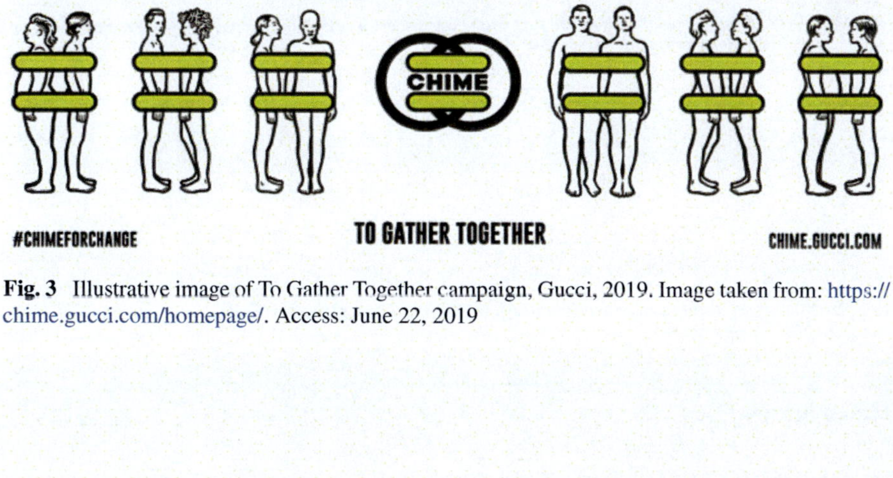

Fig. 3 Illustrative image of To Gather Together campaign, Gucci, 2019. Image taken from: https://chime.gucci.com/homepage/. Access: June 22, 2019

Fig. 4 From left to right. Metanoia V and Metanoia VI breast pin and Fazis I pendant by Darja Popolitova, 2015. Images taken from: https://popolitova.com/portfolio/androgyny-jewellery-beyond-gender-2/. Access: April 16, 2019

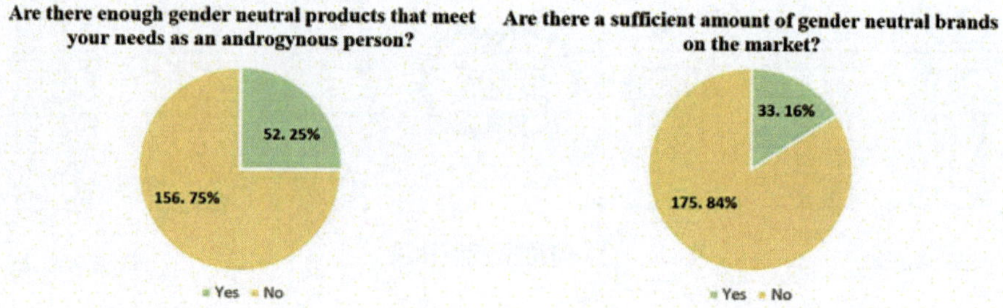

Fig. 5 Statistical data from the questionnaire survey related to genderless products and their existence on the market

References

1. Reis T (2018) Manual de comunicação LGBTI+. Núcleo de Estudos Afro-Brasileiros—Universidade Federal do Paraná, Curitiba
2. Singer J (1990) Androginia: rumo a uma nova teoria da sexualidade. Cultrix, São Paulo
3. Robinson J, Godbey G (1997) Time for life: the surprisng ways American use their time. The Pennsylvania State University Press, University Park
4. Muraro RM, Boff L (2002) Feminino e masculino: uma nova consciência para o encontro das diferenças. Sextante, Rio de Janeiro
5. Júnior JL (2009) Que nunca chegue o dia que irá nos separar: notas sobre episteme arcaica, hermafroditas, andróginos, mutilados e suas (des)continuidades modernas. Universidade Estadual de Campinas

6. Carvalho AM (2000) Nos caminhos da sexualidade: androginia na religião e na cultura portuguesa. Medicina da Beira Interior: da pré-história ao século XXI, pp 86–81
7. Garber M (1997) Vice-Versa: bissexualidade e o erotismo na vida cotidiana. Record, São Paulo
8. Butler J (2017) Problemas de Género. Orfeu Negro, Lisboa
9. Nepomuceno M (2008) O colorido cinema queer: onde desejo subverte as imagens. II Seminário Nacional Gênero e Práticas Culturais: culturas, leituras e representações, p 2
10. Barthes R (2005) Imagem e moda. Martins Fontes, São Paulo
11. Riello G (2013) História da moda: da idade média aos nossos dias. Editora Texto and Grafia
12. Lehnert G (2000) História da Moda do Século XX. Köneman, Colónia
13. Practical Androgyny, https://practicalandrogyny.com/2011/06/25/the-necker-cube-symbolfor-androgyny/. Last accessed 12 Oct 2019
14. Google Trends, https://trends.google.pt/trends/explore?q=androgyny. Last accessed 19 Nov 2019
15. Rede social mais popular do mundo está expandindo a identificação dos usuários para além da dicotomia homem/mulher, http://ladobi.uol.com.br/2014/02/56-opcoes-genero-facebook/. Last accessed 15 Jul 2019
16. Agora, Facebook permite preencher livremente o campo de gênero, https://www.amambainoticias.com.br/meio-ambiente-e-tecnologia/agora-facebook-permite-preencher-livremente-o-campo-de-genero. Last accessed 27 Jul 2017
17. Reino Unido proíbe estereótipos de género na publicidade, https://www.cartacapital.com.br/diversidade/reino-unido-proibe-estereotipos-de-genero-na-publicidade-2/. Last accessed 15 Jul 2019
18. Sweden's 'gender-neutral' pre-school, https://www.bbc.com/news/world-europe-14038419. Last accessed 18 Jun 2019
19. Generation Z looks a lot like millennials on key social and political issues, https://www.pewsocialtrends.org/2019/01/17/generation-z-looks-a-lot-like-millennials-on-key-social-andpolitical-issues/. Last accessed 08 Jul 2019
20. The Irregular Report, https://static1.squarespace.com/static/59b2b4d4a8b2b050d1cb7b9c/t/5c128c4c4d7a9c88918661c9/1544719442197/IL_04_Gender_v8.pdf. Last accessed 28 Jul 2019
21. Bauman Z (2005) Identidade. Zahar (ed), Rio de Janeiro
22. Hall S (2005) A identidade cultural na pós-modernidade. DP&A, Rio de Janeiro
23. Schweriner ME (2006) Comportamento do consumidor: identificando necejos e supérfluos essenciais. Saraiva, São Paulo
24. Canclini NG (1997) Consumidores e cidadãos: conflitos multiculturais da globalização. UFRJ, Rio de Janeiro
25. Bauman Z (1998) O mal-estar da pós-modernidade. Zahar (ed), Rio de Janeiro
26. Baudrillard J (1981) A sociedade de consumo. Edições 70, São Paulo
27. Hallock J (2003) Colour assignment. Universidade de Washington
28. What Is Gender-Neutral Design? Here's How and When to Use It, https://visme.co/blog/feminine-design-masculine-design/. Last accessed 09 May 2019
29. To Gather Together, http://equilibrium.gucci.com/people/journey-of-humanity/chime-forchange/. Last accessed 16 Oct 2019
30. Hear what a genderless AI voice sounds like—and consider why it matters, https://qz.com/work/1577597/this-ai-voice-is-gender-neutral-unlike-siri-and-alexa/. Last accessed 03 Aug 2019
31. Meet Q: The First Genderless Voice, https://www.genderlessvoice.com/about. Last accessed 12 Oct 2019
32. Popolitova D (2015) Androgyny: jewellery beyond gender. Estonian Academy of Arts

Addressing Glocalization Challenges Through Design-Driven Innovation Approaches

Rui Patrício and António Moreira

Abstract The aim of this paper is to explore a relevant conceptual limitation in the literature on Glocalization approaches, when firms deploy their innovation strategies. For that, traditional Local Global and Glocal approaches to serve different types of markets are analyzed and dealt with through a series of case studies. Although glocalization has been researched in many dimensions, the link between glocalization and the design-driven innovation approaches has not been extensively considered in the literature. For bridging the research gap, this paper adopts a design-driven glocalization framework that makes a cross-comparison of several cases of design-driven innovation approaches, which are drawn from the existing theory. Findings show that design driven approaches have the power to engage stakeholders from academia, industries and society in the globalization or/and localization strategies of products/services/business models to different markets and users. Findings also support the conclusion that design-driven approaches facilitate the expression of emotional factors in cultural products, strengthen interactions among individuals, supporting the creation of new visions that intertwine objects and the environment. As a result, design-driven approaches help deliver not only great user experiences, but also support the personalization of different concepts.

Keywords Glocalization · Design thinking · Strategy

1 Introduction

Business internationalization is becoming more important than ever as companies are trying to reach more international markets, and international management practices are becoming core activities in business practices [31]. As such, companies are using

R. Patrício (✉)
UNIDCOM/IADE - Unidade de Investigação em Design e Comunicação, IADE - Faculdade de Design, Tecnologia e Comunicação, Universidade Europeia, Lisboa, Portugal
e-mail: rui.patricio@universidadeeuropeia.pt

A. Moreira
GOVCOPP and DEGEIT, University of Aveiro, Aveiro, Portugal

E. Duarte and C. Rosa (eds.), *Developments in Design Research and Practice*, Springer Series in Design and Innovation 17,
https://doi.org/10.1007/978-3-030-86596-2_37

adaptation and standardization strategies to conquer international markets to increase their presence in abroad [41].

Globalization has forced companies to embrace a process of innovation, development and promotion of their products and brands to remain competitive in international markets [21]. One of the biggest challenges that companies have embraced is the implementation of standardization (globalization) or/and adaptation (localization) strategies of their products to the different markets where they operate [47]. However, although they have been widely accepted, controversies remain [41]. Both adaptation and standardization highlight various aspects of the international marketing strategy companies develop to tailor their products to different countries and customers, based on Localization and Globalization strategies. The adaptive behavior of consumers to corporate strategies plays a major role in understanding the complexity of the adaptation process that companies want to implement [3].

This paper attempts to overcome a relevant conceptual limitation in the literature on glocalization approaches, when firms seek Local—thinking locally, acting locally—Global—thinking globally, acting globally—and Glocal—thinking globally, acting locally—orientation to serve different types of markets as a result of their innovation strategy [37].

Design-driven approaches like design thinking promote the development of a responsive, flexible and people-centered organizational culture, which emphasize the following distinctive principles: creativity and identification of emerging needs; adoption of systematic and holistic approaches; involvement and collaboration; conception of prototypes; adoption of people first and user experience approaches [8, 17]. Given its characteristics, design thinking has the power to address market and technology changes and adapt products and services to different customer needs.

This paper seeks to ask the following research question to answer is: How are design-driven approaches addressing the challenges of Glocalization?

Although glocalization has been researched in many dimensions and contexts [22, 23, 36], the link between glocalization and the design-driven innovation approaches has not been extensively considered in the literature. For bridging this gap, the paper compares several cases of design-driven innovation approaches, which are drawn from the existing theory. This instrument implements a theoretical review-based focus on the linkages between glocalization and design innovation management approaches.

The results of the study support the idea that market entry strategies driven by design approaches deliver great user experiences, particularly in industries more threatened by market and technological changes. These findings are of added value not only to researchers but also to managers that need to create and implement entry strategies in local and global markets.

2 Literature Review

2.1 International Strategies

Globalization brought about the need for companies to clearly define their presence in international markets. Thus, companies have been trying to implement both standardization strategies of their products, trying to implement Globalization strategies, and sometimes adaptation strategies, through Localization strategies. Matanda and Ewing [21] analyzed how dynamic and continuous tensions are managed between Globalization and Localization strategies and their consequences for corporate strategy decisions and processes.

These tensions are always present among companies that want to embrace international markets. Asseraf and Shoham [2] support the need for constant adaptation when crafting international marketing strategies. They claim that there are two different situations called outside-in (OI) and inside-out (IO). Those who defend OI perspectives are strongly influenced by contextual factors and tend to focus on the market. OI perspectives ask questions such as "What do our customers need?", "What are our competitors doing?" and "What capabilities do we need to win?" and tend to be market oriented aiming to adapt to the market [2]. IO proponents are internally oriented and tend to rely on their competencies or their unique technologies. IO proponents deal with questions such as "How can we invent the future?", "How can we make innovative products?" and "What can we do with our capabilities?" and aim to change the market [2].

If continuous adaptation creates tensions among companies due to the need to focus on those two perspectives, the cultural diversity, the various target customers, the different local opportunities and geographical markets companies try to serve create a continuous unbalance between Globalization/Localization or OI/IO strategies. As such, in order to embrace their international activities across countries, the following typology has been put forward: Globalization strategies (centralized activities based on product standardization), Localization strategies (decentralized activities based on product adaptation) and Glocalization strategies (centralized decentralized activities based on product standardization and adaptation) [45].

The importance of being local stems from the differences between countries that support the idea that there are substantial worldwide differences in consumer behavior, needs, wants and preferences [9]. Those differences support the adaptation strategies to local preferences respecting the differences between markets and cultures and the needs of consumers [46]. Wafler and Badir [44] defend that being local demands the launch of unique local products in terms of brand, taste, shape, color, size and/or packaging. Thus, the more a product or service adapts to the local market conditions, the more it meets the needs of a local market [46].

Localization is ultimately the result of adapting to external factors involving, among others, consumers, distribution systems, legal and socio-cultural contexts, which force companies to adapt their product labels to local markets [43].

A global marketing strategy is implemented when a standardized product sold in international markets is identical to the product sold domestically, leading companies to standardize their product ranges, labels, brand names and positioning [24, 28]. Such global companies are capable of creating a coherent global brand for all their customers, which is the main benefit of global strategies.

The decision to go Local or Global comprises one of the most difficult and crucial decisions a company has to make as following one of these strategies has implications for all operational processes of the company. It should be noted that extreme use of either approach may lead to unsuccessful expansion into foreign markets [43]. As such, Going Glocal may be a viable option.

Increasingly those concepts are part of a continuum [46], and the simultaneous application of both approaches (Glocalization) has been advocated, so that companies can benefit from Local and Global strategies, improving their competitiveness [16, 43]. If, on the one hand, Going Local is an interesting alternative to be implemented, given the preferences of consumers and economic and legal cultural inequalities of various countries, on the other hand, Going Global enables the creation of a global, consistent brand image. Glocalization complements each other.

The idea of Glocalization is closely related to what is known as "micro-marketing", more specifically the tailoring and advertising of goods and services on a global or near-global basis to increasingly differentiated local and particular markets [27, 33]. Going Glocal means that companies seek to implement a mixture of global standardization and local adaptation.

As Wilken and Sinclair [45] put it, the global marketing perspective glocalization is an amalgam of global strategy and local adaptation in which each company, whereas at corporate or business level. With glocalization, each company embraces a particular stance depending on each organizational aspects and markets segments the company wants to address. Thus, organization, product and advertising can and often are globally aligned or locally adapted to differing degrees depending on the company, the particular point in time and other circumstances [45].

2.2 *Key Factors*

There are several factors that act as antecedents of strategic decisions that influence a company's decisions whether to be Local, Global or Glocal [31, 38].

Company size influences the company's decision to adapt or standardize its products. While large companies have financial resources, human resources, managerial capabilities, learning experience and access to information, smaller companies have specific competitive advantages because of their flexibility in providing customized solutions, responding more readily to local needs and adapting more easily to new international markets [18].

Larger companies are more likely to embrace Localization strategies [7] because of their responsiveness to foreign markets, [30].

When there are very large cultural differences between the countries of origin and destination, there is normally a heavy reliance on the parent company, which tends to concentrate all its power embracing Global strategies [42].

In countries where consumer tastes, culture and needs are very particular, there is a clear need to tailor product features to consumer characteristics, behaviors and tastes [38]. Under these circumstances, Localization strategies are mandatory.

While associating Localization strategies with highly different markets, and Global strategies with highly similar markets [5], the degree of customer acceptance also impacts the type of strategy. Thus, although there are countries that are very similar to each other, there is a clear need to adapt products to certain markets based on differences in perception and lack of homogeneity, which may open the door to a glocalization strategy. Moreover, cultural and economic differences, and different political and legal contexts are a solid obstacle to product/service standardization permeating Localization and Glocalization strategies [18, 35].

Competitors greatly influence international corporate strategies. When competition is stiff, companies tend to enter international markets through Global strategies standardizing their products to minimize risks, achieving better performances and harnessing economies of scale. Local or Glocal strategies may be embraced adapting product portfolios to find a competitive advantage, especially when companies can create a solid foothold *vis-à-vis* main competitors [42].

Product characteristics and the industry in which companies operate also influence the corporate strategy adopted. In technology-based industries, certain product types work well based on Global strategies [42], while Local and Glocal strategies are more advisable in processing industries [15].

Concerning services, the main reasons leading to standardization are cost minimization, brand image and innovation [25]. There has been a growing concern with Global-based strategies in services as a result of heterogeneity of service inputs and outputs as a result of growing trend for customized services, which have led to the implementation of Localization and Glocalization strategies [11] as they lead to greater perceived control and consequently greater consumer satisfaction.

Although Localization and Glocalization strategies improve perceived singularity and consumer experience [12, 39], competitive edges are short-lived since consumer preferences are unstable and susceptible to external influences. Thus, the added value of individually tailored offers are lower when compared to standard offerings, and do not always translate into consumer loyalty [32].

Localization and Glocalization strategies may lead to lower productivity and higher costs [14]. From this analysis it follows that while products and services have specific characteristics, as soon as markets change, they need a constant adaptation to different customer needs.

3 Methodology

Although Glocalization has been researched in many dimensions and contexts [22, 23, 36], the link between glocalization and the design-driven innovation approaches has not been extensively considered in the literature. Therefore, the research objectives of this paper are as follow: (i) to synthetize the theoretical contributions of Globalization, Localization and Glocalization; (ii) to explore the role that design-driven approaches play in the implementation of standardization (Globalization) or/and adaptation (Localization) strategies and a mixture of both (Glocalization), and (iii) to examine the outcomes of design-driven approaches to glocalization.

For bridging the research gap, the paper adopts an instrument that makes a cross-comparison of several cases of design-driven innovation approaches. The Glocalization Design-driven Innovation Framework (Fig. 1) helps to read the cases and the linkages between glocalization and design-driven approaches [26]. These linkages are established around the following dimensions: (i) types of Products/Services/Business Models (PSBMs) and sectors, e.g. industries more and less threatened by market and technological changes (ii) types of Glocalization challenges, e.g. only global, only local and both global and local; (iii) types of sustainable development goals; (iv) types of outcomes.

Using Case #1 as an example, this framework shows that ceramic artists, students and designers joined forces to develop handicraft products with local cultural identification conditions. The major challenges were related to the abandonment of the traditional craft production and the disposable consumption life culture that is spreading at a global scale. Design thinking was the driving force of the creative process of products that appeals to the emotional factor in traditional cultural. It inspired the creation of local culture handicrafts that are favorable to a more sustainable way of living.

4 Findings

4.1 Cases

The cross-comparison of existing studies about design-driven innovation approaches to Glocalization brings new and meaningful insights about the linkages between the types of PSBMs, challenges, goals and outcomes. All the different cases clearly express the value-added of design-driven approaches, which are interpreted as follows:

Case #1—Stakeholders from business (professional ceramic artists) and academia (students and design instructors) joined forces to develop handicraft products e.g. ceramics with local (Chinese) cultural identifiability through design avoiding a disposable consumption life culture [20]. The major challenges are related to abandoning of the traditional craft production and the disposable consumption life culture

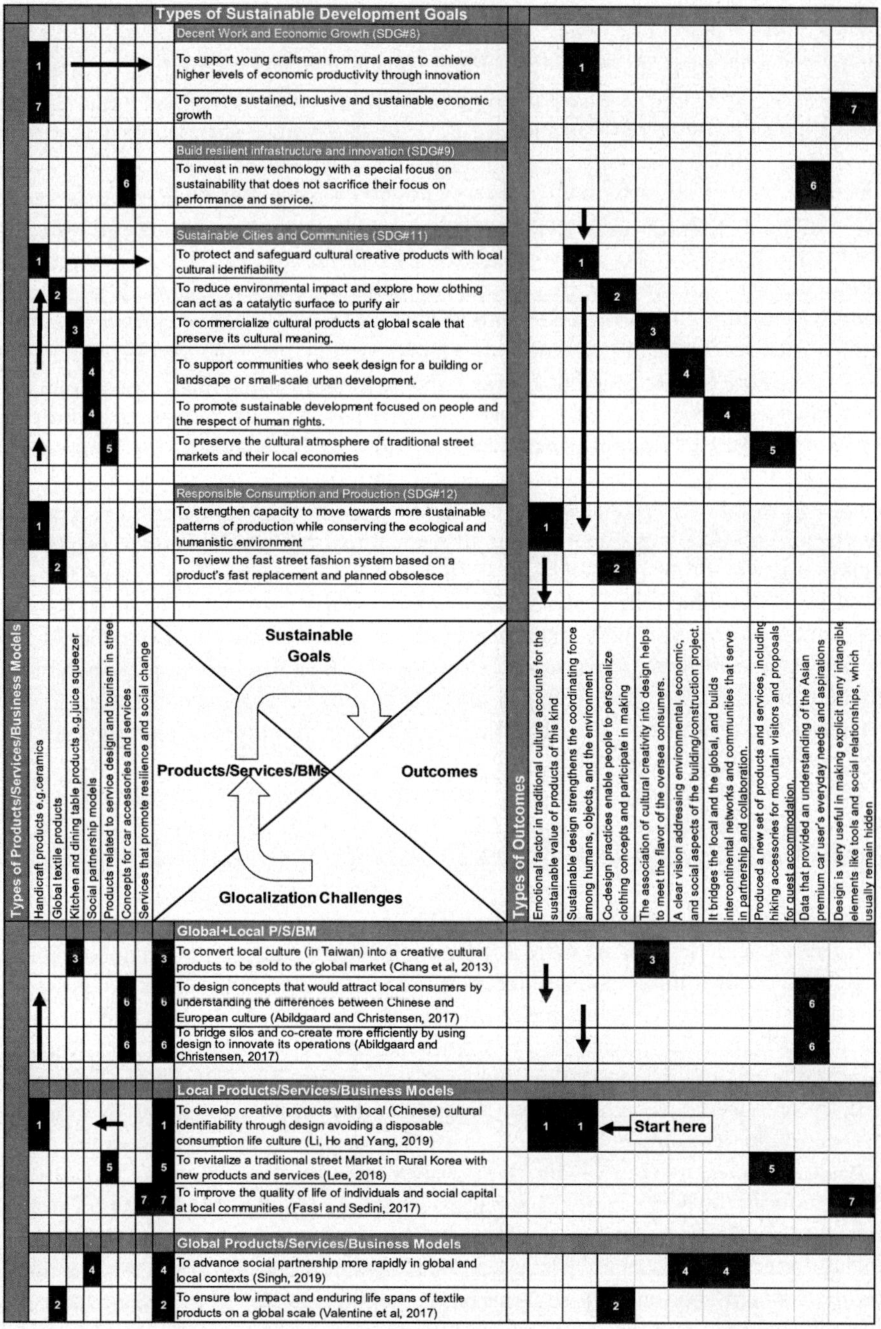

Fig. 1 Glocalization design-driven innovation framework

that is wide spreading at a global scale. Therefore, they wanted to support young craftsman from rural areas to achieve higher levels of economic productivity through innovation; protect and safeguard cultural creative products with local cultural identifiability and strengthen capacity to move towards more sustainable patterns of production while conserving the ecological and humanistic environment. The results obtained have a good fit with above-mentioned sustainable goals. It was found that sustainable design-driven approaches strengthen the coordinating force among humans, objects, and the environment. Design thinking is the driving force of the creative process of products that appeals to the emotional factor in traditional local cultural identifiability, which accounts for the sustainable value of products of this kind. Instead of duplicating traditional elements, design thinking inspired the creation of local culture handicrafts that are favorable to a more sustainable way of living.

Case #2—This case involves a partnership between textile designers and textile fashion industry to ensure low impact and enduring life spans of textile products on a global scale [40]. Globalization, sustainability and technological changes are challenging the way textiles are designed. More collaborative methods are required to ensure products align with the circular economy, e.g. the UK alone, dispose of approximately 10,000 garments every ten minutes. Sustainable goals are very much focused on the reduction of environmental impact and review of the fast street fashion system based on a product's fast replacement and planned obsolescence. Co-design practices described in the case enable people to personalize clothing concepts and participate in making. This change is supported by the development of design-driven capabilities like empathy, deep thinking, visualization, prototyping, and synthesizing in a human centered approach to innovation.

Case #3—The Taiwan National Palace Museum established a partnership with an Italian boutique design brand ALESSI to convert local culture into a creative cultural product to be sold to the global market [6]. Designers were challenged to apply domestic culture elements into global favorite style of consumers. Other stakeholders from the academia (experts and scholars) and society (group of consumers) were in line with the need to address these challenges by designing and commercializing kitchen and dining table products inspired in ancient Chinese Art that preserve its cultural meaning. The outcome of this partnership supports the idea that the association of cultural creativity into design helps to meet the flavor of the overseas consumers and enhance the cultural level of the society. The cultural creativity that is delivered through design is the key ingredient that can be applied by the industry to compete in the global market.

Case #4—The raising of creative local community-based actions around the globe requires the advance of social partnership more rapidly [34]. Typically, global changes in society often begin when people take action at a community level. These social partnership models can support local communities who seek design for a building or landscape or small-scale urban development as well as to promote sustainable development focused on people and the respect of human rights. Results from partnership cases show that design thinking can make things visible. In one of these cases, it created a clear vision that addressed environmental, economic, and social aspects of a building/construction project. Design advances partnerships, bridges the local and the global, and builds intercontinental networks and communities that serve in partnership and collaboration.

Case #5—A local foundation working to revitalize the Eonyang traditional market in Ulsan and a group of Students from Ulsan National Institute of Science and Technology (UNIST) are establishing a close collaboration to revitalize this traditional street market in Rural Korea with new products and services [19]. This type of initiative is addressing the current economic trends and efforts of traditional street markets to promote tourism and culture along with commercial offers of products and services. Thus, its main goal was to preserve the cultural atmosphere of traditional street markets and their local economies. Students applied design thinking approach to produce a new set of products and services, including fruit jams and hiking accessories for mountain visitors and proposals for guest accommodation. Design thinking facilitated the knowledge exchange between students and market merchants, which created valuable ideas for market revitalization and innovative designs.

Case #6—Designing across cultures was the key challenge addressed by a multi stakeholders' collaboration that involved an automotive manufacturer group of collaborators from different areas (design, accessories and the user Involvement departments) and countries (Scandinavia and China) and Chinese lead users. The main focus of this project was to design a "concept package" for car accessories and services that would attract local consumers [1]. To accomplish this difficult assignment, the design team collected user insights like stories and anecdotes, and spent a lot of time discussing how to navigate their own culturally influenced understandings of the differences between Chinese and European culture. They also moved into a more decentralized management structure to bridge organizational silos and co-create more efficiently with users. This automotive manufacturer is investing in new technology with a special focus on sustainability that does not sacrifice their focus on performance and service. This is very much in line with the goals of creating a reliable, sustainable and resilient transborder infrastructure to support economic development. One of the most important outcomes of this project was a much clear understanding of the Asian car user's everyday needs and aspirations. Findings show that design-driven approach through car user journeys and in-house interviews with selected premium lead users enabled a rich collection of data to map current habits and user needs.

Case #7—The case studies about participatory design and resilience jointly developed between the design department of a university and several social organizations illustrate quite well how design-driven approaches can improve the quality of life of individuals and social capital at local communities [10]. This approach addresses the need to the create desirable contextual conditions in the areas of craft and DIY, communities and social innovation, arts and cultural heritage that can promote social change and improve the quality of life of individuals. Several examples of services that promote sustained, inclusive and sustainable economic growth are provided like the improvement of deprived areas of the city and the use of kitchen gardens as educational tools. Design and participatory methods are particularly suitable to engage people and make explicit many intangible elements like tools and social relationships, which usually remain hidden.

4.2 Remarks

Glocalization challenges can be broke down into: local PSBMs, global PSBMs and a mix of both. Moreover, these categories may be slightly blurred by particular market conditions, e.g. local products that can serve global customers and consumer cultures, e.g. global products commercialized in local markets. Furthermore, the types of PSBMs discussed in these cases are mostly from industries more threatened by market, e.g. handicraft products and tourism and technology changes, e.g. textile products and car accessories.

Findings show the relevance of UN 2030 sustainable development goals (SDG) when it comes to implementation of glocalization strategies. It provides rich examples like the sustainable cities and communities (SDG#11), e.g. safeguard local cultural identifiability, preserve the cultural atmosphere of local economies and reduce environmental impact; decent work and economic growth (SDG#8), e.g. support craftsman from rural areas; build resilient infrastructure and innovation (SDG#9), e.g. invest in new technology with a special focus on sustainability; and responsible consumption and production (SDG#12), e.g. review the fast street fashion system based on a product's fast replacement and planned obsolesce.

The types of outcomes display a clear match with the outlined sustainable goals. It demonstrates what design driven approach can deliver when tackling critical glocalization challenges. Design has the power to engage quadruple helix stakeholders, typically from academia, industries and society. These stakeholders are the driving forces of economical and societal change, which are supported by a comprehensive set of design methods and tools.

5 Discussion

The classical debate between the Local/Global/Glocal perspectives has been prolific. There has been tremendous emphasis in how companies, namely multinationals, should embrace change to reach overseas markets. While being Local has been valued when companies try to focus on the differences between countries that support the idea that there are substantial worldwide differences in consumer behavior, needs, wants and preferences, going Global has been embraced by companies when a standardized product can be sold in international markets as much as the product can be sold domestically. On a different vein, Glocalization has been embraced when companies try to detach themselves from the standardization/adaptation perspective and embrace a blend of globally aligned perspective with a locally adapted solution.

If the literature on international marketing, relating product/market perspectives, and international management, or organizational needs to strategy deployment, have been extensively addressed, design-driven innovation is absent in the literature. As shown in the seven cases depicted in Fig. 1, Design thinking solutions can be important drivers in embracing Glocal solutions.

Cases #1, #2, #3 and #4 depict clear examples of inside-out (IO) perspectives. Case #1 seeks to increase cultural creative products respecting the ecological and humanistic perspective and, at the same time, seeking to increase the economic productivity. Case #3 exemplifies the desire to blend the Chinese local culture in global markets' needs, references and brands, in order to explore ancient Chinese art globally through a Glocal perspective. Case #4 depicts how design thinking supports social change and through partnerships it is possible to bridge local and global perspectives involving international communities. Case #2 shows how involving the end user in co-design activities in the fashion industry supports local product to embrace Glocal markets.

Cases #5 and #6 depict clear examples of outside-in (OI) perspectives. Case #5 shows how design thinking facilitates the adaptation of traditional street markets and local products to promote tourism through blending local culture and visitor's needs. Case #6 exemplifies how design thinking supported the understanding of the Asian car use every day needs bridging the gap between the Chinese and European culture.

Finally, Case #7 exemplifies an outside-in inside-out IO-OI perspective of how design supports the adaptation of deprived areas through participatory design that supports the improvement of life of individual at local communities.

It seems there is no one-way street in this process, as OI, IO, as well as IO-OI have been embraced to achieve their intents. As Asseraf and Shoham [2] put it, it is important to think globally and act locally regardless the main purpose—the focus on our capabilities/advantages or the focus on the market—as both perspectives can be used intertwinedly. Design-driven innovation plays an important role in the glocalization process as it can be used for blending product and market perspectives delivering great experiences, adapting products / services to markets (Cases #1, #2, #3 and #4) as well as taking advantages of the already markets served to bridge local unknown characteristics to the world (Cases #5 and #6) without falling in the tyranny of localization/globalization one street road.

Therefore, it is clear that design-driven innovation plays an important role in the glocalization process, although is it not widely covered in the literature.

6 Conclusion

This research shows that design driven approaches have the power to engage stakeholders from academia, industries and society in the globalization or/and localization strategies of products/services/business models to different markets and users. It supports the idea that design-driven market entry strategies deliver great user experiences, particularly in industries more threatened by market and technology changes.

Design driven approaches facilitate the expression of emotional factors in cultural products, strengthen the coordinating force among humans, objects, and the environment, help to create a clear vision, build networks and communities, enable people to personalize concepts, and help to understand local and global user's needs and aspirations. Moreover, taking into account their intricate interactions, design-driven approaches are very powerful in supporting Glocalization approaches as they enable stakeholders to exchange experiences from far-flung environments and deploy then in local settings underpinning the diffusion of global ideas to local environments, as well the diffusion of local ideas to global settings.

References

1. Abildgaard SJ, Christensen BT (2017) Cross-cultural and user-centered design thinking in a global organization: a collaborative case analysis. She Ji 3(4):277–289
2. Asseraf Y, Shoham A (2019) Crafting strategy for international marketing: outside-in or inside-out? Int Mark Rev 36(6):859–886
3. Baptista CS (2014) Product importance and complexity as determinants of adaptation processes in business relationships. J Bus Ind Market 29(1):75–87
4. Birkinshaw J, Hamel G, Mol MJ (2008) Management innovation. Acad Manag Rev 33(4):825–845
5. Cavusgil S, Zou S (2002) The GMS: a broad conceptualization of global marketing strategy and its effect on firm performance. J Mark 66(4):40–56
6. Chang S, Hsu C, Lin R (2013) Design thinking of the local culture and the global market. In: Proceeding of the 5th international congress of international association of societies of design research, pp 1427–1438
7. Chung HF (2005) An investigation of crossmarket standardisation strategies: experiences in the European Union. Eur J Mark 39(11/12):1345–1371
8. Deserti A, Rizzo F (2014) Design and the cultures of enterprises. Des Issues 30(1):36–56
9. Diamantopoulos A, Schlegelmilch BB, Du Preez JP (1995) Lessons for pan-European marketing? the role of consumer preferences in fine-tuning the product-market fit. Int Mark Rev 12(2):38–52
10. Fassi D, Sedini C (2017) Design actions with resilient local communities: goals, drivers and tools. Strategic Design Res J 10(1):36–46
11. Franke N, Keinz P, Steger C (2009) Testing the value of customization: when do customers really prefer products tailored to their preferences? J Mark 73(5):103–121
12. Franke N, Schreier M (2008) Product uniqueness as a driver of customer utility in mass customization. Mark Lett 19(2):93–107
13. Hamel G (2006) The why, what, and how of management innovation. Harv Bus Rev 84(2):72–84
14. Hart CW (1995) Mass customization: conceptual underpinnings, opportunities and limits. Int J Serv Ind Manag 6(2):36–45
15. Horská E, Ubreziova I, Kekäle T (2007) Product adaptation in processes of internationalization: case of the Slovak food-processing companies. Balt J Manag 2(3):319–333
16. Hultman M, Robson MJ, Katsikeas CS (2009) Export product strategy fit and performance: an empirical investigation. J Int Mark 17(4):1–23
17. Kolko J (2015) Design thinking comes of age. Harvard Bus Rev (September), 66–71
18. Lages LF, Abrantes JL, Lages CR (2008) The STRATADAPT scale: a measure of marketing strategy adaptation to international business markets. Int Mark Rev 25(5):584–600
19. Lee HK (2019) Revitalising traditional street markets in rural Korea: design thinking and sense-making methodology. Int J Art Des Educ 38(1):256–269
20. Li WT, Ho MC, Yang C (2019) A design thinking-based study of the prospect of the sustainable development of traditional handicrafts. Sustainability 11(18):1–26
21. Matanda T, Ewing MT (2012) The process of global brand strategy development and regional implementation. Int J Res Mark 29(1):5–12
22. Matusitz J (2010) Disneyland Paris: a case analysis demonstrating how glocalization works. J Strateg Mark 18(3):223–237

23. Matusitz J (2011) Disney's successful adaptation in Hong Kong: a glocalization perspective. Asia Pacific J Manag 28(4):667–681
24. Ozsomer A, Bodur M, Cavusgil TS (1991) Marketing standardisation by multinationals in an emerging market. Eur J Mark 25(12):50–64
25. Pardo-del-Val M, Martínez-Fuentes C, López-Sánchez J, Minguela-Rata B (2014) Franchising: the dilemma between standardisation and flexibility. Serv Ind J 34(9/10):828–842
26. Patrício R, Moreira AC, Zurlo F (2018) Gamification approaches to the early stage of innovation. Creativity Innovation Manag, 1–13
27. Robertson R (2012) Globalisation or glocalisation? J Int Commun 18(2):191–208
28. Rosenbloom A, Haefner JE (2009) Country-of-origin effects and global brand trust: a first look. J Glob Mark 22(4):267–278
29. Roudometof V (2016) Theorizing glocalization: three interpretations. Eur J Soc Theory 19(3):391–408
30. Schilke O, Reimann M, Thomas J (2009) When does international marketing standardization matter to firm performance? J Int Mark 17(4):24–46
31. Schmid S, Kotulla T (2011) 50 years of research on international standardization and adaptation—from a systematic literature analysis to a theoretical framework. Int Bus Rev 20(5):491–507
32. Simonson I (2005) Determinants of customers' responses to customized offers: conceptual framework and research propositions. J Mark 69(1):32–45
33. Sinclair J, Wilken R (2009) Strategic regionalization in marketing campaigns: beyond the standardization/glocalization debate. Continuum 23(2):147–157
34. Singh V (2018) Local action, global impact: From domination to partnership by design. Interdisciplinary J Partnership Studies 5(3):1–28
35. Sousa C, Bradley F (2008) Antecedents of international pricing adaptation and export performance. J World Bus 43(3):307–320
36. Svensson G (2001) "Glocalization" of business activities: a "glocal strategy" approach. Manag Decis 39(1):6–18
37. Svensson G (2002) Beyond global marketing and the globalization of marketing activities. Manag Decis 40(6):574–583
38. Theodosiou M, Leonidou LC (2003) Standardization versus adaptation of inter- national marketing strategy: an integrative assessment of the empirical research. Int Bus Rev 12(2):141–171
39. Tian KT, Bearden WO, Hunter GL (2001) Consumers' need for uniqueness: scale development and validation. J Consumer Res 28(1):50–66
40. Valentine L, Ballie J, Bletcher J, Robertson S, Stevenson F (2017) Design thinking for textiles: let's make it meaningful. Des J 20(sup1):S964–S976
41. Venaik S, Midgley DF (2019) Archetypes of marketing mix standardization-adaptation in MNC subsidiaries. Fit and equifinality as complementary explanations of performance. European J Market 53(2):366–399
42. Virvilaite R, Seinauskiene B, Sestokiene G (2011) The link between standardization/adaptation of international marketing strategy and company performance. Eng Econ 22(1):106–117
43. Vrontis D, Thrassou A, Lamprianou I (2009) International marketing adaptation versus standardisation of multinational companies. Int Mark Rev 26(4):477–500
44. Wafler B, Badir Y (2017) Global products marketing strategy of two European MNCs in Vietnam. J Product Brand Manag 26(6):573–588
45. Wilken R, Sinclair J (2011) Global marketing communications and strategic regionalism. Globalizations 8(1):1–15
46. Willis M (2006) To adapt or to standardise: Testing Asian views about purchasing Western luxury products in seven Asian locations. J East–West Bus 12(1):65–94
47. Zhang J, Beatty SE, Walsh G (2008) Review and future directions of cross-cultural consumer services research. J Bus Res 61(3):211–224